THE ORCHIDS

杜蘭

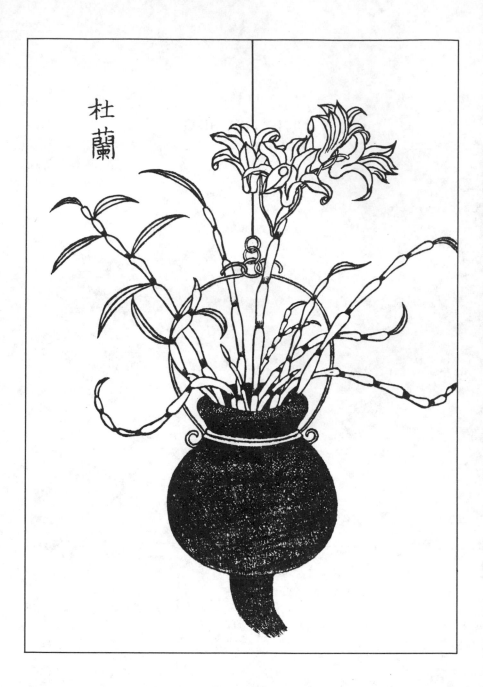

THE ORCHIDS

Scientific Studies

Edited by

CARL L. WITHNER

PROFESSOR OF BIOLOGY
BROOKLYN COLLEGE

AND

ORCHID SPECIALIST
BROOKLYN BOTANIC GARDEN

A WILEY-INTERSCIENCE PUBLICATION

JOHN WILEY & SONS, New York · London · Sydney · Toronto

Library of Congress Cataloging in Publication Data:
Withner, Carl Leslie.
    The orchids.

    "A Wiley-Interscience publication."
    Continues the editor's The orchids, a scientific
survey.
    Includes bibliographies.
    1. Orchids. I. Title.

QK495.064W53   584'.15   73-20496

ISBN 0-471-95715-1

Printed in the United States of America

10 9 8 7 6 5 4 3 2

Contributors

GARAY, LESLIE, Director, Orchid Herbarium of Oakes Ames, Botanical Museum, Harvard University, Cambridge, Massachusetts

JONES, KEITH, Jodrell Laboratory, Royal Botanic Gardens, Kew, Richmond, Surrey, England

KAMEMOTO, HARUYUKI, Professor of Horticulture, Department of Horticulture, University of Hawaii, Honolulu, Hawaii

LÜNING, BJÖRN, Docent, Institutionen för organisk kemi, Stockholms Universitet, Stockholm, Sweden

MEHLQUIST, GUSTAV A. L., Professor of Plant Science, Department of Plant Science, University of Connecticut, Storrs, Connecticut

MOREL, GEORGES, Directeur de Recherches, Station Centrale de Physiologie Végétale, Versailles, France

NELSON, PETER K., Associate Professor of Biology, Department of Biology, Brooklyn College, Brooklyn, New York

SANFORD, WILLIAM W., Professor of Botany, Department of Botany, University of Ife, Ile-Ife, Nigeria

STOUTAMIRE, WARREN, Professor of Biology, Biology Department, University of Akron, Akron, Ohio

SWEET, HERMAN, Professor of Biology, Biology Department, Tufts University, Medford, Massachusetts and Research Associate, Orchid Herbarium of Oakes Ames, Harvard University, Cambridge, Massachusetts

TANAKA, RYUSO, Professor of Botany, Botanical Institute, Hiroshima University, Hiroshima, Japan

VEYRET, YVONNE, Maître de Recherche, Institut de Botanique, Université de Paris, Paris, France

WEJKSNORA, PETER, Graduate Fellow, Biology Department, Brandeis University, Waltham, Massachusetts

WITHNER, CARL L., Professor of Biology, Biology Department, Brooklyn College, and Orchid Specialist, Brooklyn Botanic Garden, Brooklyn, New York

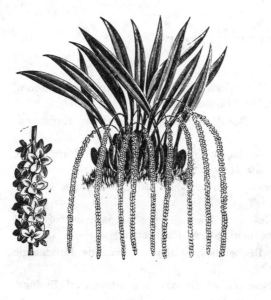

Preface

When *The Orchids, A Scientific Survey* was published by Ronald Press in 1959, it was the first volume of a proposed two-volume set designed to be part of Chronica Botanica's series of manuals on specialized plant groups. Circumstances at the time, however, prevented the completion of the second volume. There has now been a lapse of 14 years, and the publisher of this second book, *The Orchids, Scientific Studies,* is the Interscience Division of John Wiley and Sons.

It is satisfying to say that the material presented by the various specialists in the first volume is still valid and usable, and orchid enthusiasts and experts the world over refer to it for authoritative information that does not go out of date. This new book is *not* a revision of that volume, but adds to it in significant fashion. The information presented by the specialists is, again, primarily factual and will also stand the test of time. This second volume accounts for areas not reviewed previously and adds to other specialties by an increase in details and recent advances.

In assembling these new chapters and lists, each by a recognized authority in the field, I hope to present to biologists continued specialized coverage of the scientific aspects of orchidology. These writings also bridge a gap between the botanist-investigator and the orchid fancier who wants to know about the concepts and facts behind what becomes common orchid knowledge. A widely scattered literature is collated and evaluated with a thorough overview of the orchids as a result. Research areas and problems are better defined, a convenient reference record is provided, and with the first volume also at hand, there is almost a complete review of the various biological aspects of the orchid family. But even without the first volume, this second compendium will stand by itself as a useful and valuable reference source.

With the exception of the editor, the contributors to this new volume were represented in volume one only as the results of their investigations may have been recorded there. It was noted in the preface to that book that ecology and geography of orchids was not adequately covered. That new chapter in this book is one of the most significant included and gives perspective to a great deal of orchid research. Poor coverage in the former book of greenhouse automation and growth of orchids in controlled environments was noted; the same with polyploidy, evolution, speciation, and introgression. Progress in these fields is now included. The chapter on cytology details a better understanding of orchid chromosome significance.

The earlier volume had no comprehensive systematic review of orchid anatomy. This omission is now corrected, while other chapters provide more depth of detail on subjects previously discussed. This is particularly true of the greatly detailed coverage of terrestrial orchid seed germination and the orchid embryology in this volume. In orchid physiology a better understanding of flower pigments, mycorrhizal relationships, and developmental influences dominate the new information. The chapter on orchid alkaloids presents a brand new field of interest and research with no end in sight. The work on clonal propagation of orchids presents a technique, well known in botany but new in its detailed application to the orchids, that has revolutionized the commercial production of valuable awarded or otherwise desirable plants.

The number of hybrid orchid genera is now so large that a reference list is a necessity for all but the most computer-minded of orchidologists. This valuable list will be found accurate, according to the International Rules of Botanical Nomenclature, giving correct combinations and the names of the original plants for which the hybrid genera were established. The chromosome list provides a useful reference source, alphabetized by genera for easy use. It brings up to date, with a great number of additional counts, the list from *The Orchids, A Scientific Survey*, but because of space limitations includes only species counts, not those of hybrids.

Nomenclatural difficulties still cause problems in the production of a book such as this. We again have attempted to print accurately by type face and style the many orchid names used. The current botanical and horticultural rules regarding nomenclature have produced only one significant change since 1959 when *The Orchids, A Scientific Survey* was published. When making a species name from a proper name ending in -er, the *i* is no longer to be used. Thus *Vanda sanderiana* now properly becomes *Vanda sanderana*; or *Phalaenopsis schilleriana* becomes *Phalaenopsis schillerana*. This new form is followed in this volume and will explain the difference in spelling with the first.

Our appreciation and thanks are due the Missouri Botanical Garden, the Arnold Arboretum, and the Botanical Museum of Harvard University for permission to copy certain illustrations from their journals. We also wish to acknowledge the use of Fig. 15 from J. J. Barkman's *Phytosoci-*

ology and Ecology of Cryptogamic Epiphytes, published by van Gorkum, Assen; also Fig. 36 from J. P. Schulz's *Ecological Studies on Rain Forest in Northern Surinam,* published by N. V. Noord-Hollandsche Vitgevers Maatschappij Amsterdam. Certain illustrations in the anatomy chapter are published with approval from the Fedde Repertorium Beihefte, now under the Museum für Naturkunde at the Humboldt-Universität, Berlin.

My wife Pat has once again spent many hours in typing revised manuscripts and checking proof. She receives my special appreciation for her patient help and editorial expertise. John Ward receives an accolade for his careful, thorough, and most understanding work in collating and preparing the indexes for press. To the various contributors and to others who have encouraged me to continue and have worked along to bring the book finally into print, my heartfelt thanks for their gracious cooperation.

CARL L. WITHNER

Brooklyn, New York
February, 1974

Note:

Georges Morel died on the first of December, 1973. Regrettably, Professor Morel will not see the final form of his valuable chapter. The orchid world will long remember his scientific applications of meristemming to the art of orchid culture.

C. L. W

Contents

CONTENTS

THE ORCHIDS

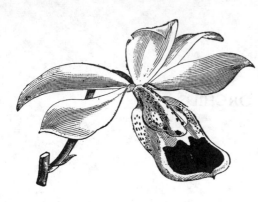

I

The Ecology of Orchids

WILLIAM W. SANFORD

Introduction

It is always well to begin by defining what we are talking about, but when we attempt to define "orchid ecology" we very nearly destroy the concept, for orchid ecology cannot be confined. It is the study of orchids in relation to their environment and so it is liable to break out and invade meteorology and soil science, when we emphasize the physical environment; genetics and taxonomy, when we look at variation in plants in relation to the environment; physiology, when we examine plant responses to environment; and climatology, geology, and geography, when we look at the way plants are distributed over large areas of the earth. Consideration of orchid ecology, then, becomes a tightrope-walking exercise: the game is to keep talking about ecology without falling into another discipline.

Talking about orchid ecology is particularly difficult because most of the available data are scattered among the observations of amateurs. [The only professional review of the subject that I know is Holttum's (1960) excellent but brief treatment of epiphytic orchid ecology.] Professional biologists tend to ignore the work of amateurs, while amateurs usually do not make the effort to evaluate their own observations by reading scientific papers related to them. I believe that the first big job to be done, therefore, is to collect relevant information from both amateurs and professionals and put it into a condensed enough form that it may serve as a manageable foundation for further work. Since the form must be manageable if it is to be useful, only a small proportion of available data can be used, and selection and synthesis will have elements of the arbitrary. Having noted this shortcoming, I can only fall back on the usual excuse and say, may readers of this review be inspired to write better ones.

Besides knowing what we are talking about, it is desirable to know why we bother to talk about it: in other words, we want to know what is the use of orchid ecology. This field of study shares the overall utility—rather, necessity—of the whole of ecology: to understand the intimate relationships between earth, atmosphere, plants, animals, and man and to direct such understanding toward eventual control of the environment so that it may continue to support human life. Besides its share in this great general aim of ecology, orchid ecology has several specific potential functions of its own.

The growing of orchids is not only an economically important industry but also an important factor in keeping many individuals sane and happy in this disturbing world. An understanding of orchid ecology is a vital tool for the successful growing of orchids: ideal culture is a controlled improvement on the natural situation, and ecology is the study of that natural situation. From a wider viewpoint, knowledge of orchid ecology is necessary for the prediction of orchid distribution. When the conditions under which various species grow are known, it becomes possible to predict where these species may be found. And, as I will show later, such knowledge can be used to predict the distribution of other plants associated with orchids and perhaps even to characterize whole vegetations and environments. Stemming from this is the possibility of using orchid ecology in the estimation of land-use potential, which is especially important in developing tropical countries.

Finally, there is a huge field of general physiological, chemical, and genetic knowledge to be gained from the study of orchid ecology. Such knowledge will apply not only to orchids—in fact it may be most useful in general application outside the orchid family, for it will form a fundamental part of our expanding knowledge of the expanding world in which we live.

Large-Scale Distribution

As a first step, let us consider the overall distribution of orchids. In such a survey we at once run into a taxonomic problem: if we are looking at orchid distribution, we must decide what orchids are. Schlechter's dichotomous, hierarchical scheme from DIANDRAE and MONANDRAE down to SYMPODIALES and MONOPODIALES is still basically followed by many, but Garay's scheme (1960) as modified by Dressler (1960) and the scheme used by Dressler and Dodson (in van der Pijl and Dodson, 1966, p. 171) may also be interesting to us here. Subfamilies of the family ORCHIDACEAE are (1) APOSTASIOIDEAE; (2) CYPRIPEDIOIDEAE; (3) NEOTTIOIDEAE;

(4) EPIDENDROIDEAE; (5) ORCHIDOIDEAE, with structural similarities suggesting genetic alliances as shown in Fig. 1-1. The most controversial point of such a scheme is whether or not to include the APOSTASIOIDEAE. Brieger (1958), for example, prefers to omit them. Hutchinson (1959) and Backer (1968) separate the APOSTASIACEAE from the ORCHIDACEAE as a distinct family. At Kew (Summerhayes and Hunt, unpublished) the family is divided into three subfamilies: APOSTASIOIDEAE, CYPRIPEDIOIDEAE, ORCHIDOIDEAE. The latter is then considered as composed of the tribes ORCHIDEAE, NEOTTEAE, EPIDENDREAE, and VANDEAE.

Referring to Fig.1-1, no group here is assumed to be an ancestor of any other group, although number 1 is thought to have the most characters similar to early "orchids" and number 5 the least. Although the first orchids probably developed monophyletically from either a BURMANNIACEAE- or LILIACEAE-like ancestor, forms living today have evolved through ancient gene exchanges between a number of secondarily differentiated groups, perhaps resembling the groups shown here (Fig. 1-1). The site of origin is probably Asia, possibly Malaysia (Garay, 1960), during the Cretaceous. Since that time, approximately 17,000 species (Hunt,

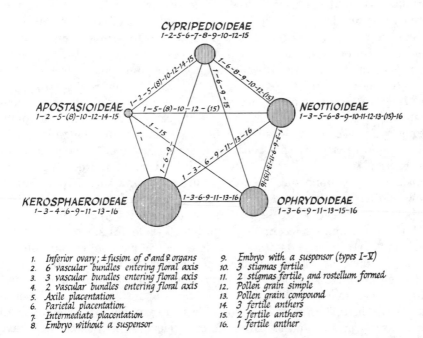

1. Inferior ovary; ± fusion of ♂ and ♀ organs
2. 6 vascular bundles entering floral axis
3. 3 vascular bundles entering floral axis
4. 2 vascular bundles entering floral axis
5. Axile placentation
6. Parietal placentation
7. Intermediate placentation
8. Embryo without a suspensor
9. Embryo with a suspensor (types I-V)
10. 3 stigmas fertile
11. 2 stigmas fertile, and rostellum formed
12. Pollen grain simple
13. Pollen grain compound
14. 3 fertile anthers
15. 2 fertile anthers
16. 1 fertile anther

Fig. 1-1. The interrelationships in the orchid family. (From Garay, 1960, p. 79.)

1967) in 750 genera (Hunt, 1969b) have come to be distributed in all the major land masses of the world except those perpetually covered with snow. The actual number of species may be debated. Whereas Hunt counted herbarium folders of "valid" species at Kew, Garay (personal communication) counted type descriptions and came up with some 30,000. He believes, however, that should the family ever be taxonomically revised, the number might drop down to about 12,000. But no matter how one looks at it, the orchids are a very large flowering plant family and account for about 7% of the species of flowering plants of the world (van der Pijl and Dodson, 1966). Their wide distribution qualifies them as a cosmopolitan family, although Good (1953, pp. 56–58) chooses rather to call them a subcosmopolitan family, because their representation is so much greater in the tropics. At any rate, we are confronted with the problems of how 12,000 to 30,000 forms developed from one or a few forms, and how these forms came to be spread over the world in the great variety of habitats where they are found today.

Considering the latter problem spatially, we may think of the following explanations: (1) gradual, short-range radiation, with each successive generation dispersing its propagules a bit farther from the center of origin; (2) sudden jumps or long-range dispersals whereby propagules are carried hundreds or even thousands of miles. I believe that both have occurred and are still occurring, although now to a much lesser extent because of greater intercontinental distances. Gradual, short-range radiation is certainly quantitatively more important and represents the infrastructure of distribution.

Many workers have not been able to see beyond this short-range radiation and so have taken advantage of various land-bridge and stepping-stone hypotheses to do away with the necessity for any long-range dispersals. Most persuasive has been van Steenis (e.g., 1962). The panacea of bridges and steppingstones has, however, been destroyed: present-day geologists and geographers almost unanimously discredit such hypotheses. A new solution has been seen by many in the revival of the continental drift hypothesis of Wegener (1915, 1937). Among recent workers, Hurley et al. (1967), for example, accept the hypothesis, while Fittkau et al. (1968) feel that the available data are insufficient to prove continental connections. On the other hand, the most recent research bears out the earlier view of Darlington (1965), who postulated that South America and Africa were joined until about the Jurassic and that Africa, South America, Australia, Antarctica, and India were grouped near the southern end of the world at Permo-Carboniferous glaciation and since then have moved northward for varying distances at different times and perhaps at varying rates.

Although it now seems logically safe to accept continental drift, the time of continental separation has been moved back to an era well before the orchids evolved. Nevertheless, it is most likely that the continents were much closer together at the time of the first great expansion of the ORCHIDACEAE (Cretaceous-Oligocene through early Miocene) than they are today.

Garay (personal communication) points out that subfamilies were probably differentiated during the early Cretaceous and so could have spread while continents were relatively close. He notes that only terrestrial genera are widely spread, that the later evolving epiphyte genera did not have time to disperse before the continents were far apart. He continues to stand by his earlier (Garay, 1964) objections to long-range dispersal. At that time, he agreed with the objections cited from Cain (1944). One of these objections is that distribution patterns are not the chance patterns expected from long-range dispersal. I do not see why "chance"—presumably meaning random—dispersal should be expected. Dispersals follow definite patterns brought about by many variables, including seasonally prevailing winds, air currents, water currents—all leading to a general east to west dispersal—land topography, bird migration, together with the final rigid patterning dictated by various establishment demands of the species.

A second objection is that endemism in some isolated regions, such as islands, is too high if long-range dispersal has been a distributional factor. My previous argument outlining pattern factors has some bearing here in that the amount of long-range dispersal can be expected to vary, and so endemism will also vary, from place to place and from time to time. Such variance does not invalidate a theory of long-range dispersal. The distance and the time by which one land mass is separated from another will affect the balance between recently dispersed species and species dispersed very long ago which are now partially altered to become endemic species.

Furthermore, the rate of evolutionary expansion (speciation) must be considered for each group of endemics before comparison is possible. And, as I will later discuss, speciation is to a great extent controlled by diversity of habitat and environmental pressure, so it will not be constant from one place to another, or from one time to another, not even for the same plant group.

Garay (1964) admits to orchid seed dispersal up to 400 miles (or even up to 500 miles; personal communication) but feels that this is rare. Rarity of a phenomenon does not mean that it is unimportant: for example, we know that successful mutations and hybridizations are rare and yet they form the basis of evolution.

Garay argues that orchid seed cannot be transported by wind across oceans because it is desiccated very rapidly, yet I have kept a number of kinds of epiphyte orchid seeds in a desiccator at tropical room temperatures (22 to 27°C) for over two years and found at least some seed viable. Knudson (1940) reported that seed in a $CaCl_2$ desiccator at 46°F remained viable for ten years—some species even longer, some not as long. He also reported that seed stored in envelopes in his office remained viable for one year. Dungal (1953) conducted limited freeze-drying tests on *Cymbidium* Sw. seeds and found that they can withstand this treatment. Burke and Northen (1948) showed that *Cattleya* Luegeae (= *C. dowiana* 'Rosita' × *C.* Enid) seed can withstand below-freezing temperatures. Thompson (personal communication), at The Jodrell Laboratories, Kew, recently found that orchid seed can be stored successfully in the deep freeze.

It appears, then, that if orchid seed is desiccated, it can remain viable for long periods, especially at cool temperatures. It may also be frozen. This means that although seed might quickly lose viability if it were near the warm, damp earth surface, if it were carried in high air currents where it would be both cool and dry—perhaps frozen—it could drift for years and still be viable when dropped down to a suitable environment.

When I presented these arguments to Professor van Steenis, his reply (personal communication) was, "Well, if orchid seeds can be so easily dispersed, show me some examples of long-range dispersal!" His point was that there is a very great difference between possible dispersal and *effective* dispersal.

Garay (in his 1964 paper, p. 180) lists 11 pantropical genera. Of these only *Vanilla* Mill., *Polystachya* Hook., and *Bulbophyllum* Thou. are normally epiphytic, and all are generally considered primitive. Two advanced genera, *Acampe* Lindl. and *Angraecum* Bory (I consider *Mystacidium* part of *Angraecum*), are distributed in Malaysia, India, Madagascar, and Africa. Whether this represents a small amount of successful long-range dispersal, or medium-range dispersal of a few early differentiated advanced genera before continental drift proceeded very far, cannot be determined. A few cases of intercontinental species distribution may be noted. Particularly interesting are the African species *Eulophidium maculatum* (Lindl.) Pfitz, and *Eulophia alta* (Linn.) Fawcet and Rendle, which are found in South America. Brieger (personal communication) has the fascinating view that these occur because they can grow on clay-loam and therefore do not have to compete with the humus-growing South American terrestrial orchids. Such distributions clearly appear to represent long-range dispersal. Admittedly, such cases are few —but some exist, thus proving the possibility.

Another possible case in point is the New World *Campylocentrum* Benth., a sarcanthine orchid quite close to such leafless angraecoid orchids of Africa as *Microcoelia* Lindl. and *Encheiridion* Summerh. It is certainly possible that this genus represents an old, fairly long-range dispersal from Africa. I am quite certain, however, that van Steenis would not consider that so few scattered cases would represent effective dispersal.

In a recent discussion, Thorne (1973) stated that continental drift was too early to help much in explaining present day distributions and that land bridges, such as Central America, Antarctica, and Malesia, were important in moderate-range dispersals. But for dispersals between continents, across ocean gaps, and to distant islands, "Only one rational explanation is left—long-distance dispersal."

Burgeff (1936) reviewed all aspects of orchid seed and found considerable buoyancy, greatest in seeds of terrestrials. Many recent workers have reported improved germination after soaking seeds in water. For example, Kano (1968) found better germination in *Cypripedium acaule* Ait. after soaking the seed from 15 to 45 days. Such observations suggest that water, as well as air, may act as a carrier of orchid seed. This is most important for medium-range dispersals along streams and rivers. It also casts doubt on alleged quick loss of viability of nondesiccated seeds. Seawater dispersal is doubtful, however, both because of distances and time involved and, more especially, because of the toxic salinity of seawater. Although a number of orchids such as *Dendrobium* Swartz, *Brassavola* R. Brown, *Schomburgkia* Lindl., and *Epidendrum* Linn. live comfortably along the sea, even in occasional spray (e.g. Wright, 1946; Anon. 1898), it is unlikely that seeds would withstand penetration of salt water. The carrying of whole green pods by the sea, on the other hand, is certainly possible. Ridley (1930) reported seeing a *Dendrobium crumenatum* Sw. plant in good condition from his ship far out in the Malacca Straits. It is thus possible that both whole plants and seed pods may occasionally be carried by sea currents from one land mass to another.

There is very little evidence either for or against the dispersal of orchids by animals—except by man who at least in recent times has consciously dispersed plants and seeds all over the world. Peter F. Hunt (personal communication) believes that in the Solomon Islands the people have, from very early times, carried orchid plants from island to island. Such early transport may have been common in many other areas of Asia as well. Ridley (1930) cites an unpublished report of orchid seeds carried inadvertently on human clothing. He also mentions insects and birds as carriers. Garay (1964) finds the chances for transport of orchid seeds on the feathers of birds very slim, but I feel the possibility cannot be ignored. Bird preening on a tree limb would seem an excellent device for planting any seeds picked up from the air during migration or from bursting pods before the journey began.

Consideration of the orchid floras of islands may help us to view dispersal more concretely. The problem of island floras in general has been well treated by MacArthur and Wilson (1967). Recently Tobler et al. (1970) considered model explanations for the floras of New Zealand and neighboring islands based on two geographical factors: the relative positions of the islands and their sizes.

Of especially interesting specific examples, that of Krakatao is most famous. Here all vegetation was destroyed by volcanic eruption in 1883. Subsequent plant colonization has been reviewed by van Leeuwen (1929, 1936). No orchids were found in 1886 but by 1896, Boerlage found three species (*Arundina speciosa* Blume, *Cymbidium finlaysonianum* Lindl., and *Spathoglottis plicata* Blume). The first epiphyte, a fern, was found in 1906. From 1908 to 1928, van Leeuwen found a total of 276 species of plants, of which 62 were cryptogams, 2 gymnosperms, 66 monocots, and 146 dicots. The list included 23 epiphytes of which 13 were orchids. By 1933, 17 epiphytic and 18 terrestrial orchids were reported. Van Leeuwen observed that until 1886 all plants found on Krakatao had seeds that could be spread by wind or ocean currents, but by 1897 the first plant with seeds spread by animals was reported. It is also important to note that no orchids were among the first arrivals even though their seed is eminently suited to wind dispersal. The major difficulty involved in orchid colonization appeared to be more a matter of other plants having prepared a suitable habitat rather than a matter of dispersal problems per se.

Java is the closest large land mass. This island has 971 orchid species distributed over 139 genera (Backer, 1968). At best, small islands such as Krakatao have only a small percentage of the flora of nearby large land masses. This is true even when the islands involved are of the same vegetational age. For example, Hunt (1969b) recently reported on the orchids of the Solomon Islands. About 230 species distributed among 70 genera occur there. These figures may be compared with those given by Good (1960) for the nearby large island of New Guinea: 2600 orchid species in 128 genera. It is also of interest to note that whereas most of the New Guinea species are endemic, only 40% of the Solomon Islands species are. [Hunt (personal communication) feels that this percentage may be considerably reduced when the orchids of New Britain, New Zealand, Louisiades, New Guinea, Fiji, and New Hebrides are better known.]

The number of species found in the smaller land mass generally is much less than the number found in the larger land mass. The underlying explanation of such reduction is probably directly related to the areas of the land masses as potential catchers of orchid propagules: if air carries dustlike seed, a large "mop" twice the size of a small one stands a chance of catching twice as many propagules. Once an orchid becomes established on a land mass, it is free to move over that territory by a series of radiating, small-range dispersals. It is also ready to evolve into a number of new species through the process of mutation, recombination, and selection by environmental differences. For both secondary dispersal and speciation, diversity of habitats is of primary importance. The large land mass normally has a much greater diversity of habitats. Thus not only may more kinds of orchids be initially established because suitable habitats are available, but also the increase in orchid species resulting from the catching of more propagules by the larger land mass is geometrically increased by speciation possibilities on the larger, more diverse land mass. Higher percentages of endemism represent a dynamic interplay between earliness of migration into the area, greater number (and therefore diversity) of immigrants, and greater selective pressure because of more varied habitats or more environmental changes.

Preston (1962) has given statistical reasons for caution in considering an Island flora depauperate. He maintains that islands are "isolates" and equal areas on the mainland are "samples" and so direct comparison is invalid. Although this is true, it is still instructive to compare speciation in "isolates" and in "samples," and his statistical caution rather misses the point.

A different type of example of dispersal from a large land mass to a smaller one has been given for Fernando Póo, 32 km off the coast of West Africa (Sanford, 1970). Here, the presence of the separating sea was found to have no more significance than a comparable strip of unfavorable land sites. A similar picture is given by Moreau's work on bird populations in this area (1966).

A still different situation is observed in the flora of the Seychelles archipelago (Summerhayes, 1931). Here only 21 species of orchids in 14 genera were reported. This represents a tremendous reduction from the East African mainland, which has about 700 species. Such a reduction probably reflects the greater distance between the Seychelles and a large land mass, their own small mass and lack of habitat diversity, and the general east to west direction of dispersal which puts them at a strong disadvantage, since the African orchid reservoir is to the west and southwest.

The Hawaiian Islands present another sort of problem, for they are separated from other land masses by very great sea distances. These islands have only 42 genera of monocots and 159 of dicots, with only 2% endemism in the monocots (Stone, 1967). Orchid species are only 3 in 3 genera (Dillon, 1953). Dispersal to these islands probably represents the longest range dispersal we encounter, since continents, before continental drift proceeded so far, were closer to each other than Hawaii has been to any major land mass. There is no reason, however, to suspect that this long-range problem put the orchids at a particular disadvantage in relation to other plants, so another explanation must be sought for the low percentage of orchids in the angiosperm flora. The explanation may lie in the scarcity of suitable mycorrhiza and in the slowness in the development of an angiosperm vegetation providing suitable orchid habitats. As the example of Krakatao illustrated, a well-developed flora must precede orchid immigration. It is also possible (Hunt, personal communication) that immigrant orchids were secondarily lost through competition from introduced nonorchidaceous plants. Such competition would probably act against epiphyte phorophytes ("hosts") rather than directly against the epiphytes.

From this very brief review of distribution, we can conclude that some long-range dispersal of orchids over water may have occurred and may still occur, although to a reduced extent. This is probably comparable to long-range dispersal over unsuitable land habitats. The size of the land mass, in the case of islands—or of suitable colonizing area, on the mainland—is important in determining the number of species found there by, first, the initial catching of propagules. Perhaps even more significantly, the greater diversity of habitats provided by the large area is important, initially, through its effect on the variety of propagules which may become established into plants and, later, through its effect on short- and medium-range dispersals. Finally, diversification into new species is largely dependent upon diversity of available habitats, which is normally greater in larger areas. Consideration of habitats leads us to our next major area of investigation: small-scale distribution and the role of environment in speciation.

Small-Scale Distribution

Our concern has been with how some 12,000 to 30,000 forms of orchids, which have developed from one or a few forms, have come to be spread over the world. We thought of this problem first in terms of space. Now we will think of it in terms of form itself; that is, in terms of the development of forms through speciation. This is intimately connected with ecology in two broad ways: (1) species can remain alive only if they are adapted to their environment, thus evolutionary change reflects an interplay between variation in environment and variation in adaptive possibilities of the plant; (2) species can survive as distinct entities only if there is some barrier—environmental or spatial, structural or chemical—preventing interbreeding with other forms.

The relationship between ecology and species is so intimate that the term "genecology" was coined to cover it partially. This was originally understood (Turesson, 1923) to be the study of infraspecific variation in relation to environment, but since infraspecific variation is a part of the continuous variation leading to interspecific variation, the concept has become considerably enlarged. In this chapter I include how dispersal and environment have affected the speciation leading to our present forms of orchids. To do this, we need to agree on what is meant by "species." Without going into arguments I consider out of date, we can accept Mayr's recent (1969) summarization of species as groups of interbreeding natural populations that are reproductively isolated from other such groups. A species by this definition is, then, a reproductive community; an ecological unit that interacts with other species with which it shares the environment; and a genetic unit consisting of an intercommunicating gene pool. (One word of caution concerning the "biological definition of species": I believe it is logically unassailable, but it can seldom be applied in practice and so has lamentably little effect on plant classification! The problem arises in proving, in the field, interbreeding on the one hand and isolation on the other.)

A convenient place to begin our discussion of species formation is to return for a moment to dispersal: when a propagule is transported to a new habitat, the first problem is establishment. With orchids there must be suitable mycorrhiza present and, in the case of epiphytes, a phorophyte, or host, as well as a suitable nonbiotic environment. Once established, a plant will either continue life or die, depending on its adaptability to the environment. This adaptability may be thought of as having two aspects: absolute adaptability to the nonbiotic environment and relative, or competitive, adaptability in regard to other plants in that environment. If the immigrant does not have the former, it will be eliminated almost at once—if, indeed, it ever became established in the first place. Its ability of the second type is comparative and so may be neutral, superior, or inferior in relation to competitor plants. Clearly, if it is inferior, the newcomer will soon be eliminated, although it may possibly hang on for a while, depending on the amount of competitive pressure.

If the immigrant is neutral, it should, we have always thought, be able to repeat its number in successive generations. Runemark (1969) has now pointed out that this is not actually so. He has calculated that the deviations from the expected numbers of a species in successive generations are not haphazard but follow frequency polygons of the Poisson series if the sample is made from an *infinite* number of objects; deviations follow approximately if made from a *large* number of objects. In orchid dispersal the number of seeds is much higher than the number of plants already established, so the number of plants in successive generations will only approximate calculated frequency and there will be appreciable random drift. In a large population the consequence of this drift would be negligible, leading to numerically small oscillations in the number of individuals of a species in following generations. But if the species is represented by a very few individuals in a large plant community, this random variation will be highly significant and Runemark has accepted the term "reproductive drift" for it.

In the new orchid immigrant we may have the case of one or a few plants in a large community of other plants. For such a situation, Runemark calculated the probabilities for the survival of neutral immigrants to 127 generations and found a 37% probability for the elimination of the first generation if one individual were the immigrant, 13% if two individuals. He concludes that a competitively neutral species introduced as a single individual or as a few individuals will in most cases be eliminated.

If the immigrant is slightly superior compared to the established species, it should theoretically increase and eliminate the species originally inhabiting the site. Runemark, however, calculates actual survival probabilities to be only slightly higher than for a neutral species. An immigrant with marked superiority, on the other hand, may be randomly eliminated but stands such a considerable chance of survival that reproductive drift is seldom an important parameter of establishment.

From these calculations we may get the impression that plants have very seldom been established in any habitat where they were not markedly superior competitively. An obvious exception to this generalization is the spread of weeds, whereby immigration into disturbed areas lacking other plant competition results in a "neutral" species being equated to a markedly superior one. In studying orchid migration, therefore, we must scrutinize orchid habitats carefully.

The epiphyte habitat appears to be equivalent to the disturbed habitat in that there are few competitors. And, in fact, epiphytic orchids have been called pioneers (Pittendrigh, 1948), and so equatable with weeds entering cultivated land. This is not, however, the case. Van Oye's early work (1921, 1924) in Java clearly demonstrated the reality of epiphyte succession, with ferns and orchids following algae, lichens, liverworts, mosses. Oliver (1930) found much the same thing in New Zealand: the first settlers are usually small lichens and mosses; ferns appear next, together with orchids. Observations in India (Dudgeon, 1923), on the basis of branch age, indicate (1) crustose lichen, (2) foliose and fruticose lichen, (3) pioneer moss, (4) climax moss, (5) ferns, and (6) flowering plants.

Fawole and Sanford (unpublished) made a number of observations in a mature, secondary, moist, lowland rain-forest along the Jamison River in southern Midwestern Nigeria. About 76 trees were chosen randomly and measured for girth; their epiphyte load was then enumerated. No significant correlation was found between tree size (hence age) and presence of algae, lichens, liverworts, and mosses. This indicated that in such a highly favorable site these epiphytes became established on very small trees and so were nearly ubiquitous. On the other hand, correlation between size and the presence of ferns and orchids was clear-cut; these macroepiphytes generally became established only on larger, older trees.

It is extremely doubtful that lichens, mosses, or liverworts are ever orchid competitors; rather they are necessary for providing a suitable environment for seed germination and subsequent seedling growth, and so even if present in considerable amount the community may be considered "open" for orchid colonization. An apparent exception to this has been reported by Teuscher (1959, 1964). He observed that *Rodriguezia* Ruiz. and Pavon, in general, and *R. batemanii* Poepp. and Endl. var. *speciosa* Mansf. and *R. teuscheri* Garay, in particular, are found growing on smooth barks without any moss. I believe that the reason for this is not that moss is a competitor to these species but rather that the *Rodriguezia* can manage without moss and so can colonize habitats free of competition from other orchids and macroepiphytes which need moss to become established. Such orchids capable of survival without a relatively massive root system adpressed to a rough substrate have been called the "air-orchids" by Dungs and Pabst (1971). Brian Morris (1970) notes that the genus *Aërangis* Rchb.f., in Malawi, is almost restricted to smoothbark trees.

Ferns and flowering plants are definitely competitors of orchids. My own observations in Africa suggest that the most serious ones are lianes, seconded by epiphytic ferns (see Plate 1). In Tropical America, bromeliads also may be serious competitors. Withner (personal communication) has pointed out that it is a common observation that trees having bromeliads do not have orchids. Although it is possible to exaggerate such demarcation of habitats [certainly orchids and bromeliads are at least sometimes found growing together on trees (e.g., Post, 1965; my own observations in Colombia, 1972) and even on old palm thatch (Foster, 1947) and lava flows (Lankester, 1920)], it is likely that general environmental preferences together with competition do serve to keep them apart.

If, on the other hand, seeds of epiphytic orchids arrive before fern and angiosperm colonization, they are comparable to weeds in disturbed ground and so without appreciable competition. Establishment will depend, then, on the nonbiotic environment itself, together with mycorrhiza and phorophyte. Reproductive drift will not be a significant factor in their survival probability. At later stages of epiphyte succession, plant competition will not only begin to act as an additional environmental pressure and so provide additional selective mechanisms but will also make reproductive drift important in the control of populations.

Plate 1.   Epiphytic ferns which have taken possession of possible orchid sites, island of Fernando Póo, near sea level, February (dry season). (Photograph by W. W. Sanford.)

In instances where forests have been destroyed and regeneration has occurred, orchid immigration tends to follow a wavelike pattern with the peak occurring when lichens are being superseded by mosses. It has been pointed out to me (John B. Hall, University of Ife, personal communication) that this may be a slow process, because many cryptogams are slow in migrating into new areas. The reason for this is probably that their propagules (most often gemmae) cannot be dispersed far and are very sensitive to unfavorable environments such as drying and overexposure. In older forests, orchids tend to be most frequent in areas least colonized by macroepiphytes—in the canopy where there is too much light and too little moisture for ferns, aroids, and so on. Elsewhere, an interplay of absolute and competitive adaptability will determine the kind and amount of orchids found.

Many terrestrial orchids are adapted to disturbed areas; perhaps a more accurate statement would be: the terrestrial orchids, of which there are many, are those adapted to disturbed areas. Certainly a number of species of temperate zone orchids are weedlike. For example, some species of the Great Lakes region in the United States have begun to decrease with the regrowth of formerly cultivated areas (Curtis, 1941). A well-known instance of association with disturbed land is the preference of *Spiranthes spiralis* (L.) Chev. for roadbanks. This orchid cannot grow in tall grass, or even in thick short grass, but is only successful where the bare earth is visible in patches (Anon., 1967). A similar preference for roadbanks is shown by *Stenorhynchus orchioides* (Sw.) A. Rich. in Florida (U.S.) (Luer, 1967). *Epipactis helleborine* Crantz, to cite another example of weedlike behavior, was not originally found in North America but was introduced there from Europe shortly before 1879 and then spread aggressively from Quebec, Canada, to at least the District of Columbia (U.S.) (Green, 1968). A number of weedlike orchids are mentioned by Correll (1950, p. 7). The original immigration of such terrestrials into cultivated or disturbed land is comparable to early epiphyte immigration from the standpoint of lack of competition and the unimportance of reproductive drift. Subsequent history also parallels what was outlined above for epiphytes.

Besides the weedlike nature of some orchid migration, it may be well to mention here the suboptimal nature of many natural orchid habitats. It is a commonplace observation that if a wild plant is removed from its natural habitat and is cultivated, it often grows bigger and better. The improvement reflects the removal of the plant from competition with other plants and the provision of a more nearly optimal level of food, water, and light. I have observed, for example, the increase in size and improved growth of *Trichocentrum panamense* Rolfe with greenhouse cultivation (Sanford, 1961). Withner (personal communication) mentions that *Oncidium lucayanum* Nash is normally much smaller in its natural habitat than in cultivation. Adderley (1965) reports considerable size variation of this species in nature in the Bahamas and suggests that this is correlated with variation in light exposure. In such cases, it is impossible to separate competition effects from those of absolute nonbiotic environmental adaptability.

The ultimate importance of competitive adaptability brings up the unfortunate concept of "ecological niche." It is commonly said that a species cannot become established unless there is a niche for it. This usage suggests an actual space as being a niche. Sometimes niche and microhabitat are used synonymously (Ashton, 1969, p. 175). This has led to an extension of the sense of niche as being an as yet uncolonized habitat—in effect, an empty space. This is a highly ambiguous usage since its meaning "empty space" disappears as soon as it becomes relevant— "filled" with some organism. On the other hand, Odum (1963, p. 27) defines ecological niche to mean "the role that the organism plays in the ecosystem"—its "profession." Mayr (1969) insists that niche was originally and rightly defined "as the requirements of a species. In other words, it was designated from the animal or plant outward, as something which the species requires in order to survive and prosper."

Richards (1969) discusses niche in relation to speciation in the tropical rain forest and concludes that "If it [a niche] can be applied in any exact sense to rain-forest trees it implies that species of each niche stand in a different relation to the resources of the environment from those of other niches, as well as to the other organisms in the ecosystem." In other words, he has adopted Mayr's definition but added as part of the meaning of niche the old and yet still debatable hypothesis that a surviving organism must have a unique niche. Connell and Orias (1964) neatly demolish most usages of niche as being illogical or irrelevant by pointing out that if we attempt to delimit niches a priori, "we find ourselves in a circular argument, since the number of niches is partially a function of the number and types of species present. We cannot then explain the number of species by the number of 'potential' niches."

In view of the considerable confusion and ambiguity surrounding the term, it would seem best to avoid it altogether. When we speak of an organism becoming established, we clearly think of it as occupying a physical space and being subjected to a complex set of environmental variables. Habitat and microhabitat seem clear enough terms for designating any such particular space and set of environmental variables, and when we wish to speak of the environmental requirements or limitations of an organism, it is fairly easy to say so. I will follow such usage in this chapter.

We have seen that although orchids may migrate into habitats of low competitive pressure, competition most often becomes a factor of environmental pressure later on. This might lead us to expect an eventual equilibrium of one or a few "most successful" species in each habitat. It is rare, however, for a single dominant orchid species to be found unless the habitat is so marginally favorable (environmental pressures so great) that few orchids are absolutely adapted to it (Sanford, 1968). Usually a branch of a tree will contain a number of species, and the number of species appears to increase with the favorableness of the environment. In southern Nigeria, for example, the most favorable sites average 14 species per tree, with a species/genera ratio of 1.6, while the least favorable sites average 4 species per tree, with a species/genera ratio of 1.0 (Sanford, 1968). In Venezuela, a felled tree in dense, montane forest was found to carry 47–48 species of orchids, with a species/genera ratio of 3.2 (calculated from Dunsterville, 1961). A similar picture seems to be true for terrestrial orchids in that wherever any one species of orchid is abundant, other orchids are likely to occur (e.g., Thomas, 1962).

There is usually no question, then, of an exclusive species in anything but a marginally favorable habitat. This means that an orchid soon after establishment is in competition with other orchids as well as with other plants: the habitat becomes an interacting community of plants. Runemark (1969) holds the view that a plant community is usually an integrated system of a number of ecological specialists inhabiting different "microniches." On the other hand, Poore (1968) finds that several species of rain forest trees appear able to occupy a single microhabitat—an area of the same specialized environment. This latter observation seems true of both epiphytic and terrestrial orchids. There are several possible explanations. The several species of a single microhabitat may be equally well adapted to the environment and so exist at a competitive equilibrium dependent upon the ratio of original propagules that became established at the habitat. This would, in effect, mean that a number of competitively neutral species had survived and, unless the original entry was into an uncolonized site, would necessitate a closer look at Runemark's reproductive drift hypothesis. It is also possible that what appears to be one microhabitat is rather a continuum of slightly different microhabitats, so that the species found there actually have different enough environmental requirements to keep them in an equilibrium controlled by slight variations of some factor(s) of the environment. It is certainly true that the human observer often oversimplifies and lumps together a number of microhabitats. For example, although it is obvious that very different environments are found as we move vertically up a single tree (temperature, light, and moisture all vary), it is easy to overlook that variation occurs horizontally along one branch of the tree as well.

Minute variations may also occur in terrestrial sites, where small differences in slope and surface contour affect water runoff, seepage, and leaching. Undersurface rocks may bring about sharp differences in soil depth with consequent differences in soil moisture and temperature. It is very likely, then, that many so-called microhabitats are collections of microhabitats, and that species occupying them reflect minute differences in environmental requirements and differences in competitive ability as well.

Even considering this, there do appear to be many instances of several species occupying the same microhabitat. This is superficially indicated by a random distribution—as opposed to regularity or clumping—of two or more species within an apparently uniform area. Such a situation has recently been reported for a few species of epiphytic bromeliads (Hazen, 1966), although the conclusions of this paper seem open to some debate. (See the following section on Moisture.) I have seen, in West Africa, tangles of *Bulbophyllum* Thou., *Polystachya* Hook., and *Calyptrochilum* Kraenzl., for example, on one branch of one tree. It is hard to believe that these physically touching species are occupying different microhabitats from the standpoint of environmental variation. It is more probable that their presence and distribution may reflect original establishment and either that the plants are now in equilibrium or that climax has not yet been reached and some of the species will eventually be eliminated.

That the vegetation piece under observation has not yet reached climax is another possible explanation for the seeming inhabitance of the same microhabitat by several species. The long life of the individual orchid plant makes elimination and the reaching of a stable or climax condition slow. Elimination of a species from a community most often comes about through its failure to produce enough propagules, or through subsequent failures of propagule establishment, so that the number of individuals of the parent generation is not reproduced. If the parent plant is long-lived, elimination will not occur for some time, even if successful propagules are not produced. (For example, Smith, 1966, draws attention to an *Epidendrum gracile* Lindl. plant in its natural habitat that is probably about 185 years old.) That single or few species dominance is not found among the orchids except in marginally favorable sites does suggest, however, that failure to reach climax is not the reason for several species being found in the same microhabitat. On the other hand, it is possible that age limits of the trees might prevent climax being reached. That is, the tree might normally die before its epiphyte load reached a stable condition. That trees may be shorter lived than their epiphyte colonies is vividly illustrated by Fowlie's (1966) report on *Broughtonia* Wall. ex Lindl.

Ashton (1969) suggests another possible mechanism for explaining equilibrium of several species in one habitat. He suggests that "intra-specific competition will always under favourable conditions give place to interspecific competition, which will evolve mutually avoiding species allowing higher total population densities and increasingly efficient utilization of habitats." The most significant corollary of this hypothesis is that we have ecosystems evolving, not separate species, and that evolution is toward increasing diversity of ecosystem components. Ashton finds considerable support for his view, such as Dobzhansky's early (1950) hypothesis that individuals that survive to reproduce are mostly those which possess combinations of traits that make them adaptable to the manifold interdependencies of the community. The concept of competitive replacement of species would apply, then, only to unstable or young ecosystems. Ashton borrows from the field of zoology to consider two *Drosophila* mutants that can evolve toward mutual avoidance to the extent that a habitat can support larger total mixed populations than either component separately (Seaton and Antinovics, 1967). To my knowledge no such situation has been experimentally shown to occur among plants, but the possibility seems very likely, and Harper and McNaughton (1962) have come close to it. They showed that if two species of *Papaver* Linn. occur together in different proportions but at the same total population density, the minority population is usually favored. This sort of minority species protection evidently occurs because the individuals of the same species, being more similar, are in stronger competition with each other than with individuals of other species in the community. Thus the minority species is protected from its chief competitor (its own kind) through increasing "outside" competition. The same mechanism might be presumed to operate in a population of mixed genera. Pielou (1970, p. 64) showed mathematically that two-species equilibria are possible. As she remarks, if such equilibria are mathematically possible, it cannot be proved that they do not in fact exist in nature; on the other hand, their existence may not be provable!

To summarize our thinking so far: orchid species may become established in sites where they do not demonstrate markedly superior adaptability over other species, because many can invade disturbed sites (e.g., cultivated land and pastures) or "open communities," sites occupied by generally noncompetitive plants (e.g., tree and shrub boles and branches in early stages of epiphyte colonization). Under such conditions reproductive drift is of little importance. In many other instances, however, and probably in most instances at later times, a number of plants come to occupy the same or adjacent habitats so that competition occurs. At many sites, a number of slightly different microhabitats exists so that species survival is controlled primarily by absolute adaptability to the environment rather than by relative or competitive ability. At others, a dynamic equilibrium of several species evolves as a community unit, with the numbers of individual species controlled by original establishment, balance between interspecific and intraspecific competition, and type of absolute environmental adaptability. In no case, however, would an orchid species not capable of absolute adaptability to the environment become established, nor would a species with inferior relative adaptability remain established over several generations, although a few plants might survive for long periods because of the great age to which many orchids live.

Environmental pressure, then, will initially act as a selective agency at the site of establishment; it will next operate at the relative level of competition with other plants; finally it will again operate absolutely with changes in the environment brought about by climatic or edaphic shifts, interference by man and other animals, and changes in the habitat brought about by the plants themselves which have developed at or near the habitat. Throughout the life of the plant there are two different types of possible response to such environmental pressure: (1) plants may differ in plastic response to the environment, even though they are genetically identical, because genes are present which are capable of variation in quantitative (and perhaps even qualitative) response to differences in the environment; (2) plants may differ in having different genotypes and thus different modes of response.

The early leading figure in the investigation of plant variation and the experimental separation, by transplantation, of the two types of variation was Gote Turesson. His basic propositions of genecology have been well summarized by Heslop-Harrison (1964). Turesson and his followers introduced a number of special terms to fit different levels of plant variation. The most widespread in use is "ecotype" (Turesson, 1922b), defined as an ecological subunit of the species, which is meant to cover the product arising as a result of the genotypical response (response type 2 as listed in the preceding paragraph) of an ecospecies to a particular habitat (where "ecospecies" is a genotype compound occurring over a "cline," that is, over a variational gradient of environmental conditions). Or, to put it as Heslop-Harrison does, an ecotype is a particular range on an ecocline. "Coenospecies" is used (Turesson, 1922b) as more or less equivalent to the larger concept of the Linnean species, which may include several ecospecies and thus a number of ecotypes.

The concepts involved with these terms are very important and useful, but it may be doubted if the splitting and resplitting of these three early terms has been a productive exercise. One tends to confuse linguistic manipulation with mechanistic explanation. The ecological aspects of species concepts as clarified by the Turesson school have helped to evolve the basic concept of "species" with which we began this section. I believe it is best here to use "species" in this biological sense and discuss it in terms of environment and genetics rather than to use extended Turessonian vocabulary.

Within a species concept, then, we normally have a great variety of forms. These forms either will be the results of plastic responses of the same genotypes to different environments, or will indicate different genotypes that have been selected for by different environmental pressures.

The question of how variable orchid species are is relevant at this point. Certainly the ORCHIDACEAE is an extremely variable group, as has been made obvious by emphasis of the tremendously high species number. A slightly different question concerns us also. How variable are the presently recognized species of the group? A number of temperate zone terrestrial orchids are famous for their variability [such as *Ophrys* Linn. (e.g., Danesch, 1970, 1969a, 1969b)]. Papers have also been presented on variation of tropical ephiphytes (e.g., Teuscher, 1967, on *Epidendrum vespa* Vell.). On the other hand, many species appear remarkably uniform; for example those Sanford (1968) has observed for West African epiphytes. Withner (personal communication) has drawn attention to the widespread *Epidendrum difforme* Jacq., which he calls *very uniform*. Perhaps both Withner and I have slightly overemphasized species uniformity and given an erroneous impression. Dunsterville and Garay (1961a, p. 114) remark that *E. difforme* is "very variable in size!" In my paper cited above, I admit to a number of slight variations among "uniform" West African species. Considerable variability was certainly recognized in most orchids by the nineteenth-century orchid hunters, growers, and collectors during that period when good forms of species rather than hybrids made up the bulk of collections. But, all things considered, it does seem remarkable that there is such a high degree of species homogeneity in so many orchids. Several possible explanations may be advanced. One is the extreme specialization of habitat preference. An orchid species often inhabits only a very limited area of fairly uniform environment; plants able to be successful outside such a specialized environment most often represent variant forms that have become another species—or are on the way to becoming another species.

A second possibility is that some species may have physiological plasticity which is not manifest in morphological variation. My own work (unpublished) has shown that secondary substances (such as the phenolics) may vary considerably from plant to plant and from time to time.

Finally, the orchids have been so extensively studied taxonomically and so highly prized horticulturally that nearly every genotypic variant is at once recognized as a species, or at least as a subspecies or variety, and this gives a false appearance of uniformity of taxa in relation to many other plant groups, where one taxon still includes a great variety of forms.

Variants resulting from plastic responses of identical genotypes will not be of taxonomic significance—if the taxonomist can recognize them for what they are—but are of great ecological and practical importance in indicating environmental differences with considerable precision and also in indicating the ability of the genotype to respond to such differences. Such plastic ability of the genotype may endow a plant with superior competitive ability and enable it to survive changes of the environment over time. One of Turesson's basic tenets (Heslop-Harrison, 1964) is that wide-ranging species show spatial variation in morphological and physiological characteristics, and that much of this variation may be correlated with habitat differences. This certainly applies to the ORCHIDACEAE as a whole.

The second group of variants, forms having different genotypes, will always be a thorn in the flesh of taxonomists because, no matter how many Turessonian terms are invented, it will always be a question whether or not to separate such forms and honor them with different names. Drawing the line between one species and another ultimately depends upon one's evaluation of discernible differences, upon his working definition of species, and upon the use put to the plant in question. To the genecologist such variants are of great importance because they either may represent incipient species or at least provide information on how speciation comes about. To the unwary, however, pigeonholing every genotypic variant as a distinct taxon may, as remarked previously, give a false impression of taxonomic tidiness and lack of variability.

Genotypic variants arise by gene and chromosome mutation and by recombinations of genes, through crossover, and recombinations of chromosomes through hybridization and introgression. Final speciation, and thus evolutionary speed and direction, will depend upon mutation and crossover rates, possibility of cross-pollination, environmental diversity, and pressure, and the formation of breeding barriers isolating new adaptive forms. Because orchids make up such a large group, it is superficially tempting to ascribe their great diversity to a basically higher rate of gene and chromosome mutation. There is no evidence for this, although a rather high incidence of polyploidy does suggest the possibility of a higher rate of this particular type of mutation. It is not yet possible to evaluate this, because so little is known about the ploidy of tropical plants. Morton (1961, 1966), studying West African plants in general, has claimed a relatively low incidence of polyploidy in the tropics, except for highland vegetation where polyploidy seems to increase with altitude, in a way reminiscent of the old concept of the "*nordfactor.*" At the time of his reports, the ploidy of so few tropical plants was known that it was impossible to evaluate his assumptions. This is still true in general, but extensive work in progress on ferns has already given some idea of the total picture (Manton, 1969).

Polyploidy seems to occur in about 60% of the species in West Tropical Africa, Ceylon, and Jamaica as compared to 53–54% in Britain, Hungary, and Southern India, while less than 50% appears in the Himalayas. (Unfortunately, South America is still largely unknown.) Manton postulates that "part of the land mass represented by Malaya, India, perhaps some of the islands in the Indian Ocean together with the Mediterranean basin which is also rich in diploids may represent an ancient tertiary flora which except for local variants such as Ceylon has been less disturbed by the climatic vicissitudes which elsewhere seem to have stimulated speciation by polyploidy in both temperate and tropical regions." The important point here is that climatic vicissitudes have placed greater environmental pressure on plants so that adaptive tolerances, which may accompany polyploidy (Barber, 1970), have been strongly selected for in certain regions. Whether this would be borne out by the ploidy of orchids is not yet known, but I would hazard a guess that for this type of mutation—and for all types of mutation—quantitative differences are not correlated with either the tropics or the temperate regions or with any particular plant groups but rather with environmental variability and consequent selective pressures. The extreme dispersibility of orchids has placed them in such a variety of habitats that mutations not only would be selected for through great variety of environmental pressures but often could be isolated easily from the parent forms and so preserved. This would superficially give the appearance of a higher mutation rate.

Hybridization is a different matter, and there can be little question that orchids are very unusual in their crossability, not only at the species level but at the genus level as well. Wallbrunn (1969a) has discussed this, suggesting that the remarkable ease of orchid hybridization might have as reasons (1) the huge number of ovules and pollen tubes present in each flower; (2) the rapidity of orchid evolution, which has resulted in the formation of a number of species not yet diverged far enough to be incompatible; (3) faulty definition of species and genera, which has resulted in a situation where so-called hybridization is not hybridization at all; or (5) that orchids are "just different." His first point is reasonable—certainly there is a greater chance for fertilization the more ovules and pollen tubes there are, and numbers may run up to the millions in orchids. Withner (personal communication) has pointed out that even if two triploids are crossed, there is a chance of some fertile seed being produced. The next two possibilities, relating to species divergence and species and genera definition, involve us in something close to circular thinking and thus are not helpful in elucidating the problem.

Wallbrunn's point that orchids are "just different" is the most useful observation; van Steenis (1969) reminds us that orchid species have evolved through the development of pollinator specificity as opposed to the incompatibility mechanisms that are usual in other plants. [Most often the specificity is achieved by a combination of flower scent and structure (van der Pijl and Dodson, 1966), although color may often play a role as well. For a brilliant example in another area see Lewis, 1969, on *Phlox*.] Thus if this breeding barrier that segregates the population, and so maintains the species, is removed by hand pollination, it is not surprising that seed is formed. The relevance of this to speciation and environmental adaptability under natural circumstances hinges on whether or not the barrier of pollinator specificity can be bridged in nature. Van der Pijl (1969) significantly pointed out that times, conditions, vegetations, and, most important to the present discussion, pollinators change, and the importance of selfing and crossing shift. Crossing in orchids may occur when new pollinators invade an area, or, more significantly, when orchids invade a new area where different pollinators may be found, and when new pollinators evolve. Since orchid pollinators are largely insects, which comprise such a mammoth and variable group, there is considerable chance of local stabilization between orchid species and insect pollinator, and considerable chance for the development of new relationships even in nearby localities. Extremely interesting examples are given by van der Pijl and Dodson (1966). It is also possible

that when orchids move to new areas which may require new adaptations, new pollinators may make possible genic recombinations leading
to structural changes that effect new pollinator specificity. A considerable
possibility of coevolution of insects and orchids exists. Recent work has
been conducted on the chemical identity of the sugars of orchid nectar
(Jeffrey and Arditti, 1969). It is possible that subsequent work will show
specific pollinator dependence on certain complex sugars or other materials specific to an orchid species.

Withner (personal communication) notes the possibility that cleistogamy has been selected for in some orchids when the species migrated to
areas where no pollinator was available. He cites examples of cleistogamous *Epidendrum* Linn. of the section *Encyclia* Hooker in a recent paper
(Withner, 1970). The possibility of this evolutionary direction is fascinating but extremely rare in orchids. A related phenomenon, the production
of parthenocarpic "seeds," is also rare, although the example of *Zygo-*
petalum intermedium Lodd. ex Lindl. (Z. "*mackayi*") is well known.

However, selfing is rare in orchids and this brings us back to the question of the role of hybridization in speciation. A general review of the
problem may help us to hypothesize on the orchid situation. Hybrids
appear to be exceedingly uncommon among rain-forest trees, and the
most likely explanation is that there is strong selective pressure against
hybridization (Ashton, 1969). On the other hand, few tropical forest species are able to colonize artificially cleared sites, while a number of examples exist of successful colonization of such areas by hybrids. (As, for
example, hybrids referred to by van Steenis, 1969, and *Hevea* Aubl., by
Seibert and Baldwin, as quoted by Stebbins, 1950, p. 264.) The reason
for this is that plants in a more or less undisturbed habitat are adapted
to that habitat or they would not have survived there. Genic recombination—as well as mutation—simply by being different gives rise to a great
preponderance of genotypes not adapted to the environment. It is also
probable that species adaptation is based to a large extent upon epi-
static gene interaction (i.e., interaction at different loci), in which case
the complex interacting system will very easily be upset by hybridization (Mayr, 1969; Stebbins, 1969). Rarely, recombination may give rise
to plants adapted to a different environment, hence the survival of hybrids perhaps at points of natural environmental transition and most certainly at points where disturbance has led to removal of most competitors as well as change in the nonbiotic environment. More rarely still,
recombinations may give rise to individuals better adapted to the same
environment inhabited by the parents (i.e., having superior competitive
adaptability).

In spite of the ease of hybridization, if the difficulty of pollinator specificity is overcome, epiphytic orchids appear to follow the general pattern, outlined above, of hybrid rarity. It is remarkable how little apparent hybridization relative to the size of the group there is among them (Sanford, 1968), although some interesting examples of hybridization have been reported (e.g., Fowlie, re. *Broughtonia* Wall. ex Lindl., 1961e; Dodson and Frymire, possible hybrid swarms of *Oncidium serratum* Lindl., 1959; Withner and Stevenson, *Oncidium tetrapetalum-pulchellum* syngameon, 1968). Apparently, then, the barrier of pollinator specificity is seldom removed in nature. On the other hand, the extreme frequency of hybridization among many terrestrials, especially those of the temperate zones, has made orchid taxonomy here almost a farcical chess game. (For excellent illustrations of hybrid forms see Nelson, 1962, 1968, Danesch, 1970, 1969a, 1969b; Sundermann, 1969; also the discussion of Reinhard, 1970). This may be due to a combination of the factors of less specificity of pollinator, together with the disturbed, noncompetitive nature of the habitat of these terrestrials.

Since terrestrial species are relatively "primitive" (i.e., like older forms), they might reasonably be expected not to have developed the complex pollinator-specificity barrier to interbreeding found in the advanced forms. The question then remains of how these terrestrial species have achieved speciation: Have terrestrial species always been so fluid and relatively short lived or have recent changes in environmental conditions broken down former barriers of the type common to most land plants? There seems to be some reason to suspect the latter. Many terrestrials have been, throughout the past, spatially-ecologically isolated. The development of agriculture in Europe probably eliminated a number of species and gave many of the surviving forms the status of weeds: plants successful mainly or exclusively in disturbed areas. [As Walter (1971, p. 2) remarks, natural biotic communities are nonexistent in Europe because of the extreme and long disturbance by man.] Such disturbed sites have become nearly continuous, both from the standpoint of open communities suitable for potential seed establishment and from the standpoint of continuous successful foraging area for the bees and flies that pollinate orchids. A great deal of study remains to be done before this situation is satisfactorily clarified.

Both the limitations and the importance of hybridization in evolution have been lucidly discussed by Stebbins (1969). How this discussion can be applied to orchids remains a largely open question. We have seen that hybrids in orchids occur easily if pollinator specificity is overcome; for the resulting hybrids to play a part in evolution and speciation rather than be only transitory variants, breeding barriers must separate the forms before they are swamped by the wild (parent) stock. This may be easier among the orchids than among most plants because, as we have emphasized, orchids exist in small populations in very specialized habitats; there is a good chance, then, that a hybrid may not be surrounded by a large number of individuals of the wild stock. Furthermore, the ready dispersibility of orchid seeds leads readily to spatial separation of the hybrids. With small, spatially scattered populations, the development of slight differences in blooming time becomes an important isolating mechanism. This has been discussed for tropical forest trees (Ashton, 1969). Evidence for perhaps over 10% of the epiphytic orchid species of West Africa having developed forms varying in blooming time has been presented recently (Sanford, 1971). Variation in blooming season might allow enough isolation for the development of subpopulations. Occasional flowering overlap or migration into new areas where environmental factors forced a shift in blooming time would allow hybridization.

The most important factor in the creation of breeding barriers is the extreme diversity and specialization of the habitat: it is quite easy for a hybrid to be spatially separated from the parent types either by emigration or through the ability to survive in a habitat different from that of the parents. Thus diversity of microhabitat acts not only as a barrier mechanism which preserves mutants and hybrids but also as the mechanism of environmental pressure which selects for variety of forms in the first place.

Environmental pressures have been recognized, at least since Darwin, as the primary selective factors guiding the course of speciation and evolution. There was a time, recently, when the environment was given a secondary position in relation to the size of the gene pool, but the environment and ecological dynamics have come back into their own again (e.g., Ehrlich and Raven, 1969). The two factors are not independent: a large gene pool will tend to be built up in a region of diverse habitats whereas a uniform environment will tend to lead to a small pool. An interesting controversy has arisen, however, concerning the role of environmental pressures in the tropical rain forest. Since orchids, although technically cosmopolitan, are primarily tropical, this argument becomes relevant here.

In 1940, Dubinin (cited by Fedorov, 1966) wrote that it was quite likely that the relatively small size of the populations of tropical organisms made possible the rapid initiation of new races and species by means of the accumulation both of adaptive characters and of diverse indifferent (i.e., neutral) characters fixed by means of automatic genetic processes as opposed to natural selection; that is, by genetic drift.

In 1966, Fedorov argued that the environment of the tropical rain forest (presumably meaning the wet, evergreen tropical forest) was too uniform and too nearly optimal to provide the environmental pressures necessary to account for the great plant diversity found there. He further pointed out that the tropical flora is remarkable in containing complete, uninterrupted series of taxa, particularly species connected by indubitable affinity. (Orchids would have provided many excellent examples for his argument.) He noted that, contrary to the usual situation in temperate zone floras, closely allied forms exist side by side, apparently maintained as separate forms largely by differences in blooming times. He concluded that in the tropical rain forest, in the absence of any considerable disturbance, there is no—or very little—competition and no species is superseded by other species closely allied to it. Diversity, then, has been built up through genetic drift, the random fixation of non-adaptive mutations and not through environmental selection. Species are maintained almost solely by differences in blooming times.

Following Fedorov, an attempt was made by Sanford (1968) to evaluate the relationship between Nigerian epiphytic orchid distribution and gross environmental pressure, particularly moisture availability. No single species of orchid predominated at favorable sites, and species of the same genera were found to occur together on the same tree. While this was in agreement with Fedorov's basic set of observations, it was noted that considerable variation occurred in the microhabitats of one tree. Thus no evidence was given by this study for lack of environmental pressure even at the most favorable sites. Neither was there any evidence for the lack of competition assumed by Fedorov; rather, the density of epiphytes at favorable sites suggested inevitable competition.

Ashton (1969) has argued against Fedorov's view. We have already noted that he suspects that several species can inhabit the same microhabitat, not because of lack of competition, but rather because environmental pressures have led to the evolution of ecosystems as units with mutual avoidance relationships and diversity protection of individual species through the minimization of intraspecific competition. On the other hand, van Steenis (1969) inclines toward Fedorov's view, pointing out how very low the selective pressure is in the tropical rain forest. Even granting this debatable point, reproductive drift should work against Fedorov's condition of the random fixation of variant forms: the variant form being represented by one or a few individuals would, if adaptively neutral—and, according to Fedorov's view of the lack of environmental and competitive pressures, new forms would necessarily be neutral—be randomly eliminated.

It appears to me that environmental pressures and the consequent selection of successful forms have probably been important in orchid evolution even in the tropics and that random fixation of characters—genetic drift—is important only in small populations on new sites where competition has not yet developed. A profitable approach to this fundamental problem would be a more detailed examination of the orchid habitat, in order to determine the presence or absence of environmental pressures, and of the interspecific and intraspecific distribution of plants within habitats, particularly whether or not distributions were random.

Habitats

Presumably the first, and so by definition most primitive, orchid habitat was the soil. And presumably the relationship between orchids and mycorrhiza developed very early evolutionarily while all orchids were still terrestrial, or possibly even before "orchids" evolved from their ancestral group. It is remarkable that this relationship is so ancient and even more remarkable that it has not been lost in any of the present-day species, terrestrial or epiphytic.

That some *Bletilla* Rchb.f. seed will develop aseptically without added sugar (Withner, 1959, p. 237) does not indicate that this genus has lost its mycorrhizal habit. Rather, it merely indicates that some of these quick developing seeds have enough stored carbohydrate to bridge the gap from water imbibition to protocorm development and greening, with consequent photosynthesis. Normally, *Bletilla* are invaded by mycorrhiza, fed with sugars digested by the fungi, and develop much more vigorously than do aseptically cultured seeds. It is true that the fully differentiated seedling of all green-leaved orchids can photosynthesize enough carbohydrate so that the plants could be independent of mycorrhiza. Nevertheless, the fungi that invaded the embryonic tissue often remain associated with the orchid throughout its life.

One interesting area of seed germination study has been neglected: I would guess that the bark-moss-liverwort environment of some tropical epiphytes contains a high enough concentration of sugar and aminonitrogen leachates to allow protocorm development without mycorrhiza. This has not yet been shown, however, and it is possible that bacteria and yeasts on the site may utilize carbohydrates more rapidly than can the orchid seed.

There can only be speculation as to how the mycorrhizal relationship was originally established. Such speculation has not been changed substantially or improved upon since Rayner's work (1928). He regarded mycorrhiza as ecological phenomena resulting directly from the inevitable competition in soil between roots of vascular plants and the mycelia of numerous soil fungi. This relationship was originally facultative, or quasi-parasitic. A stable symbiotic relationship developed by way of the evolution of control mechanisms of susceptibility, resistance, and immunity in the host plant. Resistance of the orchid to fungal destruction is apparently effected at least partly by the orchid synthesis of phytoalexins—chemical substances inhibiting fungal growth. [The first recognition of phytoalexins was, according to Fisch et al. (1972), by Noel Bernard about 1910 working on *Neottia nidus-avis* (L.) Rich. The first chemical isolation and characterization of a phytoalexin also concerned an orchid: a chemical isolated from *Orchis militaris* L. by Gäumann and Jaag in 1945. As Arditti (personal communication) has remarked, anything that can be done with plants can be done better with orchids!]

The orchid-fungal relationship, although very broadly "symbiotic," may vary a great deal, as has been carefully reviewed by Arditti (1972). That such a relationship should have evolved in the ORCHIDACEAE seems less remarkable to us when we note the frequency of soil fungi and the frequency of their having established definite relationships with plant groups other than the ORCHIDACEAE. For example, it has long been recognized that a very high proportion of tropical trees possess mycorrhiza: thus, out of 75 species examined in Java, 69 contained mycorrhiza (Richards, 1957, p. 220). Burgeff has quoted Stahl to the effect that, with a few exceptions, all land plant families that grow on dry soil contain members having mycorrhiza (Burgeff, 1959, p. 361).

True mycorrhiza, as opposed to soil fungi inhabiting the rhizosphere, may be divided into (1) obligate saprophytes showing variable substrate preference, and (2) obligate root-inhabiting forms, dependent on living tissue, and showing specific substrate demands (Harley, 1959, p. 12). The mycorrhiza of orchids are of the latter type, although specificity is not very great, the same fungi being able to penetrate several genera.

The dependence of orchids on mycorrhiza for seed germination and seedling establishment has been a major factor in habitat determination since the very beginning of the family. Any possible exception to this would be in environments rich in sugar and amino-nitrogen leachates from plant material (see above) and would constitute very special habitats. Present-day species may remain dependent on mycorrhiza throughout life or may become independent by maturity (Summerhayes, 1968, pp. 14–15). When Withner (1953) cultured orchid root sections aseptically, he found no fungi in any of them. This indicates at least a considerable degree of independence of adult plants from mycorrhiza, and raises the very interesting question of what has occurred in the internal environment of the orchid root to make it no longer hospitable to the fungus. The orchid which is dependent upon mycorrhizal association is, technically at least, partially saprophytic, since organic compounds are supplied to it via diffusion and active transport from the fungus. The term "saprophyte" is, however, generally restricted in usage to apply to those plants not able to photosynthesize and so incapable of utilizing carbon dioxide for the production of organic substrate. Thus all fungi are saprophytes and some orchids—at least the nongreen species—are. This classification becomes awkward, as do all classifications at some point, when we consider orchid species that have either only a few green leaves or green roots and so are able to synthesize only a small part of the organic materials they need. Some of the African *Eulophia* R.Br. fall into such a category. For example, while *E. galeoloides* Kraenzl. is clearly a leafless saprophyte, I find both the very similar *E. brevipetala* Rolfe and *E. warneckeana* Kraenzl. to have such rudimentary or small leaves that they also must depend to a considerable extent upon saprophytism. Another interesting example is that of *Gastrodia africana* Schltr. This peculiar species, allied to the Australian *Gastrodia*, was described by Schlechter early this century. The type specimen was lost and no record of the existence of this plant remained except for the Latin description until Letouzey collected it in Cameroon in 1971. This long-lost orchid is a leafless saprophyte, but it has green flower scapes, indicating limited photosynthetic ability.

A somewhat different situation is presented by the leafless orchids [e.g., *Chiloschista* Lindl. from Malaya, *Campylocentrum* Benth., *Harrisella* Fawcett and Rendle, *Polyrrhiza* Pfitz., some species of *Vanilla* Plum. of warm and tropical America, *Dipodium punctatum* (Sw.) R.Br. of Australia, *Cymbidium macrorrhizum* Lindl. of Sikkim, and the genera *Chauliodon* Summerh., *Encheiridion* Summerh., and *Microcoelia* Lindl. of Africa.] These have chlorophyll-containing roots, but how much organic material is synthesized and how much is absorbed via mycorrhiza or directly from leachate, drip, runoff, and decaying material of the phorophyte has not been determined (see also Kerr, 1972; Teuscher, 1972).

Epipactis purpurata Sm. presents an interesting case of variable saprophytism. The type specimen is without any chlorophyll and such completely saprophytic forms are still commonly found. Chlorophyll-containing plants are even more common, however. Although the two forms are morphologically identical, Hunt (personal communication) reports that the saprophyte is much more robust and rapid-growing. He has suggested that chlorophyll metabolism may be an inhibiting factor. Certainly this highlights the extreme efficiency mycorrhizal saprophytism may attain in the ORCHIDACEAE. On the other hand, Walter Vöth, head orchid grower, University of Vienna Botanical Gardens, tells me that the chlorophyll-containing forms are invariably larger and more vigorous!

An extension of mycorrhizal saprophytism into parasitism is found in Australia where *Gastrodia cunninghamii* Hook.f. is reported to live almost entirely underground. The tubers are covered with a network of mycelia of the fungus *Armillaria mellea* (Vahl.) Quel., which penetrates the living roots of an adjacent tree, most often *Nothofagus* Blume. This might be thought of as parasitism by proxy. Another species of *Gastrodia, G. minor* Petrie, grows only in association with *Leptospermum scoparium* Forst., a scrub plant of poor soil. *Gastrodia sesamoides* R.Br. grows among roots of *Acacia melanoxylon,* R.Br. The mycorrhiza of this orchid penetrate the living, bacteria nodule-bearing roots of the *Acacia* Mill. (Campbell, 1964).

The Australian orchids *Cryptanthemis slateri* Rupp and *Rhizanthella gardneri* R.S. Rogers live and flower 8–30 cm below the surface of the ground. *Corybas saprophyticus* Hatch. lives on the surface but is most often covered with up to 15 cm of litter (Hatch, 1953). These are completely saprophytic; nor can the possibility of some degree of parasitism be entirely excluded. The genus *Corallorhiza* (Hall.) Chat. comprises a well-known group of new world saprophytes.

Arditti and Ernst recently (1972) drew our attention to the fact that orchids logically should be considered as parasites upon saprophytic fungi. This is certainly true during at least a part of their life cycle, but we do not know much about the apparent gain of independence by many species after the seedling stage.

Summerhayes (1967) has pointed out that a seedling of the European terrestrial *Orchis ustulata* L. may live underground, in association with its mycorrhiza, for 10–15 years before producing the first leafy aerial stem. (He has since scaled downward the length of the underground period but still feels it is often very considerable; personal communication.) Withner (personal communication) reports much the same thing for the North American *Cypripedium acaule* Ait.

Terrestrial orchids then may exist rooted in the ground but with normal photosynthesizing aerial parts, or, either permanently or for variable periods of time, completely underground with no photosynthesizing parts. Trophic gradations appear to exist from mycorrhizal saprophytism throughout life, to mycorrhizal association only during the early stages of development, to the development of individuals without mycorrhiza. Saprophytism itself may grade from complete to partial, and a condition is found in a few species which can be considered parasitism (see Arditti and Ernst, 1972).

The more "normal"-growing terrestrial orchids, starting as chlorophyll-less masses of saprophytic tissue nourished via mycorrhiza which soon develop into stem- and leaf-bearing green plants, show considerable variation in habitat. These habitats have been variously classified for convenience. In general, there have been two approaches: habitat classification according to (1) nonbiotic environmental characters such as soil type and (2) biotic characters, usually the associated vegetation type. Neither classification can be exact, and considerable difficulty is encountered in meaningful definition. Fuller (1933), for the Wisconsin area of the United States, has, for example, classified orchid habitats according to dominant plant associations: prairie, conifer forest, oak-hickory forest, pine barren, mixed hardwood forest. This represents a combination of physiognomic and taxonomic definition of the vegetation types involved and is subject to the same lack of precision as all such classifications, but it is serviceable for the restricted area for which it was made. The same may be said for Summerhayes' (1968a) habitat characterizations for Great Britain: woodland, scrub, grassland—except that here the classification is more nearly physiognomic. (Characterization of vegetation is still a major problem of modern ecology and will be touched upon in the final section of this chapter.)

Case (1962) categorizes orchid habitats according to major environmental factors, both biotic and nonbiotic: (1) soil requirements, such as heavy clay, sandy or peaty soil, black humus; (2) freedom from competition; (3) mycorrhiza; (4) acidity-basicity; (5) soil temperature; (6) exposure. Precision of definition may be even less here, both because of variation from microhabitat to microhabitat within a very small area and also because of measurement difficulties. For example, it is not clear how mycorrhiza would be characterized either qualitatively or quantitatively in practice, or at what times of the day or year or at what depths soil temperatures would be taken. The measurement of competition from other plants is a highly complex parameter requiring such extensive study that it would be an impracticable means of characterizing sites.

That ancient, earth-growing orchid ancestors were able to utilize fungal attacks for their own nutrition was probably a great factor in their survival and evolutionary development. When the next peculiar step of taking to the trees as epiphytes occurred, we cannot know. The success of this evolutionary leap was in finding a habitat where competition was at a minimum. The preceding adoption of mycorrhiza, by means of which cellulose and other complex organic materials could be broken down and utilized, made life on the bark surface of trees possible. The main problem in the new perch was—and is today—water conservation. For this reason, it has often been suggested that epiphytic orchids must have evolved from xerophytic forms. Although it is true that most present-day epiphytes have some xerophytic modifications, there is no evidence that they have evolved from xerophytic ancestors. Water-conserving modifications may as well have occurred and been selected for after early orchids took to the trees in very moist, nonseasonal sites. Such modifications would then have allowed expansion into new, less competitive habitats.

Overall characterization of the epiphyte habitat presents the same sort of difficulties as does that of the terrestrial. Crude categorization according to major differences in nonbiotic environmental requirements is possible; for example: exposure, that is, sun-shade; temperature, as the familiar horticultural classes of cool, intermediate, and warm; moisture, as wet, moderate, dry. And, as for terrestrials, classification according to dominant associated vegetation types is frequently encountered, such as epiphytes of lowland semideciduous, tropical forest; foothills; montane woodland (e.g., Sanford, 1968, 1969b, 1970a). (See Plates 2, 3, 4.) Again, precision of characterization is a major problem and will be discussed in the final section of this Chapter.

Speaking very generally, the two most outstanding determining factors of the epiphyte environment are moisture availability and phorophyte character. Both of these aspects are discussed in following sections, but the relationship between phorophyte and general habitat environment is best treated here.

In what is probably still the major ecological paper on epiphytic orchids, Went (1940) reports that specific relationships between orchid and phorophyte species is usual. On the other hand, Brieger (personal communication) has remarked, apropos of the thousands of tree species in Brazil, that one of the happiest moments of his life was when he found that there was no specific relationship between host species and orchid. Several more or less casual observations seem, at first glance, to support Went. Allen (1959), working in Panama, reports an interesting situation where branches of *Licania arborea* Seem. actually intermingle with those of some oak species and yet *Laelia rubescens* Lindl. occurs almost entirely on the *Licania* Aubl. Sulit (1950, 1953), in the Philippines, writes of "the uncanny preference shown by many epiphytic orchids for certain species of trees." He states that about 80% of the

Plate 2. High-forest, mature regrowth, with mixed structure, Equatorial Guinea (Rio Muni), February (rainy season). (Photograph by W. W. Sanford.)

Phalaenopsis Blume *spp.* occur on *Diplodiscus paniculatus* Turcz.; over 95% of *Vanda sanderana* Rchb.f. on several species of dipterocarps; *Dendrobium taurinum* Lindl. never grows on anything but *Pterocarpus* L. *spp.*, and so on. Withner (personal communication) mentions *Pseudo-laelia vellozicola* (Hoehne) C. Porto and Brade believed to grow only on *Vellozia*. Vand. Lecoufle (1964) reports that a Madagascar endemic, *Cymbidiella rhodochila* Rolfe, grows only on the staghorn fern, *Platy-cerium madagascariense* Baker, and never on other *Platycerium* Desvaux species growing in the vicinity. He guesses that the fern is necessary for germination of the orchid seed. He also reports that *C. humblotii* Rolfe grows only on the raphia palm. Kennedy (1972) confirms the narrow habitat range of *C. rhodochila* but suggests the reason is largely the necessity for an acid substrate; he has had great success in growing the orchid in redwood shavings or palco wool. He has not, on the other hand, been able to grow *C. humblotii* away from its palm host.

Somewhat more generally, it has often been noted (probably first in the *Orchid Review*; Anon., 1906) that the calabash tree, *Crescentia cujete* (L.) Sesse and Moc., is a favorite host of orchids. It is interesting

Plate 3.   Montane woodland between 6000 and 7000 ft, Mt. Sta. Isabel, island of Fernando Póo, February (dry season). (Photograph by W. W. Sanford.)

that I have never been able to find a single orchid on the scattered calabash trees growing in Nigeria. Leakey (1968) remarked that *Eurychone rothschildiana* Schltr. is uncommon in Uganda, apparently being restricted to a very few host trees, most commonly *Spathodea nilotica* Seem. (= *S. campanulata* Beauv.). In southern Nigeria the orchid is, on the other hand, quite common and found on a great variety of phorophytes, most usually on small trees and shrubs in thickets; I have never found it on *Spathodea campanulata*, which occurs commonly here. A rather similar situation is provided by comparison of the mango, *Mangifera indica* Linn., as a phorophyte: Jesup (1966) reports that the leafless *Dendrophylax funalis* Hort. Ames and Nash occurs on this tree in Jamaica. In West Africa I have never found any orchids on mangoes, ex-

Plate 4. An example of Northern Guinea Savanna, Boubanjida Forest Reserve, northern Cameroon, March (late dry season). (Photograph by W. W. Sanford.)

cept at Douala—a very wet region on the coast of east Cameroon. Again, *Ansellia gigantea* Rchb.f. var. *nilotica* is said to be associated with the baobab tree (*Adansonia digitata* Linn.) in Malawi (Morris, 1968), but I have never found it on this tree in Nigeria; here it seems to favor the *Borassus* palm.

These observations indicate the importance of the totality of the habitat as opposed to a single combination of factors such as the phorophyte species. One of the few professional ecological papers that gives some attention to the total forest environment in relation to epiphytes is that of Eggeling (1947) on a rain forest in Uganda.

An extreme situation is illustrated by the frequent observation that the same orchid species may be either epiphytic or epilithic—or in some cases even terrestrial—at different sites. For example, in least favorable sites, I find the African orchid, *Tridactyle tridactylites* Schltr., only as an epiphyte. In more favorable sites, it is frequently found on rocks as well. I have seen similar versatility in several species of *Bulbophyllum* Thou. and *Polystachya* Hook., and less commonly in such species as *Calyptrochilum christyanum* (Rchb.f.) Summerh., *Aërangis kotschyana*, Schltr., and *Angraecum distichum* Lindl.

A remarkable example of many epiphytic orchids growing both terrestrially and epilithically is given by Dunsterville (1964), describing a table mountain in Guyana. Allen (1959), in Panama, mentions that *Laelia rubescens* Lindl. usually grows on *Crescentia alata* H. B. and K. (= *Parmentiera alata* Miers) but if this tree is scarce, the orchid may grow on rocks in lowland valleys. (This is the same species observed by the same author to be sometimes specific on the tree *Licania arborea* Seem.!)

The total environment of moisture, temperature, exposure, nutrients, light, and available space must, therefore, act together in determining the microhabitat where any particular orchid species may be found, and, indeed, where any particular tree serving as phorophyte may be found. Most epiphytes can be found growing epilithically or even terrestrially under certain conditions. They are more adaptable in this way than are the more primitive terrestrial orchids which are normally limited to a specifically defined soil. [Although I have seen a few cases of terrestrial *Habenaria* Willd. and *Eulophia* R.Br. growing epiphytically in humus-filled tree branch crotches or *Platycerium* Desvaux brackets; *Liparis nervosa* (Thub.) Lindl., commonly thought of as a terrestrial, is more at home as an epiphyte at some sites.]

The realization that epiphytes may be extremely adaptable does not alone solve the problem of phorophyte specificity. No one will argue that the phorophyte must be in a suitable site and must itself provide a suitable microhabitat for the orchid. Nor can it be argued that some orchids are particularly adaptable in regards to phorophytes and to different total environments. For example, *Ansellia gigantea* Rchb.f. may be found on almost every kind of tree in Rhodesia, from palm to baobab (Dickenson, 1969). It is the case of the epiphyte which is not adapted to a wide environmental range and therefore not adapted to a range of phorophytes that provides ground for argument.

The most extensive and careful studies of epiphyte-phorophyte relationship have concerned cryptogamic rather than vascular epiphytes. Barkman (1958) reviews and discusses this field, including methodology, in detail. More recently, Kershaw (1964) has developed a clever method of estimating percentage cover of lichens on trees in Britain. His impressive study has led him to conclude that while there are no specific lichen-species tree-species relationships, tree species do have characteristic lichen spectra. I believe that a similar complex relationship, reflecting total environment as well as phorophyte species, would probably be found to exist between epiphytic orchids and phorophyte.

The first methodological step in any kind of relationship study is that of gathering statistically significant data. This may be extraordinarily difficult. With the exception of the Went (1940) paper, all orchid reports thus far cited have been subjective or intuitive estimates rather than actual counts or systematic observations (and even this paper cannot be subjected to statistical validification). Sulit (1950) notes that *Diplodiscus* Turcz. is predominant in the area where *Phalaenopsis* Blume occurs mainly on this tree. He states that the percentage occurrence of the orchid on this particular tree is higher than the percentage occurrence of the tree in the area, but it is difficult to be sure of such estimates in the tropical forest. Again, when he mentions that *Vanda sanderana* Rchb.f. occurs almost exclusively on dipterocarps, the problem is that dipterocarps are very abundant in the area, and may be almost the only emergents and thus the sole or at least main providers of a special environment for factors only secondarily associated with the tree: light exposure, air movement, temperature, atmospheric humidity.

As to cryptogamic epiphytes, Barkman early concluded that there is no simple correlation between phorophyte species and epiphyte species. Instead, the relationship depends on the interplay of a number of complex environmental factors, beginning with the overall climate and site environment and possibly broken down into such categories as (1) habitat of tree, (2) light periodicity, (3) amount of light transmitted, (4) type of crown, (5) bark relief, (6) scaling of bark, (7) hardness of bark, (8) water capacity and related properties, (9) resin, (10) tannins, (11) lime content, (12) PO_4 content, (13) total salt content, (14) actual acidity, (15) buffering capacity.

It seems logical that such factors may determine the presence of orchids as well as cryptogams. The most attention has so far been given to bark relief and scaling, probably because these factors are most easily observed *in situ*. Thus Garnett (1929) early observed in South America that in "jungle" areas he sometimes encountered trees literally covered with epiphytes and adjacent or intermingling patches of trees with few or none. Trying to explain such mosaic occurrence, he found that the various orchids did not grow exclusively on one particular kind of tree, and he was at a loss for an explanation. Ultimately, bark texture provided the clue. Sulit (1950) concluded that barks are the main explanation of the correlations he noted (see above). He observed that phorophyte barks are nearly always finely canaled and made up of small scales, and that bark-shedding or peeling occurs only after many years' growth. He suggests that the fine scales are suitable for seed germination. Dickinson (1968) noted that *Laelia speciosa* Schltr. in Mexico prefers a small-leaved oak but will adjust to other rough-barked trees or even to cactus. On the other hand, *Masdevallia* Ruiz. and Pav., on the Central Cordillera, is reported to favor smooth trunks (Lager, 1932). Allen (1959) states (for Panama) that orchids generally shun trees with smooth or peeling bark. He suggests that this is due to difficulty of seed establishment. Together with these observations, it should be remembered that an "understorey" of lichens, liverworts, and small mosses provides a suitable environment for seed germination and seedling growth and that Barkman and the workers whom he cites have made it clear how important bark texture is for colonization by such cryptogams.

Went (1940) states that it is his belief that bark is responsible for orchid-phorophyte specificity, and he largely discounts both the physical structure of bark and its water-holding capacity as the variable factors leading to the specificity. Rather, he feels that it is largely the chemical composition of the bark and the composition of the water runoff and leachate from it.

Recently Sanford and students (University of Ife, unpublished) examined seven chemical fractions extracted from barks of a number of West African tree species, selected for heaviness and paucity of epiphyte load. Relatively high sugar concentrations were found, the highest being sucrose, with significant amounts of glucose, fructose, xylose, and glucuronic acid, and lesser amounts of several unidentified sugars or sugar acids. Phenolic compounds and amino acids were also determined to be present. Amounts of all these compounds varied with tree species. A possible correlation was noted between rough-barked trees in optimal locations but with few orchids and high phenolic content of the bark. Recent laboratory experiments by Frei (1973) indicate inhibition by some barks, but at relatively high concentrations and under conditions not necessarily applicable to natural situations.

A few species of orchids are occasionally able to become established in very odd microhabitats. Withner (personal communication) reports seeing small orchids—probably *Comparettia* Poep. and Endl.—growing on mango leaves. Since the leaves last about three years, this suggests that the orchid may complete a life cycle within this relatively short time. Even more bizarre, Bowling (orchid propagator, Kew; personal communication) told me that he had once seen a tiny *Microcoelia* Lindl. growing on a spider's web in Ghana! I very much doubt, however, that either leaves or webs will ever become very important orchid habitats.

In West Africa, my own observations are in agreement with the view that specific phorophyte-epiphyte species associations do not occur, but that the complex environmental factors determining the presence of tree species together with the many variable characters of the tree itself determine the presence of the orchids. Thus certain species are likely to carry definite orchid "spectra." Morris (1970) makes the same observation for Malawi. He writes that "few specific host/orchid relationships exist" and that "the majority of orchids are found on a wide variety of trees" but that some trees are "orchid prone" with 71% of the orchids in the Blantyre-Mlanje area found on five species of tree. It is apparent that the less favorable the site, the more specific is the phorophyte-epiphyte relationship because the orchid, near the margin of its adjustment range, becomes very dependent on the ameliorating characteristics of the tree, especially its bark. On the other hand, in very favorable sites, orchids can be found almost anywhere on almost any kind of tree. For example, in moist forest, orchids may be found on both smooth- and rough-barked trees, whereas in drier forests the same orchids are only on rough-barked hosts.

We have reviewed the general orchid habitats of earth and phorophyte. It has also been remarked that orchids may grow on rock surfaces (i.e., may be lithophytic or rupiculous). This habitat appears possibly ambiguous, representing an extension of terrestrial species on the one hand and of epiphytic ones on the other. Kapuler (1962), writing of Colombia, referred to a few orchids growing on exposed mountain cliffs as "true lithophytes." I doubt that such a category of orchids exists. In this case, for example, we find that *Pleurothallis velaticaulis* Rchb.f., which he lists as a true lithophyte, occurs very often in Peru upon trees (Schweinfurth, 1958). *Catasetum barbatum* Lindl. generally grows terrestrially but may be even more luxurious as an epiphyte (Dunsterville, 1967). Examples abound of orchids growing epiphytically or in humus piles on rock (e.g., Fowlie, 1961a, 1961b), as do descriptions of orchids growing on cliffs (e.g., Berkeley, 1894; Horich, 1960), but it is more realistic to look upon orchid species as either predominantly earth-growing or tree-growing, with rocks often serving as a compromise for either habitat—more often for the tree. South American observers, however, feel that some orchids such as some of the Brazilian *Laelia* Lindl. may be "truly lithophytic." (Withner; Urpia; personal communication). I must admit to being impressed by thousands of *Brassavola nodosa* (L.) Lindl. plants growing on rocks along the sea in northwestern Colombia.

Now that we have reviewed the basic habitats of orchids, we are ready to consider the major environmental factors that define these habitats. Any treatment of so complex and dynamic a system must be greatly simplified and so, to a degree, falsified. But in spite of this warning, there is, I believe, much to be gained by examining at least the most obvious variables: light, water, food, temperature, other plants, and animals.

Major Environmental Variables

Moisture.   The moisture available to the plant may be considered on the large scale of climate and weather and on the small scale of microclimate. Most available data concern the large-scale aspects. Such information provides a crude foundation for the environmental conditions under which plants grow. Moisture is expressed as daily precipitation and as monthly and annual means; as relative atmospheric humidity in the morning and early afternoon, or occasionally as saturation deficit. On a smaller scale, Piche evaporation rate is sometimes obtainable. Most often overlooked in large-scale consideration is the distribution of moisture both over the day and over the year.

The relevance of these aspects of moisture availability to the orchids of a locality is very different for epiphytes and terrestrials. For example, a foggy climate with little precipitation is drier to terrestrials than a rainy one with little fog, but to the epiphyte it is moister (Barkman, 1958). Epiphytes will be saturated after a few minutes of rain; continued precipitation will merely run off. On the other hand, the soil may require a relatively long soaking to become wet below the surface. Thus to epiphytes the distribution of rain is much more critical than its amount. High atmospheric humidity is important for water conservation and is thus much more important to the exposed epiphyte than to terrestrials.

Water may be present in four forms: (1) as a liquid film covering the particles of substrate or, if excessive, filling the spaces between substrate particles as well; (2) as fine droplets in the air, that is, mist, cloud, fog [Boynton (1969) measured cloud water in an Elfin Forest in Puerto Rico and found it to be only slightly less than 10% of the rainfall, a very appreciable amount.]; (3) as fine droplets, dew, condensed on vegetation and substrate surfaces; (4) as water vapor in the air. In all of these forms excepting the last, water is in a liquid state. There is no question but that most water used by the plant, epiphyte or terrestrial, is absorbed in the liquid form; the question of water vapor absorption is more complex. When water vapor is absorbed, both latent heat and free energy are released, but the direction of vapor transfer is determined by the potential gradient of free energy. In the plant cell, the free energy deficit is the difference between the osmotic potential energy of the vacuolar fluid and the elastic potential energy of the cell wall. Monteith (1963) terms this "diffusion pressure deficit" (a better term is "water deficit") and states that deficits are normally met by transfers from one part of the plant to another, but direct uptake of atmospheric water vapor is possible. Leaves at the wilting point (conventionally considered as 15 atmospheres water deficit) will absorb vapor from an atmosphere with a relative humidity exceeding 98.8%. For a plant surviving great water deficit, absorption is possible when relative humidity is as low as 92% (Slatyer, 1960). It is doubtful that absorption of vapor is appreciable in orchids because a situation of such extreme water deficit in the plant coupled with such high atmospheric humidity is rare. Conceivably this situation could exist with epiphytes at high altitudes in exposed positions and possibly also with epiphytes in the crowns of lowland forest trees during the dry season. The thickness of the cuticle, however, would probably prevent much foliar absorption except through the stomata. The velamen of the root, contrary to earlier popular belief, is not a structure for water absorption but rather for water conservation and is of very limited permeability (Dycus and Knudson, 1957), so probably little water vapor would be absorbed through the aerial root surfaces. Vapor absorption is made even more doubtful by the low osmotic concentrations of tissue fluids in orchids (Harris, 1918, 1934), although very limited work has been done in this area. (See also Walter, 1971, p. 105-107.)

Mention of aerial root and foliar permeability brings up the question of how liquid water is absorbed by the orchid. It is primarily the root that accounts for this. Some years ago, foliar feeding was popular for greenhouse orchids, and although both foliar absorption and leaching have been shown to be significant phenomena for plants in general (e.g.,

Tukey and Mecklenburg, 1964), the heavy cuticle of most orchids determines that it cannot be important for this group and probably least so for the epiphytes. That some small amount of absorption does take place through the leaves is likely and that it is important in thin-leaved orchids living in very moist habitats is possible, but it is the root that must assume the burden of liquid absorption. The root of the epiphyte is made impermeable to a considerable degree by the velamen (Dycus and Knudson, 1957), so that most water absorption is presumably through the portions of the root surfaces which are adpressed to the phorophyte surface. Here the root is structurally modified so that velamen is absent or modified, and absorption can occur freely (Freire, 1939). Some species in which many of the roots persistently hang free do, however, present possible exceptions requiring study (e.g., such leafless orchids as *Microcoelia* Lindl.).

Sanford and Adanlawo (1973) have studied velamen and exodermis of 76 species of epiphytic orchids occurring in West Africa from the standpoints of number of velamen layers, wall striations and hairs of the velamen, wall thickening and lignification of velamen and exodermal cells, and uniformity of shape and size of cells. High correlation was reported between the number of velamen layers and habitat tolerance: species of environments with severe dry seasons and species growing in exposed habitats had thick (i.e., many cell layered) velamens, whereas moist-growing species had only one layer of velamen or only the epivelamen and one layer of velamen. Thickness of the wall of the exodermal cells was also shown to be correlated with habitat tolerance. Modifications of the velamen and exodermis with affixation to the substrate were remarkable: the velamen became modified to one cell layer in thickness, the cells were thinner walled and usually of a different shape. Furthermore, the exodermis became modified in structure, the cells coming to resemble cortical cells. The authors postulate that the main function of the outer root layers, especially the exodermis, is conservation of water, through prevention of transpiration and evaporation from the cortical tissue.

The function of velamen postulated by Went (1940) may also be mentioned. He points out that the very first runoff from the bark contains the highest concentration of dissolved minerals and that this may be trapped by the empty velamen cells.

Another root peculiarity is found with some epiphytic species having apparently two types of root: one type goes downward toward the phorophyte surface where the roots spread out in the normal way; another type grows upward in the same direction as the shoots, forming a dense mass of tiny, spearlike roots. This is commonly observed in *Ansellia* Lindl. species and in many *Oncidium* Sw., sometimes in *Catasetum* L.C. Rich. ex Kunth and *Cycnoches* Lindl. Latif (1969) observed that *Grammatophyllum speciosum* Bl. adapts to growing terrestrially in thick humus where two kinds of root develop: short ones pointing upward and a heavy bunch of enormous adventitious roots in the humus. Such upward-growing roots might possibly be of adaptive advantage to the plant in catching debris and holding it for decay. It is doubtful whether they would be significant as supplementary photosynthesizing or gas-exchanging organs.

Sanford and Adanlawo (1973) compared free portions of "normal" roots of *Ansellia gigantea* Rchb.f. var. *nilotica* (Bak.) Summerh. which grew along the substrate with those that grew upward into the air. The former coarse roots had velamen of five to seven cell layers, whereas the latter had only one layer of velamen. Beside this difference, exodermal walls of the upward-growing roots were significantly thinner and less lignified. These differences suggest that upward-growing roots may have absorptive functions.

Upward-growing roots also present an interesting problem in root growth control. In this area there is little to go on, however, since the control of root growth in epiphytic orchids has not been seriously studied. Roots cannot be much influenced by gravity since they normally spread over the phorophyte surface in all directions, up and down, though most often horizontally. Possibly growth control is exerted through moisture and nutritional gradients; possibly tactile stimulation and negative phototropism play some part.

Rain, fog, mist, and cloud may all wet the phorophyte surface and provide water for absorption by the epiphyte. Rock surfaces are similarly easily wet for epilithic (rupiculous) plants. Terrestrials generally require rain, although some tropical woodland and forest orchids have their spongy root systems near enough the soil surface or forest litter that wetting is easy and rapid, even with fog or mist. Dew is probably not of much importance in providing water to orchids. Dew as an ecological factor has been generally reviewed by Stone (1957). Formulas for dew formation together with evidence that while formation is virtually independent of climate, it does increase with relative humidity and varies from crop to crop are given by Monteith (1963). Generally, epiphytic orchids grow in spots sheltered enough—even though high in the forest canopy— to ensure that there is not sufficient difference between leaf and air temperature for condensation to occur. Exceptions are epiliths and epiphytes of exposed habitats, especially on seasonally deciduous trees and in montane regions. Above the mist zones of tropical mountains, dew may be an appreciable factor in providing moisture, although few orchids are found here. [*Sophronitis grandiflora* Lindl. may possibly be an example of an orchid benefiting from dew. Gardner (1846) early described this plant occurring as the only orchid at about 6000 ft in the Organ Mountains in Brazil. It usually inhabits exposed locations, whether on trees or rocks.]

Barkman (1958, p. 64) discussed both large- and small-scale aspects of water economy and his Figure 15, reproduced here as Fig. 1-2, although specifically relating to cryptogamic epiphytes, gives a very good idea of the complexity of water relations for epiphytes in general.

In the case of forest-growing plants, whether terrestrial or epiphytic, the amount of water reaching the plant is variable. Some of the rain falling on a forest area falls through the canopy and may or may not strike the bole; a portion is retained by twigs and leaves where a small amount is absorbed by the plant but more evaporates; a portion is conducted along the twigs to the periphery of the crown, where—together with some water from leaf surfaces—it drips unevenly to the floor; a portion is conducted by twigs and branches back toward the trunk where it runs down the bole, not evenly but in various channels depending on exposure, tree form, bark characteristics. Measurements in an Elfin Forest in Puerto Rico (Boynton, 1969) have shown trunk flow to account for 21% of the rainfall and the canopy to have a storage capacity equal to a depth of 0.035 in. or 886 ml/m². The forest floor, then, receives somewhat variable amounts of water depending on the trees above it, while epiphyte positions on the tree vary tremendously in amount of wetting received. The amount of wetting may also depend on the leaves of the orchid itself. Many have leaves so oriented and folded that they act as troughs channeling water to the roots of the plant. This is, incidentally, the probable explanation for the efficacy of foliar feeding: the leaves drip the fertilizer down to the roots!

One study of tropical forest in the Caribbean (Clegg, 1963) showed that rainfall interception by the trees varies inversely with the amount of rain, so it is especially significant in the critically important brief rains of the dry season. This is very probably true of all forests, tropical and temperate.

A few early attempts were made to measure precipitation running down the tree bole. Voth (1939), working in tropical rain forest on Barro Colorado Island in the Canal Zone, found that collecting cups on small *Annona* L. *sp.* and *Myriocarpa* Benth. *sp.* trees collected six to seven times more water than rain gauges in the open. This means that the crown

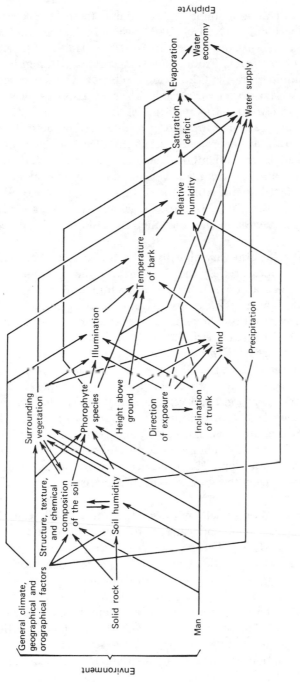

Fig. 1-2. Indirect (left) and direct (right) factors on epiphytes, as illustrated by the complex factor water. Taken from the original diagram with the permission of Van Gorkum Assen, The Netherlands, from J. J. Barkman, *Phytosociology and Ecology of Cryptogamic Epiphytes*, 1958.

47

diverted a considerable amount of the rainfall to the trunk, which conducted it to the ground. Curtis (1946), using the same type of "collar" bole collector as Voth, measured rain runoff in Haiti from an arc of 6 in. (15.3 cm) on the trunk. From a single 2-hr rain of 14 mm he collected 1060 cm³ of liquid from his 6-in. panel collector. On the other hand, I have often observed dry or nearly dry boles in dense tropical forests in Africa during rainstorms. The position of the tree in relation to other trees, as well as its own growth form, is a determining factor in rainwater conductance down the bole. Such variation seems to have been somewhat overlooked in the Boynton (1969) paper cited. Variation in the amount of precipitation striking a surface and running along a surface effects sharp distributional differences in cryptogamic epiphytes over very small areas (van Oye, 1921, 1924), which in turn, as we have seen before, effect differences in suitable microhabitats for orchid seed establishment.

Evaporation rates also vary sharply on a microscale. The daily march of evaporation as measured by the Piche evaporimeter is most extreme in unshaded clearings at 1.5 m above the ground but almost as great in shaded clearings and in the canopy of rain forest at 30 m. On the other hand, there is very little change in evaporation rate during the day in the rain forest at 10 cm above the floor. [For Surinam, see Fig. 1-3 (from Schulz, 1960, Fig. 36).] Exposure and air movement may be especially important. Barkman (1958) has shown for temperate-zone cryptogamic epiphytes that moving air, with consequent rapid evaporation from the bark surface and from the epiphyte mass itself, may be a more decisive factor than air humidity in epiphyte occurrence. Stefureac (1941) reported that on hills in Rumania, epiphytic cryptogams reached highest up tree trunks on the side facing the slope, the height being proportional to the steepness of the slope. He explained this by the fact that evaporation will decrease near the ground where there is less wind and higher air humidity.

Atmospheric humidity itself will also vary on a small scale. The contrast between forest floor and canopy in Surinam is well brought out by Schulz's work (1960). During the dry season, the relative humidity in the canopy at 2 PM may be as low as 38% with a saturation deficit of 24 mm, while at 10 cm above the ground readings only as low as 74% and 7 mm are obtained. These figures are roughly comparable to those available for Southern Nigeria in semideciduous tropical forest (Evans, 1939). Cochan (1963), working in the Ivory Coast, noted vertical stratification of microclimates in the evergreen forest.

Such microvariation also occurs in the temperate zones and undoubt-

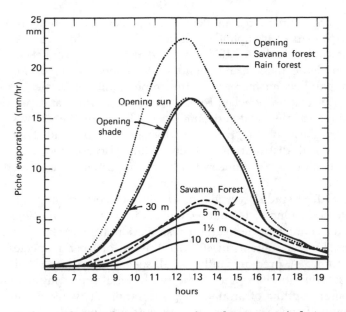

Fig. 1-3. Daily march of Piche evaporation (in arbitrary units) during part of the dry season of 1957 (28 Sept.–23 Oct.) at 10 cm, 1½ m, 5 m, and ca. 30 m height in rain forest, at 1½ m height in savanna forest, and in an extensive clearing. Meters sheltered from direct sunlight, except 1 m in the opening which was exposed to direct sunlight. Reproduced by permission of N. V. Noord-Hollandsche Uitgevers Maatchappij, Amsterdam, from J. P. Schulz, *Ecological Studies on Rain Forest in Northern Surinam*, 1960.

edly affects terrestrial orchid distribution there. Studies of temperature and humidity variation at different sites and exposures have been reviewed by Boerboom (1964). Potzger (1939), working on a ridge in Central Indiana (U.S.), reports the great effect of exposure on evaporation rates. For example, evaporation was found to be 61% higher on the southern slope than on the northern, while the surface soil had 30% more moisture on the northern slope.

Moisture considerations, together with light intensity, are the factors largely responsible for the positioning of epiphytes on the tree. That different plants are found at different vertical levels has long been recognized. Very early, Schimper (1903, p. 199) stated that even in the most pronounced rain forests one finds xerophilous types of epiphytes on upper limbs of the tallest trees, whereas lower limbs and trunks have hygrophilous plants. Shreve (1914) slightly later discussed vertical distribution of epiphytes in montane forests in Jamaica, noting that lowest level epiphytes are hygrophytes, midlevel ones somewhat drought resistant or else close to mats of bryophytes, while top-level epiphytes are xerophiles.

Hosokawa and Kubota (1957) worked with the resistance to desiccation of epiphytic mosses and found that an increase in osmotic values generally parallels an increase in the upper limits of vertical ranges on the tree. Correlation between resistance to desiccation and position was also found, with the most resistant mosses being in the crowns and the least at the base of the trunks. Culberson (1955) and Hale (1952) studied vertical distribution of lichens and bryophytes in temperate forests and van Oye's pioneer work in the tropics has already been mentioned.

Harris (1971) studied vertical positioning of corticolous lichens in the temperate zone. One of his interesting findings is that variations in optimum water content for net carbon assimilation are correlated with habitat: optimal water contents of 40% saturation for treetop species and for the top forms of predominantly basal species were recorded. Species with basal distributions and those with basal-type physiologies had water content optima of around 75% saturation. It would be interesting to determine if such variations occur in epiphytic orchids.

Shreve (1908) worked in Jamaica and remarked on the exactness of the vertical distribution of macroepiphytes. He classified the strata as follows: lower—PIPERACEAE, ferns; mid—a few water-storing PIPERACEAE, orchids; high—bromeliad. He noted structural adaptations of the higher strata species concerned with water storage and conservation. Some *Stelis ophioglossoides* Sw. leaves, for example, were observed to have water-storage tissue making up nearly two-thirds of the leaf thickness. Thick epidermis, heavy cuticle, and sunken stomata are other water-conserving modifications often noted. A good drawing of *Brassavola tuberculata* Hook. that shows thickened cuticle and a sunken stoma with a raised "collar" of cuticle around it is given by Arens and Pedroita (1948). These workers also compare water loss from shade and sun and report that as much as 66.66% of the sample weight may be lost without death occurring. I have found that excised leaves of *Calyptrochilum* Kraenzl. may remain alive in the laboratory for over one month. The nineteenth-century shipments of orchids to Europe from Asia and tropical America even more vividly demonstrate the resistance of orchids to desiccation.

Beside modifications in stomatal structure, differences in frequency and behavior may be expected. Epiphytic orchids so far observed have stomata only on the lower leaf surface, but frequencies in relation to habitat have thus far been studied only very superficially. Preliminary counts have correlated lowest frequency with most xerophytic habitat (Sanford, unpublished). Differences in stomatal opening and closing have been studied in other plants: for example, Rawitscher (1948), working on the Campos Corrados of Southern Brazil, showed the following possibilities: (1) stomata open without regulation throughout the day; (2) stomata partly closed during the driest hours; (3) stomata closed throughout the day. In the last case, only cuticular transpiration occurred. Presumably the carbon dioxide for photosynthesis is absorbed during the night. Such adaptation, shown by many succulent plants, for night absorption of carbon dioxide through open stomata, may be found in some orchids, although this has not yet been shown. Appreciable dark-fixation of carbon dioxide by various orchids has, however, long been known. (See recent papers by Nuernbergk, 1964, and Knauft and Arditti, 1969, which also briefly review past work.) This carbon dioxide fixation may be associated with stomatal opening during the night. Such a condition would be most likely in fleshy-leaved xerophytic forms such as *Oncidium cebolleta* Sw., to take an extreme example.

Water loss by transpiration has been measured for a few orchids and sharp differences between species have been found (e.g., Spanner, 1939; Kamerling, 1912). Rates have not, however, been extensively or systematically studied in orchids and not studied at all in relation to habitat, but see Walter (1971) for water loss in relation to orchid plant form. It is relevant that tree leaves in the canopy of an Elfin Forest in Puerto Rico have been shown to transpire even at relative humidities of 80–100%, with leaf temperatures probably as low as 15–25°C (Gates, 1969). Presumably orchids, too, transpire even in the very high humidities that are usual in the habitats where they live.

Transpiration of desert plants under different environmental conditions (el Rahman and Batanoung, 1965) follows evaporation rate, as measured by Piche evaporimeter, in the microhabitat where they live. Wide differences, varying with habitat, such as these workers found, could reasonably be expected to be found in orchids and give added significance to differences in evaporation rates at different forest microhabitats, as cited earlier.

Studies of the vertical distribution of angiosperm epiphytes are numerous. Went's (1940) paper is the earliest to give attention to such distribution for orchids alone. Hosokawa (1943), working in Taiwan, enumerated epiphytes as occurring on (1) trunk bases, (2) trunks, (3) crown bases, (4) crowns, and (5) leaves. His attempt at quantification at these sites is impressive, and he was able to show clear species-position correlations. Schnell (1950), reviewing the tropical forest—primarily of Africa—states that while ferns tend to be low on trunks and in the shade, orchids tend to be on upper branches where they are subject to periods of brutal desiccation, particularly during the dry season. He observes that many orchids occurring in such sites have water-storage organs. He classifies epiphytes into two groups: (1) those of the trunks and lower branches—*strate épiphytique inférieure,* and (2) those of the high branches—*strate épiphytique supérieure.*

The most detailed vertical distribution studies, however, have concerned bromeliads in Trinidad. Pittendrigh (1948) follows earlier authors' concepts of epiphytes as more or less independent vegetational units organized into definite communities (also the tendency of Hosokawa et al., 1957). This premise has perhaps led him to pay insufficient attention to the interrelationships between vascular and cryptogamic epiphytes and the host trees (phorophytes) and total environment, but he does recognize three "mechanical features" of the phorophyte affecting epiphyte position: (1) bark texture; (2) frequency of forks; (3) angle of limb orientation. He feels that these features account for nonrandom distribution of bromeliads on a particular tree species. He also recognizes that the canopy microclimate may be important, most especially in respect to light, and that the exposure a species of bromeliad can endure also depends on the rainfall of the area. Actual positions of various species are given in distances above the ground and in forest profile drawings. Such a treatment would be applicable to orchids but would prove even more cumbersome and imprecise. Accurate comparisons of such data would be impossible.

More recently, Hazen (1966), working exclusively with bromeliads, has tried, unsuccessfully I believe, to quantify epiphyte location. He converted all positions to linear values and thus obscured possible horizontal differences at different vertical heights. Furthermore, he worked with only one identified species plus one or more unidentified, immature species so has not considered species differences but rather only plant frequency along a horizontal. An analysis of four branches showed random plant distribution. This is of little general significance to epiphyte distribution, and it is based on such limited data that the conclusion is open to question. Random plant distribution is almost never encountered in nature.

Initial studies in Nigeria (Sanford, unpublished) show that position of epiphytic orchids on the tree varies with orchid species, with growth habit of the tree, with the position of the tree in relation to other vegetation, and with overall microclimate and climate of the area.

Except for extremes in temperature, moisture availability is the chief factor determining large-scale as well as small-scale orchid distribution. Sanford (1968) studied species number and species/genus ratio per tree in relation to moisture and characterized epiphyte occurrence throughout Nigeria in relation to precipitation and humidity (Sanford, 1969a). Thorold (1952), working on the blackpod fungus disease of cocoa in Nigeria, examined the relationship between incidence of this disease, epiphyte load on the cocoa trees, and rainfall. He used a clever system of epiphyte rating in which absence and presence of one and presence of two or more (a) leafy liverwort colonies, (b) moss colonies, (c) juvenile vascular plants, (d) adult vascular plants were combined in a 7-point cumulative scale, with a tree having all receiving 7 and a tree with only mosses and liverworts receiving 2. He found that the three groups of epiphytes were related to dryness and wetness, with liverworts alone at the driest sites and vascular plants at the wettest, and that correlation of mean total epiphyte rating with annual rainfall and incidence of blackpod disease was high. My treatment of his data by correlation analysis does not, however, show significant epiphyte and rainfall relationship to disease incidence. Ordination of the data suggests other factors primarily affecting the incidence of the blackpod.

Hardy (1962) brought together available information on climate, including both temperature and moisture, and incidence of *Odontoglossum* H. B. and K. species in South America. Vacin (1952) discussed the original Asian distribution of *Cymbidium* Sw. in relation to rainfall.

An elementary survey of the climates of the main tropical orchid countries is given by Volkert (1958). A vegetation map from any standard atlas is slightly more useful, and more detailed climate and vegetation maps are more useful still (e.g., Keay, 1959, for Africa; Hueck, 1955, for South America; Nuttonson, 1963, Thailand; Goldman, 1951, Mexico; Gaussen et al., 1967, southeast Asia; Meher-Homji, 1963, and Labrove et al., 1965, India). A well-detailed paper giving rainfall, relative humidity, and physiogeography in relation to vegetation in the West Indies has been published (Loveless, 1960), as has a similar work for Madagascar (Humbert and Darne, 1965). Excellent maps and climate and weather charts but little vegetation information are given for Brazil by Aubreville (1961). Curtis (1947a) related the distribution of orchids in Haiti to climate and vegetation and found that the occurrence of most species is closely correlated to a main vegetation type which is in turn largely determined by moisture availability.

The most detailed comparison of lowland and montane forest vegetation that I know has been made in Ecuador (Grubb and Whitmore, 1966). A forest at 1710 m was compared in detail with one at 380 m.

Perhaps the most interesting conclusion to come out of this work is that fog and cloud are more important than temperature in distribution of plants from lowlands to highlands. This clarifies some puzzling earlier views, including that of Richards (1957, p. 135; p. 216), who stated that change in temperature with increase in altitude rather than precipitation was primarily responsible for change in forest type on mountains. My own observations (e.g., Sanford, 1967) are that moisture is more important than temperature in orchid distribution except at very high altitudes—that is, above 7000 ft—where temperature does become critical. It is important to note, however, that the crucial factor is not precipitation but moisture in the form of cloud and mist (see Plate 5.).

Light.   We have already seen how very closely related light and moisture are in determining the distribution of orchids, especially epiphytes. We will now try to emphasize the special effects of light alone. Light may vary in spectral composition, intensity, and duration. These aspects in turn affect the plant primarily through photosynthesis, tissue heating, transpiration and evaporation, and control of periodic functions of the circadian type (e.g., many metabolic functions, oftentimes flower scent, and flower and leaf movements) and of the long-period type (e.g., vegetative growth and dormancy, flowering).

If we consider light on a large scale (climatologically), the most significant variables are duration or day length and intensity. At the equator, dark and light periods are equal the year around, but as we move north and south differences between the length of the day and night become seasonally cyclic. Since orchids are predominantly a tropical group and thus native to regions where differences in day length are slight, possible effect of these slight differences becomes a subtle area for investigation. It has been hypothesized since at least as early as 1948 (Bünning, cited by Richards, 1952, 1957, p. 192) that tropical plants are more sensitive to small differences in day length than are temperate-zone plants. Day length control of the flowering of some tropical orchids has been demonstrated, and the subject is well reviewed by Rotor (1951, 1959) and Arditti (1966, 1967a). A recent study of photoperiodism in West African orchids (Sanford, 1971) claims photocontrol of vegetative dormancy for some species. In many cases in this study, it was difficult or impossible to separate temperature effects from those of light, but in spite of this difficulty, the flowering of at least 35% of West African epiphytes and perhaps as many as 27% of the terrestrials quite clearly appears to be photo-influenced.

That light intensity may vary tremendously on both macroscales and microscales is obvious. It has remained more of a problem to determine

Plate 5. Mist layers in northern savanna highlands in the dry season, between Meiganga and Ngaoundéré, northern Cameroon, March. (Photograph by W. W. Sanford.)

55

how the quality of light may vary. There is certainly a relative increase in ultraviolet content at high altitudes, and the effect of this on alpine plants has been reviewed by Caldwell (1968), who claims that the UV content of light, under natural circumstances, does not seem a significant factor in montane environments. It is commonly held, however, that UV together with visible light near the UV range and in the green range repress plant growth (e.g., Klein, 1964; Klein et al., 1965), while various morphogenetic effects of light of different wave lengths have been noted (e.g., on fern gametophytes, by Miller and Miller, 1969). It is extremely doubtful that such effects constitute significant environmental variables *in situ*.

It is true that Bernard, in 1945 (cited by Richards, 1952, p. 148), working in the Central Congo Basin, reported deficiency in the blue and UV range of light there. Such supposed spectral differences have sometimes been popularly—and erroneously—claimed to promote the extreme lushness of plant growth which is characteristic to the tropics. On the other hand, Schulz (1960, pp. 53-56), summarizing his own work in Surinam and that of other workers in the tropics, points out that spectral measurements tend, superficially at least, to show a strong decrease in the region of maximum chlorophyll absorption. If this were true, photosynthetic efficiency might conceivably be reduced, although to such a slight extent as not significantly to affect plant growth.

The most intensive early work on the quality of light was done by Orth (1939), who showed that, in a number of African forests, the maximum excess in the red part of the spectrum is likely to fall beyond the red end of the visible spectrum. He also cites evidence, however, of an increase in the transmission of the forest canopy just beyond the red end of the visible spectrum in European beech woods (Seybold, 1936). Evans (1939) reported similar findings in southern Nigeria. Ashton (1958), working in Brazil, concluded that spectral composition in the tropical forest does not differ essentially from that elsewhere. Federer (1966), working in the temperate zone under corn, sugar maple, oak, pine, spruce, also found a very high maximum in the near infrared. Robertson (1966), for the temperate zone, reports that a closed crop canopy transmits a high proportion of far red energy. He notes that increased haziness or decreasing solar elevation increases the proportion of red and far red and that decreasing solar elevation results in decrease of both UV and blue energy. Both decrease in solar elevation and increase in space containing haze particles through which light travels account for spectral changes during a day—early morning to evening. On a wider time scale, it should be remembered that in much of the lowland tropics there is considerable moisture haze throughout the day in the wet season and considerable dust haze in the dry season, both of which would tend toward increasing red and far red transmission.

Evans (1969) very recently reviewed the technical problems involved in the qualitative measurement of light in the field. The complexity of the problem casts considerable doubt on the results reported by earlier workers who did not have the advantage of sophisticated instrumentation. Even though considerable caution is exercised, it remains quite certain that UV and blue-range light *is* relatively reduced under vegetation and that near infrared *is* increased. That slight differences may be correlated with vegetation type has been brought out by the work of Vehagen and Wilson (1969), who show that propagation of light inside a canopy depends on the structure and optical properties of the foliage.

In summary, it is likely that there is not much spectral difference between the tropics and the temperate zones if sites comparable in vegetation are chosen for measurement. An exception may be some average increase in the tropics—at least during part of the year—in the red and near infrared brought about by atmospheric haze. Slight decrease in the UV and near UV may also occur. It is doubtful that either is enough for it to be a significant factor in plant growth.

Quantitative measurements of light appear easier to handle, but the extreme variation on a microspace scale and on a minute-by-minute time scale make this problem complex also. For example, Evans (1956, 1969) in Southern Nigerian forest calculated that the contribution of continually shifting sunflecks to the noon light reaching a plant in the undergrowth may be as high as 80%. Allee (1926), in Panama, found abrupt rises in light intensity between lower treetops (12–18 m) and the upper canopy (23–25 m). The same is probably true in Brazil (Ashton, 1958). We have, then, extremely complicated variation in light intensity: (1) on an ever-shifting, mosaic-patterned horizontal plane brought about by sunflecks which are determined by the canopy above; and (2) a gradient variation on a vertical plane extending from emergents of the canopy through the canopy and down to the forest floor. Clearly, many different microhabitats—in respect to light—may occur within a very small region of forest.

Whether quantitative differences in light reaching the orchid consistently occur between the tropics and the temperate zones has been debated. Certainly in many parts of the tropics the growing season is a period of extremely low light intensity because of almost continual rain and cloud, while the bright sunny part of the year is too dry for significant growth to occur.

It has been popularly claimed that the tropical forest is a dark and dismal place because of the lushness of tree growth. Orth (1939) claims, however, that under comparable conditions of sunlight, light intensities in African forests are of the same order of magnitude as those in European forests, and he strongly denies that plants living in undergrowth in the tropics should be adapted to lower light intensities than those found in deciduous forests in temperate regions. I believe that tropical forests are too variable to make Orth's limited view generally applicable.

Orchids may be crudely sorted into those preferring open habitats, woodland habitats, or forest habitats, but we have seen that micro-habitats within these classes—especially within the latter two—may vary as much as, or more than, between classes. Furthermore, other environmental variables may modify or compensate for light effects. It is common knowledge, for example, that orchids will stand much greater exposure if root moisture is continually available and atmospheric humidity high. The relationship between light and temperature is also important. A portion of the light absorbed by the plant is transformed to heat and in this way the light intensity tolerated by an orchid is related to ambient air temperature and air movement, transpiration and general water relations, and heat tolerance of the protoplasm.

Leaves of common plants are reported (Loomis, 1965) to absorb 80–95% of the blue (400–500 nm), 60–80% green (500–600 nm), and 80–90% of the red (600–700 nm). Absorption of infrared is about 5% at 800–1200 nm and increases to nearly 100% beyond 3000 nm. Leaves tend to assume the temperature of the surrounding air. They are heated rapidly by radiant energy and cooled primarily by conduction of energy to the air—hence the importance of some air movement and one importance of high atmospheric humidity, since water vapor more readily absorbs and conducts heat than does dry air. Leaves exposed to sky radiation but shaded from sunlight are typically 1°C warmer than the air. In sun, leaves heat rapidly to equilibrium at a rate nearly inversely proportional to their mass and to a temperature 6–10°C above air temperature for thin leaves and 20–30°C higher in the case of leaves several millimeters thick. Loomis, along with most plant physiologists, feels that transpiration is relatively unimportant in cooling leaves, that conductance and convection are the major factors.

Gates (1968), however, believes that because transpiration and leaf temperature are results of the interaction of several simultaneous environmental factors, the two have not been properly related experimentally. He states, "If the plant physiologist ever has had a doubt concerning the value to the plant of transpirational cooling as it affects leaf temperature, this gross misconception should be dispelled now once and for all." He admits that there are conditions when variations in transpiration rate may make little difference in leaf temperature but insists that often the ability to transpire will make a substantial difference in temperature when the heat load on a leaf is large and thermal death is likely. He cites experimental evidence to show that a transpiring leaf may be as much as 10°C cooler than a nontranspiring leaf when the air temperature is 40°C. This appears to be true, however, only for relatively large leaves.

Too little is known about transpiration in orchids to make possible the evaluation of its role in temperature control. The epiphytic orchids with which I have done preliminary work have had transpiration rates too low to affect leaf temperature measurably. Transpiration may, however, be important in leaf temperature control of thin-leaved terrestrials growing in the open, such as American and European marsh and pasture orchids.

As for knowledge of critical temperature of orchid leaves, I know only of McDade's (1947) incomplete report on the result of dipping leaves into heated water for various time periods. He found that *Cattleya* Lindl. leaves submerged 2 hr at 110°F (43°C) or 30 min at 120°F (49°C) or 10 min at 130°F (54°C) were injured or killed.

Similarly, there is almost nothing available on light in relation to maximum photosynthesis in orchids. Davidson (1964) states that we can assume that orchids respond as other plants and so are able to utilize from 1500 to 2500 fc of light. Anderson (1967) provides an excellent review of chlorophyll content and photosynthesis under natural conditions for plant communities—excepting orchids! It is likely, however, that generalities may apply to natural orchid communities as well. This writer states that generally the normal chlorophyll content of leaves is adequate to absorb available light under most natural conditions, and that chlorophyll content of plant communities therefore is not likely to be a limiting factor in photosynthetic production. It is much more likely that light itself will be the limiting factor.

Some work has been done on the optimum light requirements of orchids under cultivation (e.g., Fairburn and Pring, 1945), but such work is generally too inexact to allow for definite conclusions; moreover, it concentrates on only one or two commercially important genera. The difficulty of taking into consideration such important parameters of environmental variation as temperature, moisture availability, ontogenetic and circadian metabolic status have also discouraged work in this area. In general, I believe that the amount of light received (intensity over the day) by orchids under cultivation in temperate zones is probably inadequate for optimum photosynthesis and flowering. This may also very often be true for plants in nature in the tropics. Increased light, higher humidity, and lower day temperatures would almost always lead to better growth and flowering.

Tests have been reported on photosynthesis in flowers and various plant parts other than leaves (Erickson, 1957, Arditti and Dueker, 1968), but these are of little significance in the overall metabolism of the plant. (The exception is photosynthesis in the roots of leafless epiphytic orchids—but no tests have been made here!)

Most orchid structural modifications related to light are correlated with moisture and temperature as well. Particularly noticeable are pigment changes of the stem and foliage in response to extreme exposure (e.g., Fowlie, 1960a, 1960b). The most striking examples I have seen are in *Bulbophyllum* (e.g., *falcatum* Rchb.f. and *schimperanum* Kraenzl.) and *Polystachya* (e.g., *subulata* Finet, *stricta* Rolfe) growing in exposed sites in the highlands as contrasted with specimens growing in less exposed sites or even in exposed sites in the lowlands. Such plants quickly lose their intense red-purple pigmentation on transplantation to semishade in the lowlands. This production of anthocyanins is certainly related to light intensity, temperature (the lower temperature of the highlands favors production), and carbohydrate metabolism (and possibly to UV content of light, and under some special circumstances, to nutritional deficiencies or excesses as well). It is possible that red pigmentation may have some adaptive value in reducing the amount of red and near far-red light absorbed, thus reducing leaf temperatures. Martin and Juniper (1970, p. 73) remark that anthocyanin is often found in the protoplasm of epidermal cells. I have never found leaf anthocyanins here, but rather in the first to third layers of cells just beneath the epidermis (mesophyll or hypodermis). In any case, the pigment does shield the majority of the chlorophyll-containing cells and may prevent chlorophyll breakdown from excess light. [A similar role has been reported for carotenoid pigments (Anderson and Robertson, 1960).]

The problem of optimal leaf size is subtle. Although the large, fleshy leaf can store considerable water, it has the great disadvantages of overheating and of being very inefficient in photosynthesis. The small, thin leaf, on the other hand, may cool rapidly and present a relatively huge area for light absorption and gas exchange, but it stores so little water that it is very easily desiccated. In general, there is a correlation between moist, shaded habitat and fast-growing, thin-leaved orchids and between dry, exposed habitats and slower growing, thick-leaved forms. For example, the orchids found growing on the dry rocky hills and ridges of Oaxaca, Mexico, include such forms as *Oncidium cebolleta* (Jacq.) Sw. (MacDougall, 1959). But, just to make the world of orchids more difficult to understand, *Scuticaria steelei* Lindl., which has similar fleshy, nearly terete leaves, is reported (Dunsterville and Garay, 1961) to grow in Venezuela principally along river banks, at moderate height and light intensity with generally damp conditions and only one month when the rainfall is negligible. A fascinating report has been published (Narodny, 1945) on how size and shape of the *Vanilla planifolia* Andr. leaf may be used as a practical indicator of light conditions in the field.

The relationship between light and nutrition via photosynthesis is obvious, but Wenger's report (1955) that light enhances mycorrhizal development in pine seedlings is suggestive of possibly unsuspected effects of light on orchid seed germination and seedling establishment, especially of epiphytes.

An effect of light intensity on root formation in *Cymbidium* Swartz. seedlings has been reported by Ueda and Torikata (1972).

Nutrition. Most aspects of nutrition are best treated in conventional physiological works (e.g., Withner, 1959; Arditti, 1967b; Ernst, Arditti, and Healey, 1970), but some ecologically directed notes may be useful.

One of the most widespread and long-lasting orchid arguments has been whether or not cultivated orchids should be fertilized. Young in 1897 warned that use of manure on orchid plants is very unwise. As late as 1957, Eigeldinger (pp. 56–57) discouraged fertilizing in the greenhouse, but Rolfe in 1897 had already said, I believe, the last word on the subject when he wrote: "Some orchids which under cultivation gradually dwindle away are literally starved to death."

It is now generally understood that conditions of light, moisture, temperature, nutrition, and growth cycle must be balanced and correlated, and few people seriously doubt the necessity, under most circumstances, of a good fertilizing program. It is often overlooked that orchids in their natural habitats may be barely surviving under nutritionally marginal conditions, so that one might, in some instances, learn more of orchid food requirements in the greenhouse than in the field. In many cases in nature, the growth rate of epiphytes at least is limited by food and moisture as well as by innate character of the species.

Several important problems of nutrition remain. One is the question of whether organic compounds in the medium are necessary for optimal growth of the adult orchid plant. A number of experiments have been conducted in hydroponic culture of orchids whereby plants are held in an inert pebble or plastic-fiber substrate and bathed with a weak inorganic salt solution. Such tests generally have had limited success, although adequate controls have often been lacking and relatively few species have been tried. Optimal growth seems to be obtained when some organic material is used, although future work on properly balanced inorganic regimes may change this.

It is well known that plants in general can and do take up complex organic compounds from the soil. It has even been shown that protein molecules may enter roots (Bradfute and McLaren, 1964; McLaren et al., 1960; Ulrich et al., 1964), but there is considerable doubt that such uptake is of any nutritional benefit. Feeding of known organic precursors of specific materials has frequently been shown to increase production of the end product in the plant (e.g., anthocyanin biosynthesis: Pacheco et al., 1966; Sugano, 1967), and a whole body of literature has grown up on the relation of the chemical composition of the growth medium and the chemical composition of the plant. Most fascinating ecologically has been Ovington's work (1953, 1956, 1957, 1958) on the effect of various tree species on the forest soil under them and the consequent feedback of soil effect on the trees. He has shown, for example, that the pH of leaves varies with the pH of the soil—and that the pH of the soil is affected by the pH of the leaf litter from the trees (see also Zinke, 1962). More directly related to orchids is the work of Zeigler et al. (1967), who studied the effects of various growing media and photoperiod on the amino acid content of orchid seedlings.

We can, then, be certain that a dynamic relationship exists on a chemical level between plant and substrate, but we do not yet know very much about this relationship. Akintola and Sanford (1971) suggest that the phenolic biochemistry of trees may vary enough from site to site to pose a problem in "biochemical conservation" if plants from only one type of site are preserved. Such extreme habitat differences as highland and lowland have been shown to affect such basic plant constituents as carbohydrates (e.g., Mooney and Billings, 1965). As to orchids specifically, almost nothing is known. Some plants are reported to particularly need humus [e.g., *Cycnoches* Lindl. (Segars, 1960); *Mormodes* Lindl. preference for rotting wood (Adderley, 1969); *Ansellia africana* Lindl. (Sanford, unpublished)], and this may have a great deal to do with their distribution in nature.

The whole area of the role of leachates from the leaves and stems of host trees has been neglected in epiphyte nutrition, except for Went's (1940) insistence that the composition of the leachates plays an important role in orchid-host tree specificity. It is well known that foliar leaching in general may be considerable and may involve almost every known type of organic compound as well as inorganic salts, although I do not know of any detailed studies in the tropical forest (Tukey, 1964, 1971).

Curtis (1946) made a crude analysis of rainwater collected as bole run-off in Haiti that is most suggestive. He found the water to be pH 5.70 and to contain 102 ppm dissolved salts including traces of K, Ca, Mg, NH_4, NO_3, PO_4, SO_4, Cl, Na, Fe. Particularly interesting for the present discussion, he found the water negative to amino acid, protein, and reducing sugar tests but positive for catechol tannins. (I believe the water must have contained both amino acids and sugars but in very low concentrations. Careful evaporation to obtain more concentrated material for analysis would be necessary.)

A more recent and detailed study (Carlisle et al., 1967) concerns stem flow and ground flora litter and leachates in temperate zone oak (*Quercus petraea* O. Schwarz) woodland. It is significant that a number of organic materials, including relatively high concentrations of polyphenols, were found in the stem flow. Polyphenolic compounds (including tannins) might be of significance as inhibitor substances, especially to seed and spore germination. (See the note concerning work by Sanford, page 40.) Differential inhibition of fern spores and orchid seeds, and of seeds of different orchid species, might possibly be a very subtle mechanism of microhabitat variation and consequent species selection. In this connection, Brown (1967) made a suggestive study of the effects of extracts from 56 plants common in *Pinus banksiana* Lamb. forests on germination of pine seed. He found that nine extracts consistently inhibited and five often stimulated germination. Allen (1969a) claims that some kinds of bark are actually toxic to most orchids. For example, he found "by experimentation" that virtually nothing except *Catasetum integerrimum* Hook. (=*C. maculatum* Kunth), *Epidendrum ionophlebium* Rchb.f. (= *Epidendrum chacaoense* Rchb.f.), and *E. nocturnum* Jacq. grow on "oak sections." Withner (personal communication) notes that in tropical America orchids do not grow on pines. It is probable that this is not because of pine bark toxicity but rather that pine trees are found in regions environmentally unsuited to epiphytes. This is true at least of Honduras and has been pointed out by Williams (1948), who states that the pine forest regions have much too long and intense a dry season for orchids. Some evidence for bark toxicity *in situ* in Mexico is presented by Frei (1973a). In another paper (1973) she reported no inhibition to orchid seed germination by pine barks as tested in the laboratory.

Some idea of the subtle extent to which ecological conditions may vary has been suggested to me by the study of Hodges and Lorio (1969) on the effect of moisture stress on carbohydrate and nitrogen content of the bark of loblolly pine. Drought was shown to markedly increase reducing sugars and total carbohydrates. I would imagine this to be true of other trees as well. It would be interesting to know if the first rains after a drought or dry season might carry an elevated concentration of sugars in the bole runoff. Such a phenomenon might be important in epiphyte orchid seed germination and the beginning of adult plant growth at the start of the rainy season in the tropics. Necessity for mycorrhiza might even be eliminated. (The effects of various sugars on seed germination has been reviewed by Arditti, 1967b.)

Nitrogen metabolism is another interesting area. A very early way of fertilizing orchids was to give them an "ammonia vapor bath" (e.g., Wrigley, 1897): a bucket of soot and half a bucket of lime were put into a tub with 12 gallons of water, mixed, allowed to stand for 3 days. The paste was then spread over the staging of the orchid house.

A point of major interest in nitrogen nutrition is that of the comparative utility of ammonium and nitrate nitrogen sources. According to Spanner in 1939 (cited by Richards, 1952, p. 115) all true epiphytes probably obtain their nitrogen as ammonium ions. This is certainly an exaggeration but may contain an element of truth. Spoerl and Curtis in 1948 investigated amino acid nitrogen supplies to orchid seed cultures and found that effects varied greatly with embryo/seedling age. In general, weight and development of *Cattleya* Lindl. and *Vanda* R.Br. embryos appeared best with a ratio of high ammonium, low nitrate, although there was some indication that ammonium sources were best for the early protocorm stage while later stages did better with higher nitrate. Withner (1959) reported that orchids grow better when both NO_3 and NH_4 are present than when NO_3 alone is supplied. This area of study has been briefly reviewed recently by Zeigler et al. (1967). Unpublished experiments at the University of Ife have borne out, for African *Angraecum* Bory embryos and young seedlings, the necessity of ammonium nitrogen sources. Such results indicate that the early embryo and protocorm do not contain the enzyme nitrate reductase, and so must rely on already reduced (ammonium) nitrogen; however, nitrate reductase production occurs at later developmental stages, so that nitrate can then be utilized. (For a recent review of nitrate reduction in higher plants, see Beevers and Hageman, 1969.) In light of this, Curtis' early finding (1946) that bark runoff contained a much higher ammonium N concentration than nitrate (35:1) is of great interest. His later results (1947b) that seedlings of 16 orchid species out of 12 genera do better with complex organic N sources probably reflect the need for already reduced nitrogen.

Specific effects of various amino acids on seed germination have been reviewed (Arditti, 1967b). The possible deleterious effects of excess amino acids in organic substrates has not been considered for orchids. That excess of some (e.g., DL-isoleucine, DL-alloisoleucine, L-leucine, D- and L-methionine) may cause diseaselike abnormalities in some higher plants is well known (Waltz and Jackson, 1961). Accumulation of free amino acids in many plants with various suboptimal environmental conditions (drought, mineral deficiency, disease, etc.) is also well known. This area of investigation should be especially interesting in epiphytic orchids because of their association with organic substrates which must continually release amino acids. [It is interesting to note tangentially, in relation to terrestrial orchids, that tropical soils, at least, contain varying amounts of water-extractable amino acids at different times during the year (Sanford and Oluwatuyi, unpublished).]

Especially intriguing is a note by H. E. Young (1938) concerning hybrid *Dendrobium* seeds flasked without N source. These developed to the protocorm stage then began to die. When the flasks were inoculated with bacteria from garden lupine nodules, the bacteria began growing on the agar surface and the orchids resumed growth. Allen (1969b) states that nitrogen-fixing bacteria are present in the growing tips of many orchid roots, but I have been able to find no further reference to such a phenomenon. It is more likely that N-fixing bacteria occur in the epiphyte environment and were mistakenly thought to be inside the orchid roots. Ruinen (1956) has reported N-fixing *Beijerinkia spp.* to be "omnipresent" on the leaves of tropical vegetation. Also intriguing is the paper of Rice (1964) on the inhibition of N-fixing and nitrifying bacteria by extracts from seed plants. He found considerable inhibitory activity by the extracts of 13 out of the 20 species tested. This might play some role in plant succession and competition, especially interesting in affecting epiphytic floras.

Possible general inhibition of plant growth by plant exudates and leachates is a controversial field. No doubt extracts from any plant at a high enough concentration would inhibit germination or seedling growth of any other plant. The testing for inhibitors which have significantly high activity in low concentrations or which demonstrate some specificity becomes difficult, and the results of many experiments are ambiguous in these respects. A historical review of the subject is given by Garb (1961). More recent reviews are of greater significance (Rovira, 1969; Whittaker and Feeny, 1971). Such papers do not even mention orchids, but I feel that the possibility of similar inhibition of both epiphyte and terrestrial orchids would be a productive area of research.

The role of pH in the distribution of terrestrials has already been mentioned as critical. The pH of natural epiphyte habitats has received little attention except in the case of cryptogamic epiphytes (e.g., Barkman, 1958; C. Young, 1938). Especially interesting is the possibility that vascular epiphytes and the debris they hold change the pH of water running down the tree bole. For example, Kolkwitz (1932) reports that in the rain forest of Java, the pH of rainwater was lowered from 5.8 to 4.4 as it passed *Asplenium nidus* L. specimens.

The great buffering capacity of cryptogams may also be important. Adderley (1965) reported outstanding buffering capacity of the mosses *Thuidium delicatulum* |Hedw.) Mitt. [= *T. recognitum* (Hedw.) Lindb. *var. delicatulum* (Hedw.) Warnst.] and *Hypnum imponens* Hedw. [= *H. cupressiforme* Hedw. *Sp. imponens* (Hedw.) Boul.].

The effects of nonmycorrhizal fungi both in the soil and on the epiphyte substrate may be considerable, as suggested by the preliminary work of Nicot on *Vanilla* Plum. (Bouriquet, 1954, pp. 364–392) and as reviewed for higher plants in general by Barber (1968). Actually, it is largely nonmycorrhizal fungi that account for whatever effects might prove statistically significant among Ruinen's (1953) and Cook's (1926) observations claiming damage to trees by epiphytic orchids. Certainly moisture-retaining roots of epiphytes and accumulated organic debris provide a more suitable environment for fungi than does a bare branch. And probably such fungi sometimes do invade tree tissue. But Ruinen's and Cook's claims have never been either very convincing in themselves or substantiated by other observational studies. More significant is a possibility pointed out by Thorold's work (1952) that conditions favoring disease—in his case blackpod of cocoa—also favor epiphyte growth. It should also be remembered that old trees tend to have a greater macroepiphyte load and it is logical to expect higher incidence of disease in old trees than in young ones.

A very strange area of ecology that may have considerably more relevance to epiphytes than to terrestrials is the study of the chemical composition of rain. Rainwater has been found to carry appreciable amounts of inorganic ions. Madgwick and Ovington (1959) compared rain falling on open and adjacent forest plots in the temperate zone and found very considerable differences, with much higher concentrations of dissolved materials falling on the forest floor. The great difference represents dust washoff and leachate from the forest canopy. Allen et al. (1968) reported on rainfall nutrients dissolved from the air and aerial dust and not coming from any contact with vegetation: they conclude that such nutrients play an important role in the nutrient cycle of ecosystems, at least in Great Britain. The field of study is reviewed by these workers but no mention is made of similar studies in the tropics. Jones (1960), however, reported on the contribution of rainwater to the nutrient economy of soil in northern Nigeria, where he found appreciable quantities of NH_4 and NO_3 as well as N, P, K, Ca, Mg, and Na. Thornton (1965) wrote on the nutrient content of rainwater in Gambia, and Attiwell (1966) published on the role of rainwater in nutrient recycling in a mature *Eucalyptus* L'Herit. forest (Australia).

The last area of nutrition we can consider here is carbon dioxide supply. It has been fashionable occasionally to believe that CO_2 supply was the limiting factor in the photosynthetic efficiency of orchids under greenhouse conditions (e.g., Wright, 1967). Tests have not, however, been rigorously enough designed to allow for firm conclusions to be drawn. I agree with Davidson's recent statement (1967) that "no one has yet shown that orchids respond advantageously to CO_2 enriched atmosphere." Rather, lack of enough light at a low enough temperature and high enough humidity and air flow is usually the limiting factor. Under tropical conditions there is some evidence for CO_2 being the limiting factor in the photosynthesis of some plants—not orchids—rather than high temperature or low light intensity (Marx, 1973).

Startlingly little is known of CO_2 relations in orchids. I have already mentioned CO_2 dark fixation by orchids and photosynthesis by plant parts other than leaves. More significant still is the possibility that many orchids are able to absorb their CO_2 at night and use it for photosynthesis during the day. This would make possible daylight closure, or semi-closure, of stomata for water conservation without disruption of photosynthetic activity and so would be an important modification, especially in epiphytes.

Sensitivity of the photosynthetic response to moisture and temperature stress might be an important factor in orchid distribution. This is the case in the altitudinal distribution of several nonorchidaceous plants (Klikoff, 1965).

Of interest also is variation in amount of carbon dioxide in the plant habitat, although, as pointed out previously, this will seldom if ever be a limiting factor in photosynthetic efficiency. Miwa (1937) treated this in a very general way for the temperate zone greenhouse but could not give precise values. Ross (1954) and Evans (1939) provide CO_2 figures for tropical forest in southern Nigeria which may be general for tropical forests: highest CO_2 concentration (0.052%) at sunrise, falling until noon when it reached a steady value of about 0.035%. Evans found little difference in secondary and mature secondary (which he incorrectly termed "primary") forest.

Hosokawa and Odani (1957) made an interesting study on the daily compensation period (time necessary for plants to recover, by photosynthesis, the carbon lost by respiration during the previous night) in relation to the favorableness of the epiphyte (cryptogamic) habitat. They found that the lower vertical levels of habitat were correlated with ability of the epiphyte to reach the compensation point under low light intensities. A similar study, and similar results, have been communicated by Harris (1971) for the algal components of temperate zone lichens. General findings would most probably apply to orchids as well as to cryptogams.

Temperature. Although temperature is the crude large-scale sorting factor in temperate-tropical orchid distribution, on a small scale, both moisture and light become more influential.

The inductive effect of temperature on flowering has been studied and the previously mentioned papers (Arditti, 1966, 1967a, Rotor, 1951, 1959; Sanford, 1971) review this area. It is of interest that tropical orchids from areas having cool periods—usually mountains—often will not flower under warm (lowland) conditions. Well-observed instances of this are *Phalaenopsis schillerana* Rchb.f. (DeVries, 1953) and *Polystachya cultriformis* Lindl. (Sanford, 1971). Such a necessity for a period of low temperature is common to most or perhaps all temperate zone orchids and seems to be the major factor that makes it impossible to cultivate them in the tropics (unless in air-conditioned greenhouses!).

Temperature may be an inductive stimulus not only for flowering but also for breaking vegetative bud dormancy. This is perhaps the more usual mechanism in low-temperature effects on temperate zone orchids, and I (1971) suggest it in the case of some African savanna orchids as well. Although it is usually low temperature, or a temperature drop, that is inductive, flowering in some plants (e.g., tulip and hyacinth; Hartsema, 1961) may be induced by high temperature. Vöth of the Vienna Botanical Gardens (personal communication) collected experimental evidence that high soil temperatures are necessary for the flowering of some Mediterranean *Ophrys* L. A sudden drop in temperature accompanying rain is the stimulus for the famous gregarious flowering of *Dendrobium crumenatum* Sw. (Coster, 1926; Smith, 1927). I (1971) suggest that the sudden drops in temperature, general with the storms at the beginning and end of the rainy season, may be important factors in causing the blooming peaks observed at these times in Nigeria by Richards in 1939.

Diurnal temperature fluctuation is often necessary for normal or optimal plant growth in general, and we know from good greenhouse practices that this is true for many orchids, but precise experimental results are lacking. Orchids growing low in a tropical forest are subject to remarkably little temperature fluctuation, whereas those growing high in the canopy normally are subjected to some diurnal variation and montane species are subject to very considerable fluctuations.

The two extremes, burning and freezing, are of some ecological interest. It has been claimed (Perkins, 1962) that annual grass burning brings about the flowering of some African terrestrial orchids. I know of no instances of apparent correlation which are not simply fortuitous. Burning may, on the other hand, eliminate some species and so be one of the profound biotic effects brought about by man. Man's burning for hunting, agriculture, and occasionally by accident has already converted thousands of acres of forest into "derived savanna" in many parts of the world. This is hardly the place to embark on a general conservation lecture, but, beside the elimination of the habitats of scores of forest orchids, destruction of forest may have extremely far-reaching effects. It is instructive—but tragic—to review the comments of Thompson (1910) on the forests of Ghana published over sixty years ago. He pointed out the importance of forests in affecting the general environment: by mitigating extremes of temperature; by markedly affecting regulation of the water supply, more especially by ensuring sustained feeding of springs and continuous flow of rivers and streams, and in reducing the danger of floods; in increasing the relative humidity of the air; in the mechanical action of the roots and stems in preventing erosion, especially on hillsides; in tending to increase precipitation; by acting as windbreaks; by serving as barriers in the spread of fungal and insect attacks. He mentions, most pointedly, that all these effects are more pronounced in the tropics.

Orchids are casually mentioned—but not identified beyond the genus—in a paper on fire effects on plant succession in Papua (Gillison, 1969). *Orchis maculata* Linn. is noted as able to regenerate from protected underground buds after burning of heath vegetation in Europe (Hansen, 1964). Most African *Eulophia* R.Br. and *Habenaria* Willd. can survive burning, although some with surface pseudobulbs, such as *Eulophia quartiniana* A. Rich., survive only by growing on rocks and in rock fissures where they are protected from fire.

The frost hardiness of a number of Himalayan dendrobiums is well known. *Laelia speciosa* Schltr., *L. albida* Batem. ex Lindl., *L. anceps* Lindl., and *L. gouldiana* Rchb.f. of Mexico can withstand freezing (Halbinger, 1941), but they cannot withstand the heat of lowland southern Nigeria for more than a year or two. Baxter (1958) lists "typical" orchids found in the Big Cypress Swamp, Florida (U.S.), undamaged after prolonged freezing weather. Dressler has written about the northernmost epiphytes of Mexico (1961) and recorded that most were severely damaged or destroyed in the severe winter of 1961–1962 excepting *Epidendrum conopseum* R.Br. (Dressler, 1964). Wallbrunn (1966b) has shown that *E. conopseum* under cultivation can easily withstand 12°F, and most F_1 hybrids of this species with *Cattleya* Lindl. and *Laelia* Lindl. can survive 23°F. That such hybrids prove intermediate in cold hardiness is of theoretical interest in suggesting possible species development at temperate-tropical and mountain-lowland boundaries. A factor in frost hardiness may be common with the major factor for water conservation: the cuticle. This was recently discussed by Martin and Juniper (1970, p. 220).

Other Plants and Animals. The two major fields for consideration in this subject area—competition of other plants and insect pollination—have been discussed previously but a few further notes on the second subject may be useful.

Much attention has been paid to orchid pollination. The most famous report is that of Darwin (1899), and one of the most famous cases is his prediction of the moth *Xanthopan morgani praedicta*, pollinator of the Madagascar orchid, *Agraecum sesquipedale* Thou. (Ramsey, 1965). Scarcely less famous is fertilization of European *Ophrys* Linn. by pseudocopulation of bees, first described by Robert Brown (1833) and later reviewed by Oakes Ames (1937).

The general field of pollination ecology is well treated by Faegri and van der Pijl (1966), and van der Pijl and Dodson treated orchid pollination and evolution specifically in their book of 1966. Dodson also published a distinguished series of well-illustrated articles on the subject (1961, with Frymire; 1962, 1965, 1966, 1967), and Dressler an important series on euglossine bees (1967, 1968). Arditti very recently published a bibliography on the subject (1969b). An excellent specific paper on *Cypripedium calceolus* Linn. has appeared since van der Pijl's review (Dawmann, 1968), as has Hayes' (1968) interesting paper on Brazilian miltonias.

Of more unusual pollinators, the mosquito has received the most attention, beginning in 1913 (Dexter) and more recently noted by several workers as the pollinator of various terrestrial species (Stoutamire, 1968; Arditti, 1968; Thien, 1969).

Some of the specialized biochemical and physiological mechanisms involved in pollination phenomena have been reviewed by Withner (1959). Of especial ecological interest are aspects of perfume and pollinator relationships (noted as early as 1894 by Rodway) and current methods of identifying the components of orchid scents (Hills et al., 1968; Dodson and Hills, 1966). Dressler (1968b) points out that euglossine bees have been a major factor in the evolution of some American orchids, and since odor of orchids is the critical factor in attracting these bees, odor difference constitutes the main breeding barrier between sympatric species. The preliminary note on sugar content of orchid nectar (Jeffrey and Arditti, 1968) also suggests an interesting area of study. If insect dependence on specific carbohydrates of nectar can be shown, it will point to fascinating possibilities of coevolution.

The number of seed pods actually formed relative to the number of flowers may be largely controlled by pollinator activity and by general environmental and physiological conditions. The discrepancy between flower and seed pod may be great. For example, Kurfess (1965) reports observing about 100 plants of *Cypripedium acaule* Ait. which nearly all flowered but produced only one pod.

Somewhat related but clearly dependent on environmental conditions, and perhaps on long-range plant periodicity as well, is fluctuation of flower production from year to year. This has been studied in *Cypripedium* Linn. by Curtis (1954).

Some strange animal-orchid relationships other than those of pollinating agents exist. Most disastrous are those of orchids and man. As Hunt (1969a) is laboring to point out, a number of things must be done or many orchids may be lost, to say nothing of many whole ecosystems being destroyed and total environments ruined for habitation by both plant and man. In this larger sense, Paul and Anne Ehrlich's (1970) and Taylor's (1970) tomes of doom may be salutory bedtime reading. The American Orchid Society has begun to take a strong interest in conservation (Moir, 1968; Fitch, 1969; Elbert, 1969; Dunsterville, 1968, 1969, Lynch, 1970, etc.) and the international journal *Biological Conservation* is eager for orchid conservation papers (Sanford, 1969b, 1970b). It is encouraging when the national journal of a developing country's science association devotes illustrated space to orchid conservation as has happened in Nigeria (Sanford, 1970c). Attention is also spasmodically brought to bear on great areas of destruction in South America, for example, the coastal plain in Boraceia, Brazil, where modern pesticides have now made formerly deserted forestland habitable by man and so have led to the destruction of a unique ecosystem (Spencer, 1969). Arango (1968) recently published, in Spanish, on the relation between man and orchids. Yet local publications still all too often fill their space with reports of collecting raids rather than conservation projects.

It is important in regard to these problems that everybody does not get carried away in a flood of sentimental good will, but rather that sound scientific research be done on total ecosystems, their comparative productivities, and controlling mechanisms so that man may determine which vegetation types must be saved, which can be sacrificed, and how best to control the total environment so that such choice-making will continue to be possible. An example of sound application of scientific sense is the program of Jodrell Laboratories, Kew Gardens, London. An orchid propagator (Mr. J. C. Bowling) has been appointed specifically to grow orchid species from seed so that a "bank" of seedlings will be available for distribution to other orchid-growing centers, where such species will be conserved even though their native habitats may become improved out of existence by civilization. (Such a "banking system" was suggested by Elbert, 1969, in his note cited above.)

Tangential to orchid conservation is extremely interesting work showing the great sensitivity of cryptogamic epiphytes to air pollution (e.g., Barkman, 1962; Jones, 1952). Intimate relationship between urban smoke, cryptogamic epiphytes, and vascular epiphytes is a thought-provoking example of the frequent delicacy of ecosystem balance. Less subtle and certainly more talked about are the effects of urban and industrial fumes on orchid buds and flowers in the greenhouse. "Dry sepal" caused by ethylene gas is probably the most common such ailment. It may be caused by ripening fruit as well as by urban fumes—but we can store our bushels of apples out of the greenhouse, whereas we have no control at all over the air we, and our orchids, breathe.

For relief, we may turn to less widespread and devastating special relationships. Hermit crabs have been reported to munch pseudobulbs of coastal mangrove orchids in the Grand Bahama (Hall, 1967). Gilbert (1939) describes flower picking by satin bower-birds and the blue wren as well as the use of whole orchid plants in nest building by the scrub wren. More bizarre is the Burmese villagers' insistence that pythons eat the new growths of *Dendrobium dalhousieanum* Paxt. (= *D. pulchellum* Roxb.) (Aldworth, 1938). It is doubtful that such activities will ever become major factors in orchid ecology.

A far-reaching case of seemingly trivial animal-plant relationship is the famous one between the myxomatosis disease of rabbits and terrestrial orchids in the British Isles. After the rabbit population had been decimated by the disease, regrowth in many waste areas became so extensive that many ground orchids were lost from overshading and consequent failure to flower (Thomas, 1962).

The apparently mutually beneficial association of ants and orchids was discussed as early as 1910 by Ridley, who noted that ants were attracted to live among the roots of *Dendrobium crumenatum* Sw. and in turn accumulated soil and debris which compensated for the nonwater retention and lack of moss growth on smooth barks. Even earlier (1897) Rolfe wrote of the famous hollow pseudobulbs of *Schomburgkia* Lindl. which house viciously stinging ants. Ants may, in some cases, act as guards to prevent damage to flowers of various orchid species by carpenter bees (*Xylocopa*) (van der Pijl and Dodson, 1966, p. 42). A number of epiphytes are now known to harbor such protectors (e.g., Fowlie, 1962a). Withner (personal communication) notes the interesting point that orchids which harbor ants are usually species growing in dry, exposed environments. It is possible that water retention by the orchid may be the attractive feature to the ants.

Perhaps of more general interest has been the discussion of whether or not termite mounds affect the pH and mineral and moisture content of tropical soils. Definite plant species associations with termite mounds is a common observation in the tropics. Such special distribution of plants in relation to mounds was discussed for the Sudan by Morison et al. (1948), who offer the following reasons: soil of the mounds is more moisture-retaining because of a higher clay content; the pH is more basic. It was also very sensibly suggested as possible that the woody plants growing there were those resistant to termite attacks! Although no orchids are mentioned in this article or in other papers, the possible effects on the distribution of savanna orchids is of interest and awaits investigation. (Up to now, however, a cursory search in Nigeria has not revealed any relationship of orchids to termite mounds.) More recent work (Hesse, 1955; Watson, 1969) indicates that the only chemical effects of termite mounds are a somewhat higher pH and a higher concentration of exchangeable bases. This might result in increased activity of nitrifying bacteria. (See Plate 6 for an illustration of a terminte community in African savanna.)

Research Methods and Areas of Study

Orchid ecology, like any other ecology, may be studied in the field or under controlled garden or greenhouse conditions. Fundamental to either approach is a precise knowledge of environment. In the field, this ultimately becomes a study of the microclimate: environmental conditions must be understood and characterized on the same small scale as variation in plant distribution. A recent basic and extensive coverage of the problem is provided by Wadsworth et al. (1968). Rieley et al. (1969) considered some statistical problems of samplings—always a difficult area in microclimate work. The paper of Poel and Stoutjesdijk (1960) presents excellent methods for comparing the microclimate of one community with that of another. Two possible methodological approaches are considered:

Plate 6. Mushroom-mimicking termite mounds in northern Cameroon, near Meiganga, in March (late dry season). (Photograph by W. W. Sanford.)

(1) compiling long series of observations with consequent statistical evaluation; (2) taking direct measurements in the two communities at the same time in the same way. These workers adopt the second approach, but the first way is often necessary in practice.

Instrumentation for measurement presents several difficulties. Rain collectors are cheap and portable and present no problem. Metal collars (as mentioned under Moisture, above) are suitable for measuring bole run-off. Boynton (1969) described a cloud-water collector. Satisfactory recording humidistats are available, as are temperature recorders. Light measurement is more difficult. Elaborate and expensive instruments do exist, for example, the Kipp solarimeter, but it is often necessary to have cheaper, more portable equipment. The simple chemical method of anthracene ($C_{14}H_{10}$) in benzene, which polymerizes into an insoluble form, on exposure to light, has been utilized by Dore (1958) and others (Pierik, 1965). For orchid work, standard glass tubes of known anthracene concentration and light absorption could be fastened at various positions on trees, collected after suitable time intervals, and read for absorption. In this way microhabitats could be characterized rather simply for light over a yearly cycle. Construction of cheap photometers suitable for measuring light under dense vegetation is described by Getz (1968) and Wiens (1967).

Soil and phorophyte bark analyses present rather routine chemical problems. Soil analysis is adequately covered by any modern soil science text. Bark analysis is more complicated—and more interesting. (Bark chemistry is well reviewed by Jensen et al., 1963.) Two approaches are obvious: (1) the analysis of water-extractable materials such as would leach out with precipitation; (2) more exhaustive extraction of powdered bark by organic solvents to determine the total potential leachates. The first approach has the great difficulty of necessitating the concentration of large amounts of liquid without breaking down sensitive compounds; the second has the disadvantage of artificiality.

Next to the problems of characterizing climate and nutritional environment on a small scale in the field comes the great problem of relating the plants to the environmental variables measured and to each other. Of particular practical importance is an approach from the other direction: the characterization of an environment by means of its vegetation. Clearly, a great deal of preliminary work has to be done relating specific plants and vegetation types to specific environmental variables. But once this is done, analysis of the vegetation of an area can be much easier, quicker, and more exact and subtle than analysis of the environment. Ecology works, then, on a circular treadmill of logic: continual environmental analyses used to predict plant distribution and continual vegetation analyses used to predict environmental variables.

Because temperate zone vegetation and environments have been studied for so much longer than have tropical conditions, the greatest problems remain in the latter. And the most difficult area is the tropical forest because of its great complexity of structure and diversity of species (see Plate 7). Particular attention has thus recently been turned to the ecology of the tropical forest. I shall direct my remarks specifically toward this area, but it is to be realized that most comments also apply to temperate zone problems.

The basic difficulty in ecology is that of characterizing vegetation accurately and with the least work possible. One approach is a taxonomic characterization. It is exceedingly difficult, if not almost impossible, to tabulate the kind and number of all plants in a very complex bit of vegetation and, should this be done, the result may be too ungainly for use: some generalization in the description is necessary. One way of simplification has been to characterize vegetation by dominant plant associations. The weakness of this is obvious: the selection of a few dominant associations may be arbitrary and their description so imprecise quanti-

Plate 7. Secondary regrowth forest near Yaoundé in Cameroon, November (end of rainy season). (Photograph by W. W. Sanford.)

tatively as not to be very meaningful on any comparative basis. Also, the most abundant species will obviously be the most tolerant in regards to habitat and thus give the least information about environmental variation. Furthermore, many tropical vegetations either lack dominant species and associations or contain such a variety of species that their quantitative estimation becomes arbitrary in the field.

Another way has been to characterize the vegetation physiognomically —by how it appears. The most apparent difficulty here is the lack of any descriptive means, other than photographs which cannot be precisely compared, to characterize the vegetations. Moreover, vegetation in very different climates may look superficially alike, while within one region where the vegetation must be described as more or less uniform in general appearance there may be much species difference, reflecting considerable environmental difference.

It is true that with the development of better statistical techniques and of the computer, it is now possible to classify and correlate so many data that total classificatory approaches to vegetation have become possible— and even fashionable in some quarters. The problem remains data collection: although computers can correlate and manipulate the data fed to them, no simple or accurate way has been developed for going into a forest and identifying and counting every tree there (and in Central America, for example, there are over 2000 species) to say nothing of ground herbs and epiphytes. And as to the epiphytes, tropical ecology is particularly noteworthy in that vegetation dripping with orchids is often studied but the orchids are seldom mentioned and never identified! Collection of data must still depend directly on the human eye and human knowledge. Thus, even with modern computers and statistics, more or less primitive short-cuts by means of main associations are often necessary (e.g., Sanford, 1969a).

After the plants in a piece of vegetation have been identified the problem of what to do with the data arises. A leading approach is ordination— the systematic arrangement of vegetation along one or more axes according to similarities (Curtis and McIntosh, 1950, 1951). Again, statistics and computers can process the data, but the observations still have to be collected in the field.

Greig-Smith has been one of the leaders in the development of a quantitative approach to plant ecology and his textbook (1964) is still basic. More recently (1967, 1969), he and others reviewed the use of numerical methods in characterizing tropical vegetation: they have been able to find relatively few instances. His recent methodological papers (e.g. 1971) are particularly applicable to the tropics and should make possible an increase in the use of quantitative approaches.

The necessity for rigorous quantitative methods is beyond question. Fundamentally connected with this is the necessity of associating a great number of environmental variables with the plants classified or ordinated: in order to know very much about either an environment or the vegetation in it, it is necessary to know quite a bit about what kind and how many plants are there and how they are distributionally related to a number of the environmental variables of the region. Such a multirelational approach is termed "the multivariate approach," and special techniques have been developed for it. A good practical introduction to such techniques which is also suitable for the nonspecialist is that of Allen (1970); more elaborate is the book by Blackith and Reyment (1971).

Orchids, being extremely specialized environmentally, and so sensitive to slight environmental variation, are potentially valuable "marker," or characterizing plants for vegetation analysis. If enough were known about them, it might be possible to characterize a whole vegetation not by working with every plant there but by considering only the orchid species, just as is now done with trees. It might even be possible to simplify further and obtain useful results by using only a part of the orchid flora. On the other hand, it is possible that the habitat specialization of orchids is so great that their distribution varies on too small a scale to allow their use in treating the larger collections of plant communities that are termed vegetation. (See below.)

While Grubb et al. (1963) remark that epiphytes are important in characterizing features of montane forests, the only attempt so far, other than my own, to use orchids for precise vegetation analysis is the critical work of Webb et al. (1967). Working in tropical rain forest, they selected 18 sites, classified into 6 environmental groups, where they enumerated 818 species of plants divided into the following subsets: (1) big trees; (2) lianes and epiphytes; (3) special life forms (i.e., tree ferns, pandans, orchids, etc.); (4) the remainder. These groups were further subdivided in subsequent analyses. The most precise and all-inclusive characterization of the sites was obtained by using the subset of 269 large trees. Most significantly, nearly as good results could be obtained by using only 65 (*ca.* 24%) of these species. Lianes were the next most successful group for characterizing the sites, and epiphytes and herbs were the least successful. (Numbers used were 49 species of epiphytes, 45 species of herbs, 133 species of lianes.)

Webb (personal communication) has kindly supplied me with a copy of the epiphyte list which was used in the analysis. Only ten orchid species were enumerated and of these, seven were incompletely determined. I agree with Dr. Webb's statement: "I do not think that any conclusion could be drawn about the distribution of epiphytes in the forest," and I believe it follows that no real test was therefore made of the usefulness of epiphytes, and orchids in particular, in vegetation and site characterization.

To test the possible use of orchids in general ecological work, Sanford recently (1972) applied ordination methods to his data on epiphytic orchids of Nigeria. He enumerated 117 species in 31 sites, most of which were forest reserve areas. Resulting ordinations provided more information on the over-all vegetation and environment of the areas studied than did ordinations using tree data. Vegetation types were sorted in more detail than can be shown on vegetation maps (e.g., Keay, 1959) and special peculiarities of sites could be inferred from site positions on the ordination. Correlations could also be seen between site position, rainfall, length of dry season, and one—as opposed to two—peak rainy seasons. The conclusion drawn was that of epiphytic orchids proved an especially efficient and sensitive way of characterizing vegetation and environment. (Sanford, 1974, 1974a).

Modern analytical methods of ecology have yet to be applied to small-scale orchid distribution. Multivariate analyses of orchids on single trees might prove very interesting. In this area, I have already mentioned the work of Hosokawa (1943), but multivariate techniques were not attempted by this worker. Hazen (1966) characterizes bromeliad position on a linear scale, which is not meaningful if more than one branch of a tree is considered, unless to show perfect randomness of distribution, which is certainly not the case for epiphytic orchids [nor, according to Pittendrigh's more extensive work (1948), for bromeliads].

Beside the use of orchids in vegetation characterization and the application of analytical techniques to orchid distribution, several more specific field problems exist. An obvious one is the relationship between phorophyte and epiphyte. The sort of work done by Kershaw (1964) on bark-cryptogam relationship could be usefully adapted to orchid work. Chemical analyses of barks and seed germination tests on bark extracts might also yield useful results. Detailed work in water relations in the field would also be productive and might well include study of osmotic concentrations of sap, transpiration rates, stomatal frequency, and cuticle-epidermis structure. (Martin and Juniper, 1970, are not able to mention orchids even once in their comprehensive book on cuticle!) I am far from satisfied with our knowledge of the role of aerial roots. In a number of species (of the *Microcoelia* Lindl.-*Encheiridion* Summerh. group at least) many of the roots remain free and not adpressed to the substrate. Furthermore, the velamen is extraordinarily thin. Water is apparently absorbed through these free roots, contrary to the usual pathway, as discussed above. Gas exchange and photosynthesis in such roots also remain to be studied.

The extremely important area of plant competition together with the question of whether two or more species may share the same microhabitat in equilibrium are fundamental problems in ecology, most particularly tropical ecology. Such problems are normally studied best under controlled conditions of artificial culture. The slowness of growth of orchids and their specialized substrate demands make this classical approach difficult. Field analysis over a number of sites over several years would seem more feasible. Here, multivariate analysis of single trees might be especially productive.

Controlled culture experiments are suitable for study of the more directly physiologically orientated problems such as nutritional needs, optimum light intensity and quality, or metabolic rates. There is not the need for new methodological approaches here that there is in field ecology; rather, it is a matter of careful but conventional experimental design together with taking advantage of recent, improved instrumentation for measurement and recording.

One major area of research, which combines more conventional ecology with physiology and biochemistry, remains to be mentioned: production ecology. Ultimately, almost all ecological research is directed toward the problem of how the ecosystems of vegetation can be made most efficient in the service of man. That is, how much do they produce? This involves estimation of their efficiencies in carbon dioxide fixation; nitrogen fixation, utilization, and recycling; and general mineral and water use and reuse. No work has yet been done on the estimation of the extent of carbon fixation by orchids in any piece of vegetation, nor upon their physiological-chemical relationship with other components of the ecosystem. One can only make rough guesses based on general knowledge of orchid growth rates and plant frequency. Certainly, terrestrial orchids seldom if ever produce an important percentage of the fixed carbon of any extended area. Epiphytic orchids, on the other hand, although far below the woody plants of a forest, may exceed floor-covering herbaceous plants in total productivity. In many forests and woodlands, then, orchids become quantitatively important in carbon fixation, oxygen-carbon dioxide balance, and mineral recycling. The latter is perhaps of some particular importance. The nutritional elements utilized by orchids come from rain, water dripoff, bark and leaf leachate, dust, animal droppings, decaying bark, and humus. It is quite probable that in the usual tropical forest with its heavy rainy season much of such nutritional material, unless trapped by epiphytes, would be lost through rapid soil leaching and water runoff. Because of this, it is possible that orchids play a necessary role in the ecosystem. Experiments testing this remain to be carried out.

Perhaps nitrogen utilization and recycling are especially interesting. We have seen that orchids, particularly in the early stages of development, directly utilize the ammonium-nitrogen released from organic materials by decay and leaching. At the same time, all but the very young plants can also utilize any nitrates fixed by nitrogen-fixing bacteria and blue-green algae or released by nitrifying bacteria, all of which find a suitable home on the phorophyte. (Algae on leaf and bark surfaces should especially be considered.) The slow growth and very long life of most orchids lead to the immobilization of the nitrogen, along with other nutrients, and for their slow release upon leaf shedding and back-bulb and rhizome decay, and thus make possible reuse by other members of the ecosystem.

As for special accumulations of inorganic nutrients, available plant analyses (Chin, 1966; Bouriquet, 1954b, p. 346) do not suggest anything unusual. Nothing is yet known of possible special production and accumulation of organic materials except that at least traces of alkaloids are very widespread (Lüning, 1966, 1967, 1969). Chromatography (Sanford et al., 1965; Withner and Stevenson, 1968) has revealed the presence of many phenolic compounds but identification and detailed analyses have not yet been carried out except for some preliminary work on pigments (Arditti, 1969a).

One is forced to conclude that we do not yet know the role of orchids in ecosystem balance and productivity, but that it may not be as important, at least quantitatively, as that of other vegetation components. An exception may be the epiphytes in tropical forests and woodlands, which may play an important role in overall productivity. I would, however, like to emphasize in closing that it is the height of folly to destroy orchids and their habitats without knowing exactly what their role may be—to say nothing of the psychological scar that destruction of natural beauty leaves permanently upon man.

Acknowledgments

The author wishes to express his gratitude to the University of Ife for granting him time to work on this paper. Thanks are also due to many other institutions and people for their help, particularly to the staff in the Library and the Herbarium of the Royal Botanic Gardens, Kew, and Peter F. Hunt, who read the manuscript and made many valuable suggestions, additions, and corrections. Any remaining errors, however, are the author's own. Thanks are also due to the staff of the British Museum of Natural History, who have helped in bibliographic search; to Dr. Carl L. Withner, who read the manuscript, offered many suggestions and additions, and helped in collecting literature for it; to Professors L. A. Garay and C. G. G. J. van Steenis for helpful and stimulating conversations on ecology and orchids; and to Dr. John B. Hall, of the University of Ife, who also read the manuscript. Finally, to Mr. Augustine Akoshile for typing and retyping the manuscript.

Bibliography

ADDERLEY, L. 1965. Two species of moss as culture media for orchids. *Am. Orchid Soc. Bull.* **34**: 967–968.

———. 1966. *Oncidium lucayanum. Am. Orchid Soc. Bull.* **35**: 370–372.

———. 1969. The various ecological conditions under which tropical orchids grow. *Orchidata* **8**: 171–176.

AKINTOLA, G. A., and SANFORD, W. W. 1971. The problem of biochemical conservation: A preliminary study of phenolic compounds in the leaves of *Afzelia africana* Sm. as an example. *Biol. Conservation* 1:233–255.

ALDWORTH, R. M. 1938. Orchid hunting in Burma. *Austr. Orchid Rev.* **3**: 103–104.

ALLEE, W. C. (1926). Measurement of environmental factors in the tropical rainforest of Panama. *Ecology* **7**: 273–302.

ALLEN, P. H. 1959. Orchid hosts in the tropics. *Am. Orchid Soc. Bull.* **28**: 243–244.

———. 1969a. Orchids of Middle America, Part II. *Orchidata* **9**: 99–110.

———. 1969b. Orchids of Middle America, Part III. *Orchidata* **9**: 131–138.

ALLEN, S. E., CARLISLE, A., WHITE, E. J., and EVANS, C. C. 1968. The plant nutrient content of rainwater. *J. Ecol.* **56**: 497–504.

ALLEN, T. F. H. 1970. Multivariate techniques for use in ecology and taxonomy. *Ife Univ. Herb. Bull.* No. 4.

AMES, O. 1937. Pollination of orchids through pseudocopulation. *Bot. Mus. Leaflets, Harvard* **5**: 1–30.

ANDERSON, I. C., and ROBERTSON, D. S. 1960. Role of carotenoids in protecting chlorophyll from photodestruction. *Plant Physiol.* **35**: 531–534.

ANDERSON, M. C. 1967. Photon flux, chlorophyll content, and photosynthesis under natural conditions. *Ecology* **48**: 1050–1053.

ANON. 1898. Orchids in Guiana. *Orchid Rev.* **6**: 267–269.

ANON. 1906. *Laelia rubescens. Orchid Rev.* **14**: 41–42.

ANON. 1967. *Spiranthes spiralis*, Autumn Ladies' Tresses. *Orchid Rev.* **75**: 366–367.

ARANGO, S. M. 1968. La orquidea y su relacion con la activadad humana. *Orquideologia* **3**: 11–15.

ARDITTI, J. 1966. Flower induction in orchids. *Orchid Rev.* **74**: 208–217.

———. 1967. Flower induction in orchids, II. *Orchid Rev.* **75**: 253–256.

———. 1967. Factors affecting the germination of orchid seeds. *Bot. Rev.* **33**: 1–97.

———. 1968. Mosquitoes as pollinating agents for orchids. *Orchid Rev.* **76**: 22–24.

———. 1969a. Floral anthocyanins in some orchids. *Am. Orchid Soc. Bull.* **38**: 407–413.

———. 1969b. Annotated selected references on orchid pollination. *Orchid Rev.* **77**: 249–251.

ARDITTI, J., and DUCKER, J. 1968. Photosynthesis by various organs of orchid plants. *Am. Orchid Soc. Bull.* **37**: 862–866.

ARDITTI, J., and ERNST, R. 1972. Reciprocal movement of substances between orchids and their mycorrhizae. *Proceedings of the Seventh World Orchid Conference, Medellin* (in press).

ARENS, K., and PEDROITA, M. 1948. Noticia ecologica sobre *Brassavola tuberculata* Hook. *Orquidea* **10**: 164–170.

ASHTON, P. S. 1958. Light intensity measurements in rain forest near Santarem, Brazil. *J. Ecol.* **46**: 65–70.

———. 1969. Speciation among tropical forest trees: Some deductions in the light of recent evidence. *Biol. J. Linn. Soc.* **1**: 155–196.

ATTIWELL, P. M. 1966. The chemical composition of rainwater in relation to cycling of nutrients in mature *Eucalyptus* Forest. *Plant and Soil* **24**: 390–406.

AUBREVILLE, A. 1961. *Etude écologique des principales formations végétales du Brésil.* Centre Technique Forestier Tropical, Nogent-sur-Marne.

AUSTIN, M. P., and GREIG-SMITH, P. 1968. The application of quantitative methods to vegetation survey, II. Some methodological problems of data from rain forest. *J. Ecol.* **56**: 827–844.

BACKER, C. R., and DAKHUIBEN VAN DEN BRINK, R. C. 1968. *Flora of Java.* Walters-Noordhoff N.V., Groningen.

BARBER, D. A. 1968. Microorganisms and the inorganic nutrition of higher plants. *Ann. Rev. Plant Physiol.* **19**: 71–88.

BARBER, H. N. 1970. Hybridization and the evolution of plants. *Taxon* **19**: 154–160.

BARKMAN, J. J. 1958. *Phytosociology and ecology of cryptogamic epiphytes.* Van Gorkum, Assen. (Pages 1-202 reprinted in *Belmontia*, 1958, fasc. 2.)

———. 1962. De verarming van de cryptogamenflora in ons land gedurende de laatste honderd jaar. *Belmontia*, 8. (*Natura* K.N.N.V. Jaargang **58** nr. 10, Nov. 1961: 141–151.)

BAXTER, J. F. 1958. The Big Cypress Swamp. *Am. Orchid. Soc. Bull.* **27**: 767–768.

BEEVERS, L., and HAGEMAN, R. H. 1969. Nitrate reduction in higher plants. *Ann. Rev. Plant Physiol.* **20**: 495–522.

BERKELEY, E. S. 1894. Notes on orchids in the jungle. *Orchid Rev.* **2**: 12–13; 331–333.

BLACKITH, R. E., and REYMENT, R. A. 1971. *Multivariate morphometrics.* Academic Press, London and New York.

BOERBOOM, J. H. A. 1964. Microklimatologische waarnemingen in de wassenaarse duinen (Microclimatological observations in the Wassenaar dunes). *Belmontia* II. Ecology **10**: paper 67, 1965. (*Mededelingen van de landbouwhogeschool te Wageningen. Nedl.* **64/3**, 1964.)

BOURIQUET, G. 1954a. *Le Vanillier et la Vanille dans le monde.* Editions Paul Lechevalier, Paris.

————. 1954b. Données écologiques complémentaires concernant *Vanilla fragrans*, in G. Bouriquet, Ed., *Le Vanillier et la Vanille dans le monde*. Editions Paul Lechevalier, Paris, pp. 335–350.

BOYNTON, H. W. 1969. The ecology of an elfin forest in Puerto Rico, 3. Hilltop and forest influences on the microclimate of Pico Del Oeste. *J. Arnold Aboretum* **50:** 80–92.

BRADFUTE, O. E., and McLAREN, A. D. 1964. Entry of protein molecules into plant roots. *Physiol. Plant.* **17:** 667–675.

BRIEGER, F. G. 1958. On the phytogeography of orchids. *Proceedings of the Second World Orchid Conference*, pp. 189–200.

BROWN, R. 1833. On the organs and mode of fecundation in Orchidaceae and Asclepiadaceae. *Trans. Linn. Soc.* **16:** 685–733.

BROWN, R. T. 1967. Influence of naturally occurring compounds on germination and growth of jack pine. *Ecology* **48:** 542–546.

BURGEFF, H. 1936. *Samenkeimung der Orchideen*. Verlag Gustav Fischer, Jena.

————. 1959. Mycorrhiza of orchids in C. L. Withner, Ed., *The Orchids: A Scientific Survey*. Ronald Press, New York, pp. 361–398.

BURKE, D., and NORTHEN, H. T. 1948. Effects of exposure to low temperatures on the subsequent germination of *Cattleya* Luegeae seed. *Am. Orchid Soc. Bull.* **17:** 252.

CAIN, S. A. 1944. *Foundation of Plant Geography*. Harper Bros., New York.

CALDWELL, M. M. 1968. Solar ultraviolet radiation as an ecological factor for alpine plants. *Ecol. Monogr.* **38:** 243–268.

CAMPBELL, E. O. 1964. Non-green orchids in New Zealand. *Proceedings of the Fourth World Orchid Conference*. Straits Times Press (M) Ltd., Singapore, pp. 291–295.

CARLISLE, A., BROWN, A. H. F., and WHITE, E. J. 1967. The nutrient content of tree stem flow and ground flora litter and leachates in a sessile oak (*Quercus petraea*) woodland. *J. Ecol.* **55:** 615–627.

CASE, F. W., JR. 1962. Growing native orchids of the Great Lakes region. *Am. Orchid Soc. Bull.* **31:** 437–445.

CHIN, T. T. 1966. Effect of major nutrient deficiencies in *Dendrobium phalaenopsis* hybrids. *Am. Orchid Soc. Bull.* **35:** 549–554.

CLEGG, A. G. 1963. Rainfall interception in a tropical forest. *Caribbean Forester* **24:** 75–79.

COCHAN, P. 1963. Signification écologiques des variations microclimatiques verticales dans la forêt sempervirente de Basse Côte d'Ivoire. *Annls. Fac. Sci. Univ. Dakar* **8:** 5–87; 89–155.

CONNELL, J. II., and ORIAS, E. 1964. The ecological regulations of species diversity. *Am. Naturalist* **98:** 399–414.

COOK, M. T. 1926. Epiphytic orchids as a serious pest on citrus trees. *J. Dept. Agric. Puerto Rico* **10:** 5–9.

COOMBE, D. E. 1957. The spectral composition of shade: light in woodlands. *J. Ecol.* **45:** 823–830.

CORRELL, D. S. 1950. *Native Orchids of North America.* Chronica Botanica Co., Waltham, Mass.

COSTER, C. 1926. Periodische Blüteerscheinungen in den Tropen. *Ann. Jard. bot. Btzg.* **35**: 125–162.

CULBERSON, W. L. 1955. The corticolous communities of lichens and bryophytes in the upland forests of Northern Wisconsin. *Ecol. Monogr.* **25**: 215–231.

CURTIS, J. T. 1941. Some native orchids of the Lake Superior region. *Am. Orchid Soc. Bull.* **9**: 191–194.

——. 1946. Nutrient supply of epiphytic orchids in the mountains of Haiti. *Ecology* **27**: 264–266.

——. 1947a. Ecological observations on the orchids of Haiti. *Am. Orchid Soc. Bull.* **16**: 262–269.

——. 1947b. Studies on the nitrogen nutrition of orchid embryos. *Am. Orchid Soc. Bull.* **16**: 654–660.

——. 1954. Annual fluctuation in rate of flower production by native cypripediums during two decades. *Bull. Torrey Bot. Club.* **81**: 340–352.

CURTIS, J. T., and McINTOSH, R. P. 1950. The interrelation of certain analytic and synthetic phytosociological characters. *Ecology* **31**: 434–455.

——. 1951. An upland forest continuum of the prairie-forest border region of Wisconsin. *Ecology* **32**: 476–496.

CURTIS, J. T., and SPOERL, E. 1948. Studies on the nitrogen nutrition of orchid embryos, II. Comparative utilization of nitrate and ammonium nitrogen. *Am. Orchid Soc. Bull.* **17**: 111–114.

DANESCH, O., and DANESCH, E. 1969a. Über eine neue *Ophrys speculum*-Sippe aus Portugal. *Die Orchid.* **20**: 18–26.

——. 1969b. Eine neue *Ophrys fusca*-Sippe aus Frankreich. *Die Orchid.* **20**: 254–259.

——. 1970. Drei neue *Ophrys fuciflora*-Sippen aus Italien. *Die Orchid.* **21**: 17–22.

DARLINGTON, P. J. 1965. *Biogeography of the Southern End of the World.* Harvard University Press, Cambridge, Mass.

DARWIN, C. 1899. *The Various Contrivances by Which Orchids Are Fertilized by Insects.* John Murray, London.

DAVIDSON, O. W. 1964. Advances in orchid environment control. *Proceedings of the Fourth World Orchid Conference.* Straits Times Press (M) Ltd., Singapore, pp. 147–157.

——. 1967. Question Box. *Am. Orchid Soc. Bull.* **36**: 811.

DAWMANN, E. 1968. Zur Bestäubungsökologie von *Cypripedium calceolus* L. *Österr. Bogor.* **1**: 61–76.

DeVRIES, J. T. 1953. On the flowering of *Phalaenopsis schilleriana* Rchb.f. *Ann. Bogor.* **1**: 61–76.

DEXTER, J. S. 1913. p. 867 in *Science* **37**, quoted by Anon. Mosquitoes pollinating orchids. *Orchid Rev.* **21**: 263.

DICKENSON, C. R. 1969. *Ansellia* notes. *Am. Orchid Soc. Bull.* **33**: 491–492.

DICKINSON, S. 1968. Mexico's "Lirio or Flor de Mayo"—*Laelia speciosa. Am. Orchid Soc. Bull.* **37**: 1062–1064.

DILLON, G. W. 1953. Hawaii's native orchids. *Am. Orchid Soc. Bull.* **22**: 736.

DOBZHANSKY, T. 1950. Evolution in the Tropics. *Am. Scient.* **38**: 209–221.

DODSON, C. H. 1962. The importance of pollination in the evolution of the orchids of tropical America. *Am. Orchid Soc. Bull.* **31**: 525–534; 641–649; 731–735.

———. 1965. Studies in orchid pollination. The genus *Coryanthes*. *Am. Orchid Soc. Bull.* **34**: 680–687.

———. 1966. Studies in orchid pollination. *Cypripedium, Phragmopedium* and allied genera. *Am. Orchid Soc. Bull.* **35**: 125–128.

———. 1966. Studies in orchid pollination. The genus *Anguloa*. *Am. Orchid Soc. Bull.* **35**: 624–627.

———. 1967. Studies in orchid pollination. The genus *Notylia*. *Am. Orchid Soc. Bull.* **36**: 209–214.

DODSON, C. H., and FRYMIRE, G. P. 1959. *Oncidium serratum*. *Am. Orchid Soc. Bull.* **28**: 107–110.

———. 1961. Natural pollination in orchids. *Mo. Bot. Gardens Bull.* **49**: 133–152.

DODSON, C. H., and HILLS, H. G. 1966. Gas chromatography of orchid fragrances. *Am. Orchid Soc. Bull.* **35**: 720–725.

DORE, W. G. 1958. A simple chemical light-meter. *Ecology* **39**: 151–152.

DRESSLER, R. L. 1960. A review: On the evolution of the Orchidaceae. *Am. Orchid Soc. Bull.* **29**: 759–760.

———. 1961. Tropical orchids near the Texas border. *Am. Orchid Soc. Bull.* **30**: 961–965.

———. 1967. Why do euglossine bees visit orchid flowers? *Atas do Simposio sobre a Biota Amazonica* Vol. 5 (*Zoologia*): 171–180.

———. 1968. Pollination by euglossine bees. *Evolution* **22**: 202–210.

———. 1968. Observations on orchids and euglossine bees in Panama and Costa Rica. *Rev. Biol. Trop.* **15**: 143–183.

———. 1964. Another natural hybrid in *Epidendrum*. *Am. Orchid Soc. Bull.* **33**: 289–291.

DUBININ, N. P. 1940. Darwinism and genetics of populations (In Russian). *Unsp. sovrem. Biol.* **13**: 257, as quoted by Fedorov, An. A. in *J. Ecol.* **54**: 1–11 (1966).

DUDGEON, W. 1923. Succession of epiphytes in the *Quercus incana* forest at Landour, Western Himalayas. *J. Ind. Bot. Soc.* **3**: 270–272.

DUNGAL, N. 1953. Preservation of orchid pollen and seeds by dry freezing. *Am. Orchid Soc. Bull.* **22**: 863–864.

DUNGS, F., and PABST, J. F. J. 1971. Habitats, living conditions and culture for orchids in the tropics. *Orchid Fig.* **35**: 6–10.

DUNSTERVILLE, G. C. K. 1961. How many orchids on a tree? *Am. Orchid Soc. Bull.* **30**: 362–363.

———. 1964. Auyantepui, home of fifty million orchids. *Am. Orchid Soc. Bull.* **33**: 678–689.

———. 1967. *Catasetum barbatum:* A collector's dream, taxonomist's nightmare. *Orchid Rev.* **75**: 386–390.

———. 1968. Please don't strip the forests—Even from your armchairs. *Am. Orchid Soc. Bull.* **37**: 397–401.

DUNSTERVILLE, G. C. K., and DUNSTERVILLE, E. 1969. Orchids of Caracas. *Am. Orchid Soc. Bull.* **38**: 493–496.

DUNSTERVILLE, G. C. K., and GARAY, L. A. 1961a. *Venezuelan Orchids Illustrated,* Vol. II. André Deutsch, London.

——. 1961b. Venezuelan orchids: *Scuticaria steelei* Lindl. *Orchid Rev.* **69**: 52–53.

DYCUS, A. M., and KNUDSON, L. 1957. The role of the velamen of the aerial roots of orchids. *Bot. Gaz.* **119**: 78–87.

EGGELING, W. J. 1947. Observations on the ecology of the Budongo rain forest, Uganda. *J. Ecol.* **34**: 20–87.

EHRLICH, P., and EHRLICH, A. 1970. *Population, Resources, Environment.* W. H. Freeman, London.

EHRLICH, P. R., and RAVEN, P. H. 1969. Differentiation of populations. *Science* **165**: 1228–1232.

EIGELDINGER, O. 1957. *Orchids for everyone.* John Gifford, London.

ELBERT, G. A. 1969. On the conservation of orchids. *Am. Orchid Soc. Bull.* **38**: 497–498.

ERICKSON, L. C. 1957. Respiration and photosynthesis in *Cattleya* roots. *Am. Orchid Soc. Bull.* **26**: 401–402.

ERNST, R., ARDITTI, J., and HEALEY, P. L. 1970. The nutrition of orchid seedlings. *Am. Orchid Soc. Bull.* **39**: 599–605.

EVANS, G. C. 1939. Ecological studies on the rain forest of southern Nigeria, II. The atmospheric environmental conditions. *J. Ecol.* **27**: 436–482.

——. 1956. An area survey method of investigating the distribution of light intensity in woodlands, with particular reference to sunflecks. (Including an analysis of data from rain forest in southern Nigeria.) *J. Ecol.* **44**: 391–428.

——. 1969. The spectral composition of light in the field, I. Its measurement and ecological importance. *J. Ecol.* **57**: 109–125.

EVANS, G. C., WHITMORE, T. C., and WONG, Y. K. 1960. The distribution of light reaching the ground vegetation in a tropical rain forest. *J. Ecol.* **48**: 193–204.

FAEGRI, K., and VAN DER PIJL, L. 1966. *The Principles of Pollination Ecology.* Pergamon Press, Toronto.

FAIRBURN, D. C., and PRING, G. H. 1945. Light requirements for orchids in the midwest. *Mo. Bot. Gard. Bull.* **33**: 33–56.

FEDERER, C. A. 1966. Spectral distribution of light in the forest. *Ecology* **47**: 555–560.

FEDOROV, AN. A. 1966. The structure of the tropical rain forest and speciation in the humid tropics. *J. Ecol.* **54**: 1–11.

FISCH, M. H., SCHETER, Y., and ARDITTI, J. 1972. Orchids and the discovery of phytoalexins. *Am. Orchid Soc. Bull.* **41**: 605–607.

FITCH, C. M. 1969. Orchid conservation in the tropics. *Am. Orchid Soc. Bull.* **38**: 476–485.

FITTKAU, E. J., ILLIES, J., KLINGE, H., SCHWABE, G. H., SIOLI, H., Eds. 1968. *Biogeography and Ecology in South America.* Vol. I. Dr. W. Junk N. V., Publishers, The Hague, pp. 25–53.

FOSTER, M. B. 1947. Roof gardens in Colombia. *Am. Orchid Soc. Bull.* **16**: 652–653.

FOWLIE, J. A. 1960a. Ecology notes: *Comparettia falcata*. *Am. Orchid Soc. Bull.* **29**: 752–754.

———. 1960b. Ecology notes: *Oncidium pulchellum*. *Am. Orchid Soc. Bull.* **29**: 849–850.

———. 1961a. Ecology notes: *Chondrorhyncha aromatica*. *Am. Orchid Soc. Bull.* **30**: 951–952.

———. 1961b. Ecology notes: *Oncidium jamaica*. *Am. Orchid Soc. Bull.* **30**: 223–224.

———. 1961c. Ecology notes: *Broughtonia sanguinea*. *Am. Orchid Soc. Bull.* **30**: 379–380.

———. 1961d. Ecology notes: Aberrancy in *Broughtonia sanguinea*. *Am. Orchid Soc. Bull.* **30**: 641–643.

———. 1961e. Ecology notes: Natural hybridization in the genus *Broughtonia*. *Am. Orchid Soc. Bull.* **30**: 707–710.

———. 1961f. Ecology notes: *Cochleanthes flabelliformis*. *Am. Orchid Soc. Bull.* **30**: 797–799.

———. 1962a. Ecology notes: *Huntleya burtii*. *Am. Orchid Soc. Bull.* **31**: 741–744.

———. 1962b. Ecology notes: Observations on coloration defects in nature. *Am. Orchid Soc. Bull.* **31**: 830–832.

———. 1966. The enigmatic broughtonias. *Orchid Rev.* **74**: 349–353. (Reprinted in *Lasca Leaves* **16**: 1.)

FREI, Sister JOHN KAREN. 1973. Effect of bark substrate on germination and early growth of *Encyclia tampensis* seeds. *Am. Orchid Soc. Bull.* **42**: 701–708.

———. 1973a. Orchid ecology in a cloud forest in the mountains of Oaxaca, Mexico. *Am. Orchid Soc. Bull.* **42**: 307–314.

FREIRE, C. V. 1938–1939. Estudo histologico de orchidaceas. *Orchidea* **1**: 51–52.

FULLER, A. M. 1933. Studies on the flora of Wisconsin. Part XXX 1. The orchids; Orchidaceae. *Bull. Publ. Mus. City of Milwaukee, Wisc.*

GARAY, L. A. 1960. On the origin of the Orchidaceae. *Bot. Mus. Leaflets, Harvard* **19**: 57–87.

———. 1964. Evolutionary significance of geographical distribution of orchids. *Proceedings of the Fourth World Orchid Conference*, Straits Times Press (M) Ltd., Singapore, pp. 170–187.

GARB, S. 1961. Differential growth-inhibitors produced by plants. *Bot. Rev.* **27**: 422–443.

GARDNER, G. 1846. The vegetation of Organ Mountains of Brazil. Communication to the Horticultural Society; reprinted in *The Orchid World* **1**: 125–126 (1910–1911).

GARNETT, C. S. 1929. Orchidological research and collecting expedition, Brazil, 1928. *Orchid Rev.* **37**: 40–49.

GATES, D. M. 1965. Energy, plants, and ecology. *Ecology* **46**: 1–13.

———. 1968. Transpiration and leaf temperature. *Ann. Rev. Plant Physiol.* **19**: 211–238.

———. 1969. The ecology of an elfin forest in Puerto Rico, 4. Transpiration rates and temperatures of leaves in a cool humid environment. *J. Arnold Arborteum* **50**: 93–98.

GÄUMANN, E., and JAAG, O. 1945. Über induzierte Abwehrreaktionen bei Pflanzen. *Experientia* 1: 21–22.

GAUSSEN, H., LEGRIS, P., BLASCO, F. 1967. *Bioclimats du Sud-est Asiatique.* Imprimérie de la Mission Pondichéry.

GETZ, L. L. 1968. A method for measuring light intensity under dense vegetation. *Ecology* 49: 1168–1169.

GILBERT, P. A. 1939. Birds and orchids. *Aust. Orchid Rev.* 4: 51–52.

GILLISON, A. N. 1969. Plant succession in an irregularly fired grassland area—Doma Peaks region, Papua. *J. Ecol.* 57: 415–428.

GOLDMAN, E. A. 1951. Biological investigations in Mexico. *Smithsonian Miscellaneous Coll.,* Vol. 115, Washington, D.C.

GOOD, R. 1953. *The Geography of the Flowering Plants.* Longmans, Green and Co., London.

———. 1960. On the geographical relationships of the angiosperm flora of New Guinea. *Bull. Br. Mus. (Nat. Hist.)* 2: 203–226.

GREEN, E. I. 1968. Wild orchids of the Northeastern States. *Am. Orchid Soc. Bull.* 37: 297–305.

GREIG-SMITH, P. 1964. *Quantitative Plant Ecology,* 2nd ed. Butterworths, London.

———. 1971. Application of numerical methods to tropical forests. In G. P. Potil, E. C. Pielou, and W. E. Waters, Eds., *Statistical Ecology.* Pennsylvania State University Press. pp. 195–206.

GREIG-SMITH, P., AUSTIN, M. P., and WHITMORE, T. C. 1967. The application of quantitative methods to vegetation survey. 1. Association-analysis and principal component ordination of rain forest. *J. Ecol.* 55: 483–503.

GRUBB, P. J., LLOYD, J. R., PENNINGTON, T. D., and WHITMORE, T. C. 1963. A comparison of montane and lowland rain forest in Ecuador, I. The forest structure, physiognomy and floristics. *J. Ecol.* 51: 567–601.

GRUBB, P. J., and WHITMORE, T. C. 1966. A comparison of montane and lowland rain forest in Ecuador, II. The climate and its effects on the distribution and physiognomy of the forests. *J. Ecol.* 54: 303–333.

HALBINGER, C. 1941. Hardy and beautiful Mexican laelias. *Am. Orchid Soc. Bull.* 10: 31–32.

HALE, M. E. 1952. Vertical distribution of cryptogams in a virgin forest of Wisconsin. *Ecology* 33: 398–406.

HALL, J. M., and HALL, MRS. J. M. 1967. Hermit crabs and orchids in the Bahamas. *Am. Orchid Soc. Bull.* 36: 809–810.

HANSEN, K. 1964. Studies on the regeneration of heath vegetation after burning-off. *Bot. Tids.* 60: 1–41.

HARDY, R. D. 1962. A climatological study of the *Odontoglossa. Orchid Dig.* 26: 121–124; 216–221; 253–260; 301.

HARLEY, J. L. 1959. *The Biology of Mycorrhiza.* Leonard Hill Books, London.

HARPER, J. L., and McNAUGHTON, J. H. 1962. The comparative biology of closely related species living in the same area, VII. Interference between individuals in pure and mixed populations of *Papaver* species. *New Phytol.* 61: 175–188.

HARRIS, J. A. 1918. On the osmotic concentrations of the tissue fluids of phanerogamic epiphytes. *Am. J. Bot.* 5: 490–506.

——. 1926. The specific electrical conductivity of the leaf tissue fluids of phanerogamic epiphytes. *Bull. Torrey Bot. Club* 53: 183–188.

——. 1934. *The physico-chemical properties of plant sap in relation to phytogeography.* Minneapolis. (Cited by Curtis, 1946.)

HARRIS, G. P. 1971. The ecology of corticolous lichens. II. The relationship between physiology and the environment. *J. Ecol.* 59: 441–452.

HARTSEMA, A. M. 1961. Influence of temperature on flower formation and flowering of bulbous and tuberous plants, in W. Ruhland, Ed., *Handbuch der Pflanzenphysiologie.* Springer-Verlag, Berlin, pp. 123–167.

HATCH, E. D. 1953. Orquideas subterraneas. *Orquidea* 15: 4–15.

HAYES, A. B. 1968. The morphological effects of pollination in the Brazilian miltonias. *Am. Orchid Soc. Bull.* 37: 705–707.

HAZEN, W. E. 1966. Analysis of spatial pattern in epiphytes. *Ecology* 47: 634–635.

HERMESSEN, J. L. 1916. Notes on the ecology of orchids. *Orchid Rev.* 24: 77–79.

HESLOP-HARRISON, J. 1964. Forty years of genecology, in J. B. Cragg, Ed., *Advances in Ecological Research,* Vol. 2. Academic Press, London and New York.

HESSE, P. R. 1955. A chemical and physical study of the soils of termite mounds in East Africa. *J. Ecol.* 43: 449–461.

HILLS, H. G., WILLIAMS, N. H., and DODSON, C. H. 1968. Identification of some orchid fragrance components. *Am. Orchid Soc. Bull.* 37: 967–971.

HODGES, J. D., and LORIO, P. L., JR. 1969. Carbohydrate and nitrogen fractions of the inner bark of loblolly pines under moisture stress. *Can. J. Bot.* 47: 1651–1657.

HOLTTUM, R. E. 1960. The ecology of tropical epiphytic orchids. *Proceedings of the Third World Orchid Conference.* The Royal Horticultural Society, London, pp. 196–204.

HORICH, C. KL. 1960. Orchid raid to Eucador. *Am. Orchid Soc. Bull.* 29: 838–843.

HOSOKAWA, T. 1943. Studies on the life-forms of vascular epiphytes and the epiphyte flora of Ponape, Micronesia. *Trans. Nat. Hist. Soc. Taiwan* 33: 35–55; 71–89; 113–141.

HOSOKAWA, T., and KUBOTA, H. 1957. On the osmotic pressure and resistance to desiccation of epiphytic mosses from a beech forest, Southwest Japan. *J. Ecol.* 45: 579–591.

HOSOKAWA, T., and ODANI, N. 1957. The daily compensation period and vertical ranges of epiphytes in a beech forest. *J. Ecol.* 45: 901–915.

HOSOKAWA, T., OMURA, M., and NICHIHARA, Y. 1954. Social units of epiphyte communities in forests. *Proc. VIII Congr. Internat. Bot.,* pp. 11–16.

HOWELL, F. C., and BOURLIERE, F., Eds. 1964. *African Ecology and Human Evolution.* Methuen and Co., London, pp. 28–42.

HUECK, K. 1955. *Nouvelles cartes de la végétation sud-Américaine et leur signification pour l'agriculture et la sylviculture.* Paris.

HUMBERT, H., and COURS DARNE, G. 1965. *Carte Internationle du tapis végétal Madagascar.* Pondichéry.

HUNT, P. F. 1967. Size of the orchid family. *Orchid Rev.* 75: 229.

———. 1968. Conservation of orchids. *Orchid Rev.* 76: 320–327.

———. 1969a. Conservation of orchids. *IUCN Bull.* 2: 76.

———. 1969b. Orchids of the Solomon Islands. *Phil. Trans. Roy. Soc. B.* 255: 581–587.

HURLEY, P. M., DE ALMEIDA, F. F. M., WELCHER, G. C., CORDANI, U. G., RAND, J. R., KAWASHITA, K., VANDOROS, P., PINSON, W. H., JR., FAIRBAIRN, H. W. 1967. Test of continental drift by comparison of radiometric ages. *Science* 157: 495–500.

HUTCHINSON, J. 1959. *The families of flowering plants,* Vol. II. *The Monocotyledons.* Clarendon Press, Oxford.

JEFFREY, D., and ARDITTI, J. 1968. Sugar content of orchid nectar. *Orchid Rev.* 76: 315–316.

———. 1969. The separation of sugars in orchid nectars by thin layer chromatography. *Am. Orchid Soc. Bull.* 38: 866–868.

JEFFREY, D. C., ARDITTI, J., and KOOPOWITZ, H. 1970. Sugar content in floral and extrafloral exudates of orchids: Pollination, myrmecology and chemotaxonomy implications. *New Phytol.* 69: 187–195.

JENSEN, W., PREMER, K. E., SIERILA, P., and WARTIOVAARA, V. 1963. In B. L. Browning, Ed., *The Chemistry of Wood.* Interscience, New York and London, p. 587.

JESUP, H. P. 1966. Preliminary cultural experiences with the leafless species. *Am. Orchid Soc. Bull.* 35: 731–733.

JONES, E. 1960. Contribution of rainwater to the nutrient economy of soil in northern Nigeria. *Nature* 188: 432.

JONES, E. W. 1952. Some observations on the lichen flora of tree boles, with special reference to the effect of smoke. *Rev. Bryol. Lich.* 21: 96–115.

KAMERLING, Z. 1912. De verdamping van epiphyte Orchideen. *Natuurk. Tijdschr. Ned. Ind.* 71: 54–72.

KANO, K. 1968. Acceleration of the germination of so-called "hard to germinate" orchid seeds. *Am. Orchid Soc. Bull.* 37: 690–698.

KAPULER, A. M. 1962. In search of orchids in Colombia. *Am. Orchid Soc. Bull.* 31: 701–707.

KEAY, R. W. J. 1959. *Vegetation Map of Africa South of the Tropic of Cancer.* Oxford University Press.

KENNEDY, G. C. 1972. Notes on the genera *Cymbidiella* and *Eulophiella* of Madagascar. *Orchid Sig.* 36: 121–122.

KERR, A. D. 1972. The leafless orchids. *Am. Orchid Soc. Bull.* 41: 307–309.

KERSHAW, K. A. 1964. Preliminary observations on the distribution and ecology of epiphytic lichens in Wales. *The Lichenologist* 2: 263–276.

KLEIN, R. M. 1964. Repression of tissue culture growth by visible and near visible radiation. *Plant Physiol.* 39: 536–539.

KLEIN, R. M., EDSALL, P. C., and GENTILE, A. C. 1965. Effects of near ultraviolet and green radiations on plant growth. *Plant Physiol.* 40: 903–906.

KLIKOFF, L. G. 1965. Photosynthetic response to temperature and moisture stress of three timberline meadow species. *Ecology* 46: 516–517.

KNAUFT, R. L., and ARDITTI, J. 1969. Partial identification of dark $^{14}CO_2$ fixation products in leaves of *Cattleya* (Orchidaceae). *New Phytol.* 68: 657–661.

KNUDSON, L. 1940. Viability of orchid seed. *Am. Orchid Soc. Bull.* **9**: 36–38.

KOLKWITZ, R. 1932. Urwald und epiphyten. *Ber. Deutsch. Bot. Ges.* **50**: 110–116.

KURFESS, J. F. 1965. Through the letter slot. *Am. Orchid Soc. Bull.* **34**: 914.

LABROVE, L., LEGRIS, P., VIART, M. 1965. *Bioclimats du sous-continent Indien.* Imperimerie de la Mission Pondichery.

LAGER, J. E. 1932. Reminiscences of an orchid collector. *Am. Orchid Soc. Bull.* **1**: 40–48.

LANKESTER, C. H. 1920. Orchids in Costa Rica. *Orchid Rev.* **28**: 129–132.

LATIF, S. M. 1969. The territory of the giant orchid *Grammatophyllum speciosum* Bl. *Orchid Rev.* **77**: 82–83.

LEAKEY, C. L. A. 1968. The orchids of Uganda. Various species of horticultural merit. *Am. Orchid Soc. Bull.* **37**: 1052–1055.

LECOUFLE, M. 1964. Notes about *Cymbidiella. Orchid Rev.* **72**: 233–236.

VAN LEEUWEN, W. M. 1929. Krakatou's new flora, in *Fourth Pacific Science Congress,* Part II, Krakatou. Java, pp. 57–79.

———. 1936. *Krakatou 1883–1933.* A. Botany. E. J. Brill, Leiden. (pp. 452–468, Orchidaceae.)

LEWIS, D. A. 1969. The effect of corolla color and outline on interspecific pollen flow in *Phlox. Evolution* **23**: 444–455.

LOOMIS, W. E. 1965. Absorption of radiant energy by leaves. *Ecology* **46**: 14–17.

LOVELESS, A. R. 1960. The vegetation of Antigua, West Indies. *J. Ecol.* **48**: 495–527.

LUER, C. A. 1967. Observations on *Stenorrhynchus orchioides. Am. Orchid. Soc. Bull.* **36**: 381–382.

LÜNING, B. 1966. Chemotaxonomy in a *Dendrobium* complex. *Proceedings of the Fifth World Orchid Conference,* Long Beach, Calif.

———. 1967. Studies on Orchidaceae alkaloids, IV. Screening of species for alkaloids. *Phytochem.* **6**: 857–861.

———. 1969. Orchid alkaloids. *Orchid Rev.* **77**: 52–55.

LYNCH, J. J. 1970. Conservation of America's native wild orchids. *Am. Orchid Soc. Bull.* **39**: 1060–1065.

MACARTHUR, R. H., and WILSON, E. O. 1967. *The Theory of Island Biogeography.* Princeton University Press, Princeton, N. J.

MACDOUGALL, T. 1959. The Nizanda "Rock Gardens" of Oaxaca. *Am. Orchid Soc. Bull.* **28**: 728–730.

MACHATTIE, L. B., and MCCORMACK, R. J. 1961. Forest microclimate: A topographic study in Ontario. *J. Ecol.* **49**: 301–323.

MADGWICK, H. A. I., and OVINGTON, J. D. 1959. The chemical composition of precipitation in adjacent forest and open plots. *Forestry* **32**: 14–22.

MANTON, J. 1969. Evolutionary mechanisms in tropical ferns. *Biol. J. Linn. Soc.* **1**: 219–222.

MARTIN, J. T., and JUNIPER, B. E. 1970. *The Cuticles of Plants.* Edward Arnold, London.

MARX, JEAN L. 1973. Photorespiration: key to increasing plant productivity? *Science* **179**: 365–367.

MAYR, E. 1969. The biological meaning of species. *Biol. J. Linn. Soc.* **1**: 311–320.

McDADE, E. 1947. Leaf burn: Its results, causes and prevention. *Am. Orchid Soc. Bull.* 16: 448–449.

McLAREN, A. D., JENSEN, W. A., and JACOBSON, L. 1960. Absorption of enzymes and other proteins by barley roots. *Plant Physiol.* 35: 549–556.

MEHER-HOMJI, V. M. 1963. *Les bioclimats du sub-continent indien et leurs types analogues dans le monde.* Institut Français, Pondichéry.

MILLER, J. H., and MONZ MILLER, P. 1964. Blue light in the development of fern gametophytes and its interaction with far-red and red light. *Am. J. Bot.* 51: 329–334.

MIWA, A. 1937. Carbon dioxide content of the atmospheric air of the greenhouse. *Orchid Rev.* 45: 146–152.

MOIR, W. W. G. 1968. Conservation and orchids. *Orchid Rev.* 76: 327–331.

MONTEITH, J. L. 1963. Dew: Facts and fallacies, in A. J. Rutter and F. N. Whitehead, Eds., *The Water Relations of Plants.* A Symposium of the British Ecological Society London, 5–8 April 1961. Blackwell Scientific Publications, Oxford.

MOONEY, H. A., and BILLINGS, W. D. 1965. Effects of altitude on carbohydrate content of mountain plants. *Ecology* 46: 750–751.

MOREAU, R. E. 1966. *The Bird Faunas of Africa and its Islands.* Academic Press, New York and London.

MORISON, C. G. T., HOYLE, A. C., and HOPE-SIMPSON, J. F. 1948. Tropical soil-vegetation catenas and mosaics. A study in the southwestern part of Anglo-Egyptian Sudan. *J. Ecol.* 36: 1–84.

MORRIS, B. 1968. Preliminary checklist of the epiphytic orchids of Malawi. *Am. Orchid Soc. Bull.* 37: 887–892.

——. 1970. *The Epiphytic Orchids of Malawi.* The Society of Malawi, Blantyre.

MORTON, J. K. 1961. In *Recent Advances in Botany,* Vol. 9. Toronto, pp. 900–903.

——. 1966. The role of polyploidy in the evolution of a tropical flora, in C. D. Darlington and K. R. Lewis, Eds., *Chromosomes Today,* Vol. 1. Oliver and Boyd, Edinburgh, pp. 73–76.

NARODNY, L. H. 1945. Vanilla leaves used as light indicators in Dominica, B.W.I. *Am. Orchid Soc. Bull.* 14: 269–273.

NELSON, E. 1962. *Gestaltwandel und Artbildung erörtert am Beispiel der Orchidaceen Europas und der Mittelmeerländer.* Crernix-Montreaux, Switzerland.

NICOT, J. 1954. Microflore fongique des sols de vanilleraies, in Gilbert Bouriquet, Ed., *Le Vanillier et la Vanille dans le monde.* Editions Paul Lechevalier, Paris, pp. 364–392.

NUERNBERGK, E. L. 1964. On the carbon dioxide metabolism of orchids and its ecological aspect. *Proceedings of the Fourth World Orchid Conference,* Straits Times Press (M) Ltd., Singapore, pp. 158–169.

NUTTONSON, M. Y. 1963. *The Physical Environment and Agriculture of Thailand.* Am. Inst. Crop Research, Washington, D.C.

——. 1963. *The Physical Environment and Agriculture of Burma* and *Supplement on Climate, Soils and Rice Culture of Burma.* Am. Inst. Crop Ecology, Washington, D.C.

————. 1963. *The Physical Environment and Agriculture of Vietnam, Laos and Cambodia*. Am. Inst. Crop Ecology, Washington, D.C.

ODUM, E. P. 1963. *Ecology*. Holt, Rinehart and Winston, New York.

OLIVER, W. R. B. 1930. New Zealand epiphytes. *J. Ecol.* **18:** 1–50.

ORTH, R. 1939. Zur Kenntnis des Lichtklimas der Tropen und Subtropen sowie des tropischen Urwaldes. *Gerl. Beitr. z. Geophys.* **55:** 52–102. (Cited by Schulz, 1960.)

OVINGTON, J. D. 1953. Studies of the development of woodland conditions under different trees, I. Soils pH. *J. Ecol.* **41:** 13–34; 1956. IV. The ignition loss, water, carbon and nitrogen content of the mineral soil. *J. Ecol.* **44:** 171–179; 1956. V. The mineral composition of the ground flora. *J. Ecol.* **44:** 597–604; 1957. VI. Soil sodium, potassium and phosphorus. *J. Ecol.* **46:** 127–142; 1958. VII. Soil calcium and magnesium. *J. Ecol.* **46:** 391–405.

OYE, P. VAN. 1921. Influence des facteurs climatiques sur la surface des troncs d'arbres a Java. *Rev. Gen. Bot.* **33:** 161–176; 1924. Sur l'écologie des épiphytes de la surface des troncs d'arbres à Java. *Rev. Gen. Bot.* **36:** 12–30; 68–83.

PACHECO, H., PLA, J., and VILLE, A. 1966. Comparison of the incorporation of *trans*-cinnamic-3-^{14}C acid and ($\pm$) shikimic-1-2 ^{14}C acid in violinin, an anthocyanin. *Compt. Rend. Ser. D.* **262:** 926–929.

PERKINS, B. L. 1962. *Eulophia cucullata* and other orchids in Kenya. *Orchid Rev.* **70:** 281–282.

PIELOU, E. C. 1969. *An Introduction to Mathematical Ecology*. John Wiley and Sons, New York.

PIERIK, R. L. M. 1965. Integrating photochemical light measurement, an ecological study in the middachten woodland in the Netherlands. *Mededelingen van de landbouwhogeschool. Wageningen, Nedl.* **65:** (10). (Reprinted in *Belmontia II. Ecology*, fasc. 12, paper 78, 1966.)

PIJL, L. VAN DER. 1969. Evolutionary action of tropical animals on the reproduction of plants. *Biol. J. Linn. Soc.* **1:** 85–96.

PIJL, L. VAN DER and DODSON, C. H. 1966. *Orchid Flowers: Their Pollination and Evolution*. University of Miami Press, Coral Gables, Fla.

PITTENDRIGH, C. S. 1948. The bromeliad-anopheles-malaria complex in Trinidad, I. The bromeliad flora. *Evolution* **2:** 58–89.

POEL, A. J. VAN DER, and STOUTJESDIJK, P. 1959. Some microclimatological differences between an oak wood and a *Calluna* heath. *Mededelingen van de landbouwhogeschool te Wageningen Nedl.* **59:** 1–8. (Reprinted *Belmontia* fasc. 5: 1–8, 1960.)

POHLSSON, L. 1966. Vegetation and microclimate along a belt transect from the Esker Knivsas. *Bot. Notiser.* **19:** 401–418.

POORE, M. E. D. 1968. Studies in Malaysian rain forest, I. The forest on triassic sediments in Jengka Forest Reserve. *J. Ecol.* **56:** 143–196.

POST, R. E., JR. 1965. Building blocks of the *Cattleya* genus: *Cattleya dowiana*. *Am. Orchid Soc. Bull.* **34:** 807–810.

POTZGER, J. E. 1939. Microclimate and a notable case of its influence on a ridge in Central Indiana. *Ecology* **20:** 29–37.

PRESTON, F. W. 1962. The canonical distribution of commonness and rarity. Part II. The depauperate faunas and floras of oceanic islands. *Ecology* **43:** 410–432.

RAHMAN, A. A. ABD EL, and BATANOUNG, K. H. 1965. Transpiration of desert plants under different environmental conditions. *J. Ecol.* **53**: 267–272.

RAMSEY, C. T. 1965. Darwin and the Star Orchid of Madagascar. *Am. Orchid Soc. Bull.* **34**: 1056–1062.

RAWITSCHER, F. 1948. The water economy of the vegetation of the 'Campos Corrados' in Southern Brazil. *J. Ecol.* **36**: 237–268.

RAYNER, M. C. 1928. Note on the ecology of mycorrhiza. *J. Ecol.* **16**: 418–419. (Ref. Rayner, M. C. *Mycorrhiza*. Wheldon and Wesley, London, 1927.)

REINHARD, H. R. 1970. Diskussionsbeitrag zum Problem der Bastarde europäischer Orchideen. *Die Orchid.* **21**: 167–173.

RICE, E. L. 1960. The microclimate of a relict stand of sugar maple in Devils Canyon in Canadian County, Oklahoma. *Ecology* **41**: 445–453.

——. 1964. Inhibition of nitrogen fixing and nitrifying bacteria by seed plants (1). *Ecology* **45**: 824–837.

RICHARDS, P. W. 1939. Ecological studies on the rain forest of Southern Nigeria, 1. The structure and floristic composition of the primary forest. *J. Ecol.* **27**: 1–61.

——. 1952, 1st ed.; 1957, 2nd ed. *The Tropical Rain Forest: An Ecological Study.* Cambridge University Press.

——. 1969. Speciation in the tropical rain forest and the concept of the niche. *Biol. J. Linn. Soc.* **1**: 149–153.

RIDLEY, H. N. 1910. Symbiosis of ants and plants. *Ann. Bot.* **24**: 457–483.

——. 1930. *The Dispersal of Plants throughout the World.* L. Reeve and Co., Ashford, Kent.

RIELEY, J. O., MACHIN, D., and MORTON, A. 1969. The measurement of microclimate factors under vegetation canopy—Reappraisal of Wilm's method. *J. Ecol.* **57**: 101–108.

ROBERTSON, G. W. 1966. The light composition of solar and sky spectra available to plants. *Ecology* **47**: 640–643.

RODWAY, J. 1894. Quoted in anonymous review. Notes on the fertilization of orchids in the tropics. *Orchid Rev.* **2**: 338–340.

ROLFE, R. A. 1897. Orchids abroad and at home. *Orchid Rev.* **5**: 105–109.

ROSS, R. 1954. Ecological studies on the rain forest of southern Nigeria, III. Secondary succession in the Shasha Forest Reserve. *J. Ecol.* **42**: 259–282.

ROTOR, G., JR. 1951. Daylength and temperature in relation to flowering in orchids. *Am. Orchid Soc. Bull.* **20**: 210–214.

——. 1959. The photoperiodic and temperature responses of orchids, in Carl L. Withner, Ed., *The Orchids.* Ronald Press, New York, pp. 397–418.

ROVIRA, A. D. 1969. Plant root exudates. *Bot. Rev.* **35**: 35–57.

RUINEN, J. 1953. Epiphytosis: A second view on epiphytism. *Ann. Bogor.* **1**: 101–157.

——. 1956. Occurrence of *Beijerinckia* species in the "phyllosphere." *Nature* **177**: 220–221.

RUNEMARK, H. 1969. Reproductive drift, a neglected principle in reproductive biology. *Bot. Notiser.* **122**: 90–129.

SANFORD, W. W. 1961. Orchid favorites, II. *Trichocentrum panamense. Orchid Rev.* **69**: 364–365.

————. 1967. An introduction to the climate and vegetation of Nigeria. *Am. Orchid Soc. Bull.* **36**: 963–969.

————. 1968. Distribution of epiphytic orchids in semideciduous tropical forest in southern Nigeria. *J. Ecol.* **56**: 697–75.

————. 1969. The distribution of epiphytic orchids in Nigeria in relation to each other and to geographic location and climate, type of vegetation and tree species. *Biol. J. Linn. Soc.* **1**: 247–285.

————. 1969b. Conservation of West African Orchids, I. Nigeria. *Biol. Conserv.* **1**: 148–150; 1970. II. La République du Caméroun. *Biol. Conservation* **3**: 47–50.

————. 1970a. The orchid flora of Equatorial Guinea in relation to that of West Africa. *Proceedings 7th Congress of A.E.T.F.A.T.*, Munich, Sept. 1970. *Mitt. Bot. Staatssamme.* München **10**: 287–298.

————. 1970b. Conservation of West African orchids, II. La République du Caméroun. *Biol. Conservation* **3**: 47–50.

————. 1970c. Practical conservation of orchids in Nigeria. *J. Nig. Sci. Assoc.* **4**: 49–57.

————. 1971. Flowering time of West African Orchids. *Bot. J. Linn. Soc.* **64**: 163–181.

————. 1972. Epiphytic orchids as characterizers of vegetation in West Africa. *Seventh World Orchid Conference,* Medellin. (In press.)

————. 1974. An ecological analysis of the genus *Bulbophyllum* in West Africa. *Am. Orchid Soc. Bull.* (In press.)

————. 1974a. The use of epiphytic orchids to characterize vegetation in Nigeria. *Bot. J. Linn. Soc.*(In press.)

SANFORD W. W., and ADANLAWO, I. 1973. Velamen and exodermis characters of West African epiphytic orchids in relation to taxonomic grouping and habitat tolerance. *Linn. Soc. Bot. Jour.* **66**: 307–321.

SANFORD, W. W., FOURAKIS, E., KRALLIS, A., and XANTHAKIS, A. 1965. Paper partition chromatography as an aid to orchid taxonomy. *Orchid Rev.* **73**: 178–181.

SCHIMPER, A. F. W. 1903. *Plant Geography upon a Physiological Basis.* Oxford.

SCHLECHETER, R. 1914. *Die Orchidaceen von Deutsch-Neu-Guinea.* Dahlem bei Berlin.

SCHNELL, R. 1950. *La forêt dense.* Paul Lechevalier, Paris.

SCHULZ, J. P. 1960. *Ecological Studies on Rain Forest in Northern Surinam.* N.V. Noord-Hollandsche Uitgevers Maatschappij, Amsterdam.

SCHWEINFURTH, C. 1958. *Orchids of Peru.* Fieldiana: Botany. Chicago National History Museum.

SEATON, A. P. L., and ANTINOVICS, J. 1967. Population interrelationships, I. Evolution in mixtures of *Drosophila* mutants. *Heredity,* London **22**: 19–33.

SEGARS, W. E. 1960. Notes on the culture of *Cycnoches. Am. Orchid Soc. Bull.* **29**: 582–584.

SEYBOLD, A. 1936. Über den Lichtfaktor photophysiologischer Prozesse. *Jahrb. Wiss. Bot.* **82**: 741–795. (Cited by Schulz, 1960.)

SHANKS, R. E., and NORRIS, F. H. 1950. Microclimate variation in a small valley in Eastern Tennessee. *Ecology* **31**: 532–539.

SHREVE, F. 1908. Transpiration and water storage in *Stelis ophioglossoides* Sw. *Plant World* **2**: 165–172.

————. 1914. *A Montane Rain-forest. A Contribution to the Physiological Plant Geography of Jamaica.* Carnegie Inst. of Washington, Washington, D.C.

SLATYER, R. O. 1960. Absorption of water by plants. *Bot. Rev.* 26: 331–392.

SMITH, J. E. 1966. Desirable orchids of the Bahamas. *Am. Orchid Soc. Bull.* 35: 970–975.

SMITH, J. J. 1927. Ephemeral orchids. *Orchid Rev.* 35: 13.

SPANNER, L. 1939. Untersuchungen über den Wärme- und Wasserhaushalt von *Myrmecodia* und *Hydnophytum. Jb. wiss. Bot.* 88: 243–283.

SPENCER, R. W. 1969. Boraceia—A unique Brazilian orchid habitat: "Where time stood still." *Orchid Dig.* 33: 125–128.

SPOERL, E., and CURTIS, J. T. 1948. Studies on the nitrogen nutrition of orchid embryos, III. Amino acid nitrogen. *Am. Orchid Soc. Bull.* 17: 307–312.

STEBBINS, G. L., JR. 1950. *Variation and Evolution in Plants.* Columbia University Press, New York.

————. 1969. The significance of hybridization for plant taxonomy and evolution. *Taxon* 18: 26–35.

STEENIS, C. G. G. J. VAN 1962. The land-bridge theory in botany (with particular reference to tropical plants). *Blumea* 11: 235–542.

————. 1969. Plant speciation in Malesia, with special reference to the theory of non-adaptive saltatory evolution. *Biol. J. Linn. Soc.* 1: 97–133.

STEFUREAC, T. I. 1941. Recherches synecologiques et sociologiques sur les bryophytes de la forêt vierge de Slatioara (Bucovine). *Anal. Acad. Roman. ser. III, 16,* Mem. 27, pp. 1–197. (Cited by Barkman, 1958.)

STEHLE, H. 1954. Ecologie, in G. Bouriquet, Ed., *Le Vanilier et la Vanille dans le Monde.* Editions Paul Lechevalier, Paris, pp. 291–334.

STONE, B. C. 1967. A review of the endemic genera of Hawaiian plants. *Bot. Rev.* 33: 216–259.

STONE, E. C. 1957. Dew as an ecological factor, I. A review of the literature. *Ecology* 38: 407–413.

STOUTAMIRE, W. P. 1968. Mosquito pollination of *Habenaria obtusata* (Orchidaceae). *Mich. Bot.* 7: 203–212.

SUGANO, N., and HAYASHI, K. 1967. Possible scheme for the biosynthesis of anthocyanin and phenolic compounds in a carrot aggregen as indicated by tracer experiments. *Bot. Mag. Tokyo* 80: 481–486.

SULIT, M. D. 1950. Field observations on tree hosts of orchids in the Philippines. *Philippine Orchid Rev.* 3: 3–8.

————. 1953. Field observations on tree hosts of orchids in Palawan. *Philippine Orchid Rev.* 5: 16.

SUMMERHAYES, V. S. 1931. An enumeration of the angiosperms of the Seychelles Archipelago. *Trans. Linn. Soc. London* (2nd Ser. Zool.) 19: 261–299.

————. 1967. Cited in an anonymous review, ref. *Orchis ustulata* in *Orchid Rev.* 75: 264.

————. 1968a. *Wild Orchids of Britain, 2nd ed.* Collins, London.

————. 1968b. Orchidaceae, in F. N. Hepper, Ed., *Flora of West Tropical Africa,* Vol. III Part I. (revised ed.). Crown Agents for Oversea Governments, London.

SUNDERMANN, H. 1969. *Problem der Orchideengattung Ophrys.* Brucke-Verlag, Kurt Schmersow, Hannover.

TANSLEY, A. G. 1968. *Britain's Green Mantle.* George Allen and Unwin, London.

TAYLOR, G. R. 1970. *The Doomsday Book.* Thames and Hudson, London.

TEUSCHER, H. 1959. Collector's item: *Rodriguezia teuscheri. Am. Orchid Soc. Bull.* 28: 421–424.

———. 1964. *Rodriguezia batemannii* var. *speciosa. Am. Orchid Soc. Bull.* 33: 835–837.

———. 1967. *Epidendrum vespa. Am. Orchid Soc. Bull.* 36: 203–206.

———. 1972. *Microcoelia guyoniana* and other leafless epiphytic orchids. *Am. Orchid Soc. Bull.* 41: 497–501.

THIEN, L. B. 1969. Mosquitoes can pollinate orchids. *Morris Arbor. Bull.* 20: 19–23.

———. 1969. Mosquito pollination of *Habenaria obtusata* (Orchidaceae). *Am. J. Bot.* 56: 232–237.

THOMAS, A. S. 1962. Botany and geology of the downs of Southern England. *Nature, London* 193: 214–217.

THOMPSON, H. N. 1910. *Forests of the Gold Coast.* Col. Rep. Misc. No. 66. (Summary in *Kew Bull.* 1910, pp. 60–64.)

THORNE, ROBERT F. 1973. Floristic relationships between Tropical Africa and Tropical America. In Betty J. Meggers, Edward S. Ayensu, and W. Donald Duckwerth, Eds., *Tropical Forest Ecosystems in Africa and South America: A Comparative Review.* Smithsonian Institution Press, Washington. pp. 27–47.

THORNTON, J. 1965. Nutrient content of rain water in the Gambia. *Nature, London* 205: 1025.

THOROLD, C. A. 1952. The epiphytes of *Theobroma cacao* in Nigeria in relation to the incidence of blackpod disease (*Phytophthora palmivora*). *J. Ecol.* 40: 125–142.

TOBLER, W. R., MIELKE, H. W., and DETWYLER, T. R. 1970. Geobotanical distance between New Zealand and neighboring islands. *BioScience* 20: 537–541.

TUKEY, H. B., JR. 1970. Leaching of metabolites from foliage and its implication in the tropical rain forest. In H. T. Odum and R. F. Pigeon, Eds., *A Tropical Rain Forest.* Nat. Tech. Info. Service, Springfield, Va. pp. 155–160.

TUKEY, H. B., JR., and MECKLENBURG, R. A. 1964. Leaching of metabolites from foliage and subsequent reabsorption and redistribution of the leachate in plants. *Am. J. Bot.* 51: 737–742.

TURESSON, G. 1922a. The species and the variety as ecological units. *Hereditas* 3: 100–113.

———. 1922b. The genotypical response of the plant species to the habitat. *Hereditas* 3: 211–350.

———. 1923. The scope and import of genecology. *Hereditas* 4: 171–176.

———. 1930–1931. The selective effect of climate upon the plant species. *Hereditas* 14: 99–152.

UEDA, H., and TORIKATA, H. 1972. Effects of light and culture medium on adventitious root formation by cymbidiums in aseptic culture. *Am. Orchid Soc. Bull.* 41: 322–327.

ULRICH, J. M., LUSE, R. A., and McLAREN, A. D. 1964. Growth of tomato plants in presence of proteins and amino acids. *Physiol. Plant.* **17:** 683–696.

VACIN, E. F. 1952. Climatological studies of the original habitats of *Cymbidium. Am. Orchid Soc. Bull.* **21:** 517–532.

VERHAGEN, A. M. W., and WILSON, J. H. 1969. The propagation and effectiveness of light in leaf canopies with horizontal foliage. *Ann. Bot.* **33:** 711–727.

VOLKERT, G. R. 1958. The orchid climate. *Am. Orchid Soc. Bull.* **27:** 831–836.

VÖTH, P. D. 1939. Conduction of rainfall by plant stems in a tropical rain forest. *Bot. Gaz.* **101:** 328–340.

WADSWORTH, R. M. et al., Eds. 1968. *The Measurement of Environmental Factors in Terrestrial Ecology.* A symposium of the British Ecological Society, Reading, 29–31. Blackwell Scientific Publications, Oxford and Edinburgh.

WALLBRUNN, H. M. 1966a. Concerning the inability of obtaining hybrids between certain species. *Am. Orchid Soc. Bull.* **35:** 633–634.

———. 1966b. Breeding for cold tolerance. *Am. Orchid Soc. Bull.* **35:** 831–833.

WALTER, H. 1971. *Ecology of Tropical and Subtropical Vegetation.* Oliver and Boyd, Edinburgh.

WALTZ, S. S., and JACKSON, C. R. 1961. Production of yellow strap leaf of *Chrysanthemum* and similar disorders by amino acid treatment. *Plant Physiol.* **36:** 197–201.

WATSON, J. P. 1969. Water movement in two termite mounds in Rhodesia. *J. Ecol.* **57:** 441–451.

WEBB, L. J., TRACEY, J. G., WILLIAMS, W. T., and LANCE, G. N. 1967. Studies in the numerical analysis of complex rain-forest communities, II. The problem of species-sampling. *J. Ecol.* **55.** 525 500.

WEGENER, A. 1915. *Die Entstehung der Kontinente und Ozeane.* Sammlung Cieweg, Brunswick, No. 23.

———. 1937. *La genése des continents et des océans.* Librairie Nizet et Bastard, Paris.

WENGER, K. F. 1955. Light and mycorrhiza development. *Ecology* **36:** 581–520.

WENT, F. W. 1940. Soziologie der Epiphyten eines tropishen Urwaldes. *Ann. J. Bot. Buitenzorg* **50:** 1–98.

WHITTAKER, R. H., and FEENY, P. P. 1971. Allelochemics: Chemical interactions between species. *Science* **171:** 757–770.

WIENS, J. A. 1967. An instrument for measuring light intensities in dense vegetation. *Ecology* **48:** 1006–1008.

WILLIAMS, L. O. 1948. A glimpse of Honduras. *Am. Orchid Soc. Bull.* **17:** 645–649.

WITHNER, C. L. 1953. Orchid roots *in vitro.* Abstracts of papers. B. S. A. annual meeting.

———. 1959. Orchid Physiology, in Carl L. Withner, Ed. *The Orchids: A Scientific Survey.* Ronald Press, New York.

———. 1970. Die kleistogamen *Epidendrum*-arten der Sektion *Encyclia. Die Orchidee* **21:** 4–7.

WITHNER, C. L., and STEVENSON, J. C. 1968. The *Oncidium tetrapetalum* syngameon. *Am. Orchid Soc. Bull.* **37:** 21–32.

WRIGHT, D. 1967. Carbon dioxide enrichment for cymbidiums. *Orchid Rev.* **75**: 120–122.

WRIGHT, H. D. 1946. Orchid hunting on Guadalcanal. *Am. Orchid Soc. Bull.* **15**: 106–116.

WRIGLEY, O. O. 1897. Ammonia for orchids. *Orchid Rev.* **5**: 43–44.

YOUNG, C. 1938. Acidity and moisture in tree bark. *Proc. Ind. Acad. Sc.* **47**: 106–115.

YOUNG, H. E. 1938. Nitrogen bacteria and orchid seedlings. *Aus. Orchid Rev.* **3**: 109.

YOUNG, W. H. 1897. Calendar of operations for August. *Orchid Rev.* **5**: 247–251.

ZEIGLER, A. R., SHEEHAN, T. J., and POOLE, R. T. 1967. Influence of various media and photoperiod on growth and amino acid content of orchid seedlings. *Fla. Agric. Expt. Stations Journal Ser. No.* 2369. (Reprinted in *Am. Orchid Soc. Bull.* **36**: 195–202.

ZINKE, P. J. 1962. The pattern of influence of individual forest trees on soil properties. *Ecology* **43**: 130–133.

2

Terrestrial Orchid Seedlings

WARREN STOUTAMIRE

Terrestrial orchid species, occurring in most of the climatic regions of the earth, are adapted to growth in a correspondingly wide range of conditions. In many parts of the world the terrestrials are subjected to seasonal changes with accompanying annual fluctuations in temperature, light, water and mineral supply, and fungal activity. The small and often slow-growing seedlings are strongly affected by these changes and are quickly eliminated by unsuitable conditions. To survive these cyclic changes terrestrial seedlings as well as the mature plants developing from them have become specifically adapted to particular ranges of environmental factors.

Terrestrial seedlings differ in morphology and developmental patterns from the seedlings of the more geographically restricted epiphytes, which are adjusted to entirely different environmental conditions and are also less diverse in form. A study of terrestrial seedling adaptations is being pursued in this laboratory and discussions of particular species are, in part, based on this work.* This paper also reviews some aspects of the ecology of terrestrial orchids, with emphasis on the early stages of growth. Terrestrial orchids are a highly diverse group of organisms and have received much less attention from all aspects of their biology than the number of species warrants.

Flowering periods for terrestrial species are distributed throughout the year but seed maturation and dispersal is in many cases restricted to the end of vegetative growth and the beginning of dormancy. In orchids of Mediterranean climates in which periods of cool, moist weather alternate

* All material in this laboratory is being germinated and grown in either sterile distilled water, Knudson C, or Burgeff N_3f media modified to 8 g agar and 1 g peptone/liter, under 12-hr light at 200–350 fc, Gro-Lux or white fluorescent tubes, 20–23°C.

This work was supported by National Science Foundation Grant GB 5784X.

with extended warm, dry periods many species flower after vegetative development and shortly before the onset of the dry season (some *Ophrys, Orchis, Pterostylis, Microtis, Thelymitra* species). Seed maturation quickly follows anthesis, and seeds are usually shed at the beginning of or during the period of dormancy. With the return of rains, the seeds presumably germinate and a new cycle of growth begins. Other terrestrials contrast with this in blooming during the period of vegetative dormancy, as in some Mexican *Bletia* species and in Australian *Spiculea ciliata*. Species of temperate regions with a more evenly distributed rainfall flower over a greater period of time, but in most cases seed maturation and dispersal is still delayed until the beginning of vegetative dormancy. This is as true of north temperate species flowering in late spring and early summer (i.e., *Cypripedium, Calypso, Dactylorhiza* and *Platanthera* species) as it is of some of the autumn-flowering *Corallorhiza* and *Spiranthes* species.

Dormancy in many terrestrial orchids is not as easily defined as it is in more familiar plants. The absence of above-ground vegetative organs does not necessarily mean that underground growth has stopped, and this is especially true for such genera as *Corallorhiza, Epipogium, Hexalectris,* and *Neottia* with subterranean stem systems of which little is known of growth cycles because of the difficulty of observation. Species of some genera characteristically produce leaves that persist through the winter (*Calypso, Aplectrum, Tipularia*), and there is presumably no cessation of photosynthetic activity as long as temperature permits it. These species do have marked alterations between vegetative and reproductive growth, however. The only north-temperate and boreal North American species that do not have complete annual changes in foliage belong to the genus *Goodyera,* the plants producing leaves indefinitely along the creeping rhizomes until terminated by flowering. This growth is atypical of temperate orchids generally.

The reproductive stage is terminated by capsule formation and seed production. Capsules of European and North American species cultivated here usually remain closed until growth ceases and dormancy begins, at which time capsules begin to open and seeds are shed at rates differing with the species. Sexual and apomictic races of the autumn-flowering *Spiranthes cernua* release seeds soon after anthesis. *Cypripedium* species often retain quantities of seeds during the dormant period and release them slowly. *Cypripedium arietinum* is particularly tardy in its capsule dehiscence, with some capsules remaining closed until early in the following season of growth. Since dehiscence determines the earliest time at which germination can begin, this temporal control is undoubtedly adjusted for the maximum germination potential of the seeds. Those seeds retained in capsules through the dormant period in temperate and boreal climates are subject to prolonged low temperatures, and this may fulfill low-temperature requirements for germination if such requirements exist. It can be argued that seeds at the soil surface would also be subjected to severe cold treatment, but field measurements indicate that the surface temperatures under a protective snow blanket may be considerably higher than temperatures a few inches above the snow (Getz, 1961).

In species of regions with strongly fluctuating annual temperature and rainfall cycles, the timing of germination will be determined by several factors, including the time of seed release and the period of optimal growth conditions. Although there has been little observation under natural conditions, it is probable that seedling development of temperate species is delayed until the next growth cycle, since most seeds are dispersed before or during vegetative dormancy of the adult plant. Possible exceptions to this occur in *Zeuxine strateumatica* and *Listera cordata,* both of which release seeds very soon after the early flowering period. Seedling development could be well advanced in these species during the current season of growth. There has been no successful germination of either in this laboratory, however, and the natural germination time of *Listera* is unknown. The essentially tropical *Zeuxine* has been reported to germinate only during hot summer months (Porter, 1942), and Vickers (1968) reported quick growth and flowering in less than a year from seed dispersal. Nakamura (1962, 1964) found that *Galeola septentrionalis* seeds would germinate well only at relatively high temperatures. Seeds of this species probably germinate in nature during succeeding growing seasons. They may be subjected to 0°C temperatures in native habitats during the winter, but this cold treatment is not required for germination. Harvais and Hadley (1967b) also found that seeds of several European terrestrials germinated best at 23°C or higher, suggesting that these also do not start growth until the following season.

Mature seeds of tropical *Vanilla* species and hybrids survive at least three years on synthetic media, germination occurring only when appropriate high temperature and low light levels are present (Knudson, 1950; Lugo, 1955), although these physical requirements are not present in immature seeds of *Vanilla* (Withner, 1955). Similarly, immature seeds of temperate-growing *Cypripedium acaule* germinate more readily than seeds from ripe capsules (Withner, 1953), although the temperature requirements are very different from those of *Vanilla* species.

In tropical and subtropical areas with little annual temperature change, temperature variation is probably of little significance in determining the start of seedling development. Unless prevented by genetic control mechanisms, which may vary from species to species, seedling development may begin immediately upon release of the seeds. Seeds of some tropical species, such as *Arundina bambusifolia, Bletia stenophylla, B. purpurea, Spathoglottis plicata,* and *Calanthe* sp. become photosynthetic soon after planting and early growth is rapid, no dormancy mechanism being apparent. Others, such as species of *Habenaria, Paphiopedilum* (Knudson, 1950), and *Cymbidium* usually develop at a much slower rate, sometimes after prolonged seed dormancy.

In the laboratory, fresh seeds of some species of temperate regions such as *Calopogon tuberosus* (Liddell, 1952), *Spiranthes cernua* (Curtis, 1936), *S. sinensis, Bletilla hyacinthina, Disa uniflora, Microtis unifolia,* and *Habenaria radiata* are capable of rapid germination and growth immediately upon release from the capsule, and in these species no dormancy mechanism seems operative. Seeds become photosynthetic in sterile distilled water in two to three weeks and the embryos soon die if not transferred to the greenhouse or to nutrient media for further growth. In such species the entire seed crop either germinates and grows during the next season or does not survive.

Seeds of other terrestrials behave differently. Some germinate irregularly, with young plants in all stages of development at the end of a year in culture. *Habenaria* and *Platanthera* species are particularly irregular in this regard, with some subtropical and tropical species giving good germination within two to three months of sowing (*H. radiata, H. quinquiseta, H. miersiana,* and *H. tridactylites*), whereas temperate-growing species may not reach full germination until remaining a year on media (*P. ciliaris, P. blephariglottis, P. sparsiflora*). The latter examples do not appear to be simple cases of after-ripening, since seeds begin the process of germination throughout the period. McIntyre et al. (1972) found irregular germination in Australian species. Cultures of *Dactylorhiza purpurella* and *D. majalis* have started development after a year on distilled water, and Vermeulen (1947) also found some dactylorchids viable after remaining in water for extended periods. Curtis (1936) discussed delayed germination in a number of North American species. Immediate or delayed germination under laboratory conditions does not necessarily reflect the timing under natural conditions since we usually do not know what dormancy-breaking factors may be operative in the field. Simple inhibition mechanisms involving diffusible organic compounds within the seeds do not appear to be significant in regulating germination in the species cultivated here, since even prolonged soaking in distilled water does not hasten the process when seeds are subsequently transferred to agar media. Seeds of *Epipactis latifolia, Epipactis palustris, Spiranthes or-chioides, Corallorhiza maculata,* and *Gymnadenia alba* swell but do not germinate with methods used here. When seeds of these species are mixed with growing protocorms of *Calopogon tuberosus, Spiranthes sinensis,* and *Habenaria radiata* there is no inhibition of the latter group. This suggests lack of germination may be due to inviable seeds rather than germination-controlling compounds, or, if such compounds are present, they are specific and do not affect the growth of other orchids. The first alternative is probably the most significant since it can be demonstrated that seed viability decreases rapidly after release from the capsule in some terrestrial species. *Disa uniflora* seeds are largely invi-able one month after release (Lindquist, 1965) and *Spiranthes cernua* seeds also are short-lived in this laboratory.

A dormancy-controlling device found in many plants but not convinc-ingly demonstrated in orchids is seed coat impermeability. The very deli-cate nature of most orchid seed coats suggests that they probably are not effective in controlling germination of the embryo, and this is supported by observations of many species which quickly absorb water when placed on media but are chary of further development. The embryo swells but true growth may not follow for many months. The seed coat is not the limiting factor in these cases, nor is it the effective device in species of *Cypripedium* and other terrestrials whose seeds resist wetting. Grinding of *Cypripodium* seeds with sterile sand abrades the seed coats but in this laboratory it does not accelerate germination. In South Australia, John Devlin (personal communication) has germinated seeds of *Cypripedium acaule* and *C. reginae* after grinding them in sand, but factors other than seed coat removal are likely to be the reason for his success. His cultures developed roots several centimeters long but were later apparently killed by high temperatures (30°C). In this laboratory seedlings of these spe-cies have been grown from intact seeds but germination is best at temperatures of 10–20°C in darkness.

Knudson's studies of seed viability in epiphytes (1933, 1940, 1953) indicate that many species lose viability rapidly. A few species remained viable for surprising lengths of time, although laboratory storage ad-mittedly is not equivalent to natural conditions. Knudson reported that one sample of *Cattleya* seeds dessicated at 45°F for 10 years retained 75% viability. Among the terrestrials, *Calopogon tuberosus* stored under similar conditions for five years also remained partially viable (Stout-amire, 1964b), but seeds of this species germinate quickly under appro-priate conditions and it is not a likely candidate for delayed germination in nature. Curtis (1936) reported that *Spiranthes* seeds retain their viability for three years when refrigerated and *Cypripedium* seeds could be stored four years without affecting germination. It is possible that species of genera other than *Habenaria* and *Vanilla* are capable of prolonged dormancy and begin germination in natural habitats two or more years after being shed but direct observation of this is lacking. The ability of seeds to persist throughout one or more growing seasons in the dormant condition could have survival value, allowing the seeds to survive environmental extremes that might otherwise eliminate an entire crop of seedlings. Delayed germination as a functional mechanism is well known in many nonorchidaceous plants, but its distribution among the orchids has not been explored.

Summerhayes (1951) commented on the reappearance of terrestrials in areas several years after the original populations had disappeared because of habitat disturbance. He suggested that mature plants of some species can return to a subterranean mycorrhizal existence during periods of adverse conditions, later to reappear when conditions improve. Case (1964) was skeptical of this, at least in regard to North American species. The later reappearance of a species could result from the development of seeds persisting after the original population has been destroyed or from the reseeding of an area from outside sources. Houzean de Lehaie (1910) described Belgian orchids colonizing a disturbed quarry site, some of the species probably arising from colonies 10 miles away, and Richardson (1957) documented the colonization of British claypits by *Dactylorhiza* species and hybrids. The rapid spread of *Epipactis helleborine* and *Zeuxine strateumatica* in North America (Correll, 1950) and the appearance of mature plants of *Arundina, Phaius,* and *Spathoglottis* species on Krakatao shortly after the eruption (Ridley, 1930) are further examples of efficient seed dispersal in terrestrial orchids, at least over short distances. There is ample evidence for rapid seed dispersal in nature but little support for delayed germination, even though it has been observed in the laboratory. Determining the fate of orchid seeds in their native environments should be high on the priority list of research problems in orchid ecology.

The only orchid species producing sclerotic seed coats belong to the essentially terrestrial genera *Selenipedium, Vanilla, Galeola,* and *Epistephium* (illustrated in Garay, 1960; Beer, 1863). Of these, species of *Vanilla* (Knudson, 1950; Lugo, 1955; Withner, 1955) and *Galeola* (Nakamura, 1962, 1964) have been experimentally germinated and the seed coats have not been strong barriers to germination. Nakamura found that *Galeola septentrionalis* seeds float for prolonged periods on water but quickly sink in KCl solutions. The KCl-treated seeds also germinate more rapidly, suggesting that the treatment may increase the rate of water absorption by affecting seed coat permeability. The fruits of *Selenipedium* and *Vanilla* are fleshy, aromatic, and tardily dehiscent; fruits of some *Neuwiedia* are baccate (Vermeulen, 1966; Ridley, 1896; Holttum, 1964); and those of *Galeola septentrionalis* are baccate and orange to red in color (Ohwi, 1965), characters often associated with animal dispersal. Since the seeds of these genera appear to be adapted for such dispersal, the indurate and sometimes winged seeds are probably adaptive in this context. Garay (1960) suggested that indurate seed coats may represent a primitive condition within the family.

Secondary adaptation for animal dispersal is to be expected in a family as large as the ORCHIDACEAE and the dispersal mechanism may have arisen more than once. Seeds of *Neuwiedia veratrifolia* (APOSTASIOIDEAE) resemble seeds of the majority of orchid species in being small and non-sclerotic, although this and other members of the subfamily are often regarded as being primitive orchids. We do not have enough information as to what the ancestral orchids were like to state what the original mechanisms of dispersal were, and sclerotic seeds alone are not convincing evidence of relationship (Dressler and Dodson, 1960).

The early stages in seedling development have been documented in terrestrials by Bernard (1899, 1904, 1909), Burgeff (1909, 1911, 1932, 1936, 1954), Downie (1940, 1941, 1943a, 1943b, 1949a, 1949b, 1959), Eiberg (1969), Fabre (1853), Fuchs and Ziegenspeck (1922, 1926), Montefort and Kusters (1941), Prillieux and Riviere (1856), Salisbury (1804), Stoutamire (1963, 1964a, 1964b), Vermeulen (1947), and Veyret (1965). Most of the early reports of seedling development were concerned with plantlets discovered in natural habitats, usually in association with mature individuals of the species. The classic works of Bernard and Burgeff not only deal with seedlings in such habitats, but these researchers were also the first to experimentally grow orchid seedlings asymbiotically.

The seeds of some terrestrial species will begin growth in distilled water (Downie, 1940, 1941, 1943a; Vermeulen, 1947). In the course of water germination, embryos of species in the genera *Dactylorhiza*, *Habenaria*, and *Ophrys* swell to several times their original volume, protrude from the split testa, and produce long protocorm hairs. Seedlings remain in this state until transferred to a carbohydrate source on which further development occurs. Illuminated protocorms of these genera usually do not produce chlorophyll during this first stage of development. Other terrestrial species undergo an initial swelling accompanied by chlorophyll production if exposed to light and become bright green in distilled water (*Spiranthes cernua, S. sinensis, S. romanzoffiana, Pogonia ophioglossoides, Bletia purpurea, Bletilla hyacinthina, Disa uniflora,* and *Calopogon tuberosus*). These could presumably develop autotrophically in nature and would not be dependent on the immediate establishment of mycorrhizae, although this has not been experimentally demonstrated. These immediately photosynthetic species generally have less differentiation of protocorm hairs during initial stages of growth than do their non-photosynthetic brethren. The extensive development of hairs in non-photosynthetic protocorms suggests that one of their functions may be mycorrhizal establishment. Bernard (1902) described the penetration of fungal hyphae through such hairs and Burgeff (1909, 1911, 1936) and Harvais and Hadley (1967b) documented the same process.

A factor probably responsible for the slow seedling development of many orchids is the necessity for establishment of mycorrhizal associations in those species incapable of photosynthetic activity at the beginning of development. The necessity for the activation of enzyme systems or the supplying of growth substances by fungal associates may be responsible for the prolonged dormancy or the slow development in some terrestrial species in asymbiotic culture. The latter possibility is supported by Hadley and Harvais' (1968) finding that IAA and kinetin when used together are synergistic and stimulate growth of *Dactylorhiza purpurella* protocorms in asymbiotic cultures. The difficulties of establishing stable symbiotic relationships in experimental cultures are discussed in the papers of Harvais and Hadley (1967a, 1967b). Numerous papers report attempts to isolate endophytes and reestablish them with orchid seed cultures. Some Australian neottioid orchids appear to have specific fungal associates, while the association is less specific for other species (Warcup, 1971, 1973; Warcup and Talbot, 1967, 1971). The range from highly specific to nonspecific associations probably occurs in other geographic and taxonomic orchid groups as well. Hijner and Arditti (1973) have suggested that orchids and their fungi may have coevolved with respect to vitamin requirements, and that exchanges of vitamins or their components are important aspects of the symbiotic relationship.

Symbiotic cultures of seedlings are difficult to control and procedures are much more complicated than asymbiotic culture methods because of the introduction of another organism which is itself temporally unstable. Growth studies using asymbiotic methods have the disadvantage of creating a very artificial environment but the advantage of providing a controllable and reproducible milieu for young plants. Asymbiotic methods may affect seedling morphology tó some extent, but the influences are minor. When field-collected seedlings are compared with those artificially grown, there are no significant morphological differences other than those

Fig. 2-1. *Cypripedium* seedlings (subfamily CYPRIPEDIOIDEAE) photographed in culture bottles or freshly removed from bottles. All grown in darkness. White bar = 3 mm. (A) *Cypripedium reginae*, 8 months. (B) *Cypripedium reginae*, 17 months. (C) *Cypripedium acaule*, 9 months. (D) *Cypripedium acaule*, 19 months. (E) *Cypripedium irapeanum*, 9 months. (F) *Cypripedium calceolus* var. *pubescens*, 6 months.

due to differences in nutrition and growth rate, the variation being quantitative. Arditti (1967) has reviewed the extensive literature on this complex subject.

Given sufficient water to begin growth, seeds may develop either at the soil surface or in deeper layers. Photosynthesis would be possible only in those seedlings exposed at the soil surface, and these must be protected from drying during the several months required for the first season of growth. Most species with photosynthetic protocorms are plants of marshes, bogs, or soils which are usually well watered and brightly lighted during the growing season. On the other hand, nonphotosynthetic protocorms would have no advantage in surface development but could presumably develop in any of the deeper layers of soil providing the necessary fungi are present.

There appears to be a close correlation of this type of protocorm and the habitat of the species. Most of the terrestrials producing nonphotosynthetic protocorms which have been grown here occur in well-drained forest soils, in open grasslands, or in seasonally dry soils. The unpredictable water supply at the surface in such habitats would make it very unlikely that surface-developing protocorms could survive the first growth cycle. Some European species of *Dactylorhiza* (Fig. 2-2C), *Gymnadenia* (Fig. 2-2E, F), and *Ophrys* as well as American species of *Cypripedium* (Fig. 2-1A to F), and *Goodyera* (Fig. 2-3B) are plants of soils which are usually well drained during part of the growing season, at least over part of their ranges, and subterranean development of protocorms would seem to be advantageous. Protocorms of these species lack, or are slow to develop, chlorophyll in experimental cultures, and they would be obligately mycorrhizal in natural conditions.

The seedling development of several terrestrial species grown in this laboratory is unique in that the first growth is positively geotropic with the developing stems consistently growing downward into the medium rather than upward. *Cymbidium virescens* (Fig. 2-4A) develops branched, photosynthetic stem systems bearing scale leaves growing downward to a depth of several centimeters before normal erect growth in the apical meristem is initiated. *Calochilus robertsonii* (Fig. 2-3H), *Prasophyllum australe* (Fig. 2-3C), *Pterostylis rufa* (Fig. 2-3E), and *P. falcata* (Fig. 2-3F) also develop branched or unbranched positively geotropic growth for periods of several months before negatively geotropic leafy stems are formed from apical meristems.

A modification of this occurs in *Microtis media* (Fig. 2-3G) and *M. unifolia* in which the apical meristem is carried into the downward-growing leaf sheath by elongation of part of the protocorm and the leaf base

Fig. 2-2. Seedlings of subfamily ORCHIDOIDEAE in culture. All grown in darkness to this stage. White bar = 5 mm. (A) *Platanthera leucophaea*, 9 months. (B) *Platanthera ciliaris*, 23 months. (C) *Dactylorhiza maculata*, 5 months. (D) *Coeloglossum viride*, 9 months. (E) *Gymnadenia conopsea*, 6 months. (F) *Gymnadenia conopsea*, 14 months. (G) *Platanthera bifolia*, 8 months. (H) *Barlia longibracteata*, 8 months.

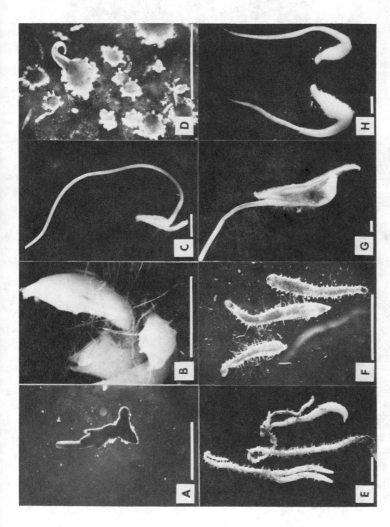

Fig. 2-3. Seedlings of subfamilies Epidendroideae (A) and Neottioideae in culture bottles or freshly removed. White bar = 3 mm. (A) *Pogonia ophioglossoides*, 14 months. Protocorm is at lower left, first root is at lower right, and shoot is above. (B) *Goodyera pubescens*, 18 months. (C) *Prasophyllum australe*, 7 months. First root and leaf emerging from base of protocorm. (D) *Pterostylis curta*, 4 months. First leaf appearing. (E) *Pterostylis rufa*, 24 months. Plants are growing as branching, positively geotropic stem systems at this stage. (F) *Pterostylis falcata*, 13 months. Plants growing as positively geotropic stems. Right-hand seedling is inverted, with protocorm at base. (G) *Microtis unifolia*, 20 months. The seedlings of this species grew downward for 1–2 cm, producing the tubular leaf from the lower level. Roots emerge from the area near the protocorm. (H) *Calochilus robertsonii*, 11 months. Protocorms grow downward into medium, producing the first leaf at a lower level.

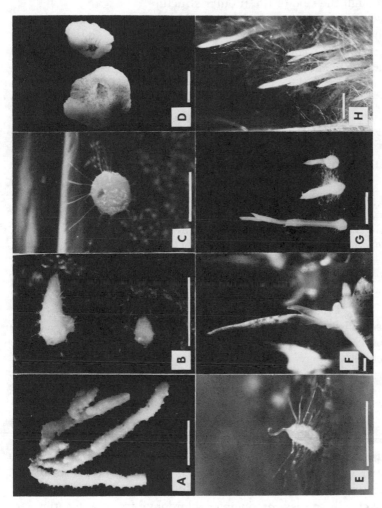

Fig. 2-4. Seedlings of EPIDENDROIDEAE (A-D) and ORCHIDOIDEAE (E-H) in culture or freshly removed from culture. All seedlings except A, F, G are grown in darkness. White bar = 3 mm. (A) *Cymbidium virescens*, 14 months. Positively geotropic, branching stem systems bearing scale leaves. (B) *Aplectrum hyemale*, 6 months. (C) *Tipularia discolor*, 9 months. (D) *Geodorum pictum*, 14 months. (E) *Corycium carnosum*, 8 months. (F) *Corycium orobanchioides*, 20 months. The protocorm enlarges to produce a large, tuberlike structure before shoot development. (G) *Satyrium carneum*, 13 months. (H) *Holothrix villosa*, 8 months.

with the result that the meristem is 1-2 cm lower than the point at which germination began. The first roots arise at the level at which stem differentiation began, producing a strangely inverted seedling. These species are also unique in producing a single tubular leaf which persists from seed germination through flowering in pot culture. The inflorescence emerges through the side of the tubular structure.

The positively geotropic growth exhibited by these Australian species will effectively bury the apical meristem deeper in the soil during the first cycle of growth. These *Microtis* species are native to areas where bush fires are frequent and such buried meristems may be an adaptation for survival in such habitats. Another mechanism that also lowers the apical meristem involves contractile roots, and this is well known in other plant families. Beer (1863) reported them in *Orchis tridentata* and Fuchs and Ziegenspeck (1922) reported them in *Dactylorhiza traunsteineri*.

Most of the NEOTTIOIDEAE germinated in this laboratory have colorless protocorms, as Vermeulen (1966) pointed out, but *Spiranthes cernua* and *S. sinensis* are exceptions with green, rapidly growing protocorms. Terrestrial species reported to grow as chlorophyll-free protocorms for prolonged periods or until leaves are initiated are *Goodyera repens* (Downie, 1940, 1949a; Montefort and Kusters, 1941), *G. pubescens* (Ames, 1922a), *Cypripedium* species (Curtis, 1943), species of *Caladenia, Calochilus, Microtis, Pterostylis, Thelymitra,* and *Diuris* (Stoutamire, 1963), *Goodyera, Listera,* and *Habenaria* (*Platanthera*) (Stoutamire, 1964). In the EPIDENDROIDEAE *Liparis loesellii* consistently produces chlorophyll-free protocorms while those of the tropical *Liparis nervosa* are green. The presence of chlorophyll is less indicative of phylogenetic relationship than of seedling ecology.

Seed germination in many plants is affected by light, being either stimulated or inhibited depending on the wave length and the plant used. Light is apparently necessary for the normal development of the green-protocorm terrestrials mentioned previously, but it inhibits germination and protocorm growth in *Gymnadenia conopsea* (Fig. 2-2E,F), *Cypripedium irapeanum* (Fig. 2-1E) and other species (Table 1). Most of the species listed germinate in dark-maintained cultures but poorly or not at all under lights. This sensitivity to light decreases when leaves are initiated. Not all species producing chlorophyll-free protocorms are equally affected by light, some developing on agar under fluorescent lights as rapidly as dark-maintained controls. More than one mechanism is involved, and the production of colorless protocorms does not automatically mean that the species requires darkness for development at this stage. We have no information as to what portions of the spectrum are involved in this inhibition.

TABLE 2-1
Comparison of Seed Germination in Light (12 hr Light, 12 hr Dark) and in Dark-Maintained Cultures of Selected Species

Species, Culture Number, Seed Source, and Media	Age of Seedlings (months)	Growth in 12 hr Light, 12 hr Dark	Growth in Continuous Dark
CYPRIPEDIOIDEAE			
Cypripedium reginae 6391, Switzerland, KC, N₃f	12	None	1–4 mm
Cypripedium irapeanum 6660, Oaxaca, Mex., KC, N₃f	4	None	1–5 mm
Cypripedium californicum 6071, California, KC, N₃f	14	None	1–3 mm
Cypripedium acaule 6059, Michigan, KC	14	None	3 mm
Cypripedium acaule 6134m Michigan, KC, N₃f	14	None	1–10 mm
Cypripedium candidum 6057, Michigan, KC	14	None	10 mm
ORCHIDOIDEAE			
Coeloglossum viride 6623, Scotland, KC, N₃f	4	None	2–3 mm
Gymnadenia conopsea 5645, Romania, KC	24	None	3 mm
Gymnadenia conopsea 6253, E. Germany, KC, N₃f	12	None	3–10 mm
Gymnadenia conopsea 6460, Sweden, KC, N₃f	9	None	1–3 mm
Gymnadenia conopsea 5754, France, KC, N₃f	9	None	5–10 mm
Gymnadenia conopsea 5788, Netherlands, KC	6	None	8 mm
Orchis militaris 6119, Switzerland, KC, N₃f	14	None	3 cm leaves
Dactylorhiza maculata 6120, Switzerland, KC	10	2 mm	1 cm leaves
Dactylorhiza papilionacea 6158, Italy, KC	13	None	2–3 cm leaves
Barlia longibracteata 6164, France, N₃f	13	None	2–5 cm leaves
Platanthera hyperborea 6214, Michigan, KC	10	None	4 cm leaves
Corycium orobanchioides 6116, S. Africa, KC, N₃f	13	None	2 cm leaves

Withner was successful in growing seedlings of *Cypripedium acaule* (1953), his cultures being kept in the dark during the first eight months of development. *Cypripedium acaule* and *Cypripedium reginae* seedlings have been dark-grown in this laboratory for 24 months with shoot development beginning only after this period. Others who have maintained cultures in the dark were Knudson (1950), working with *Vanilla;* Burgeff (1936), working with *Cymbidium, Nervilia,* and *Platanthera;* Downie (1959a), who germinated species of *Dactylorhiza, Coeloglossum, Goodyera,* and *Listera;* and Harvais and Hadley (1967b), working with *Gymnadenia, Coeloglossum* (Fig. 2-2D), and *Dactylorhiza.* Species of these genera are characteristically plants of well-drained woodland or open grassy areas, although individual species vary in their edaphic preferences.

Light-inhibited germination probably is part of the protective mechanism discussed previously, making it impossible for seedlings to develop at the soil surface where they would be subjected to drying during the growing period. With light control operative, only buried seeds could begin growth and these would at the same time be protected from water loss and would be obligately mycorrhizal. Those species grown here which seem to be light inhibited, will, when transferred to light, continue to grow if leaf initiation has started. Harvais and Hadley (1967b) believe that light inhibits the initial germination stages, but this mechanism is not operative once the protocorm is developing. Detailed studies of the period of inhibition in relation to morphological development are necessary before many species of terrestrial orchids can be consistently germinated and grown.

Terrestrial seedlings in agar cultures have high mortality rates at two stages of development, dying either shortly after the protocorm emerges from the seed coat and reaches 1–2 mm in length or shortly after the first root appears from the developing shoot. Early protocorm death occurs in members of all subfamilies of the ORCHIDACEAE and may be due to inappropriate cultural conditions such as illumination of light-sensitive seeds, improperly balanced nutrients, lack of proper growth-stimulating substances, temperature extremes, or improper gas mixtures (we know nothing of the CO_2-O_2-N_2 supply to subterranean orchid protocorms, and these gases may be present in very different proportions above and below the surface of highly organic soils). The second period of high mortality, associated with rapid shoot elongation and the development of the first root, is most characteristic of the subfamily ORCHIDOIDEAE, species of *Dactylorhiza* (Fig. 2-2C), *Platanthera* (Fig. 2-2A, B), *Ophrys, Gymnadenia* (Fig. 2-2F), and *Satyrium* (Fig. 2-4G) being especially difficult to grow beyond this stage. Necrosis is indicated by darkening of root and shoot tissue, diffusion of pinkish-brown substances into the agar, and the later appearance of necrotic tissues throughout the plantlet. There is some indication that this is caused by improper nutrient conditions. Death often occurs soon after the first root penetrates the agar of standard media.

In nature this developmental stage involves a changeover from almost completely mycorrhizal nutrition to the partially autotrophic stages. The nutrient requirements may change rapidly during this phase of growth and the relatively high sugar and mineral content of standard media may be toxic to such seedlings. Developing seedlings transferred to Knudson C containing only 10–25% of the normal mineral and carbohydrate supply survive longer and death is delayed. Curtis (1946a) found that the nutrients flowing down the bark of Haitian trees were present in extremely low quantities, and he ascribed the luxuriant growth of epiphytes there to the continuous irrigation of root systems by very dilute solutions. Something similar to this may occur in terrestrial seedlings. Many North American species are restricted to soils very low in macronutrients, and it should not be surprising that they do not survive in surroundings much richer in these compounds, both inside and outside the culture bottle.

The soil temperature parameters for the majority of terrestrial species are unknown. Seeds of some species begin development if the cultures are refrigerated at 5–10°C for several months before being kept at higher temperatures, while little or no development occurs in controls kept at 20°C for the same period. This response occurs in several temperate-boreal North American species including *Cypripedium reginae* and *Goodyera pubescens,* whose seeds would be subjected to prolonged near-freezing temperatures in their native habitats. Soil temperatures in orchid habitats of southern Michigan have been recorded (Fig. 2-5), and seeds and seedlings of the local species will be subjected to these temperatures at the root level under field conditions. Many factors can influence surface temperature, but the subsurface temperatures of contrasting habitats in a restricted climatic region are less variable.

In temperate regions with pronounced annual cycles, soil temperatures usually lag behind air temperatures (Heinselman, 1963), as Fig. 2-5 suggests. At this latitude the 6-in. soil temperatures are at or above freezing from mid-March to December while the temperature during winter may be slightly below freezing. The highest summer temperature at root level in southern Michigan woodland and bog habitats is approximately 20°C. Contrasting with this, the soil temperatures in south Florida in December are this high. Optimal growth of orchid species adapted to these and to other temperature regimes can occur only when the temperature of native habitats is approximated.

Fire is an environmentally imposed factor which stimulates flowering

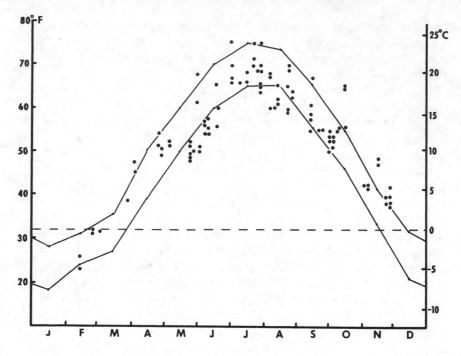

Fig. 2-5.   Dots are soil temperatures at the 6 in. depth, taken from orchid habitats in southern Michigan, chiefly in Livingston and Oakland counties, 1962-1967. Lower solid line represents the monthly mean temperature for Chatham, Michigan, and the upper line gives monthly mean air temperature for Sandusky, Ohio, north and south of the area sampled. (Data from *World Weather Records 1951-1960*, Vol. 1, *North America*. U. S. Department of Commerce.)

of terrestrial species in certain areas. Some orchid species of bush country in South Africa and Australia flower profusely after fires, but more rarely in its absence. Australian orchids stimulated by fire are *Leptoceras* and *Lyperanthus* species (Erickson, 1951), some *Caladenia* and *Elythranthera* species (K. Newbey, personal communication), *Microtis unifolia* and *Caladenia latifolia* (R. Nash, personal communication), as well as *Calochilus saprophyticus* (Jones, 1968). Fire is not necessary for *Microtis* flowering, since it blooms readily in the greenhouse, but fire apparently increases flowering in its native habitat. South African species growing in similarly summer-dry climates also are stimulated to flower following bush fires. *Disa obtusa, D. racemosa, Monadenia ophrydea, Bartholina burmanniana,* and *Penthea patens* behave in this manner (H.

Hofmann, M. Lawder, personal communication), as do species of *Satyridium* and *Forficaria* (Schelpe, 1966).

Fire could have a relatively direct effect in making available an increased supply of minerals that are otherwise in limited supply. This mechanism is operative when wet savannahs are burned in the southeastern United States, stimulating *Sarracenia* growth and flowering (Plummer, 1963), and heavy flowering of orchids follows bog fires in southern Mississippi (S. B. Jones, Jr., personal communication). Vegetative growth of *Cypripedium acaule* is stimulated by brush fires in Rhode Island (Stuckey, 1967). *Isotria verticillata* may respond in the same manner (Baldwin and Wieboldt, 1968). Increased light and decreased competition from other plants may be equally significant in such responses, but the subject has remained largely unexplored. Colonies of *Platanthera blephariglottis*, when sprayed with the weedkiller 2–4–D, increased in vigor (Stuckey, 1967) as a result of either direct stimulation by the compound or the reduction of competition from other plants. The experimental use of herbicides and fire on orchid colonies, although startling, could provide us with valuable ecological information if limited to abundant species.

Disturbances to the habitat other than by fire also result in increased vegetative growth. Case (1964), Correll (1950), Curtis (1946b), Curtis and Greene (1953), Houzean de Lehaie (1910), Richardson (1957), and Stuckey (1967) have discussed the appearance of terrestrial orchids in areas within a few years of disturbance. Some orchids are capable of colonizing such sites as clay and sand pits, road embankments and abandoned roadbeds, old fields, city dumps, and lawns. Orchid species noted for this colonizing ability are capable of relatively fast development in such open habitats, and they flourish until increased competition from other plants reduces population size. The colonizing species maintain their vigor only as long as the competition for light, growing space, and nutrients remains low.

Seedling mortality in stable communities is high and only mature plants survive (Tamm, 1948). Salisbury (1942) stated that the very small and nutritionally poorly equipped seeds of orchids avoid competition by wide dispersal while the larger seeds of other plants, better supplied with growth substances and perhaps not as well equipped for long-distance dispersal, meet the competition. Subsurface mycorrhizal development in many species is another step in this direction. Such development frees the plant from normal light requirements and direct competition with other vegetation. Species of *Platanthera, Orchis, Epipactis, Ophrys,* and

Neottia discussed by the foregoing authors are examples of rapid colonizers having subterranean protocorms.

The length of time terrestrial orchid species require to develop from protocorm to the reproductive stage varies widely. Curtis (1943) observed native populations of *Cypripedium* species and reported that plants reached flowering size eight or more years after germination. Such prolonged vegetative growth may be due to inherent patterns of development or in part to unfavorable environmental conditions. Some terrestrial species, contrasting with *Cypripedium*, have some of the shortest maturation periods reported for orchids. Becker (1964) reported that *Zeuxine strateumatica* and *Habenaria repens* are fast growing and that seedling development in *Eulophidium maculatum* is rapid. Plants of the latter reached the size of the seed parent 6 months after germination. *Zeuxine strateumatica* is reported to flower during the first year of growth (Vickers, 1968). Harbeck (1964) flowered *Dactylorhiza maculata* 3.5 years after germination and he has raised *Disa uncinata, Disa uniflora, Monadenia micrantha,* and *Stenoglottis fimbriata* to flowering in 20–26 months (personal communication). *Pterostylis curta* flowers in 2 years (R. Nash, personal communication), as does *Bletia purpurea* (J. Devlin, personal communication). Wisniewski (1965) reported that *Epipactis palustris* flowers in nature in 4 years, and Vermeulen (1947) flowered *Dactylorhiza* species in 3–4 years. Burgeff did the same in 5 years, and the classic record is that for *Disa* hybrids which were grown to flowering in 18–21 months (Anon., 1893). *Disa uniflora* is flowered at Kirstenbosch in as short a time as 20 months (Hofmann, personal communication), at Gothenburg in 33 months (Lindquist, 1958), and it has flowered here in 3.5 years. Volunteer seedlings of *Erythrodes querceticola* develop to maturity and flower within 12 months in the Akron University greenhouses.

Ames (1921) indicated that *Spiranthes cernua* flowers in its second year of growth, although it has required 3 years in this laboratory (Table 2). *Microtis unifolia* probably blooms in its native Australian range in 2 years given proper growing conditions. Raymond Nash has found the species flowering in the bottom of ditches the second year after excavation. He further states that this species flowers best after bush fires, suggesting that it is a species of quick growth. The development of these and other short-cycle species would be favored in those areas in which plant succession is accelerated by disturbance. Long cycle species would be less likely to survive to the reproductive stage under rapidly changing conditions.

Terrestrials adapted to climates with strongly contrasting seasons are for the most part provided with special resting organs which serve to

TABLE 2-2

Growth Period (Seed Germination to Anthesis) Observed in
Greenhouse Culture of Terrestrial Orchids, University of Akron

Species and Culture Number	Months to Flowering
Calopogon tuberosus, 4123	41
Corycium orobanchioides 6116	26
Dactylorhiza incarnata 5150	40
Dactylorhiza sesquipedalis 3849	36
Disa uncinata 6750	40
Disa uniflora 4132	41
Platanthera blephariglottis × P. ciliaris 6086 F$_1$	40
Microtis unifolia 4167	23
Spiranthes cernua 5142	35
Spiranthes sinensis 4791	29
Thelymitra pauciflora 4161	39
Thelymitra rubra 4199	23

carry the plant through the dormant period. Temperate exceptions to this are the species of *Goodyera* whose leaf rosettes, borne on thickened rhizomes, persist through all seasons without either the production of morphologically distinctive resting bodies or the loss of foliage. Most terrestrial species, however, produce during their first year of growth either a short, thickened axis (corm, pseudobulb) or highly modified lateral buds which have been called droppers or sinkers. Propagative structures can be formed either from the stem or root systems, from a combination of the two, or, in the case of *Malaxis paludosa*, from foliar embryos (Taylor, 1967).

Primary roots are not produced by members of the ORCHIDACEAE and root propagation always involves the secondary system. Adventitious buds develop from the roots of species of *Pogonia* (Carlson, 1938), *Cleistes* and *Isotria* (Ames, 1922b; Burgeff, 1932; Holm, 1900), *Neottia* (Prillieux and Riviere, 1856), *Listera, Cephalanthera,* and *Epipactis* (Summerhayes, 1951), and *Spiranthes* (Correll, 1950). Root buds also occur in the Australian *Chiloglottis reflexa, Lyperanthus suaveolens,* and *Corybas* species (Nash, 1968) and in some Mexican species of *Cranichis* and *Ponthieva* (Pollard, personal communication). The formation of root buds is an efficient method of vegetative propagation in *Pogonia ophioglossoides* (Fig. 2-3A), *Listera convallarioides* and *L. cordata* in eastern North America, where the wide-ranging root systems often develop numerous shoots, each in turn developing its own root system. Colonies consisting of hundreds of plants can arise this way. Seedlings of *Pogonia* in culture develop one or more root buds in the first year of growth. The genera producing adventitious root buds belong for the most part to the subfamilies NEOTTIOIDEAE and EPIDENDROIDEAE. These structures have not been described in the subfamilies APOSTASIOIDEAE and CYPRIPEDIOIDEAE. *Habenaria repens* (ORCHIDOIDEAE) cultivated here develops adventitious shoots from the root system, but this is the only reported member of the subfamily propagating in this manner.

Direct modification of the stem is a more common device in developing seedlings. *Calopogon, Liparis, Cremastra, Cymbidium, Eulophia, Eulophidium,* and *Geodorum* seedlings develop enlarged nodes and internodes in conjunction with, or above, the protocorm, and these corms survive the dormant period. The next growth develops from lateral buds in the axils of the previous year's foliage or from the lateral and terminal meristems if leaves were not formed. Some species grow indefinitely as branching stem systems, developing roots and leafy shoots tardily, if at all. This is carried to the extreme in *Corallorhiza* and *Hexalectris* species in which roots apparently never develop. Enlarged protocorms and branching leafless rhizomes are produced by *Cymbidium virescens* (Morel and Champagnat, 1966), *Cremastra appendiculata, Eulophidium maculatum,* and *Geodorum pictum* (Fig. 2-4D) in asymbiotic culture and these are apparently able to grow slowly in this juvenile form for indefinite periods. Such indeterminate juvenile growth also occurs in asymbiotic cultures of the epiphyte *Polyrrhiza lindenii* in which the protocorm itself has been grown as a branching, thalloid structure for 10 years. The protocorm is devoid of the normal node-internode differentiation characteristic of stem tissues and is not morphologically equivalent to surface or subterranean stems of terrestrials, but it exhibits a similarly protracted juvenile growth in culture.

Structurally complex sinkers or droppers are formed in seedling and later stages of the ORCHIDOIDEAE. These are formed by the transport of an apical or lateral meristem away from the plant axis by lateral growth of tissues derived from both the leaf base and the stem, and the meristem involved may be carried anywhere from a few millimeters to 10 cm away from the plant axis. The new organ which is formed consists of the evaginated leaf base and associated stem tissue, the transported bud, and usually a tuberous structure distal to the bud consisting of adventitious root tissue. Absorbing hairs are produced by portions of this new structure both distal and proximal to the transported meristem. Fabre (1853) illustrated the structures in *Orchis* and *Dactylorhiza,* and more recent workers (White, 1907; Arber, 1925; Sharman, 1939; Ogura, 1953; Kumazawa, 1956, 1958) have made more detailed studies. In all cases the bud which will develop the second year is preformed in the seedling stem axis, it is not a *de novo* development from root tissues as appears to be the case in the previously mentioned genera of the NEOTTIOIDEAE and EPIDENDROIDEAE. The elongation of the dropper carries the bud deeper into the soil and laterally away from the original plant axis. The thickened structures formed distally to the transported bud take several forms varying from fusiform (some platantheras) to branched fusiform (dactylorchids) to ovoid or spherical (*Ophrys, Orchis,* other *Habenaria* species). Superficially similar tuberlike structures are produced by *Pterostylis, Diuris, Caladenia,* and *Thelymitra* of the NEOTTIOIDEAE, but we do not know the ontogeny of these.

Enlarged propagative structures (tubers, corms, storage roots) are produced along with the first seedling leaves or soon thereafter in many terrestrial species. These structures are fungus-free in many cases and their maturation is followed by the disintegration of part or all of the original root system. Tubers are all that remain of most *Orchis, Dactylorhiza,* and *Ophrys* and many *Habenaria* species during the periods of dormancy. Each annual growth cycle results in a complete new root system which is reinvaded by fungi. The resting organs were first suspected by Bernard to contain or produce fungistatic compounds which prevent infection of the organ, and these defense reactions have been demonstrated by Braun (1963), Cruickshank et al. (1964), and Nüesch (1963). Such fungistatic compounds have not yet been clearly demonstrated in orchids with perennial root systems.

The resting organs are subjected to a variety of environmental extremes during the dormant period. Species of colder regions may have these structures surrounded by ice through most of the winter. On the other hand, plants of Mediterranean climates, such as *Orchis* and *Ophrys* species of southern Europe, *Satyrium, Corycium* (Fig. 2-4E, F), some *Disa* species of South Africa, and some *Thelymitra, Diuris, Monadenia,* and *Pterostylis* species of Australia are subjected to prolonged warm, dry soil conditions. The varied temperatures, the water relations, and the soil types of these climatically diverse regions are difficult to approximate in experimental studies, and the mortality of experimental plants is usually high when climatic information is not available.

The propagative structures of young terrestrial orchids are especially subject to injury through failure to provide the necessary conditions during dormancy. Plants of cold areas usually do not survive the next growth cycle unless their resting structures are refrigerated for several months, and the species of Mediterranean climates must be kept dry for similar periods. The dormant structures do not tolerate changing conditions and they are in part responsible for the restricted habitats of many species.

Bibliography

AMES, O. 1921. Notes on the New England orchids, I. *Spiranthes*. *Rhodora* **23**:73.

———. 1922a. Notes on New England orchids, II. The mycorrhiza of *Goodyera pubescens*. *Rhodora* **24**:37–46.

———. 1922b. A discussion of *Pogonia* and its allies in Northeastern United States, with reference to extra-limital genera and species. *Orchidaceae* **7**:3–38.

ANON. 1893. *Disa* × *kewensis*. *Orchid Rev.* **1**:13.

ARBER, A. 1925. *Monocotyledons, a Morphological Study*. Cambridge University Press.

ARDITTI, J. 1967. Factors affecting the germination of orchid seeds. *Bot. Rev.* **33**:1–97.

BALDWIN, J. T., and WIEBOLDT, 1968. A dense population of *Isotria verticillata* in Virginia. *Am. Orchid Soc. Bull.* **37**:988.

BECKER, J. 1964. *Eulophidium maculatum*. *Am. Orchid Soc. Bull.* **33**:1066.

BEER, J. G. 1863. *Beiträge zur Morphologie und Biologie der Familie der Orchideen.* Verlag von Carl Gerold's Sohn, Wien.

BERNARD, N. 1899. Sur la germination du *Neottia Nidus-avis*. *Compt. Rendu. Acad. Sci. France* **128**:1253–1255.

———. 1902. Etudes sur la tuberization. These Doct. Sc. Nat. Paul Dupont Ed., Paris.

———. 1904. Recherches experimentales sur les Orchidées. *Rev. gén. Bot.* **16**:405–451, 458–476.

———. 1909. L'evolution dans la symbiose. Les orchidées et leur champignons commensaux. *Ann. Sci. Nat. Bot.* **9**:1–96.

BRAUN, R. 1963. Orchinol, in *Moderne Methoden der Pflanzenanalyse*, Vol 6. Springer Verlag, Berlin, pp. 130–134.

BURGEFF, H. 1909. *Die Wurzelpilze der Orchideen, ihre Kultur und ihr Leben in der Pflanze*. Verlag Gustaf Fischer, Jena.

———. 1911. *Die Anzucht tropischer Orchideen aus Samen.* Verlag Gustav Fischer, Jena.

———. 1932. *Saprophytismus und Symbiose*. Verlag Gustav Fischer, Jena.

———. 1936. *Samenkeimung der Orchideen*. Verlag Gustav Fischer, Jena.

———. 1954. *Samenkeimung und Kultur europäischer Erdorchideen, nebst versuchen zu ihrer Verbreitung*. Verlag Gustav Fischer, Jena.

CARLSON, M. C. 1938. Origin and development of shoots from the tips of roots of *Pogonia ophioglossoides*. *Bot. Gaz.* **100**:215–225.

CASE, F. 1964. *Orchids of the Western Great Lakes Region*. Bull. 48, Cranbrook Inst. of Sci., Bloomfield Hills, Mich.

CORRELL, D. S. 1950. *Native Orchids of North America North of Mexico.* Chronica Botanica 24, Waltham, Mass.

CRUICKSHANK, I. A. M., and PERRIN, D. R. 1964. Pathological function of phenolic compounds in plants, in *Biochemistry of Phenolic Compounds.* Academic Press, New York, pp. 511–544.

CURTIS, J. T. 1936. The germination of native orchid seeds. *Am. Orchid Soc. Bull.* 5:42–47.

———. 1943. Germination and seedling development in five species of *Cypripedium* L. *Am. J. Bot.* 30:199–205.

———. 1946a. Nutrient supply of epiphytic orchids in the mountains of Haiti. *Ecology* 27:264–266.

———. 1946b. Use of mowing in management of white Lady's-slipper. *J. Wildlife Management* 10:303–308.

CURTIS, J. T., and GREENE, H. C. 1953. Population changes in some native orchids of southern Wisconsin, especially in the University of Wisconsin Arboretum. *Orchid J.* 2:152–155.

DOWNIE, D. G. 1940. On the germination and growth of *Goodyera repens.* *Trans. Bot. Soc. Edinb.* 33:36–51.

———. 1941. Notes on the germination of some British orchids. *Trans. Bot. Soc. Edinb.* 33:94–103.

———. 1943a. Source of the symbiont of *Goodyera repens.* *Trans. Bot. Soc. Edinb.* 33:383–390.

———. 1943b. Notes on the germination of *Corallorhiza innata.* *Trans. Bot. Soc. Edinb.* 33:380–382.

———. 1949a. The germination of *Goodyera repens* (L.) R. Br. in fungal extract. *Trans. Bot. Soc. Edinb.* 35:120–125.

———. 1949b. The germination of *Listera ovata* (L.). *Trans. Bot. Soc. Edinb.* 35:126–130.

———. 1959. *Rhizoctonia solani* and orchid seed. *Trans. Bot. Soc. Edinb.* 37:279–285.

DRESSLER, R. L., and DODSON, C. H. 1960. Classification and phylogeny in the Orchidaceae. *Ann. Mo. Bot. Gard.* 47:25–68.

EIBERG, H. 1969. Keimung europäischer Erdorchideen. *Die Orchidee* 5:266–270.

ERICKSON, R. 1951. *Orchids of the West.* Paterson Brokensha Pty. Perth.

FABRE, J. H. 1853. De la germination des Ophrydées. *Ann. Sci. Nat. Bot.*, ser 4. 5:163–186.

FUCHS, A., and ZIEGENSPECK, H. 1922. Aus der Monographie des *Orchis traunsteineri.* *Bot. Arch.* 2:238–248.

———. 1926. Entwicklungsgeschichte der Axen der einheimischen Orchideen und ihre Physiologie und Biologie. Teil 1. *Bot. Arch.* 14:165–260.

GARAY, L. A. 1960. On the origin of the Orchidaceae. *Bot. Mus. Leaflets, Harvard* 19:57–96.

GETZ, L. L. 1961. Temperatures in different vegetation types in southern Michigan. *Jack Pine Warbler* 39:132–147.

HADLEY, G., and HARVAIS, G. 1968. The effect of certain growth substances on

126   THE ORCHIDS: SCIENTIFIC STUDIES

asymbiotic germination and development of *Orchis purpurella*. *New Phytol.* **67:** 441–445.

HARBECK, M. 1964. Anzucht von *Orchis maculata* von Samen bis zur Blute. *Die Orchidee* **15:**57–61.

HARVAIS, G., and HADLEY, G. 1967a. The relation between host and endophyte in orchid mycorrhiza. *New Phytol.* **66:**205–215.

———. 1967b. The development of *Orchis purpurella* in asymbiotic and inoculated cultures. *New Phytol.* **66:**217–230.

HEINSELMAN, M. L. 1963. Forest sites, bog processes and peatland types in the glacial Lake Agassis region, Minnesota. *Ecol. Monogr.* **33:**327–374.

HIJNER, J. A., and ARDITTI, J. 1973. Orchid mycorrhiza: vitamin production and requirements by the symbionts. *Am. J. Bot.* **60:**829–835.

HOLM, T. 1900. *Pogonia ophioglossoides* Nutt., a morphological and anatomical study. *Am. J. Sci.* **9:**13–19.

HOLTTUM, R. E. 1964. *A Revised Flora of Malaya*, Vol. 1. *Orchids of Malaya*. Govt. Printing Office, Singapore.

HOUZEAN DE LEHAIE, M. J. 1910. Observations pour servir a l'étude de la dissémination des Orchidées indigènes en Belgique. *Bull. Soc. Roy. Bot. Belg.* **47:**45–52.

JONES, D. L. 1968. Bearded orchids. *Australian Plants* **4:**353.

KNUDSON, L. 1933. Storage and viability of orchid seed. *Am. Orchid Soc. Bull.* **2:**66.

———. 1940. Viability of orchid seed. *Am. Orchid Soc. Bull.* **9:**36–38.

———. 1950. Germination of seeds of *Vanilla*. *Am. J. Bot.* **37:**241–247.

———. 1953. Viability of orchid seed. *Am. Orchid Soc. Bull.* **22:**260–261.

KUMAZAWA, M. 1956. Morphology and development of the sinker in *Pecteilis radiata* (Orchidaceae). *Bot. Mag. Tokyo* **69:**455–461.

———. 1958. The sinker of *Platanthera* and *Perularia*—Its morphology and development. *Phytomorphology* **8:**137–145.

LIDDELL, R. W. 1952. Germinating native orchid seed. *Am. Orchid Soc. Bull.* **12:** 344–345.

LINDQUIST, B. 1958. A greenhouse culture of *Disa uniflora* Berg. in Gothenburg. *Am. Orchid Soc. Bull.* **27:**652–657.

———. 1965. The raising of *Disa uniflora* seedlings in Gothenburg. *Bull. Am. Orchid Soc.* **34:**317–318.

LUGO, H. L. 1955. The effect of nitrogen on the germination of *Vanilla planifolia*. *Am. J. Bot.* **42:**679–684.

MCINTYRE, D. K., VEITCH, G. J., and WRIGLEY, J. W. 1972. Australian terrestrial orchids from seed. *Am. Orchid Soc. Bull.* **41:**1093–1097.

MONTEFORT, G., and KÜSTERS, E. 1941. Saprophytismus und Photosynthese, I. Biochemische und physiologische Studien an Humus-orchideen. *Bot. Arch.* **40:**571.

MOREL, G., and CHAMPAGNAT, M. 1966. Wachstum und Entwicklung von *Cymbidium virescens*. *Die Orchidee* **17:**250–251.

NAKAMURA, S. I. 1962. Zur Samenkeimung einer chlorophyllfreien Erdorchidee *Galeola septentrionalis* Reichb. f. *Zetischr. Bot.* **50:**487–497.

———. 1964. Einige Experimente zur Samenkeimung einer chlorophyllfreien Erdorchidee *Galeola septentrionalis* Reichb. f. *Mem. Coll. Agr. Kyoto Univ.* Bot. ser. 4. 86:1–48.

NASH, R. C. 1968. Terrestrial orchids in a limited space. *Australian Plants* 4:309–310.

NÜESCH, J. 1963. Defense reactions in orchid bulbs, in *Symbiotic Associations*. Cambridge University Press, pp. 335–343.

OGURA, Y. 1953. Anatomy and morphology of the subterranean organs in some Orchidaceae. *J. Fac. Sci. Univ. Tokyo. Sect. 3, Bot.* 6:135–157.

OHWI, J. 1965. *Flora of Japan*. English transl. Smithsonian Inst., Washington, D.C.

PLUMMER, G. L. 1963. Soils of the pitcher plant habitats in the Georgia Coastal Plain. *Ecology* 44:727–734.

PORTER, J. N. 1942. Mycorrhiza of *Zeuxine strateumatica*. *Mycologia* 34:380–390.

PRILLIEUX, E. 1860. Observations sur la germination du *Miltonia spectabilis* et de diverses autres orchidees. *Ann. Sci. Nat. Bot.*, ser. 4. 13:288–296.

PRILLIEUX, E., and RIVIERE, A. 1856. Germination et le developement d'une orchidee. *Ann. Sci. Nat.*, ser. 4. 5:119–136.

RICHARDSON, J. A. 1957. The development of orchid populations in claypits in County Durham. *Proc. Bot. Soc. Br. Isles* 2:354–361.

RIDLEY, H. 1896. The Orchidaceae and Apostasiaceae of the Malay Peninsula. *J. Linn. Soc. London* 32:213–416.

RIDLEY, H. N. 1930. *The Dispersal of Plants Throughout the World*. L. Reeve & Co., Ashland, Kent.

SALISBURY, E. J. 1942. *The Reproductive Capacity of Plants*. G. Bell and Sons, London.

SALIODUNY, R. A. H. 1804. On the germination of the seeds of the Orchideae. *Trans. Linn. Soc. London* 7:29–32.

SCHELPE, E. A. C. L. E. 1966. *An Introduction to the South African Orchids*. Macdonald, London.

SHARMAN, B. C. 1939. The development of the sinker of *Orchis mascula* Linn. *J. Linn. Soc. London. Bot.* 52:145–158.

STOUTAMIRE, W. P. 1963. Terrestrial orchid seedlings. *Australian Plants* 2:119–122.

———. 1964a. Terrestrial orchid seedlings II. *Australian Plants* 2:264–266.

———. 1964b. Seeds and seedlings of native orchids. *Mich. Bot.* 3:107–119.

STUCKEY, I. H. 1967. Environmental factors and the growth of native orchids. *Am. J. Bot.* 54:232–241.

SUMMERHAYES, V. S. 1951. *Wild Orchids of Britain with a Key to the Species*. Collins, London.

TAMM, C. O. 1948. Observations on reproduction and survival of some perennial herbs. *Bot. Notiser* 1948:305–321.

TAYLOR, R. L. 1967. The foliar embryos of *Malaxis paludosa*. *Can. J. Bot.* 45:1553–1556.

VERMEULEN, P. 1947. *Studies on Dactylorchids*. Drukkerij Fa. Schotanus & Jens, Utrecht.

———. 1966. The system of the Orchidales. *Acta Bot. Neerlandica* 15:224–253.

VEYRET, Y. 1965. *Embryogénie Comparée et Blastogénie chez les Orchidaceae-Monandrae.* Office de la Recherche Sci. et Tech. Outre-Mer, Paris.

VICKERS, G. T. 1968. Zeuxine strateumatica, annual or perennial. *Am. Orchid Soc. Bull.* **38**:311–312.

WARCUP, J. H. 1971. Specificity of mycorrhizal associations in some Australian terrestrial orchids. *New Phytol.* **70**:41–46.

———. Symbiotic germination of some Australian terrestrial orchids. *New Phytol.* **72**: 387–392.

WARCUP, J. H., and TALBOT, P. H. B. 1967, 1971. Perfect states of rhizoctonias associated with orchids. *New Phytol.* (1967), **66**:631–641; (1971), **70**:35–40.

WHITE, J. H. 1907. On polystely in roots of Orchidaceae. *Univ. Toronto Stud. Biol.* **6**:1–20.

WISNIEWSKI, N. 1965. Zur Entwicklungsdauer von *Epipactis palustris* (Miller) Crantz vom Samenflug bis zur Blüte. *Mitt. des Arbeitskreises jur Beobachtung und zum Schutz heimischer Orchideen* **2**:31–32.

WITHNER, C. L. 1953. Germination of "Cyps." *Orchid J.* **2**:473–477.

———. 1955. Ovule culture and growth of *Vanilla* seedlings. *Am. Orchid Soc. Bull.* **24**: 380–392.

3

Developments in Orchid Physiology

CARL L. WITHNER

Advances in the field of orchid physiology involve seed germination and seedling development, flower physiology, particularly flower pigment studies, and a better understanding of mycorrhizal nutritional relationships in nature. Along with a knowledge of carbon dioxide fixation at night—indications that orchids may have the same metabolic patterns as many succulents—have come more advanced attempts at controlling the orchid environment in cultivation. Finally, the ability to grow orchid tissues *in vitro* comes as an extension of general plant tissue culture techniques, thus permitting the commercial development of moricloning. This technique has achieved such importance and generated such interest that it is covered here in a separate chapter (see Morel).

Seed Germination, Seedling Nutrition, and Development *in vitro*

The germination of orchid seed has changed little over the years; basal media remain the same, and there are still many orchids that cannot readily be grown from seed. Ovule or green pod culture (Withner, 1943) is used increasingly as a way to save time and obtain seedlings of hard-to-grow species and hybrids. Valmayor and Sagawa (1967) and Sagawa and Valmayor (1966) have presented data on the time intervals between pollination, fertilization, and growth from ovule cultures of various species and genera so that one can plant at the earliest times. Stort (1972a) has made a detailed study of ovule development in *Eulophidium maculatum*.

Vitamins and Other Growth Factors. Arditti and his students have investigated closely the role of the B vitamins during germination and have shown that niacin, especially, is effective with *Cattleya* and *Laelio-*

cattleya hybrids (Lawrence and Arditti, 1974; Arditti, 1965; Arditti and Bils, 1965). This extends the earlier work of Schaffstein (1941) and Withner (1942) and particularly Mariat (1952). The vandophytin originally described by Schaffstein was confirmed as niacin by Burgeff (1959). The positive effects of niacin led Arditti (1966c, 1967a, 1967b) to investigate further the possible synthesis of this molecule in the orchid seedling. Since orchids can use 3-hydroxyanthranilic acid, kyneurenine, and quinolinic acid as niacin sources, it is likely that they have the same synthetic pathways for this vitamin as do other plants, particularly fungi, as well as birds and mammals. The niacin is probably formed from tryptophan degradation. Though this amino acid inhibits the development of very young seedlings, older ones with developed leaves show less effects. Since tryptophan also acts as a precursor to auxin, it is difficult to assess the latter results.

This variable effect on seedlings, depending upon their stage of development, brings out a point, however, that may be emphasized in other orchid nutritional work concerning vitamins. Although the vitamin may not be stimulating, greatly increasing growth as such, it may aid in the formative or morphogenetic processes of the seedling. For instance, Mariat early thought that B_1 (thiamin) favorably affected the differentiation of *Cattleya* protocorms and later (1952) found that B_2 (riboflavin) and B_6 (pyridoxine) were also helpful. Rao and Avadhani (1963) and Rao (1963) made a specific study of the effects of additives on histogenesis in *Vanda* seedlings and demonstrated the varied effects of plant extracts, amino acids, auxins, and compounds inducing cell division. Tomato juice in small quantities had a distinct influence causing a higher rate of germination, cell division, maximum expansion of cells, and early differentiation, forming organs three times faster than the controls. Media with auxin additives, such as pollinium extract, showed callus proliferative effects at high concentrations and were not especially useful. Inhibitory or no particular effects were noted with the gibberellins or other compounds tested. Kotomori and Murashige (1965) found coconut milk was inhibitory during germination of *Dendrobium* but stimulatory to older seedlings.

Lucke (1969, 1971) also demonstrated formative effects with vitamins. He found that biotin caused better development of *Paphiopedilum* protocorms and also increased their ability to form chlorophyll. The seeds were planted from immature pods of eight months. Niacinamide was not effective.

Ueda and Torikata in a series of experiments (1969a, 1969b, 1972) showed a unique correlation between light and vitamins in the development of *Cymbidium* seedlings. They find two patterns of seedling development based on the earlier work of Terikata, Sawa, and Shisa (1965) showing vitamins B_2 (riboflavin,), C (ascorbic acid), and biotin affected their growth. Group I, consisting of tropical cymbidiums such as *C. insigne, C. pumilum, C. tracyanum,* which are mostly epiphytic, has two reactions. In the light the protocorm stage is directly followed by shoot and root formation, resulting in whole plantlets. In the dark, the protocorm stage is followed by shoot formation only, no roots. In Group II, composed of terrestrial *C. goeringii (virescens), C. kanran, C. ensifolium,* and others, in the light or dark, protocorms are followed by rhizome formation with neither shoots nor roots produced readily. In Group I shoot formation was enhanced by vitamin C; in Group II rhizome formation was enhanced by biotin. Addition of kinetin, a growth hormone, and L-arginine supported shoot growth well and induced root formation in most cultures of Group II in the dark. Kinetin alone caused only shoot formation, the arginine being necessary for roots. Withner (1955) previously demonstrated a beneficial effect of arginine (also lysine) on *Vanilla* seedlings.

Hardley's work (1970) with north temperate orchids shows an example of another interplay between growth factors and temperatures in affecting germination and growth. Kinetin, 1 to 10 ppm with or without 1 ppm indole acetic acid (IAA), can delay germination of *Dactylorhiza (Orchis) purpurella* and other terrestrial species, but the same materials enhance the growth of protocorms when they begin to form leaves. Chlorophyll formation was independent of these factors and took place quickly with temperatures under 15°C. The cold also encouraged the subsequent development of the young greened seedling. Hardley has also shown that with symbiotic culturing of seedlings, the growth does not require the hormones, and it occurs more rapidly than on the sterile media.

This only begins to scratch the surface of the formative effects and balance among various vitamins, growth regulators, and other organic additives and how they interact in the intact plant and with the environment to control growth and development. Morel (chapter 4 of this book) discusses some of these instances in more detail as related to meristem growth *in vitro*. When orchid seed first germinates, and before all the usual metabolic patterns may be fully developed and functioning—a kind of biochemical ontogeny—the seedling growth or differentiation may indeed be limited by the lack of a particular compound. Not only may we find out by trial and error what those compounds may be; they may also give us clues as to their role in orchid physiology. Certainly in nature, as Burgeff and others including Hardley (see above) have shown previously, mycorrhizal associations promote and speed germination. In fact, that is the only way to germinate some terrestrial or chlorophylless orchids that will not ordinarily grow from seed *in vitro* (also see Chapter 2 by Stoutamire). The use of fungal extracts, the addition of banana, which has been shown to be a source of natural cytokinin, as has coconut water (endosperm), or the use of honey, pineapple juice, or similar substances, adds a mixture of vitamins and many other compounds in what often seems to be beneficial natural proportions.

Arditti studied banana (1968) and previously tomato juice (1966a), emphasizing development and differentiation rather than increase in size as such. He noted enhanced root number and better growth with *Cattleya*. Ernst (1967a) did the same for *Phalaenopsis*. Attempts to fractionate chemically the organic supplements indicate that the active materials are in the water- or alcohol-insoluble fractions but are still unidentified. Other workers have shown banana contains cytokinins, auxins, gibberellins, biotin and other vitamins, mineral nutrients, and amino acids. Banana may also act as a natural chelating agent providing iron or other necessary elements in soluble form. The alcohol-insoluble fraction recalls the early work of Downie (1949), who found a similar extract of fungus stimulated the germination of *Goodyera* seed. Karasawa (1964) developed a banana-honey medium based on various inorganic nutrient combinations that seems effective for *Cymbidium, Paphiopedilum, Dendrobium,* and other genera. The use of banana seems to result from the original research of Graeflinger (1950) in Brazil, who devised a successful medium using dried banana powder (see Withner, 1959, for recipe). Though similar to coconut in effect, the banana, being more available in temperate areas, has been more readily used and investigated so far. In spite of many other organic supplements used on an empirical basis, it remains the additive of choice. Ernst, Arditti, and Healy (1970) have also confirmed the value of 5% pineapple juice, added to 15% banana, for additional stimulative effects.

Of course germination may not be held up by lack of nutrients. Veyret (1969) showed that the cuticle on the seeds of terrestrial orchids may inhibit penetration of necessary nutrients, thus producing a sort of mechanical dormancy. She feels that one of the functions of *Rhizoctonia* species in nature is to lyse this cuticular material so that the embryo may be penetrated. Kano (1968) feels that sealing culture flasks may help promote germination of some types of seed by preventing aeration. He also used a soaking treatment prior to planting.

Carbohydrate Sources. The nature of carbohydrate for orchid seedling nutrition has been reinvestigated and extended by Ernst (1967b) and Ernst, Arditti, and Healy (1970, 1971b), confirming that only D-series sugars are useful during germination, particularly fructose, and even xylose. Galactose, as had been demonstrated earlier, was inhibitory to germination, as were disaccharides containing it. Sucrose was still satisfactory and, as has been realized for some time, from the early experiments by Burgeff on *Paphiopedilum* germination and by later work of Ball (1953), is mostly converted to glucose and fructose during the autoclaving of culture medium.

Ernst et al. (1970, 1971b) found evidence that the roots of orchid seedlings may themselves *in vitro* produce an extracellular invertase that can hydrolyze sucrose to glucose and fructose. In addition, more complex sugars could also be split. These workers also found by electron-microscopic studies that the galactose caused rupture of vacuolar membranes in the cell as well as abnormalities of the chromatin and the nuclear membrane, which accounts for its toxic effects to the young seedlings. The total effect seemed to be one on membrane integrity and permeability, both necessary for continuation of life. *Phalaenopsis* and *Cattleya* protocorms were used for these experiments.

Nitrogenous Compounds. Raghavan and Torrey (1963, 1964) made a study with *Cattleya labiata* seedlings of the comparative utilization of ammonium and nitrate nitrogen. Young seedlings could germinate and grow well on ammonium nitrate, showing an increase in their total nitrogen, but they failed to grow well on only nitrate, even with the addition of vitamins or other supplements. These seedlings, after 60 days, finally developed nitrate reductase activity in their tissues and were then able to use nitrate as a sole nitrogen source, another fine example of biochemical ontogeny.

Ziegler, Sheehan, and Poole (1967) showed that Edamin, a lactalbumin hydrolysate containing 18 amino acids, could increase the growth and dry weight of orchid seedlings. The seedlings also contained greater concentrations of amino compounds, particularly glutamine, α-amino butyric acid, and asparagine, though none of these is in the Edamin. Amino acids, particularly L-arginine, have also been discussed previously.

Seeds and Pods. In a series of papers Ito (1960, 1964, 1966, 1968) described experiments developing the technique of ovary culture under partially sterile conditions from cut flowers of *Dendrobium nobile*. If the stigma is protected during the sterilizing process, viable seed may be obtained, even though the processes of embryo formation may take over 100 days after pollination. The flowers had been pollinated on the plants about one week before the culturing procedure began. The technique involves removing the sepals and petals, cleaning the ovaries, and inserting their disinfected stalks through rubber caps into bottles with sterile nutrient solution. Ito was able to grow the ovaries at a faster rate in culture than they grew on the plant during their early stages of development, but the ultimate growth was only one-third of the volume of a normally grown fruit on the plant. Seed, reduced in amount and concentrated near the stalk end of the ovary, appeared normal, however, and germinated satisfactorily. Carbohydrate was best supplied as 6% maltose, though sucrose worked well too; raffinose was inferior, and monosaccharides caused death of the peduncle tissue. Peptone at 50–100 ppm stimulated ovary growth, and coconut milk at 20% definitely promoted development. The same technique has been used to start *Phalaenopsis* stem propagations (Urata and Iwanaga, 1965).

Israel (1963) cultured *Dendrobium* Jaquelyn Thomas ovaries *in vitro* at weekly intervals after pollination. They were planted upside down after washing and sterilization by inserting the stigmatic end into agar medium. The addition of napthalene acetic acid (NAA) up to 20 ppm was effective in the culture medium, and without it death of the cultured ovaries occurred within five days. Fertilization of ovules and seed formation took place after culturing, and viable seedlings were grown from placental samples taken 11 weeks after the date of pollination. These techniques of ovary culture might be used to decrease the strain of pod production of a parent plant or to overcome certain sterility factors in the production of hybrid seed. Moreover, they tell us something of the correlation between plant and pod as nutrients and hormones work their influences on growth of fruit, ovule, and embryo.

Unusual seedlings have now been germinated *in vitro*, particularly *Galeola*, a chlorophylless orchid related to *Vanilla*, by Nakamura (1962). A study of *Taeniophyllum aphyllum* seedlings from nature by Mutsuura, Ito, and Nakahira (1962) shows that the plant up to seven years of age, at least, is composed of a modified leaflike body that is a mixture of cotyledon and hypocotyl tissue. Such seedlings reach only 9–12 mm length in seven years when the first root and a reduced scalelike leaf are produced.

Tree Bark Constituents.   It has been the observation of many that orchids prefer certain trees over others as hosts for germination and growth. Frei (1973a, 1973b) has begun an investigation relating oak bark constituents to orchid distribution patterns near Oaxaca, Mexico. The oaks all had the same sort of ridged bark, so that the surface or texture could not explain the differences. Bromeliad, fern, moss, and lichen growth was also greater on the trees with more orchids. By using extracts of the bark in germination experiments with *Epidendrum tampense* and *Odontoglossum reichenheimii,* it was found that *Quercus castanea* and *Q. vicentensis* lacked inhibitors and bark of *Q. magnoliaefolia* contained gallic and ellagic acids, both of which are growth inhibitors. Barks of *Q. peduncularis* and *Q. scytophylla* contained some inhibiting substances, all examples correlating with the orchid plant frequency on the trees. Frei feels that the moss-lichen growth on the bark acts as a physical and chemical buffer zone in supporting the germination and early growth of orchid seedlings, and she is extending her work on epiphytic symbiosis. She believes these orchids may be humus epiphytes, in contrast to being bark epiphytes, a distinction described by Went (1940). Humus epiphytes germinate only after a layer of moss and lichen is built up on the bark.

Surface Active Agents.   Arditti (1971) and Ernst, Arditti, and Healy (1971a) published information on the relative toxicity of surfactants used as drenches for wetting potting mixes or spreading sprays. Their conclusions are to use low concentrations, 10–100 ppm of low-toxicity materials. They also give procedures for purifying the agents, since they are generally less toxic in pure form. These workers believe that most of the toxic influences occur because of possible emulsification of cell membrane lipids and also because of destructive effects on proteins.

Mycorrhizal Relationships

The term rhizosphere encompasses the soil components and microorganisms immediately surrounding the root system of a plant, and thus composes the medium in which many complex interactions between root, soil, and microorganism occur. This region includes exudates or leachable materials from the roots, products of the fungi or bacteria, and the constituents of the soil. Burgeff (1959) summarized the early work with orchids and their rhizospheres and defined three types of endotrophic mycorrhiza: tolypophagy, ptyophagy, and thamiscophagy as well as types of ectotrophic mycorrhiza. The relationship between host and fungus conventionally has been called symbiotic, but it is really a delicate balance among root, mycorrhizal fungus, and the rhizosphere with its various components.

Mycorrhizal Fungus versus Pathogen.   The difference between a mycorrhizal fungus and a root pathogen may not be great. *Rhizoctonia solani* attacks the root of *Vanilla* and produces various types of infections.

Cortical cells invaded with strongly pathogenic mycelia may lie adjacent to cells in which mycorrhizal interaction is apparent. A cell in which a mycorrhizal association is present may be invaded a second time by pathogenic mycelia. There is a constant fluctuation in the kind of association and the degree of pathogenicity. The association may be advantageous, neutral, or disadvantageous and parasitic (Alconero, 1969).

Many of the endotrophic fungal forms are also saprophytic in habit. *R. solani* exhibits cellulolytic activity, the kind of activity that characterizes the saprophytic mode of existence (Harvais and Hadley, 1967a). *R. repens*, another typical orchid symbiont, is known to produce large quantities of pectic enzymes (Hadley and Perombelom, 1963). It is clear that the host-endophyte complex must in its interaction maintain certain factors or conditions that bring the pectic enzyme activity of the fungus under control. If control is not maintained, pathogenic conditions may result. The control of the pectic enzyme activity of the fungus is due to the interaction of the metabolic processes of the endophyte and its host. But beside being destructive to the host, the cellulolytic activity of the endophytic fungus may be of great value. The fungus is able to break down soil cellulose into nutritive compounds that can be readily absorbed by the fungus, then used by the host.

Harvais and Hadley (1967b) found that in *Orchis* (*Dactylorhiza*) *purpurella,* once the photosynthetic shoot of the orchid has developed, the plantlet exhibits a relative immunity to destruction by parasitic fungi. Before attaining the photosynthetic level of development, the existence of the protocorm has been likened to the running of an obstacle course. The most successful protocorms are consistently reinforced by symbiotic rather than parasitic varieties of invading fungi.

Harvais and Hadley (1967a) state that the simple acceptance of an endophytic fungus by a host should not lead one to the conclusion that symbiosis is the result of this union. The endophytic fungus anchored in the roots of the adult plants *Orchis* (*Dactylorhiza*) *purpurella* have not been proven to be beneficial to their host. It is even possible that the adult photosynthetic orchid would be more successful at this stage of its development in the absence of infection. "The evidence available at present suggests that because of its partial inefficiency in controlling the infection of its roots, *O. purpurella* is doomed to harbor a variety of *Rhizoctonia* species and strains that may be or may not be beneficial to it (or to its protocorms)."

The virulence of a fungus may vary both with the specific host that is invaded and with the root-soil environment of that host. Many species of *Rhizoctonia* that exhibit mycorrhizal associations with orchid roots consistently interact with other plants as parasites. Thus the physiological constitution of the host is of primary importance in determining whether an association will be of the mycorrhizal type. The rhizosphere, which may contain varying quantities of organic substances, inorganic substances, and soil microorganisms, is also important. Weinhold, Bowman, and Dodman (1969) demonstrated the effects of exogenous nutrition on pathogenic interaction. When the external carbon and nitrogen sources were deficient, the mycelia of *Rhizoctonia* attacked the stems of cotton seedlings much less than when substantial quantities of these elements were available. In a second paper (1972), it was concluded that an increased nutrient concentration in the environment of a pathogenic fungus will increase the organism's virulence. The pathogen with an abundant nutrient supply forms a cushion type of infection and produces an adequate quantity of metabolites to destroy its host. The fungus may find a sufficient quantity of nutrients to cause host damage in the seed, the soil solution, or even in the host exudates. Diminished production of any of these nutrient sources reduces virulence and disease severity.

Mycorrhizal Specificity. A still controversial question concerns the specificity of the orchid-fungus association. Harley (1959) regards the evidence for or against the specificity of the host-endophyte relationship as being inconclusive. "Amongst the green and partially saprophytic orchids there is a tendency for certain of the fungal species to be most effective with certain taxonomic or ecological groupings of orchids, but any generalization which can be made would be highly imperfect." Burgeff (1959) believed that though the actual physiological germination of the orchid is independent of the symbiont, a good deal of specificity was present. Kerr (1972) listed the saprophytic orchid species of the world.

Fungi that were isolated from a host of orchid and nonorchid species have been found to be effective in stimulating orchid seed germination. Parasitic forms have been isolated from crop plants which, in germination tests, have stimulated the growth of *Orchis (Dactylorhiza) purpurella* embryos. Downie (1957) found that *Corticum solani* from wheat straw and *R. solani* from both cauliflower and tomato plants were effective symbionts when associated with *O. purpurella;* parasitic *R. solani* isolates formed typical peletons in the orchid cells and were effective in stimulating embryo growth. Yet other varieties of known mycorrhizal fungi do not initiate seed germination at all (Arditti, 1967b). Many endophytic fungi isolated from adult orchid plants exhibit no symbiotic potential with the germinating seeds of their host. Although present in the adult, they do not symbiotically associate with the seed. It has been established that the orchid protocorm often exhibits a greater degree of selectivity than the mature plant (Harvais and Hadley, 1967a).

Some orchid endophytes form nonaggressive associations with a variety of different host orchids. *Tulasnella calospora* was observed to participate in such nonaggressive associations, and Hadley (1970) suggested that this fungus is a ubiquitous orchid endophyte. He further postulates that the host-fungus interaction may be limited by the process by which the fungus is either permitted or prevented from entering the cells of the host.

Many orchids have been found to interact symbiotically with a variety of fungi. Williamson and Hadley (1970) showed that eight different fungi were able to invade the roots of *Dactylorhiza (Orchis) purpurella*. The associations were mycorrhizal. Hadley, testing photosynthetic orchids in their protocorm stage of development, found that the protocorms of *Epidendrum radicans* were able to form associations with several different fungi. The other photosynthetic orchids tested only associated symbiotically with the most benign fungi. But there was no evidence of specificity between a single host and a single strain of fungus (Hadley, 1967a).

Harvais and Hadley found that several *Rhizoctonia* species were capable of symbiotically associating with *Orchis purpurella*. They also found that *R. solani* strains, isolated from the soil of orchid-free regions, were able to function as orchid endophytes. Strains pathogenic to certain non-orchid hosts could associate symbiotically with orchid protocorms. Thus *R. solani* can exist as a soil saprophyte, as a facultative parasite, or as an orchid symbiont. The finding that *R. solani* is an effective saprophytic fungus indicates that it is capable of cellulolytic activity. Garrett (1962) showed *R. solani* to be an active cellulose decomposer.

Soil and climatic conditions are often crucial in determining which endophyte will be found associated with the roots of the orchid plant. The tolerance of a specific fungus to a particular habitat is more important in certain situations than the specificity of the fungus for a host. It has been discovered that certain fungi, such as *R. repens*, survive well in wet areas. The presence of this fungus in various orchid species (those found in very wet regions) may be more indicative of the organism's ability to survive in an extremely wet environment than of its specificity for the organism in which it is harbored (Harvais and Hadley, 1967b). This idea could be extended, and one could raise the possibility that the nutrient content of the soil may also restrict the number and variety of fungi that live in it. The orchid-fungus association in various types of soil may be indicative of the soil nutrient supply, rather than of a particular host-fungus specificity.

Recently Warcup (1971) studied the specificity of Australian terrestrial orchids. Approximately 94% of the fungi isolated from *Caladenia* were found to be *Sebacina vermifera*. Of these isolates, 103 were *Sebacina vermifera,* and seven were other fungal species. Warcup, using 28 infected *Diuris* orchids, made 30 endophyte isolations and showed all of the isolates to be *Tulasnella calospora*. As a final bit of research, Warcup decided to view the effect of various endophytes on the seeds of *Diuris maculata*. He used *Sebacina vermifera* and five species of *Tulasnella* including *Tulasnella calospora*—the endophyte that was found 100% of the time in the adult orchid. The seed infected with *Tulasnella calaspora* after three months produced large protocorms with a short green shoot, whereas the seeds infected with the other fungi made very little progress in their growth.

Fungal Penetration.   The mode by which a fungus penetrates its host has been found to vary (Dodman, Barker, and Walker, 1968a). Penetration by *R. solani* has been scrutinized carefully in both orchid and nonorchid hosts. *Rhizoctonia solani,* as a parasite, usually pierces the roots and stems of its host from dome-shaped infection cushions, while it enters the leaf tissue of the host by means of lobate appressoria. The hyphae of the pathogen may also enter the body of the host through openings such as stomata or lenticels. Upon contact with the protocorms of *Dactylorhiza* (*Orchis*) *purpurella,* the hyphae of the fungus began to pass through epidermal hairs. The time between hyphal contact with the host's epidermal hairs and the penetration of these hairs was most frequently 10–16 hours (Hadley and Williamson, 1971). Cytoplasmic streaming was visible in the infected hairs, and cytoplasmic particles were seen to collect around the hyphal apex. The living epidermal hairs of this orchid were pierced by single hyphae and were the only site of penetration.

The morphology of fungal penetration has consistently been found to differ for pathogenic and symbiotic interaction. Whether the morphology and kind of infection are determined at the surface of the host (by specific surface factors), and whether those hosts capable of effectively reducing the pathogenic activity of the fungus all exhibit similar surface factors, remain uncertain. Flentje and Stretton (1964), also studying the morphology of fungal penetration, found it likely that surface exudates, plus other undefined characteristics of the host's surface, are the essential factors in the determination of whether infection cushion formation occurs. They suggest that these exudates or metabolites are continually produced by the host and propose that they serve to retard or inhibit the apical growth of the fungus: such inhibition would result in the formation of an infection cushion (see section on Phytoalexins). According to these ideas, exudates from orchid protocorms would not contain apical growth-inhibiting factors. The possibility may even be raised that the epidermal hairs of the orchid protocorm contain some factor or a group of factors which facilitate their penetration by a single hypha.

Once the fungal hyphae pierce through the epidermal cells of an orchid root, there is further penetration through passage cells of the exodermis. Alconero (1969) observed these passage cells in the *Vanilla* root and found them to be thin-walled and more active metabolically than the other cells of the exodermis. Two distinct infection layers are often present in the root. The peripheral, or outer layer of infection is comprised of live cells of both the fungus and host. Usually no digestion takes place in this layer. In the adjacent inner layer, digestion of the hyphae by the host parenchyma is apparent. Masses of hyphae, called peletons, fill the lumena of these inner cortical cells. Three types of peletons have been observed in the orchid protocorm. In the subepidermal parenchyma one finds small peletons composed of loosely coiled hyphae. Large peletons with firmly compressed, thin-walled hyphae are also apparent. Finally, there are heterogeneous peletons composed of hyphae that vary in diameter from 3 to 17 μ. The swollen portions of hyphae from this third kind of peleton probably correspond to the storage hyphae described by Burgeff (Hadley and Williamson, 1971).

Alconero (1969), working with *R. solani* and the *Vanilla* orchid, found that separation into a fungal-host cell layer and a digestion layer was not evident. Hyphal digestion occurred in cells scattered throughout the root, but it most frequently occurred in the peripheral cells. Williamson and Hadley (1970) observed the patterns of infection formed in the cells of protocorms of *D. purpurella*. The subepidermal cells contained peletons consisting of loosely coiled hyphae. They usually remained undigested. In agreement with Burgeff's observations, digestion most frequently occurred in the innermost parenchyma cells. The digestive or lytic process lasted for under 24 hours. A single parenchyma cell could be infected more than once, and secondary lysis was observed.

Burgeff's generalizations concerning the layered pattern of infection, in accord with the observations of Hadley and Williamson, are in discord with Alconero's observations. But both sets of observations strongly suggest that the orchid does gain nutrients from the hyphae of the fungus through the process of digestion. The two-layered infection pattern that Burgeff described is normal for orchid protocorms when a plentiful quantity of nutrients is available. Under conditions of starvation, all of the peletons are lysed. Hadley and Williamson suggested that the lytic process is not a mechanism expressly for obtaining food. Rather, they believe the process, except during starvation conditions, to be part of the defense mechanism of the protocorm. The possibility also exists that beside supplying the host protocorm with food, the endophyte may stimulate meristem development and cell differentiation.

Carbohydrate Movement and Nutrient Requirements. The exact role of the fungus as an intermediate in the usage of exogenous nutrients by the orchid host is still not clearly defined. Smith's (1966) work has helped explain the nature of the orchid-fungus interaction. She found that orchid seedlings grew extensively if the endophytic fungus was exposed to an external cellulose source. Working with *Dactylorhiza (Orchis) purpurella,* she showed that when the only connection between the carbohydrate source and the orchid was hyphae of *Rhizoctonia repens,* growth of the seedling occurred. The nutrient supply was then labeled with phosphorus-32 and carbon-12 and made available only to the fungal component of the fungus-orchid association. Radioactivity was soon detected in the seedling.

In another experiment, Smith (1967) supplied the hyphae protruding from the infected orchid seedling with labeled glucose. Over the seven-day experimentation period, there was a continuous movement of the label into the seedling. Both soluble and insoluble substances were found to contain the label. In the mycelia of the fungus, during the early stages of the experiment, most of the label was contained in the compound trehalose, a disaccharide of glucose found in yeast and fungi. Smith suggests that trehalose is the carbon compound that the fungus translocates, though it is possible that the large quantity of labeled trehalose reflects the specific nutrient source administered to the hyphae of the endophyte. Eventually the amount of labeled trehalose declined while the amount of labeled sucrose increased. This rise in the labeled sucrose quantity was thought to be a result of the movement of the label from the fungus to the orchid. Smith also suggests that trehalase breaks down the trehalose. The simple sugar glucose then enters the tissue of the orchid where it may be rapidly used for growth and differentiation and as an energy source for warding off the invading endophyte. Although a concentration gradient for glucose is thus established, the possibility remains that, in addition, a specific mechanism exists which facilitates the transport of sugar from the fungus to the orchid.

Hadley (1969) also found that cellulose can be used by the orchid mycorrhiza. Infected tropical photosynthetic orchid protocorms were used in his studies. When cellulose was used as a carbon source for the fungus-orchid association, there was a very large increase in the orchid growth rate. Hadley, viewing the mycorrhiza as a viable unit existing in nature, realized that a cellulose-rich medium might be more advantageous to the complex than a glucose-rich medium. To test this idea, he supplied one group of developing protocorms with glucose and a second group with cellulose. The carbon sources were administered in one dose or in small intermittent doses. Both carbon sources caused increased growth of the protocorms, and the photosynthetic apparatus of the organisms developed. But after three months of development, many of the protocorms maintained on the dextrose medium were being parasitically invaded. The young plants protruding from the dextrose medium were turning brown, while the plants grown on the cellulose medium had extended shoots. These latter organisms were not being parasitized. Hadley suggests that the more complex carbohydrates retard the endophytes' trend toward parasitism, and thereby allow for the survival of a larger proportion of the protocorms.

Endophyte Growth.   Factors affecting the growth and development of the endophyte should also be carefully researched. Such knowledge might aid in the revelation of the mechanism whereby fungal hyphae are transformed from symbiotic to parasitic organs. Stephen and Fung (1971a) recently studied the effects of nitrogen compounds and various vitamins on the growth of endophytes isolated from *Arundina chinensis* (the bamboo orchid). They found that yeast extracts enhanced the growth of the fungal components of the mycorrhiza and hypothesized that a primary factor affecting the extent of endophyte growth was the quantity of nitrogen available to the infected root. They found that ammonium nitrogen allowed only limited growth and that several of the simple organic compounds they tested did not prove to be good nitrogen sources.

Harley (1959), in his review of information concerning mycorrhiza and nitrogen absorption, states that nitrates were not efficiently absorbed by the fungus-orchid complex, while both simple nitrogen-containing organic compounds and ammonium salts were. Stephen and Fung also tested five amino acids, and nine vitamins were also added to the basal medium. Vitamins alone did not stimulate the growth of the endophyte, nor did methionine or proline, but with the addition of glutamic acid there was a marked increase in growth. This increase nearly equalled the increase obtained when the basal medium was furnished with yeast extract. Arginine and asparagine also stimulated the growth of the fungal mycelia, but neither was as effective as glutamic acid. Thus it appears that the yeast extract-stimulated endophyte growth is at least partially due to the amino acid content of the extract. The authors suggested that glutamic acid is able to enter a metabolic pathway, leading to the synthesis of proteins, through the transamination reaction.

Stephen and Fung (1971b) conducted another study concerning the vitamin requirements of the endophytes of *Arundina chinensis*. They discovered that the growth of endophytes R14 and R29 (two *Rhizoctonia* fungi) was most augmented when thiamin was added to the medium in the presence of *p*-aminobenzoic acid. The addition of thiamin and *p*-aminobenzoic acid to the basal medium resulted in R29 mycelial yields, which were almost equivalent to those yields obtained when yeast extract was added to the medium. This suggests that the R29 endophyte has at least a partial thiamin and *p*-aminobenzoic acid deficiency. Growth of the R14 strain in the *p*-aminobenzoic acid and thiamin-enriched basal medium was only about 72% of that obtained when the yeast extract-enriched medium was employed. When other vitamins (biotin, inositol, pyridoxine, and pantothenic acid) were added to the medium, the growth of the fungus increased, and the mycelial yields were nearly as large as those produced on the yeast extract medium. So, like R29, the R14 *Rhizoctonia* has partial deficiencies for thiamin and *p*-aminobenzoic acid. In addition to these deficiencies, it seems to be partially in want of at least one other vitamin. In nature, the orchid may be the important contributor of several substances that the fungus cannot normally manufacture. Research revealing both the endophyte's and the host's nutritional requirements will aid in the search for the exact definition of the physiological relationship that exists among the orchid, the endophyte, and the rhizosphere.

Defense Mechanisms of Orchids.   Arditti (1966b) declares that fungistatic compounds, as a group now called phytoalexins, which develop in the infected root check the spread of the endophyte. The defense action of these compounds allows for the maintenance of the symbiotic association and the effective resistance of the host tissue to stray fungal pathogens. Campbell (1962, 1963, 1964) and Hamada and Nakamura (1963) described the delicate balance between saprophytism and parasitism that exists between *Gastrodia* and *Galeola* and their fungal symbionts.

Bernard (1911) was the first to demonstrate the defense reaction by placing tuber sections from *Loroglossum hircinum* near its own fungus, isolated and grown in culture. The fungus was unable to invade the tissue, and there was a zone of fungal growth inhibition around it. Norbecourt (1923) confirmed Bernard's observations and showed that dead tubers could not produce the fungistatic effect. A series of papers by Gäumann and his associates (1945, 1959a, 1959b, 1960) have reconfirmed the findings of these earlier workers, using *Orchis militaris* tissues and *Rhizoctonia repens* isolated from it. A compound, orchinol, was then isolated from the plant tissue and later synthesized chemically by these workers. It was a 2,4-dimethoxy-7-hydroxy-9,10-dihydrophenanthrene (Hardegger et al., 1963a, 1963b, 1963c). This compound is formed in the plants about 36 hours after infection and may be produced in roots, tubers, or stems. Twenty-four other species of European terrestrial orchids were also shown to be able to produce the orchinol or other similar compounds. Urech et al. (1963) eventually described hircinol from *Loroglossum hircinum*, the species used in the original work by Bernard; they also described loroglossol from the same species.

These compounds, all trisubstituted dihydrophenanthrenes, have a broad spectrum of activity against soil bacteria, mycorrhizal fungal isolates from orchids, and semiparasitic or saprophytic soil fungi. But the soil fungi do not normally induce the orchid plants to produce their protective compounds. They are formed only in response to the mycorrhizal fungi and certain of the bacteria. Research indicates that the phytoalexins are host-specific rather than fungus-specific (Fisch et al., 1973). Since the compounds are degraded slowly by the fungi, a slowly degradable phytoalexin would be ideal, keeping the fungus under control without destroying it completely.

Flower and Fruit Physiology

Knowledge of flower and fruit physiology has increased on several fronts, and Hans Fitting, as Arditti (1971) points out, was the first to apply the term "hormone" to plant phenomena as a result of his orchid investigations.

Postpollination Changes. Withner (1959) detailed the basic information relative to this topic and discussed the early work of Fitting which disputed the ideas of Müller, who believed that orchid pollen was "poisonous" to the flowers, since it brought about a wilting of the sepals and petals after pollination. Fitting concluded from his experiments that orchid pollen contained a substance, *Pollenhormon*, that caused postpollination effects. Later research showed that many of these were due to the auxin (indole acetic acid, IAA) content of the pollen, but that still other compounds beside IAA might be involved.

Arditti (1969c) and his co-workers (Arditti and Knauft, 1969, Arditti et al., 1971) studied these postpollination effects in detail for *Cymbidium* flowers. They found that application of naphthalene acetic acid (NAA) to the *Cymbidium* stigma will reproduce the effects of pollination, bringing about anthocyanin formation in columns and lips, increased size of column with stigmatic closure, and wilting of the sepals and petals. The pigment formation and wilting could be inhibited by using agents such as actinomycin D which have effects on nucleic acid and protein synthesis, but the swelling of the column and closure of the stigmas were independent of such action. Abscisic acid (ABA) could raise anthocyanin levels and produce wilting but could not cause column changes. ABA plus gibberellin or kinetin mixtures generally caused the same effects as ABA alone, but with lowered pigment production as compared to ABA alone. ABA plus NAA did not permit the increase in anthocyanin that each could induce separately, but all other changes did take place.

Since IAA or ethylene can initiate most postpollination changes, and the other hormones only initiate certain ones, Arditti et al. (1971) made the suggestion that several control sites or mechanisms may be involved. The auxin may thus trigger other processes, such as ethylene production, rather than causing the changes directly. In their experiments with *Vanda*, Burg and Dijkman (1967) found that either pollination or auxin treatment could cause ethylene evolution that in turn brought about the fading and wilting of the flowers.

Oertli and Kohl (1960) showed the movement of nitrogenous compounds from sepals, lip, and petals into the column and ovary, extending the work of Gessner (1948). Poole and Sheehan (1967) also investigated nitrogenous compounds, that is, amino acids and amides, in developing *Cattleya* Trimos fruit. They correlated these studies with growth in diameter and cytological events and hoped to determine which amino acids might be required for germinating immature embryos. Fifteen different compounds from the fruits were assayed weekly by paper chromatography. Alanine, serine, and α-aminobutyric acid reached peak concentrations before fertilization; glutamine, aspartic acid, and glutamic acid reached a first peak soon after pollination, decreased until fertilization (13 weeks in *Cattleya* Trimos), and then reached a maximum peak afterward. The initial rapid increase in amino acid content of ovaries could be correlated with transport from the sepals and petals after pollination. When pod tissue was separated from developing seeds, and each was analyzed separately, serine, glutamine, aspartic, and glutamic acids were high in seeds and low in the pod.

The greening and swelling of the sepals and petals of some orchid flowers after pollination provides photosynthetic tissue that helps support pod growth. *Phalaenopsis mariae* can be added to that list (Ringstrom, 1968), and Matsumoto (1966) actually extracted the chlorophyll from green *Cymbidium* flowers. Dueker and Arditti (1968) demonstrated actual photosynthetic carbon dioxide fixation, which now confirms the process. Some *Miltonia* species show persistent sepals that turn green and become fleshy at the base after pollination (Hayes, 1968).

Alvarez (1968) worked with peroxidase changes during fruit growth in *Epidendrum tampense*. Pollination initiates its production, and activity becomes highest in the ovules as the lignification of their integuments occurs in the later phases of development. Disk electrophoresis procedures showed the enzyme to be only a single isozymic form. Earlier studies by Alvarez and Sagawa (1965a, 1965b) involved histochemical studies in developing embryo sac and embryos of *Vanda* to show changes in proteins, nucleic acids, histones, and insoluble polysaccharides.

Wild-Altamirano (1969) investigated enzyme changes during ripening and curing of *Vanilla* pods, concentrating on proteinase, glucosidase, peroxidase, and polyphenoloxidase. The proteinase, determined by using casein or hemoglobin substrates, decreased as the pods aged. The other enzymes increased with the ripening. The proteinase and polyphenoloxidase were partially purified, and their reactions were studied in an attempt to learn any information relative to the curing process and the development of the typical *Vanilla* aroma and color.

Pollen Studies.   Klass (1964) in our laboratory made a study following Müller's (1953) isolation of IAA from mixed orchid pollen. *Cattleya* Enid pollen and pooled pollen from many standard hybrid *Cymbidium* flowers were fractionated and analyzed by chromatography, spectrophotometric assay, color reagents, and *Avena* internode growth tests. After chromatography, three spots promoting the growth of *Avena* were found on the plates of both samples. They had Rf's at .2 to .3, .5 to .6, and .8 to 1. The .5 to .6 zone was identified as IAA; the other two remain to be identified in our future work.

Ito (1965) studied the ultralow-temperature storage of orchid seeds and pollen using vacuum bottles and −79°C produced with pure ethyl alcohol and dry ice. The seed germinated after as much as 465 days in storage, and pollen of comparable age could be used to produce fruit with fertile seed. Some drying of seed or pollen was necessary before freezing.

Meeyot and Kamemoto (1969) made a thorough study of storing pollen that does not involve the ultralow temperatures used by Ito. They found that silica gel and calcium chloride drastically reduced the viability of pollen, probably by excessive dehydration, and that satisfactory storage for up to a year can be obtained without dehydrants at a temperature of 45°F in a refrigerator. Pollen of different species varies much in its storability, especially at higher temperatures. The viability was checked by growing it on a synthetic medium, adapted from Miwa (1937) and Curtis and Duncan (1947), that contained 5 g sucrose, 1 g agar, and 100 ml distilled water. The medium was sterilized in Petri dishes and the cultures kept at 72°F.

Stort (1972b) observed pollen germination in *Cyrtopodium* and found supernumerary nuclei, as many as eight, from divisions of the generative nucleus. She was not able to tell what happened to them at the time of fertilization.

Nectar.   Both floral and extrafloral nectaries have been described (see Chapter 6 on orchid anatomy) for orchids. Baskin and Bliss (1969) analyzed the sugar content of many orchid nectars by gas chromatography, and no differences could be detected from intrafloral or extrafloral sources. Jeffrey et al. (1970) performed additional sugar analyses of nectar by a thin layer chromatography technique (Jeffrey and Arditti, 1969) and find no correlation between types of sugars and pollinators of the flowers involved. This suggests that scent, color, and form are more important than sugar in attracting insect vectors.

The nectars all contain glucose, fructose, and sucrose. In addition, certain ones contain raffinose, and fewer still, stachyose. Cellobiose, gentiobiose, lactose, maltose, melibiose, melezitose, and a few large oligosaccharides can occasionally be found. Jeffrey et al. (1970) found it difficult to make any chemotaxonomic correlations with the presence of particular sugars. Raffinose was generally absent from *Epidendrum* and *Cattleya* nectar but more common in *Laelia*. It was also in all nectar from *Oncidium* and *Angraecum* and in almost all *Laeliocattleya* hybrids studied. Stachyose was present in *Cymbidium* and *Mormodes*. Melezitose was found only in *Oncidium* nectar.

Flower Fragrance.   That *Satyrium* flowers produce a goaty odor (as a result of caproic acid) has been observed for ages and accounts for the plant's ancient name (Arditti, 1972). Vogel (1966) described scent glands in orchids for the first time (see Chapter 6); and Dodson and Hills (1966) and Hills, Williams, and Dodson (1968) recently investigated scent production in orchids and detected a number of compounds by using gas chromatography. Some orchids may produce a mixture of 18 compounds, and these investigators sampled over 150 species of orchids in 26 genera. Maximum odor production can vary from the first to the fifth day after the flower opens, depending on the species. Most fre-

quently identified compounds were α-pinene, β-pinene, 1,8-cineole, linalool, methyl benzoate, benzyl acetate, d-carvone, citronellol, methyl sali-
cylate, methyl cinnamate, and eugenol. These are, in most cases, the
major component of the fragrances in which they occur. More than 30
other compounds were involved, however, only some of which have been
tentatively identified.

Brassavola nodosa produces cineole with its slightly medicinal odor as
its major compound, and B. digbyana forms citronellol (roselike) and
linalool (resembling lily-of-the-valley). Stanhopea, Cycnoches, and Catasetum species form benzyl acetate that has jasminelike odor, but in
Catasetum many additional compounds are also present. The rye bread-
caraway fragrance of C. discolor results from d-carvone, and the cinna-
mon smell of C. roseum comes from methyl cinnamate, which is also
present in Stanhopea and Gongora fragrances. Eucalyptol (1,8-cineole) is
the medicinal odor in Stanhopea cirrhata. These are some of the examples
from the variety of species analyzed. Adams (1968) studied the effects of
various artificial fragrance mixtures in the field to show that specific pollinator bees could be attracted by duplicating the natural smell of a given
species with synthetic chemicals. Dodson et al. (1969) further described
the habits of the bees. The flowers attract only male bees, and the role
of the fragrances in their life cycles is still specifically unknown.

Flower Pigments.   Several more orchid flower pigments have been determined; Sanford and his students (1964a, 1964b) and Arditti (1969a)
have added numbers of new compounds to the list, though several remain chemically unidentified. The identified types are listed by Arditti
(1969a, 1969b), and one should consult his papers for the additional
references and the more detailed organic chemical information not
presented here.

Aside from the purely descriptive value of this work, the presence or
absence of particular pigments may show taxonomic relationships, assum-
ing that plants with common ancestry, or hybrids with parents in common, will show the inheritance of these compounds from their parents.
Arditti (1969a, 1969b) showed that Broughtonia sanguinea and B. negril-
ensis have different anthocyanins and may therefore be considered as
separate species, all other taxonomic evidence being considered as well.
There is indication of relationship between the genera Broughtonia and
Cattleyopsis, but they are different as far as what they actually produce,
the Cattleyopsis producing a single compound, the Broughtonia, four
different compounds.

In the hybrid Brassotonia John Miller, two of the Broughtonia pigments are found, but the single pigment, raphanusin C, from the Brassa-

vola nodosa parent is not. Four additional pigments not found in either parent are also present, so that interpretations of such data in relation to taxonomy must be carefully considered.

Sanford and his co-workers (1964a) used root tips instead of flowers for their assays. Of orchids tested that were not in the LAELIINAE, only *Vanda* showed pigmentation. This group included *Oncidium, Brassia, Dendrobium, Vanilla, Rodriguezia, Aspasia, Notylia, Orchis,* and *Cymbidium.* Their continued work (Sanford et al., 1964b) involved a few flower pigments as well. This work confirmed the close taxonomic relationship between Brazilian *Laelia* species and the genus *Cattleya,* as well as the distant relationship of Mexican *Laelia* species from their Brazilian counterparts. Another point from their research is that anthocyanins may be present in flowers but completely absent from roots or leaves of the same plants; also, when pigments are present in the roots, they are sometimes the same as, and sometimes different from, the floral anthocyanins. These patterns are not yet understood. Arditti (1969b) points out that there may be different combinations of anthocyanins in the lips of some flowers as compared to those in the sepals and other petals, accounting for different or darker colors.

As part of a survey of flowers of *Cattleya* and their relatives and hybrids, Stefanski (1965) in our laboratory has also noted this latter point. In *Cattleya* species (*lawrenceana, percivaliana, labiata,* and *amethystoglossa*) two or occasionally three pigments, with one usually in dominance, were present in sepals and petals. In some species, however, there was only a single compound (*C. mossiae, walkeriana, aurantiaca*). Lips, with the exception of *C. mossiae* and *C. aurantiaca,* had at least two anthocyanins, and they often had Rf's different from the sepal and petal pigments. The same sorts of patterns were noted in *Laelia* species. We concluded that the darker color of lips is often the result of different compounds being present rather than from a stronger concentration of the same pigments found in sepals and petals.

In flowers like *C. aurantiaca, lawrenceana, skinneri,* and *L. gouldiana,* which show little obviously visible difference in coloration of the lip, a match of compounds was found among lips, sepals, and petals. A series of hybrids of *Cattleya, Laeliocattleya, Brassolaelia, Sophrolaelia,* and *Sophrolaeliocattleya* mostly showed greater numbers and varieties of pigments as compared to the species. Flowers of *Slc.* Beverly Salmon had four pigments, for example, and *Lc.* Arthur Miles and *Lc.* Adolph Hecker both had five. Three pigments per flower in hybrids were common. We also found three in flowers of *S. coccinea* (*grandiflora*) that would help account for the multiple pigments in *Sophro* hybrids. The anthocyanins of

Epidendrum phoeniceum and *E. kranzlinii*, with the exception of one spot, were different from those in *Laelia* and *Cattleya*.

Assaying a number of the blue forms of various species (*L. purpurata werkhauseri, C. labiata coerulea, mossiae coerulea,* and *percivaliana coerulea*) produced interesting results. The pigments, except for one compound in *C. labiata* and its *coerulea* form, were all different from those in the normally colored flowers. *Lc.* Blue Boy 'Gainsborough' showed only a single pigment, and it was the same in both lip and the other petals and sepals. *C.* Portia 'Coerulea' showed a single pigment in the lip and another one in the sepals and petals. What the relationship of these blue flower pigments may be to the usual ones no one yet understands, but they appear to be a distinctly separate spectrum of anthocyanins. Whitlow (1970) has written a series of papers on the blue cultivars and is recording their breeding behavior.

Stefanski (1965) also found that after yellow, red, or orange flowers were extracted for anthocyanin content, a large amount of yellow or orange pigmentation was left in the tissues. Spectrophotometric analysis and other tests showed the material to be carotenoids. Such analyses were done on *C. aurantiaca, S. coccinea (grandiflora), L. cinnabarina,* and *Lc.* Eva.

Sophronitis flowers were examined microscopically to show the presence of yellow granular plastids containing the carotenoids. They were more concentrated in the lip, and the same pattern could be noted in the other flowers as well. Curtis and Duncan (1942) previously described such plastids in the yellow "eyes" on the lips of *Cattleya* and its hybrids. They also reported crystalline anthoxanthin in *C. dowiana* flowers, but none of the yellow plastid pigments examined in this study proved to be anthoxanthinlike. Much interesting work remains to be done in this field.

Environmental Factors and Orchid Development

Of the various external influences affecting orchid growth, light is among the most influential. Withner (1961, 1962, 1964a, 1964b) wrote a general review of light in relation to orchid growth, and Davidson (1967a, 1967b) has reviewed light effects as well as other environmental factors on orchids.

Photosynthesis.   The path of carbon dioxide fixation of orchids is beginning to be better understood, but many details must be worked out by more research. As a result of the paper by Nuernbergk (1963), who worked with *Bryophyllum* and other succulents and then investigated some thick-leaved tropical orchids with respect to their carbon dioxide metabolism, we began better to realize that they possessed "crassulacean" metabolic patterns. It would also seem likely that some orchids, at least some of the thick-leaved tropical epiphytic types tested so far, can belong to the C4 group of higher plants that differ from the C3 types in their capacity to assimilate carbon dioxide with light from a normal atmosphere with relatively high oxygen content and low carbon dioxide concentration (Devlin and Barker, 1970; Coombs, 1973). Other evidence besides CO_2 fixation must be sought, however, to coordinate with these ideas.

Warburg (1886) listed several orchid species as showing the same diurnal variations in malic acid as various succulents, the amount of malic acid increasing as much as 60% during the night and decreasing during the day. Bendrat (1929) continued Warburg's work and found that *Paphiopedilum* did not follow the crassulacean pattern. At this point, it may be presumed to be a C3 plant with usual photosynthetic CO_2 fixation in the light and respiratory CO_2 release at night. Nuernbergk has added *Calanthe, Catasetum, Cymbidium,* and *Thunia* to this list, but this must be confirmed experimentally. Hatch, Slack, and Johnson (1967) were also not able to show a dark fixation of CO_2 with *Cymbidium* leaves.

Discussing the crassulacean pattern, Nuernbergk (1963) describes what happens as follows:

Under natural conditions, i.e., a daylength of about 14 hours and higher day and lower night temperatures all these succulent plants take up CO_2 mainly during the night and only little, if at all in the late hours of the afternoon, just before dusk. In the early morning the uptake of CO_2 which is increasing steadily until about midnight begins to decrease and then stops in order to make place to a small release of CO_2 during the daytime. Sometimes the plants absorb a little bit of CO_2 during the whole day, but the rate of CO_2 which is taken up in the night is always much higher than the whole amount of CO_2 usually taken up in the light period.

The adaptive value of this de Saussure effect, as it was originally termed, to succulents or epiphytic orchids with fleshy leaves is easy to understand. During daylight hours temperatures are high, bark and roots are dry, and humidity is low. The stomata of the plants remain closed, and water is conserved within the tissues. At night stomata open, and temperature-moisture relationships do not induce stress; in addition, more CO_2 is available. But the temperature changes have even more profound effects on the plant than to increase or decrease transpiration.

If plants are kept experimentally under constant conditions of light and temperature, no de Saussure effect is observed. In fact, CO_2 may be released by the plants instead of being fixed into malate. That means the plants are not fixing any carbon for the manufacture of carbohydrates or other compounds and eventually will show this by a slowing or cessation of growth. As soon as high or low temperatures alternate, the effect will proceed, with CO_2 being fixed at night under natural conditions. Light is of little influence in the reaction, and CO_2 absorption can also occur in the light if the temperature is kept low. CO_2 may be produced, even in the dark, if temperatures are high, causing a lack of reserve accumulation as mentioned above.

Light is important, however, for another reason. It provides the proper precondition for the cells to store the CO_2 at night. The amount of CO_2 absorbed at night was found to be directly proportional to the intensity and duration of light during the day. ATP and TPN produced as by-products of photosynthesis during the day provide the energy-rich compounds and reducing power necessary for the fixation to occur at night. The more they are formed, the greater the amount of CO_2 that may be absorbed. If plants are kept in the dark, even after bright, sunny days, the rate of CO_2 incorporation begins to decrease after several hours.

These observations correlate nicely with a paper by Murashige et al. (1967) showing that seasonal flower production of *Vanda* Miss Joaquim was not controlled by photoperiod or other external factors besides the availability of sunlight, and therefore the amounts of reserves in the plants. Disbudding during the fall, thus conserving reserves, enabled the plants to flower in winter when light intensities were lower and flowering would not usually occur.

The papers by Dueker and Arditti (1968) and Arditti and Dueker (1968) indicate that green *Cymbidium* flowers can fix CO_2 in the light, especially in the sepals, and also fix CO_2 in the dark. This CO_2 fixation at night does not correlate with the noncrassulacean status for *Cymbidium* leaves mentioned above. Different organs of the plant are likely to have completely different metabolic patterns. These investigators also found that *Cattleya* leaves can maintain their photosynthetic efficiency for as long as three years before the rate begins to fall off. The experiments by Knauft and Arditti (1969) showed labeled malate after exposing *Cattleya* leaves to radioactive carbon dioxide in the dark. Citrate and an unidentified compound were also labeled. Borriss (1967) had previously shown acid increase in leaves of *Phalaenopsis, Cattleya,* and *Epidendrum* at night. Who will be the first to investigate fertilizing orchids with CO_2 at night instead of the usual daytime regimen?

Photosynthetic Efficiency. Studies of the efficiency of leaves under various light or temperature conditions are possible with the use of the infrared CO_2 analyzer. Esser (1973) determined the rates in leaves of *Odontioda* Lippstadt, and the data may be used directly to indicate optimal growing conditions for the plants under cultivation. In the dark, as measured by CO_2 production, respiration using up plant reserves is ten times greater at 35 than at 10°C. Keeping the plants at 10°C and gradually illuminating the leaves, a balance point is reached at 200 Lux where photosynthesis or food formation is just equal to respiration. At light intensities above 200 Lux, up to 5000 Lux, the rate of photosynthesis increases to its maximum. The curve then falls off as the leaf is saturated.

By studying the balance between the light and temperature, Esser found that the higher the temperature the more intense the light had to be for the plant just to compensate—for 20°C it was 500 Lux, for 25°C it was 1200 Lux, for 30°C it was 2100 Lux—and at 35°C the compensation point was never reached, even with 20,000 Lux. The maximum light to reach peak efficiency for any of the lower temperatures was 10,000 Lux.

From these results it is possible to say that *O*. Lippstadt will grow most efficiently at 15°C and 5000–10,000 Lux. Since sunlight can reach over 100,000 Lux in summer, these plants must be shaded to produce the 10,000 Lux, and this in turn helps keep temperatures lower. In winter the plants are best kept at 10°C so that the poorer winter light of 500 Lux will produce at least a balance of photosynthesis with respiration in the leaves. If the balance is only just maintained or is negative, the plants will grow very little, progressively become smaller, produce a few weak flowers, and be susceptible to disease. If a positive balance is maintained, then growth and flower production will be as usually desired. One could only wish that such data were readily available for a variety of species and hybrids, especially those that are difficult to grow. Knowledge of native habitats and past experiences in growing tend to give us the equivalent information by empirical process, but think of how many plants could have been saved if such data had been at hand.

Hormone, Fungicide, or other Treatments.   Brewer, Gradowski, and Meyer (1969) treated *Cymbidium* plants with abscisic acid in an effort to delay flowering. Since Bivins (1968) found that gibberellic acid could speed up flower development of *Cymbidium* with little increase in flower size and since abscisic acid is antagonistic to gibberellin, the idea presented a possibility. But it did not affect flower spike development. It did, however, effectively defoliate the plants at 250 or 500 ppm concentration, speeding up the senescence of leaves and delaying new growths at the higher concentration.

Not for orchid plants, but for the diminution of *Oxalis* weeds that grow with them all too often, Gripp (1969), Bivins and Kofranek (1965), and Murashige, Sheehan, and Kamemoto (1963) recommend a Simazine treatment. Simazine has a wide margin of safety required to give *Oxalis* emergence control and still not cause damage to orchids. It is also effective on algae, mosses, and liverworts, though the chemical may take two weeks for action. Generally one application, or no more than two, should be sufficient per year. *Oxalis,* beside being a pest, also harbors red spider. Bivins, Sachs, and Debie (1968) have also studied Simazine on *Cymbidium* plants. They found the orchids were resistant to its toxic effects, possibly because it is converted by a nonenzymatic detoxification factor into hydroxy-Simazine, which is metabolically degradable.

Nutrients. Davidson made several good points in his paper on orchid nutrition (1960) that are still valid. In his long-term experiments he found that orchids are not different from other plants in their requirements, but they may take longer to show deficiencies because of their growth habits. In general, the faster the growth rate, the greater the nutrient demands of a plant. Also, different species and genera, because of their varied environmental origins, might have slightly different requirements. Nitrogen and phosphate deficiencies were most apparent with *Cattleya* seedlings, for instance, with lesser requirements for potassium, calcium, and magnesium. Chin (1966) demonstrated that potassium is especially important for *Dendrobium phalaenopsis* hybrids, as are phosphate and magnesium. Sheehan's (1960) work relating nutrients to types of bark for potting media showed that nitrogen levels definitely affected the growth and numbers of flowers produced by *Cattleya,* while phosphorus and potassium levels did not correlate as well.

Minor elements seldom caused difficulty in experiments, whereas they would have with faster-growing plants. The minor elements—iron, zinc, and boron—can cause toxicity symptoms, however, when in excessive supply. Sodium may also. The water supply may contain these ions, so that demineralizing or deionizing (not water softening) may be necessary. Davidson finds that the pH of the water may vary greatly with no particular significance of itself in the growth of orchids. It is the mineral content of water that can react with added nutrients at particular pH's, or supply others in excess, that is critical. He has critically discussed water quality and other nondisease orchid ailments in two fine articles (Davidson, 1967a, 1967b).

Davidson (1960) found that osmunda fiber plus water provided adequate fertility for orchids. Except for the lower potassium content of osmunda fiber, it generally resembles orchid plants in its mineral composition. Skilled growing in osmunda may therefore produce plants that do not respond to fertilizing. When response does occur, it may imply that the nutrient content of the osmunda is leached by age and repeated waterings or that, since the plants usually complete seasonal growth within a relatively short period of time, the nutrients may temporarily be inadequate, especially the amounts of nitrogen and phosphate that can be either absorbed through roots or translocated from older plant parts to growing regions. Of all the nutrients required by plants, nitrates are the most easily washed out of potting mixes by watering.

Comparative chemical analyses of plants growing in osmunda fiber and in bark—the bark fertilized regularly with a complete nutrient combination, the osmunda not—showed the same composition. There were no significant differences. The nutrient combinations, however, showed that higher nitrogen levels were necessary for proper growth in bark, since the bark fungus (*Sphaerobolus stellatus*) also has a high nutrient requirement for nitrogen. "Over a period of 18 to 28 months, the dry weight fungus grown per pot of bark was found to be approximately 80 per cent as great as the dry weight of the orchid tissue, other than flowers, produced at the same time. The mass of fungus, moreover, contained about twice as much nitrogen as the orchid tissue produced in that period." A 20-20-20 soluble fertilizer is recommended for growing of plants in bark, 1 level teaspoon to 5 gallons of water, plus 1 rounded teaspoon of ammonium nitrate, giving a final ratio of 6-1-1 for nitrogen, phosphorus, and potassium.

Sheehan (1958, 1960) gave systematic reviews of various potting media and related them to fertilization, using *Cattleya* and *Phalaenopsis*. Some kinds of bark do not break down as rapidly as others and therefore require less nitrogen in nutrient mixtures. Deterioration related to the percentage of fine-sized particles in the bark mixes and consequently how rapidly they leached and how much liquid they retained. Adderley (1965) used sheet moss successfully as a potting medium, species of *Thuidium* and *Hypnum*, and this procedure is much used in the New York Area.

Sheehan (1966) demonstrated foliar uptake of phosphate to answer the question of whether foliar feeding of orchids is a possibility. By applying isotopic phosphate to a two-year-old leaf, he was able to show definite uptake and then translocation to pseudobulbs and leaves on either side of the treated one. Pseudobulbs and roots contained more of the activity than other plant parts. Another result was to show that *Cattleya* roots three or four years old were still actively absorbing nutrients and water and should therefore not be removed when repotting. Young roots on the same plants did not absorb any faster than the older ones.

The question of leaf-tip die-back in *Cattleya* has been studied by

Poole and Sheehan (1973) and it was hypothesized that it may result from temporary calcium deficiency during periods of rapid leaf elongation prior to maturation. This would occur especially, they report, if temperatures around the root system were high, resulting in defective absorption. Frequent watering may supply more calcium if the water is hard, but it helps even more by its cooling effect on the potting medium.

Brubaker (1969) makes the interesting point that many of the fertilizer recommendations for orchids may be too high, and he has consistently used 20 ppm of nitrogen without his plants suffering from starvation or deficiency symptoms. He gives a convenient table for the appropriate amounts of the common fertilizers to use with various mixer-proportioners to keep a level of 50 ppm of nitrogen in the fertilizing solution. He feels orchids "should never be housebroken" and that between fertilizing leaching should proceed until there is a copious flow from the bottom of each pot. Davidson (1967a) also described symptoms of over-fertilization.

Experiences with balanced slow-release fertilizers are beginning to appear. MagAmp is one such fertilizer, and its use is described by Kock (1972).

Temperature. Davidson (1963) has extended our knowledge of orchid leaf temperatures in relation to watering practices and growth, particularly with *Cymbidium* and *Cattleya*. The relationship between photosynthesis and temperature was referred to previously (Esser, 1973). Davidson found that *Cattleya* leaves could not tolerate temperatures of 130°F for more than a few minutes and that the position of the leaves with respect to incident sunlight was an important determinant of the amount of solar energy absorbed versus the amount that was reflected away. Areas near the ends of leaves did not show burns as readily as mid-regions, since heat was better dissipated than from central regions of the leaf. The practical application of fans for air movement to help prevent heat buildup, as well as increasing the cooling effects from increased transpiration, is easily understood. The temperature of leaves is almost always higher than that of the surrounding air, particularly in enclosed structures such as greenhouses, where direct radiation from the structure and its contents is also a factor in heat buildup.

Misting for cooling is definitely effective, both from the cool initial temperature of the water and from its evaporative effects from leaf surfaces. Davidson's data on *Cymbidium* leaves show that the plants may be grown under high summer light intensities, as long as foliage temperature is controlled by mist cooling. He says that "when the northern grower re-

fers to the light required or the light tolerated by his plants, he is referring to his practice of controlling the temperature of his orchid greenhouse." Davidson also makes a point that reflected light is as yet little
used for growing, even though it does not have the high heat buildup
qualities of direct solar radiation. The characteristics of fiberglass greenhouses, which produce considerably more reflected light, must be more
completely investigated. The fewer the long wave lengths of heat energy
that are re-radiated from glass, dark walls, or objects in the greenhouse,
the easier the job of growing will be. An east-west orientation for the
green house will give the best light conditions, especially in winter, for
growing plants in northern temperate regions, particularly when internal
temperatures and air circulation and humidity are properly controlled.

Virus and Fungal Infection.   When Thornberry and Phillippe (1964)
reported on the infectious nature of *Cattleya* blossom brown necrotic
streak, they found that both *Cymbidium* mosaic virus (CyMV) and the
orchid strain of tobacco mosaic virus (TMV-O) were isolatable from the
plants. Thornberry and his associates (1968) believed that neither of
these viruses alone would cause the flower necrosis, but Corbett (1967)
felt that the flower problems described by Thornberry in the north were
seldom observed in Florida, though many plants infected with CyMV
and TMV-O were to be found. Lawson (1967) then wrote that *Odontoglossum* ringspot virus (ORsV) alone could cause color-break in flowers,
and when combined with CyMV would cause the flower necrosis. These
two viruses could infect leaves, roots, and flowers, often producing visible
symptoms, but sometimes being symptomless in their infection. Lawson
later (1970a) decided that CyMV alone would produce necrotic flower
streak in *Cattleya* and that the presence of the TMV-O was only inci-
dental, not causal to the disease. The TMV-O could, however, cause color
break in lavender cattleyas (1970b), and this should not be confused
with the necrotic flower streak. The flower necrosis symptoms may resemble senescent changes in the flowers, but they occur much sooner
than would otherwise be expected, a few days after the flowers open.
The CyMV can also infect *Phalaenopsis* plants.

Lawson (1967) tested the inactivation of CyMV and ORsV by various
reagents used to clean pots and tools. He found that 2% Clorox (5.25%
sodium hypochlorite) completely inactivated CyMV, although ORsV was
slightly more resistant, especially if fresh orchid tissue was added to the
mixtures. This indicates that fresh solutions should always be used for
maximum effectiveness. If a pseudobulb is inoculated with CyMV it will
quickly spread to roots, leaves, and flowers within a few days.

Testing for CyMV was done by application of the virus-containing tissue brei onto the cotyledons of *Cassia occidentalis*. Brown virus lesions would develop after four days. The ORsV could be detected by applying samples of the virus onto leaves of *Chenopodium amaranticolor*. Five or six days after inoculation pinpoint yellow spots are produced on the blades. For TMV assays, plants of the *Chenopodium* are used, also *Nicotiana tabacum* var. Xanthi and *Gomphrena globosa*. The tissue extract to be tested is placed on the leaf along with some carborundum powder and lightly spread about; further details are given in Lawson's (1970b) paper.

Lawson found that it is possible to free *Cattleya* mericorms of CyMV by repeated excision and subculturing of the tissue; but if TMV-O is present, it will remain, even in very small sections, and cannot be eliminated. The elimination of all viruses by meristem culture is thus not yet possible and presents a great problem for future research.

Certain fungal diseases such as *Pseudomanas* infection of *Phalaenopsis* have been investigated in detail (McCorkle et al., 1969), and McCorkle recommends organic mercurial sprays to prevent the damage. Burnett (1973) studied the "black rots," *Pythium ultimum* and *Phytophthora cactorum*, on a variety of orchid plants, recommending Truban as an effective control. Burnett (1965) has also written an account of orchid diseases in general that is recommended reading for all those interested in greater detail. Benomyl or Benlate (Burnett, 1971) is described as a wide-spectrum fungicide with systemic action that is effective against a variety of fungi without harmful effects, even on *Broughtonia* or other often sensitive species. Research on petal spotting caused by *Botrytis* has also been done (Peterson et al., 1968), but a good control spray or fumigant has not yet been found. A *Fusarium* causing root rot of *Cattleya* was isolated and characterized by Good and Jackson (1965, 1966). Its pathology was increased with cool temperatures and overwatering.

Irradiation Effects. Radiation generally produces injury, deformity, or lethal effects, depending on dosage. It may also cause mutation of genes which bring about a desirable change. Now that mericloning techniques can produce quantities of genetically similar protocorms for irradiation, it is possible to distinguish systematically possible mutant effects. Harn (1970) has irradiated *Cymbidium* protocorms and slices and finds optimum dosage to be 2 kR for slices, 2–3 kR for younger whole protocorms, and 3–4 kR for old whole protocorms. The latter are less sensitive than the young, and the meristematic cells were most susceptible of all the protocorm cells.

Pollution Effects. The earlier discussion relating floral responses to postpollination effects shows that ethylene, produced in plant tissues, can cause the wilting of sepals and petals. An external supply of ethylene as an air pollutant can increase the rate of the aging process in flowers without benefit of pollination at all. The syndrome is called sepal wilt or dry sepal. Davidson (1976b) pictured this nondisease ailment of *Cattleya* orchid flowers. He also described *Phalaenopsis* bud drop. If the pollution is particularly severe, it may cause flower sheath deterioration and the loss of leaves, particularly the older ones on a plant. There is no simple treatment to counteract the effects of ethylene.

Tissue and Organ Culture *in vitro*

Only certain aspects of this topic need be discussed here. For other details consult Chapter 4, which emphasizes propagating *in vitro*. No one has yet described single cell cultures from orchids, but no doubt they will occur one day when our technical knowledge of nutrients and hormones is sufficient. It has so far been possible to grow leaf, meristem, flower stalk, and protocorm tissue by tissue culture techniques. Root tips have been cultivated *in vitro*, but they do not callus or proliferate to form plantlets (Churchill, Ball, and Arditti, 1972).

In a paper presented before the Botanical Society of America, Withner (1953) reported on growing several species of orchid roots *in vitro*, using Bonner medium with various vitamin, auxin, and amino acid supplements. Most experiments were done with roots of *Oncidium sphacelatum* and *Epidendrum* Obrienianum. Growth was followed *in vitro* for 20 to 25 weeks, through three to four transfers to new media, and a comparison with growth rates of roots remaining on the plants was possible. The growth rate of individual roots may vary, whether *in vitro* or *in vivo*, but in no case did the best growth *in vitro* amount to more than one-third to one-half of intact root growth on the plant. Solid medium worked better than liquid. Further, only cultures left in the light would grow, indicating that photosynthesis probably played a role in the survival and growth *in vitro*. Considerable research remains to be done in this area.

Tse, Smith, and Hackett (1971) reviewed the methods of propagating *Phalaenopsis* from nodal cuttings from the flower spikes, and they were able to induce callus formation and shoot production. By slicing or puncturing the bud and using NAA in the culture medium, they promoted the numbers of positive responses, but varietal differences were noticed. Some *Phalaenopsis* would not form callus or adventitious buds.

Acknowledgment

I am indebted to Arthur Grossman, who assembled the material covered in the section on mycorrhiza in a Research Tutorial, Brooklyn College, January 1972.

Bibliography

ADAMS, R. M. 1968. The attraction of *Euglossini* to fragrance compounds of orchid flowers. Ph. D. thesis. University of Miami.

ADDERLEY, L. 1965. Two species of moss as culture media for orchids. *Am. Orchid Soc. Bull.* **34**:967–968.

AKAMINE, E. 1963. Ethylene production in fading *Vanda* orchid blossoms. *Science* **140**:1217–1218.

ALCONERO, R. 1969. Mycorrhizal synthesis and pathology of *Rhyzoctonia solani* in *Vanilla* orchid roots. *Phytopathology* **59**:426–430.

ALVAREZ, M. R. 1968. Temporal and spatial changes in peroxidase activity during fruit development in *Encyclia tampensis* (Orchidaceae). *Am. J. Bot.* **55**:619–625.

ALVAREZ, M. R., and SAGAWA, Y. 1965a. A histochemical study of embryo sac development in *Vanda* (Orchidaceae). *Caryologia* **18**:241–249.

———. 1965b. A histochemical study of embryo development in *Vanda* (Orchidaceae). *Caryologia* **18**:251–261.

ARDITTI, J. 1965. Studies in growth factor requirements and niacin metabolism of germinating orchid seeds and young tissues. Ph. D. dissertation. University of Southern California.

———. 1966a. The effect of tomato juice and its fractions on the germination of orchid seeds and on seedling growth. *Am. Orchid Soc. Bull.* **35**:175–182.

———. 1966b. The production of fungal growth regulating compounds by orchids. *Orchid Digest* **30**:88–90.

———. 1966c.The effects of niacin, adenine, ribose, and niacinamide coenzymes on germinating orchid seeds and young seedlings. *Am. Orchid Soc. Bull* **35**:892–898.

———. 1967a. Niacin biosynthesis in germinating x *Laeliocattleya* orchid embryos and young seedlings. *Am. J. Bot.* **54**:291–298.

———. 1967b. Factors affecting the germination of orchid seeds. *Bot. Rev.* **33**:1–97.

———. 1968. Germination and growth of orchids on banana fruit tissue and some of its extracts. *Am. Orchid Soc. Bull.* **37**:112–116.

———. 1969a. Floral anthocyanins in some orchids. *Am. Orchid Soc. Bull.* **31**:407–413.

———. 1969b. Floral anthocyanins in species and hybrids of *Broughtonia, Brassavola,* and *Cattleyopsis* (Orchidaceae). *Am. J. Bot.* **56**:59–68.

———. 1969c. Post-pollination phenomena in orchid flowers. *Austral. Orchid Rev.* **34**:155–158.

———. 1971. Orchids and the discovery of auxin. *Am. Orchid Soc. Bull.* **40**:211–214.

———. 1972. Caproic acid in *Satyrium* flowers: Biochemical origins of a myth. *Am. Orchid Soc. Bull.* **41**:298–300.

ARDITTI, J., and BILS, R. 1965. The germination of an orchid seed. *Orchid Soc. Southern Calif. Rev.* **7**:5–6.

ARDITTI, J., and DUEKER, J. 1968. Photosynthesis by various organs of orchid plants. *Am. Orchid Soc. Bull.* **37**:862–866.

ARDITTI, J., and ERNST, R. Floral pigments in orchids. *Orchid Digest* **33**:129–131.

ARDITTI, J., FLICK, B., and JEFFREY, D. 1971. Post-pollination phenomena in orchid flowers, II. Effects of abscisic acid and its interactions with auxin, gibberellic acid, and kinetin. *New Phytol.* 70:333–341.

ARDITTI, J., HEALEY, P., and ERNST, R. 1971. Use of surface active agents in orchid culture. *Am. Orchid Soc. Bull.* 40:317–318.

———. 1972. The role of mycorrhiza in nutrient uptake of orchids, II. Extra-cellular hydrolysis of oligosaccharides by asymbiotic seedlings. *Am. Orchid Soc. Bull.* 41:503–509.

ARDITTI, J., and KNAUFT, R. 1969. The effects of auxin, actinomycin D, ethionine, and puromycin on post-pollination behavior in orchid flowers. *Am. J. Bot.* 56: 620–628.

BALL, E. 1953. Hydrolysis of sucrose by autoclaving media, a neglected aspect in the technique of culture of plant tissues. *Bull. Torrey Bot. Club* 80:409–411.

BASKIN, S., and BLISS, C. 1969. Sugar content in extrafloral exudates from orchids. *Phytochemistry* 8:1139–1145.

BENDRAT, M. 1929. Ein Beitrag zur Kenntnis des Säurestoffwechsels sukkulenter Pflanzen. *Planta* 7:508–584.

BERNARD, N. 1911. Sur la fonction fungicide des bulbes d'ophrydees. *Ann. Sci. Nat. Bot., Ser.* 9 14:221–234.

BIVINS, J. L. 1968. Effect of growth regulating substances on the size of flower and bloom date of *Cymbidium* Sicily "Grandee." *Am. Orchid Soc. Bull.* 37:385–387.

BIVINS, J. L., and KOFRANEK, A. 1965. Chemical weed control in *Cymbidium* beds and pots. *Am. Orchid Soc. Bull.* 34:397–398.

BIVINS, J. L., SACHS, R., and DEBIE, J. 1968. Penetration, transportation, and metabolism of C^{14} Simazine in *Cymbidium*. *Am. Orchid Soc. Bull.* 37:989–991.

BORRISS, H. 1967. Kohlenstoff–Assimilation und diurnaler Säurerhythmus epiphytischer Orchideen. *Die Orchidee* 7:396–406.

BREWER, K., GRADOWSKI, C., and MEYER, M. 1909. Effect of abscisic acid on *Cymbidium* orchid plants. *Am. Orchid Soc. Bull.* 38:591–592.

BRUBAKER, M. M. 1969. Fertilizer proportioners. *Am. Orchid Soc. Bull.* 38:774–776.

BURG, S., and DIJKMAN, M. 1967. Ethylene and auxin participation in pollen-induced fading of *Vanda* orchid blossoms. *Plant Phys.* 42:1648–1650.

BURGEFF, H. 1959. Mycorrhiza of orchids, in C. Withner, Ed., *The Orchids: A Scientific Survey.* Ronald Press, New York, pp. 361–395.

BURNETT, H. C. 1965. Orchid diseases. St. of Fla., Dept. of Agriculture.

———. 1971. Benomyl (Benlate), a new systemic fungicide found effective in controlling certain orchid diseases. *Am. Orchid Soc. Bull.* 40:325.

———. 1973. Controlling *Pythium* and *Phytophthora* black rots of orchids. *Am. Orchid Soc. Bull.* 42:326.

CAMPBELL, E. 1962. The mycorrhiza of *Gastrodia cunninghamii* Hook. *Trans. Royal Soc. New Zealand* 1:289–296.

———. 1963. *Gastrodia minor* Petrie, an epiparasite of Manuka. *Trans. Royal Soc. New Zealand* 2:73–81.

———. 1964. The fungal association in a colony of *Gastrodia sesamoides*. R. Br. *Trans. Royal Soc. New Zealand* 2:237–246.

CHIN, T. T. 1966. Effect of major nutrient deficiencies in *Dendrobium phalaenopsis* hybrids. *Am. Orchid Soc. Bull.* 35:549–554.

CHURCHILL, M., BALL, E., and ARDITTI, J. 1972. Tissue culture of orchids, II. Methods for root tips. *Am. Orchid Soc. Bull.* 41:726–730.

COOMBS, J. 1973. β-carboxylation, photorespiration, and photosynthetic carbon assimilation in plants. *Curr. Adv. Plant Sci.* 2:1–10; Commentaries in Plant Sci. No. 1.

CORBETT, M. K. 1967. Some distinguishing characteristics of the orchid strain of tobacco mosiac virus. *Phytopathology* 57:164–172.

CURTIS, J. T., and DUNCAN, R. 1947. Studies in the germination of orchid pollen. *Am. Orchid Soc. Bull.* 16:594–597, 616–619.

DAVIDSON, O. W. 1960. Principles of orchid nutrition. *Proc. Third World Orchid Conf.*, pp. 224–233.

———. 1963. Advances in orchid environmental control. *Proc. Fourth World Orchid Conf.*, pp. 147–157.

———. 1967a. Orchid ailments not caused by insects or diseases, I. *Am. Orchid Soc. Bull.* 36:464–475; 1967b. II. *ibid.*, 564–574.

DEVLIN, R. M., and BARKER, A. V. 1971. *Photosynthesis.* Van Nostrand Reinhold Co., New York.

DODMAN, R. L., BARKER, K. R., and WALKER, J. C. 1968a. Modes of penetration of different isolates of *Rhizoctonia solani*. *Phytopathology* 58:31–33.

———. 1968b. A detailed study of the different modes of penetration of *Rhizoctonia solani*. *Phytopathology* 58:1271–1272.

DODSON, C., DRESSLER, R., HILLS, H., ADAMS, R., and WILLIAMS, N. 1969. Biologically active compounds in orchid fragrances. *Science* 164(3885):1243–1249.

DODSON, C., and HILLS, H. 1966. Gas chromotography of orchid fragrances. *Am. Orchid Soc. Bull.* 35:720–725.

DOWNIE, D. G. 1949. The germination of *Goodyera repens* (L.) R. Br. in fungal extract. *Trans. Proc. Bot. Soc. Edinburgh* 35:120–125.

———. 1957. *Corticum solani*—an orchid endophyte. *Nature* 179:160.

DUEKER, J., and ARDITTI, J. 1968. Photosynthetic $^{14}CO_2$ fixation by green *Cymbidium* flowers. *Plant Physiol.* 43:130–132.

ERNST, R. 1967a. Effect of select organic nutrient additives on growth *in vitro* of *Phalaenopsis* seedlings. *Am. Orchid Soc. Bull.* 36:694–704.

———. 1967b. Effects of carbohydrate selection on the growth rate of freshly germinated *Phalaenopsis* and *Dendrobium* seed. *Am. Orchid Soc. Bull.* 36:1068–1072.

ERNST, R., ARDITTI, J., and HEALEY, P. 1970. The nutrition of orchid seedlings. *Am. Orchid Soc. Bull.* 39:599–605, 691–700.

———. 1971a. Biological effects of surfactants, I. Influence on the growth of orchid seedlings. *New Phytol.* 70:457–475.

———. 1971b. Carbohydrate physiology of orchid seedings, II. Hydrolysis and effects of oligosaccharides. *Am. J. Bot.* 58:827–835.

ESSER, G. 1973. Photosynthesemessung mit dem URAS. *Die Orchidee* 24:12–14.

FISCH, M., FLICK, B., and ARDITTI, J. 1973. Structure and antifungal activity of hircinol, loroglossol, and orchinol. *Phytochemistry* 12:437–441.

FISCH, M., SCHECHTER, Y., and ARDITTI, J. 1972. Orchids and the discovery of phyto-alexins. *Am. Orchid Soc. Bull.* 41:605–607.

FLENTJE, N. T., and STRETTON, H. M. 1964. Mechanism of variation in *Thanatephorus cucumeris* and *T. praticolus*. *Austral. J. Bio. Sci.* 17:686–704.

FORSYTH, W., and SIMMONDS, N. 1954. A survey of anthocyanins of some tropical plants. *Proc. Royal Soc. London, Ser. B* 142:549–564.

FREI, J. K., SR. 1973. Orchid ecology in a cloud forest in the mountains of Oaxaca, Mexico. *Am. Orchid Soc. Bull.* 42:307–314.

FREI, J. K., SR., and DODSON, C. 1973. The chemical effect of certain bark substrates on the germination and early growth of protocorms of epiphytic orchids. *Bull. Torrey Bot. Club.*

GARRETT, S. D. 1962. Decomposition of cellulose in soil by *Rhizoctonia solani*. *Trans. Brit. Mycol. Soc.* 45:115–120.

GASCOIGNE, R., RITCHIE, E., and WHITE, D. 1949. A survey of anthocyanins in the Australian flora. *J. Proc. Royal Soc. N.S.W.* 82:44–70.

GÄUMANN, E., and JAAG, O. 1945. Über induzierte Abwehrreaktionen bei Pflanzen. *Experientia* 1:21–22.

GÄUMANN, E., and KERN, H. 1959a. Über die Isolierung und die chemischen Nachweis des Orchinols. *Phytopath. Z.* 35:347–356.

——. 1959b. Über chemische Abwehrreaktionen bei Orchideen. *Phytopath. Z.* 36: 1–26.

GÄUMANN, E., NUESCH, J., and RIMPAU, R. 1960. Weitere Untersuchungen über die chemischen Abwehrreaktionen der Orchideen. *Phytopath. Z.* 38:274–308.

GESSNER, F. 1948. Stoffwanderungen in bestäubten Orchideenblüten. *Biol. Zentralbl.* 67:457–477.

GOOD, H. M., and JACKSON, R. 1966. Effects of some environmental factors on disease development. *Am. Orchid Soc. Bull.* 35:22–27.

——. 1965. Studies on a root rot of *Cattleya*, I. Isolation of the causal organism. *Am. Orchid Soc. Bull.* 35:715–718.

GRAEFLINGER, B. 1950. Repicagem precoce de orquideas sobre musceneas. *Orquidea* 12:131–134.

GRIPP, P. 1969. *Oxalis,* the orchid growers' pest: What to do about it. *Am. Orchid Soc. Bull.* 38:1079–1081.

HADLEY, G. 1969. Cellulose as a carbon source for orchid mycorrhiza. *New Phytol.* 68:933–939.

——. 1970. Nonspecificity of symbiotic infection in orchid mycorrhiza. *New Phytol.* 69:1015–1023.

HADLEY, G., and HARVAIS, G. 1968. The effect of certain growth substances on asymbiotic germination and development of *Orchis purpurella*. *New Phytol.* 67:441–445.

HADLEY, G., and PEROMBELOM, M. 1963. Production of pectic enzymes by *Rhizoctonia solani* and orchid endophytes. *Nature* 200:1337.

HADLEY, G., and WILLIAMSON, B. 1971. Analysis of the post-infection growth stimulus in orchid mycorrhiza. *New Phytol.* 70:445–455.

HAMADA, M., and NAKAMURA, S. 1963. Wurzelsymbiose von *Galeola altissima* Rechb. f. *Sci. Rept. Tohaku Univ., 4th Ser.* **29**:227–238.

HARDEGGER, E., SCHELLENBAUM, M., and CORRODI, H. 1963a. Über induzierte Abwehstoffe bei Orchideen, II. *Helv. Chim. Acta* **46**:1171–1180.

HARDEGGER, E., BILAND, H., and CORRODI, H. 1963b. Synthese von 2,4-dimethoxy-6-hydroxyphenanthren und Konstitution des Orchinols. *Helv. Chim. Acta* **45**:1354–1360.

HARDEGGER, E., RIGASSI, N., SERES, J., EGLI, C., MULLER, P., and FITZI, K. 1963c. Synthese von 2,4-dimethoxy-6-hydroxy-9,10-dihydroxyphenanthren. *Helv. Chim. Acta* **46**:2543–2551.

HARDLEY, G. 1970. The interaction of kinetin, auxin, and other factors in the development of north temperate orchids. *New Phytol.* **69**:549–555.

HARLEY, J. L. 1959. *The Biology of the Mycorrhiza.* Leonard Hill Books, London; Interscience Publishers, New York (Plant Science Monographs, ed. Nicholas Polunin), 233 pp.

HARN, C. 1970. Radiosensitivity of *Cymbidium* protocorms. *Am. Orchid Soc. Bull.* **39**: 499–505.

HARVAIS, G., and HADLEY, G. 1967a. The relationship between host and endophyte in orchid mycorrhiza. *New Phytol.* **66**:205–216.

———. 1967b. The development of *Orchis purpurella* in asymbiotic and inoculated cultures. *New Phytol.* **66**:217–230.

HATCH, M., SLACK, C., and JOHNSON, H. 1967. Further studies of a new pathway of photosynthesis. Carbon dioxide fixation in sugar cane and its occurrence in other plant species. *Biochem. J.* **102**:417–

HAYES, A. B. 1968. The morphological effects of pollination in the Brazilian miltonias. *Am. Orchid Soc. Bull.* **37**:705–707.

HEALEY, P., ERNST, R., and ARDITTI, J. 1971. Biological effects of surfactants, II. Influence on the ultrastructure of orchid seedlings. *New Phytol.* **70**:477–482.

HILLS, H., WILLIAMS, N., and DODSON, C. 1968. Identification of some orchid fragrance components. *Am. Orchid Soc. Bull.* **37**:967–971.

INTUWONG, O., and SAGAWA, Y. 1973. Clonal propagation of sarcanthine orchids by aseptic culture of inflorescences. *Am. Orchid Soc. Bull.* **42**:209–215.

ISRAEL, H. 1963. Production of *Dendrobium* seedlings by aseptic culture of excised ovaries. *Am. Orchid Soc. Bull.* **32**:441–443.

ITO, I. 1958. Culture of seedlings from cut flowers and its possibilities. *Japan Orchid Soc. Bull.* **4**:14–15.

———. 1960. Culture of orchid seedlings by way of completing the growth of ovaries of cut flowers. *Japan Orchid Soc. Bull.* **6**:4–7.

———. 1964. A device for culture *in vitro* of orchid ovary. *Japan Orchid Soc. Bull.* **10**: 18–19.

———. 1965. Ultra-low temperature storage of orchid pollinia and seeds. *Japan Orchid Soc. Bull.* **11**:4–15.

———. 1966. *In vitro* culture of ovary in orchids, I. Effects of sugar, peptone, and coconut milk upon the growth of ovary of *Dendrobium nobile*. *Sci. Rpts. Kyoto Pref. Univ., Agri.* **18**:38–50.

JEFFREY, D., and ARDITTI, J. 1969. The separation of sugars in orchid nectars by thin layer chromatography. *Bull. Am. Orchid Soc.* 38:866–867.

JEFFREY, D., ARDITTI, J., and KOOPOWITZ, H. 1970. Sugar content in floral and extra-floral exudates of orchids: pollination, myrmecology, and chemotaxonomy implication. *New Phytol.* 69:187–195.

KANO, K. 1968. Acceleration of the germination of so-called "hard-to-germinate" orchid seeds. *Am. Orchid Soc. Bull.* 37:690–698.

KARASAWA, K. 1964. Banana-honey media for seed germination and growth of orchids. *Japan Orchid Soc. Bull.* 10:20–24.

KERR, A. D. 1972. Look, Ma—No chlorophyll! *Am. Orchid Soc. Bull.* 41:756–788.

KHALIFAH, R. 1966. Gibberellin-like substances from the developing banana fruit. *Plant Physiol.* 41:771–773.

KLASS, C. S. 1964. The extraction and identification of free auxin from orchid pollinia. Master's Thesis, Brooklyn College.

KNAUFT, R. L., and ARDITTI, J. 1969. Partial identification of dark $^{14}CO_2$ fixation products in leaves of *Cattleya* (Orchidaceae). *New Phytol.* 68:657–661.

KOCK, W. E. 1972. Surprising success with a little-known fertilizer. *Am. Orchid Soc. Bull.* 41:338–340.

KOTOMORI, S., and MURASHIGE, T. 1965. Some aspects of aseptic propagation of orchids. *Am. Orchid Soc. Bull.* 34:484–489.

LAWRENCE, G. D., and ARDITTI, J. 1964. A new medium for the germination of orchid seeds. *Am. Orchid Soc. Bull.* 33:766–768.

LAWSON, R. 1967. Chemical inactivation of *Symbidium* mosaic and *Odontoglossum* ringspot viruses. *Am. Orchid Soc. Bull.* 36:998–1001.

———. 1970a. Flower necrosis in *Cattleya* orchids. *Am. Orchid Soc. Bull.* 39:306–312.

———. 1970b. Virus-induced color-breaking in *Cattleya* orchid flowers. *Am. Orchid Soc. Bull.* 39:395–400.

LUCKE, E. 1969. Biotin bei der Samenkeimung von *Paphiopedilum. Die Orchidee* 20:270–271.

———. 1971. The effect of biotin on sowings of *Paphiopedilum. Am. Orchid Soc. Bull.* 40:24–26.

MARIAT, F. 1952. Recherches sur la physiologie des embryons d'Orchidées. *Rev. Gén. Bot.* 59:324–377.

MATSUMOTO. K. 1966. Determination of the chlorophyll content of *Cymbidium* blooms. *Cymb. Soc. News.* 20:11–14.

McCORKLE, J., REILLY, L., and O'DELL, T. 1969. *Pseudomonas* infection of *Phalaenopsis. Am. Orchid Soc. Bull.* 38:1073–1078.

MEEYOT, W., and KAMEMOTO, H. 1969. Studies on storage of orchid pollen. *Am. Orchid Soc. Bull.* 38:388–393.

MIWA, A. 1937. Test of the germinating power of orchid pollen. *Orchid Rev.* 45:345–359.

MOREL, G. 1960. Producing virus-free *Cymbidium. Am. Orchid Soc. Bull.* 29:495–497.

———. 1965. Clonal propagation of orchids by meristem culture. *Cymb. Soc. News* 20:3–11.

————. 1971. The principles of clonal propagation of orchids. *Proc. Sixth World Orchid Conf.*, pp. 101–106.

MÜLLER, R. 1953. Zur quantitativen Bestimmung von Indolylessigsäure mittels Papier-chromatographie und Papierelektrophorese. *Beitr. Biol. Pflanzen* **30**:1–32.

MURASHIGE, T., KAMEMOTO, H., and SHEEHAN, T. 1967. Experiments on the seasonal flowering behavior of *Vanda* Miss Joaquim. *Am. Soc. Hort. Sci.* **91**:672–679.

MURASHIGE, T., SHEEHAN, T., and KAMEMOTO, H. 1963. Controlling weeds in orchids with herbicides. *Am. Orchid Soc. Bull.* **32**:521–526.

MUTSUURA, O., ITO, I., and NAKAHIRA, R. 1962. Studies on the germination and the development of seedlings of *Taeniophyllum aphyllum* (Makino) Makino. *Sci. Rep. Kyoto Pref. Univ. (Nat. Sci. and Liv. Sci.)* **3**:189–194.

NAKAMURA, S. 1962. Zur Samenkeimung einer chlorophyllfreien Erdorchidee *Galeola septentrionalis* Reich. f. *Zeitsch. Bot.* **50**:487–497.

NORBECOURT, P. 1923. Sur la production d'anticorps par les tubercles des ophrydees. *Compt. Rend. Acad. Sci., Paris* **177**:1055–1057.

NUERNBERGK, E. 1963. On the carbon dioxide metabolism of orchids and its ecological aspect. *Proc. Fourth World Orchid Conf.*, pp. 158–169.

OERTLI, J., and KOHL, H. 1960. Der Einfluss der Bestaeubung auf die Stoffbewe-gungen in *Cymbidium* Blüten. *Die Gartenbauwiss.* **25**:107–114.

PETERSON, J., DAVIS, S., and DAVIDSON, O. W., 1968. *Botrytis* flower spot control in orchids. *Am. Orchid Soc. Bull.* **37**:227–230.

POOLE, R. T., and SHEEHAN, T. 1967. Growth, amino acid and amide content of developing *Cattleya* Trimos fruit. *Am. Orchid Soc. Bull.* **36**:985–972.

————. 1973. Leaf tip die-back of *Cattleya*—What's the real cause? *Am. Orchid Soc. Bull.* **42**:227–230.

RAGHAVAN, W., and TORREY, J. 1963. Inorganic nitrogen nutrition of the embryos of the orchid *Cattleya*. *Am. J. Bot.* **50**:617 (abstract).

————. 1964. Inorganic nitrogen nutrition of the embryos of the orchid *Cattleya*. *Am. J. Bot.* **51**:264–274.

RAO, A. N. 1963. Organogenesis in callus cultures of orchid seeds, in Plant Tissue and Organ Culture—A Symposium, pp. 332–343.

RAO, A. N., and AVADHANI, P. 1963. Some aspects of *in vitro* culture of *Vanda* seeds. *Proc. Fourth World Orchid Conf.*, pp. 194–202.

RINGSTROM, S. 1968. The response of *Phalaenopsis mariae* to pollination. *Am. Orchid Soc. Bull.* **37**:512–

SAGAWA, Y., SHOJI, T., and SHOJI, T. 1967. Clonal propagation of dendrobiums through shoot meristem culture. *Am. Orchid Soc. Bull.* **36**:856–859.

SAGAWA, Y., and VALMAYOR, H. 1966. Embryo culture of orchids. *Proc. Sixth World Orchid Conf.*, pp. 99–101.

SANFORD, W. W., KRALLIS, A., XANTHAKIS, A., FOURAKIS, F., and KAPRI, K. 1964a. Anthocyanins in orchids. *Orchid Digest* **28**:362–367.

————. 1964b. Anthocyanins in orchids. *Orchid Digest* **28**:405–410.

SCHAFFSTEIN, G. 1941. Die Avitaminose der Orchideenkeimlinge. *Jahrb. Wiss. Bot.* **90**:141–198.

SESHAGIRIAH, K. 1941. Physiology of pollination in Orchidaceae. *Curr. Sci.* 10:30–32.

SHEEHAN, T. 1958. Orchid potting media. *Proc. Second World Orchid Conf.*, pp. 226–230.

——. 1960. Effects of nutrition and plotting media on growth and flowering of certain epiphytic orchids. *Proc. Third World Orchid Conf.*, pp. 211–218.

——. 1966. Fertilization of orchids. *Proc. Fifth World Orchid Conf.* pp. 95–97.

SMITH, S. 1966. (Formerly S. Harley.) Physiology and ecology of orchid mycorrhizal fungi with reference to seedling nutrition. *New Phytol.* 65:488–499.

——. 1967. Carbohydrate translocation in orchid mycorrhizas. *New Phytol.* 66:371–378.

STEFANSKI, T. 1965. A study on the anthocyanins in orchids. Master's thesis, Brooklyn College.

STEPHEN, R. C., and FUNG, K. K. 1971a. Nitrogen requirements of the fungal endophytes of *Arundina chinensis*. *Can. Jr. Bot.* 49:407–410.

——. 1971b. Vitamin requirements of the fungal endophytes of *Arundina chinensis*. *Can. J. Bot.* 49:411–415.

STORT, M. 1972a. Ovule development after pollination in *Eulophidium maculatum*. *Am. Orchid Soc. Bull.* 41:23–28.

——. 1972b. The pollen tubes of *Cyrtopodium* species. *Am. Orchid Soc. Bull.* 41:426–427.

TERIKATA, H., SAWA, Y., and SHISA, M. 1965. Non-symbiotic germination and growth of orchid seeds. *J. Jap. Soc. Hort. Sci.* 34:63–70.

THORNBERRY, H. H., and PHILLIPPE, M. R. 1964. Orchid disease: *Cattleya* blossom brown necrotic streak. *Plant Disease Reptr.* 48:936–940.

THORNBERRY, H. H., THOMPSON, M., IZADPANAH, K., and CANARES, P. 1968. Orchid viruses: Causality of *Cattleya* infectious blossom necrosis. *Phytopath. Z.* 62:305–310.

TSE, A., SMITH, R. J., and HACKETT, W. 1971. Adventitious shoot formation on *Phalaenopsis* nodes. *Am. Orchid Soc. Bull.* 40:807–810.

UEDA, H., and TORIKATA, H. 1969a. Organogenesis in meristem cultures of cymbidiums, II. Effects of growth substances on the organogenesis in dark culture. *J. Jap. Soc. Hort. Sci.* 38:188–193.

——. 1969b. Organogenesis in meristem cultures of cymbidiums, III. Histological study on the shoot formation from the rhizome tips of *Cymbidium goeringii* cultured *in vitro*. *J. Jap. Soc. Hort. Sci.* 38:262–266.

——. 1972. Effects of light and culture medium on adventitious root formation by cymbidiums in aseptic culture. *Am. Orchid Soc. Bull.* 41:322–327.

URATA, U., and IWANAGA, E. 1965. The use of Ito-type vials for starting vegetative propagation of *Phalaenopsis*. *Am. Orchid. Soc. Bull.* 34:410–413.

URECH, J., FECHTIG, B., NUESCH, J., and VISHER, E. 1963. Hircinol eine antifungisch wirksame Substanz aus Knollen von *Loroglossum hircinum* (L.) Rich. *Helv. Chim. Acta* 46:2758–2766.

VALMAYER, H., and SAGAWA, Y. 1967. Ovule culture in some orchids. *Am. Orchid Soc. Bull.* 36:766–769.

VEYRET, Y. 1969. La structure des semences des Orchidaceae et leur aptitude à la germination in vitro en cultures pures. *Travaux du Lab. de "La Jaysinia"* 3:89–98.

VOGEL, S. 1966. Scent organs of orchid flowers and their relation to insect pollination. *Proc. Fifth World Orchid Conf.*, pp. 254–259.

WARBURG, O. 1886. Über die Bedeutung der organischen Säuren für den Lebenprocess der Pflanzen (speziell der sog. Fettpflanzen) *Unters. Bot. Inst., Tubingen* 2:53–150.

WARCUP, J. H. 1971. Specificity of mycorrhizal association in some Australian terrestrial orchids. *New Phytol.* 70:41–46.

WEINHOLD, A., BOWMAN, T., and DODMAN, R. 1969. Virulence of *Rhizoctonia solani* as affected by nutrition of the pathogen. *Phytopathology* 59:1601–1605.

WEINHOLD, A., DODMAN, R., and BOWMAN, T. 1972. Influence of exogenous nutrition on virulence of *Rhizoctonia solani*. *Phytopathology* 62:278–281.

WENT, F. W. 1940. Sociologie der Epiphyten eines tropischen Urwaldes. *Ann. Jardin Bot. Buitenzorg* 50:1–98.

WHITLOW, C. 1970. Blue cattleyas—hopes, observations, and aspirations. *Orchid Digest* 34:292–295.

WILD-ALTAMIRANO, C. 1969. Enzymic activity during growth of *Vanilla* fruit, 1. Proteinase, glucosidise, peroxidase, and polyphenoloxidase. *J. Food Sci.* 34:235–238.

WILLIAMSON, B. 1970. Induced DNA synthesis in orchid mycorrhiza. *Planta* 96:347–354.

WILLIAMSON, B., and HADLEY, G. 1970. Penetration and infection of orchid protocorms by *Thanatephorus cucumeris* and other *Rhizoctonia solani* isolates. *Phytopathology* 60:1092–1096.

WITHNER, C. L. 1942. Nutrition experiments with orchid seedlings. *Am. Orchid Soc. Bull.* 11:112–114.

———. 1943. Ovule culture: A new method for starting orchid seedlings. *Am. Orchid Soc. Bull.* 11:261–263.

———. 1953. Orchid roots *in vitro*. Abstracts of papers. B.S.A. annual meeting.

———. 1959. Orchid physiology, in C. L. Withner, Ed., *The Orchids: a Scientific Survey.* Ronald Press, New York, pp. 315–360.

———. 1961. The importance of light for orchid growth. I. The intensity of light. *Orchidata* 1:6–8; reprinted *Am. Orchid Soc. Bull.* (1964) 33:218–220.

———. 1962. The importance of light for orchid growth, II. Use of light by the orchid. *Orchidata* 2:189–190; reprinted *Am. Orchid Soc. Bull.* (1964) 33:284–285.

———. 1964a. The importance of light for orchid growth, III. Phototropism and etiolation. *Orchidata* 3:230–232; reprinted *Am. Orchid Soc. Bull.* (1964) 33:372–373.

———. 1964b. The importance of light for orchid growth, IV. Phototoperiodism, hormones, and effects on flowering. *Orchidata* 4:11–13; reprinted *Am. Orchid Soc. Bull.* (1964) 33:579–581.

ZEIGLER, A., SHEEHAN, T., and POOLE, R. 1967. Influence of various media and photoperiod on growth and amino acid content of orchid seedlings. *Am. Orchid Soc. Bull.* 36:185–202.

4

Clonal Multiplication of Orchids

GEORGES M. MOREL

The length of time taken for clonal propagation by the conventional method of backbulb culture has been a serious drawback in the orchid industry. According to J. W. Blowers (1964), depending on the kind, it takes about ten years to cultivate from six to a dozen good-sized propagations. Up to now, it has been almost impossible to raise a clone of uniform plants of well-known characteristics for industrial cultivation, as is done with almost all other plants used for the cut flower market, such as roses, gladioli, or carnations.

The cut flower industry increasingly is programmed in such a manner that the right flower will be ready on the exact day it is needed. Only a pure line or clone can be used for that purpose, since only the individual of a pure line or clone will respond exactly in the same manner to the cultural treatment. It is only when working with such plants that the grower will be able to evaluate, months in advance, the exact yield of his greenhouse in white or red cattleyas, green or yellow cymbidiums, or odontoglossums, exactly like a crop of roses, tomatoes, or cucumbers.

All the cultivated orchids, being complex hybrids, are highly heterozygous, and there is very little hope of breeding pure lines out of them in the near future. On the other hand, the new technique of meristem tissue culture makes it possible to multiply unlimited numbers of any desirable clone exactly as is done for carnations, roses, or chrysanthemums. We shall, in this chapter, describe the principles of this new technique and the practical way to apply it to orchids.

The Problem of Bud Regeneration in Plants

Despite years of extensive studies, the general problem of regeneration in plants is not yet fully understood. By regeneration, we mean the phe-

169

nomenon by which adult cells dedifferentiate and form new meristems leading to new organs, such as roots and buds.

It is a fact of common observation that stems or leaves of a good number of species, when excised from the plant, will regenerate roots. In some cases, pieces of roots or leaves will also regenerate buds or stem primordia. These properties have been applied since time immemorial in horticultural practice.

This phenomenon is generally interpreted as the consequence of the disruption of correlation between the excised organ and the plant. For example, a new bud is never formed *de novo* on an attached leaf; the leaf has to be detached from the plant. The buds often observed on the attached leaves of many CRASSULACEAE, such as *Kalanchoe*, are not formed *de novo* from an adult cell but derive from a preexisting meristem localized on the margin of the leaf.

If the regeneration of roots from the pericycle of cuttings is a common phenomenon, that of bud regeneration is much less frequent.

Two different types of bud regeneration have been observed in two kinds of tissues (1) the epidermis and (2) the inner parenchyma.

Regeneration from the Epidermis. In the leaves of *Begonia, Peperomia, Saintpaulia,* and many LILIACAE, a single epidermal cell starts to divide. It forms a small clump of meristematic tissue which quickly organizes itself into a growing point differentiating leaf primordia. The only requirement for this bud induction is the excision of the leaf by which the correlations between the leaf and the plant are broken. Regeneration in begonia leaves has been well studied by Prevot (1939). He showed that this property is a hereditary characteristic which is dominant in the hybrids. It may be enhanced by some factors, like a low oxygen tension or a supply of an exogenous cytokinin, but so far it has been impossible to induce bud neoformation in the leaves of species which do not have this genetic property. The buds are localized on the basal part of the leaf and are more numerous along the main leaf nerves.

Regeneration from Inner Parenchyma. This kind of bud formation by the inner tissue of the stem has been extensively studied by tissue culture technique, mostly by Gautheret (1942, 1944) on *Cichorium intybus* and by Skoog, Miller, and other co-workers (1944, 1950, 1955, 1957) on tobacco. It occurs in all the parenchymatous cells of phloem, xylem, or pith; in all cases it is highly polarized. The apical end of the stem only regenerates bud primordia. Some 6 substituted purines, such as adenine or kinetin, strongly stimulate this regeneration. On the other hand, auxin is inhibitory even at a very low concentration, but since in tobacco this substance is necessary for cell division in any case, buds are formed only when a proper balance between exogeneous auxin and cytokinin is realized in the nutrient medium.

In carrot, Steward et al. (1958, 1963) showed that free cells derived from the cambium (when pieces of this tissue were cultivated in a liquid medium on a rotating device) form small nodules bounded by a cambiumlike spherical layer of growing cells. When transferred to a stationary agar medium containing coconut milk, they differentiate shoot-growing points and form embryolike structures from which plants develop. Adventive embryo formations in carrot tissue culture have been extensively studied by Halperin and Wetherel (1964, 1965, 1966).

This phenomenon seems to be widespread among umbellifers. It has also been observed in other species such as eggplant by Yamada et al. (1967), *Ranunculus sceleratus* by Konar and Nataraja (1965), and in *Macleya cordata* by Kohlenbach (1965).

In orchids, very few cases of bud regeneration have been observed, although two have been known for a long time. The first example is *Neottia nidus-avis,* described by Irmisch (1853) and also studied by Keller and Schlechter (1930, 1940). After flowering, the main axis of the rhizome decays, but the densely packed roots stay alive and regenerate a new stem primordium at their apical end. A histological study was recently done by Champagnat (1971). The other example is the leaves of *Malaxis paludosa,* which also regenerate bulbils near their apical end, even when still attached to the plant (Keller and Schlechter, 1930, 1940). Another case has been observed by Thomale (1957). Pieces from the bulb of *Orchis maculata,* aseptically cultivated on nutrient medium, soon regenerated stems and roots (Fig. 4-1). These three cases are very exceptional.

Some years ago Morel (1960) investigated the potentiality of the apical meristem and discovered that, in many species, this organ is able to regenerate new plants. This regeneration is of the epidermal type, analogous to the begonia leaf, but in orchids it is limited to the stem apical meristem. But before describing the processes of regeneration, we should first analyze the general structure of the apical meristem.

The Apical Meristem

Most vascular plants have an unlimited growth. Their stems begin in a small group of embryonic cells, which make new tissues and organs as long as the plant lives. These groups of embryonic cells constitute what

Fig. 4-1. Regeneration of roots and shoots occurring on a piece of tuber of *Orchis maculata*. (After Thomale.)

we call apical meristems or growing points. In higher plants the limited embryogenesis which occurs during the maturation of the seed is prolonged in the apical meristem during the plant's whole life. Not only the stem elongation but the entire ontogeny of the plant is due to the growth and differentiation of these organs.

We would like to point out here the dual character of the meristem. This organ is self-perpetuating and, at the same time, it ensures the growth of the stems and the differentiation of new leaves. The cells divide in such a way that part of their offspring remains in the meristem and perpetuates it when the other part forms the new differentiated organs.

In many plants, the growth of the apical meristem never ceases; their structure is called monopodial. In others, after a while the main growing point stops dividing, and it is then a lateral meristem which starts to make a new bud. This type of growth is called sympodial. Among orchids one finds both types of growth: *Phalaenopsis, Vanda, Angraecum,* and *Renanthera* are monopodial; *Odontoglossum, Dendrobium,* and *Cattleya* are sympodial.

Since the whole ontogenesis of the plant is determined by the development of the growing points, it is easy to understand why so many morphologists have studied their structure and function in the last 100 years or so. As a matter of fact, the various schools do not seem to agree yet on this structure. I shall not discuss here in detail the different points of view. I will only give a general idea of the interpretation of the apex.

The classical conception presented in most of the textbooks derives from the theory of Schmidt, published in 1924. In the apex of an angiosperm, Schmidt (1924) recognizes two regions, different by their cytology and their histology. In the middle part of the apex, one finds a group of isodiametric cells, disposed at random, with a large vacuole. This group of cells has been called the corpus. The corpus is surrounded by two or more parallel layers of small cells with smaller vacuoles and a denser cytoplasm. These layers have been called the tunica (Fig. 4-2).

The main difference in the histology of the tunica and the corpus is in the orientation of the mitoses. These are anticlinal in the tunica everywhere but in the leaf primordia. This means that the cell walls are formed at a right angle to the surface of the growing points. They are mainly periclinal in the corpus.

The number of layers of the tunica may vary, according to the species, from two up to eight or nine, and in each species it may also vary during the ontogeny of the plant and under the influence of the seasonal growth changes. Thus it is often difficult to draw a limit between the tunica and the corpus.

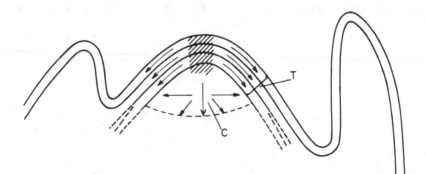

Fig. 4-2.   Apical meristem (according to Schmidt, 1924). T = tunica; C = corpus.

There are also marked cytological differences between these two regions of the apex: the vacuoles are always larger in the corpus cells and
the cytoplasm much less dense and chromophilic. Many people think that
the main difference in the tunica and corpus is of a cytological nature.

The study of periclinal chimaeras of *Datura* by Satina, Blakeslee, and
Avery (1940) seems to provide an experimental proof of Schmidt's
theory. These authors examined the apex of *Datura* after colchicine treatment, and they always found three concentric layers of cells. The chromosome numbers might be different in each of these layers, but it was
always uniform in each one. The conclusion of Satina and Blakeslee was
that each of these germ layers divides independently of the other and
produces a specific tissue; the first one produces the epidermis, the second the cortical parenchyma and the mesophyll, and the inner one the
pith parenchyma, and the vascular system.

Many other chimaeras have been studied since, and unfortunately the
case of *Datura* is unique. In cranberry, for example, Dermen (1947)
found that, after a while, the layers III and IV are more or less mixed,
so it has been impossible to attribute to each of these layers a specific
histogenic function. One cannot say, for example, that the mesophyll
always derived from layer II.

The tunica-corpus theory has been widely accepted by almost all
botanists for years. Nevertheless, the careful studies of Foster (1941)
showed that this concept was too rigid and could not be applied to all
the angiosperms. In monocotyledons, for example, one often finds periclinal mitoses in the tunica layers with consequent change in the number
of these layers. These observations led him to a wider interpretation of
the apex. Foster distinguishes (Fig. 4–3):

1. The apical initials localized at the top of the apex and producing
anticlinal as well as periclinal mitoses, the anticlinal mitoses giving rise to
the meristematic cells of the peripheral layers,

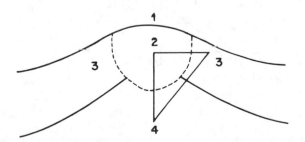

Fig. 4-3.   Interpretation of the apex of a gymnosperm (according to Foster). 1 =
apical initials; 2 = central mother cells; 3 = peripheral layers; 4 = rib meristem.

2. Immediately below, the central meristematic cells issued from the initial apical cells,

3. Around the central mother cells, the peripheral layers deriving from I and II,

4. The rib meristem.

Foster opposes the ideas of cells localized on the axis, larger, often vacuolized, as compared to the lateral cells, smaller and with a denser cytoplasm. For the first time, the notion of zonation of the apex, opposing a central zone to a lateral meristem, was introduced. The interpretation of Foster has been accepted by most morphologists for the gymnosperms and modified by Popham (1951) for the angiosperms.

After the theoretical consideration of Plantefol (1946) on the foliar helices, a new dynamic theory of the apex has been proposed by Buvat (1955). The study of the frequency of mitoses in the apex led this author to conclude that most of the mitoses occur during vegetative growth in a ring located inside the apical dome, which he called the initial ring. This ring might be considered as a lateral meristematic zone on the side of the meristematic dome comprising the old tunica, as well as the lateral portion of what used to be called the corpus, the central core being more or less inert as long as floral induction has not occurred. The concept of the apex, according to the theory of Buvat, might be summarized in the schema of Fig. 4-4.

One finds on the outside several surface layers where the mitoses are mostly, but not exclusively, periclinal, the number of parallel periclinal layers varying from one species to another and during ontogeny. In the central part, this "tunica" is reduced to a single layer of large cells dividing very rarely; these correspond to the initial apical of Foster. In these central parts, immediately below the initial apicals, one finds a zone of large vacuolated cells dividing also very rarely during vegetative growth. This central zone has been called by Buvat "méristème d'attente," or promeristem. It becomes functional only after floral induction.

This zone is surrounded by a ring of small cells with denser cytoplasm where most of the mitoses occur; this is the initial ring. These two areas are linked gradually by a transition zone. Below the central zone, one finds the long rows of parallel cells of the rib meristem that will produce the pith and, in the prolongation of the leaf primordia, the procambium precursor of the vascular system.

A great variety of structures may be found in the apices of the different orchids. Most of the time the zonation is less marked than in other plants, especially in *Cymbidium* (Fig. 4-5), where these different regions are difficult to recognize.

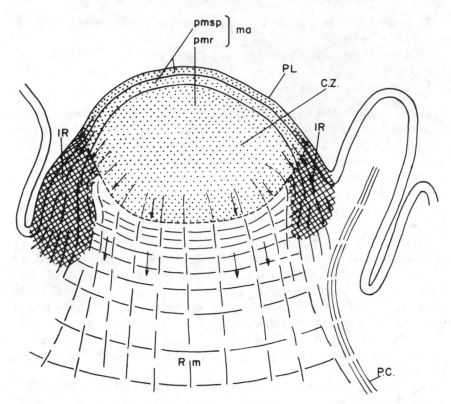

Fig. 4-4. Interpretation of the apical meristem of *Lupinus* (according to Buvat, 1955). ma = promeristem; IR = initial ring; PL = peripheral layers; C.Z. = central zone; Rm = rib meristem.

The Growth of the Apical Meristem Cultivated *in vitro*

Meristem Culture and Organogenesis. It has proved very time consuming and difficult to grow the apical meristem *in vitro* on a nutrient medium, and even now, despite numerous attempts, the meristems of only a limited number of species have been grown with success.

The first attempts made by Robbins (1922) with pea, cotton, or corn and by White in 1933 with *Stellaria media* were unsuccessful. White used very small explants of about .1 mm and grew them in hanging drops in the medium that he had just devised for root culture. In this medium the meristems stayed alive for several weeks. Some grew very little, making small calluses, but none of them differentiated into a normal stem. Ball

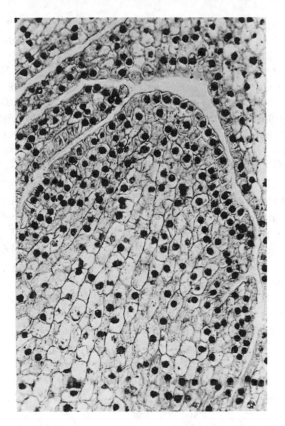

Fig. 4-5. Longitudinal section of an apical meristem of *Cymbidium*. The zonation is very little marked. (Photographed by M. Champagnat.)

(1946) made a new trial with stem tips of *Tropaeoleum* and *Lupinus*. The explants measured about 400 μ long. The results were very erratic. Only a small percentage of the explants was able to regenerate a normal plant.

The first positive results were obtained by Wetmore and Morel (1949) and Wetmore, (1954) with vascular cryptogams, such as *Adiantum*, *Osmunda*, or *Selaginella*. Growing points of these plants, from 100 to 150 μ long, grew well on a very simple medium: Knop solution with 3% surose. But this medium proved inadequate to support the growth of the meristem of higher plants. Morel, by 1960, was able, however, to grow *Cymbidium* meristems, and in 1963 Wimber grew clonal propagations from rotating liquid meristem cultures of *Cymbidium* apices, using a mineral medium with sucrose and tryptone supplements.

Working with dahlia and potatoes, Morel and Muller (1964b) found gibberellin to be an essential growth factor for these meristems. In the first experiments, they found that on a medium without gibberellin most of the explants stayed alive for months, made a callus or distorted stems looking like teratomas, but very rarely developed a normal stem. When gibberellic acid was added to the nutrient medium, the differentiation was perfectly normal but the growth was very weak and stopped after two or three months when the explants were still less than 1 cm long.

Apparently another factor was lacking. We found later (Morel and Muller, 1964a, 1965a, 1965b) that the other requirement was a high ionic content in potassium and ammonium. By adding 1 g per liter of potassium chloride and 1 g per liter of ammonium sulfate to the Knop solution, we had a perfectly normal growth. Other media with high potassium and ammonium content, such as the medium devised by Murashige and Skoog (1962) for optimal growth of tobacco callus, gave equally good results when .1 ppm of gibberellic acid was added. On such a medium, explants formed by the apical dome of the meristem of many herbaceous species, such as potato, dahlia, and chrysanthemum, will resume their growth at once and quickly form a perfectly normal stem.

On the other hand, explants of the same size taken from the meristem of most of the orchids do not grow. They stay alive for a while. The explant swells up by cell enlargement, but most of the time it dies. Only with *Cymbidium* and *Vanda* have we been able to raise plants from the apical dome and then very rarely.

In contrast, if instead of the apical dome a much larger explant is taken—a piece of .5 to 2 mm including at least two or three leaf primordia—a different and very striking phenomenon occurs (Fig. 4-6). First, the growth is not restricted to the apex as in buds which develop *in vivo*. Diffuse growth occurs by random cell divisions and mostly cell enlargement. Then, on the outer part, small spherical bulges are formed. They develop rhizoids at their bases and a few scales on top. They are morphologically identical to the protocorm (Morel, 1960, 1963a, 1964b).

Morphological Study of the Orchid Stem Apex.   We shall now briefly describe the type of development occurring in two very different genera: *Cymbidium* and *Cattleya,* all the other types observed being more or less intermediate.

With the *Cymbidium* (Champagnat et al., 1966) the explants generally used for these cultures consist of the apical dome with two or three leaf primordia and a small amount of pith parenchyma. They are cultivated

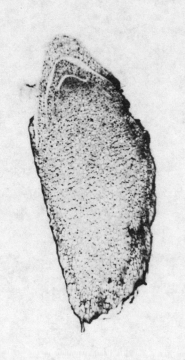

Fig. 4-6.   Section through an explant of *Cymbidium* ready to be cultivated. (Photograph by M. Champagnat.)

on a very simple medium such as Knudson. The development starts by a swelling of the part of the bud included between the two largest leaf primordia, and after a while it becomes impossible to recognize the apex itself. Sometimes this swelling occurs on one side only, giving an asymmetrical body. So the growth pattern *in vitro* is very different from the growth of the bud *in situ*. The apex stops making new leaf primordia on a short tuberized axis as it does on the plant when a new bulb is formed. Sometimes the apex disappears entirely; nevertheless, new organs are often formed: scales, very small, distorted leaves, numerous rhizoids.

A little later small areas of intense cell division can be recognized on the surface of the explant, either in the axis or on the abaxial part of the leaf primordia. At these points some epidermal cells start to divide. Mitosis can be localized at one point or may be more or less diffuse (Fig.

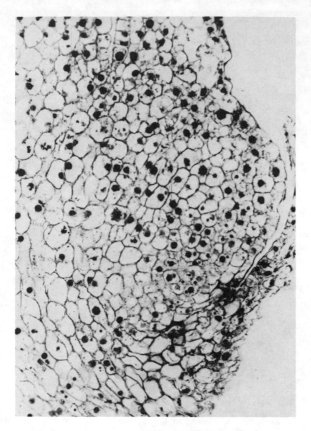

Fig. 4-7. Longitudinal section through an explant of *Cymbidium* after three weeks of cultivation. Notice the swelling at the abaxial bases of the leaf primordia. (Photograph by M. Champagnat.)

4-7). They are periclinal and give rise to long rows of cells producing the protocorm.

Protocorm is a term coined by the French botanist Bernard to designate a stage of development in the orchid embryo. The orchid seeds contain an embryo which is extremely minute, consisting of a few hundred cells only. Its development goes on when the seeds are shed under the influence of mycorrhizal fungi or when planted in a nutrient medium where the protocorm is formed. We can consider the new protocorms appearing on the bud in culture as adventive embroys analogous to the embryos formed in culture on various umbellifers or on *Ranunculus sceleratus* tissue.

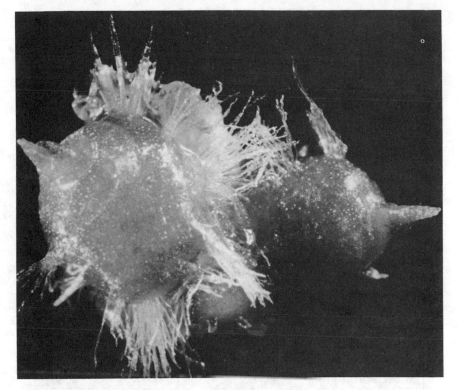

Fig. 4-8. Aggregate of several protocorms of *Cymbidium*. (Photograph by J. F. Muller.)

On the explant these protocorms appear as round, glistening swellings, white at first, then greenish, very often with long rhizoids on their equatorial planes. After a while, at the top of this mass of proliferating cells a new bud is regenerated. An aggregate of up to 10 mm protocorms can be formed from a single meristem (Fig. 4-8). They slowly develop into plantlets.

In very few instances, the original meristem, after a rest period of several weeks, starts growing again and makes new leaf primordia. But the phyllotaxy of these first leaves is always highly abnormal: instead of being distichous or at 180°, they are more often at a right angle between 90 and 180°. This abnormal development is always limited to the first leaves, the others being perfectly distichous. On the other hand, the development of a plantlet from a lateral protocorm not disturbed by excision is always normal.

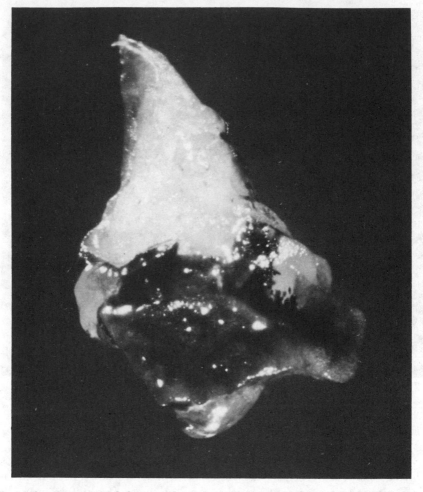

Fig. 4-9. Beginning of the growth on an explant of *Cattleya* after two months of culture. The swelling of the tissues at the base of the leaf primordia cracks the dark sheet of dead material. (Photograph by J. F. Muller.)

The *Cattleya* bud shows an entirely different type of development. The best results are obtained when fairly large buds up to 1.5 mm long are excised. They also require a more complex medium than *Cymbidium*. It has to be supplemented with an auxin and a cytokinin or coconut milk. On this medium the whole bud enlarges slowly. After two months of culture the swelling of the tissues at the base of the leaf primordia cracks the dark sheet of dead material and the newly formed green tissue appears (Fig. 4-9).

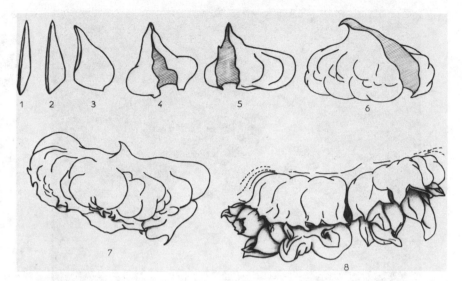

Fig. 4-10. Swelling and formation of new buds at the base of a *Cattleya* leaf primordium cultivated *in vitro*.

The reactions of the bud are quite complex (Champagnat and Morel, 1969). First, cell division occurs on the cut surface of the leaf primordia leading to a palisade tissue observed by Kuster during the last century. A few centers of meristematic cells appear at the same time, mostly near the vascular system.

The main reaction is from the leaf scales. Their bases swell and enlarge progressively and become corrugated and covered with various outgrowths (Fig. 4-10). A few typical protocorms with rhizoids can be formed on these leaf bases (Fig. 4-11), but most of the time numerous new meristems develop leading directly to buds (Fig. 4-12). Thus in *Cattleya* the cultivation leads to an intense production of buds at the base of the leaf scales. These buds are used for propagation. Rutkowski (1971) has also observed these differences between *Cymbidium* and *Cattleya*, and some of the stages are portrayed in color in his paper.

Leaf Culture

This abnormal organogenesis of the bud led us to investigate the potentialities of isolated leaves (Champagnat and Morel, 1969). For that purpose, entire leaves from 10 to 15 mm long are detached from seedling or plantlet and planted in a mineral solution such as Murashige and Skoog's containing 2% saccharose and 10^{-7} IAA plus 10^{-7} kinetin (isopentenyl-

Fig. 4-11. Development of *Cattleya*. One month later, the leaf primordia were taken apart and new protocorms with rhizoids were found on their bases. (Photograph by J. F. Muller.)

adenine). After two weeks a callus is formed on the cut surface; little by little differentiation occurs on this callus. One can observe two different types: numerous buds may appear on the callus as well as on the lamina (Fig. 4-13), but in most cases protocorms are formed on the callus and give rise to plantlets (Fig. 4-14). The same proliferation can occur on detached leaf primordia taken inside the bud of an adult plant, but this phenomenon is exceptional. Churchill, Ball, and Arditti (1970, 1971,

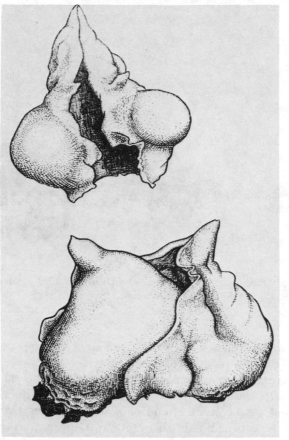

Fig. 4-12. Development of *Cattleya*. One can see the formation of numerous buds on the bases of leaf primordia.

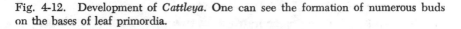

1973) showed that with *Cattleya* and also with *Epidendrum* not only the leaves but also the leaf tips can proliferate to form protocorms and plant-lets. Similar results have been found by Wimber (1965) in rotating liquid cultures.

Propagation of the Protocorms Cultivated *in vitro*

Cymbidium and Cattleya.   The clonal propagation of orchids is based on the phenomenon of regeneration of newly formed protocorms by pro-tocorm sections. When a new bud has been formed on a protocorm, its growth is absolutely identical to the growth of a seedling. The bud pro-

Fig. 4-13. Buds formed on the callus at the base of a *Cattleya* leaf and on the lamina after 20, 25, and 59 days of cultivation.

duces the first leaf. When it is about 1 cm long a root appears in the base of the bud and a new plant is regenerated.

Early experiments have shown that when the protocorm is sectioned into several parts and transferred into a new medium, it does not differentiate a bud but regenerates a clump of new protocorms (Morel, 1963b). It is thus possible to maintain the growth of the protocorm for an indefinite period of time and to increase the stock at a fantastic rate.

In the case of *Cymbidium*, for example, if each protocorm is sliced into four and each piece regenerates only two protocorms (and this is a minimum), the clone can be increased every month by a factor of eight. This means that it is possible to obtain more than one billion plants from a single bud in only nine months.

To determine the nature of these proliferations, adult protocorms 2.5–3 mm in diameter were sectioned in three or four slices in a plane at a right angle with the main axis. Then these slices were peeled by taking off a sheet of cells comprising the epidermis and three or four layers of subepidermal cells (Fig. 4-15 and 4-16). When the central core and the pieces of the epidermis were transferred into a new medium, only the epidermal fragment started to grow to make new protocorms. The central parenchyma remained alive a few months, but no cell divisions were observed.

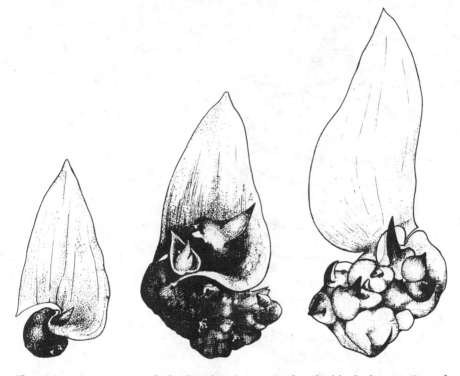

Fig. 4-14. Protocorms and plantlets found on a *Cattleya* leaf bud after 34, 52, and 67 days of cultivation.

One can question the reason of this failure to divide and differentiate. Nagl and Rucker (1972) and Nagl (1972) investigated the DNA content of the nuclei of the various cells of regenerated *Cymbidium* protocorms: only the epidermal stay normal. The others show very strong endoploidization and a disproportionate increase in nuclear DNA content due to an unscheduled increase of the DNA of some chromocenters. The DNA value may be as high as 1024 C. It is then easy to understand why such abnormal nuclei cannot divide.

An anatomical study showed that the cell divisions in these explants are limited to the outside layers. Cells generally start to divide in the subepidermal layer, sometimes below, but very rarely in the epidermis itself. These periclinal divisions give rise to a small aggregate of two to six cells appearing at random on the protocorm surface. These aggregates of small meristematic cells are very active, and in a few days small protocorms are formed.

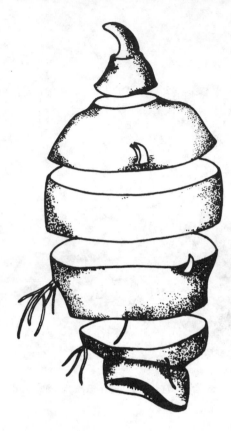

Fig. 4-15. Protocorm of *Cymbidium* sectioned in several slices.

The first differentiation to appear on these protocorms is the formation of a leaf primordium. Immediately afterward a procambial strand is differentiated in the tissue below this leaf primordium, but at that time the apex is not yet differentiated (Fig. 4-17).

Several leaf primordia, each with an axillary meristem and a procambial strand, are usually formed, but each one remains isolated and does not develop. After a while, one of these forms a growing point with leaves in a distichous phyllotaxy. Usually the others are then inhibited or start their development only when excised, but they may sometimes develop and then a large protocorm with several buds is formed.

The Oriental species of *Cymbidium,* such as *C. virescens* and *C. farreri* (studied by Champagnat, Morel, and Gambade, 1968), have a very unusual type of development. In these species, the embryo does not produce the usual protocorm but instead a long, branched rhizome. This rhizome

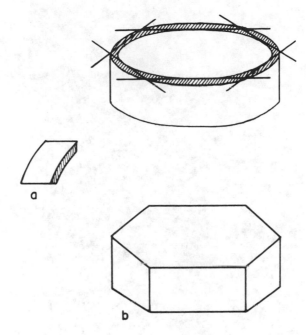

Fig. 4-16. Removal of the peripheral layers of a slice of *Cymbidium* protocorm from the central core. Only these peripheral layers are able to regenerate new protocorms. No cell division has been observed in the central core.

cultivated *in vitro* on Knudson C medium may become 5 or 6 cm long before differentiating a leaf bud (Fig. 4-18). The apical meristem of *C. virescens* or *C. farreri,* when excised and cultivated *in vitro,* develops in the same manner. After the usual swelling, a bud is formed from the epidermal cells. It elongates slowly and branches to produce a long, ramified rhizome, exactly like the one coming from the embryo. Ueda and Torikata (1972) found that certain vitamins and amino acids, as well as light, have an influence on this development.

The various clones of the *Cattleya* alliance can be increased just as fast as *Cymbidium.* Well-individualized protocorms are seldom formed, and the propagation is mostly made by the budding process that we have described on the scales of adult buds or on young leaf bases. When these buds are sectioned, their leaves or leaf primordia form calluses and regenerate new primordia leading to plantlets. Here again, the process can go on forever.

Other Genera.   In addition to *Cymbidium* and *Cattleya,* most of the

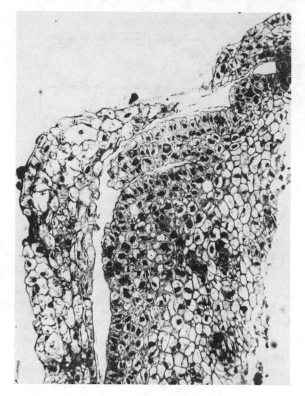

Fig. 4-17. Beginning of the differentiation of a slice of protocorm. Notice a new leaf primordium and the formation of a procambial layer. (Photograph by M. Champagnat.)

cultivated orchids have been propagated by the meristem culture technique. The development *in vitro* of *Miltonia* and of the *Odontoglossum* alliance, *Odontonia, Odontioda,* and *Vuylstekeara* is relatively simple. The explant makes a fairly large callus (Fig. 4-19), (Morel and Champagnat, 1969; Morel, 1970) on which numerous small protocorms are formed, each leading to a plantlet. The budding of the callus can go on for a long time. Thus hundreds of plantlets can be obtained from a single tip. The propagation of *Dendrobium* has been studied by Sagawa and Shoji (1967) and Kim et al.)1970).

For the monopodial orchids such as *Vanda* and *Phalaenopsis* studied by Morel (1970), Kunisaki et al. (1972), Vajrabhaya and Vajrabhaya (1970), and Teo et al. (1973), the meristems have a different structure. The buds in the leaf axils are entirely covered by circular scales like

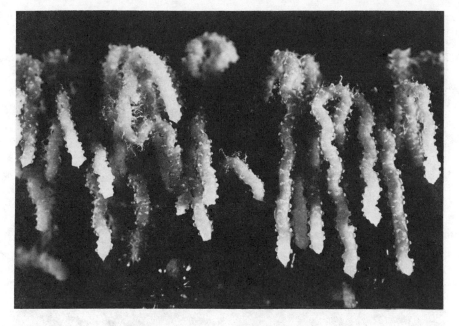

Fig. 4-18.   Long branched rhizome produced by the germination of a seed of *Cymbidium virescens*. (Photograph by J. F. Muller.)

onion scales. When the leaf is removed, only a pore making a hole through these scales is visible. The apex itself (Fig. 4-20) is very little individualized, a single-layered tunica overlying a vaguely defined meristematic corpus. The explant is removed by a circular incision into the cortex at this pore, going down to the vascular fibers.

In our experiments with terete vandas the growth of the explant is at first more or less diffuse. It enlarges slowly in many cases. The apex develops into a leafy shoot (Fig. 4-21), and the callus formation stops. This shoot has to be excised to promote further growth. After two to four months, protocorms start to differentiate on the callus (Fig. 4-22). They can be propagated like *Cattleya* or *Cymbidium* meristems.

In *Phalaenopsis* (Morel, 1960) we have used the buds at the base of the inflorescence. In some cases the buds, cultivated in the same conditions as the axillary bud of *Vanda*, develop in similar manner forming callus, protocorms, and plantlets (Fig. 4-23). But so far we have not been able to induce at will the protocorm formation.

Recently Intuwong and Sagawa (1973) found that sarcanthine orchids such as *Vascostylis*, *Neostylis*, and *Ascofinetia* can be propagated by the

Fig. 4-19. Large callus produced by a meristem of *Odontioda*. It differentiates numerous protocorms. (Photograph by J. F. Muller.)

culture of very young inflorescences, less than 1.5 cm long. When excised and planted in the mineral medium of Vacin and Went, supplemented with coconut milk, the hypodermal layer of the rachis forms clusters of meristematic cells developing into protocorms. In that case the production of protocorms is not restricted to the juvenile tissues surrounding the apex.

It seems to be the same with the tubers of various OPHRYDEAE studied by Champagnat and Morel (1972). Pieces of mature tuber parenchyma of various *Ophrys* taken in July during the rest period also make calluses on which new protocorms are formed (Fig. 4-24).

Thus far, only the members of the subfamily DIANDRAE, the CYPRIPEDI-OIDEAE, have given entirely negative results. We have attempted to grow

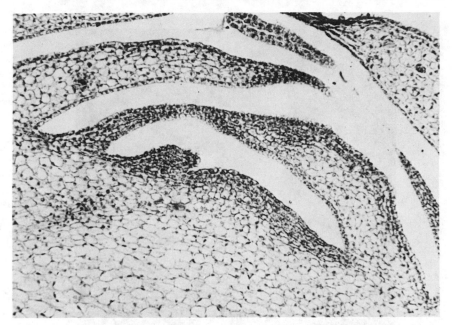

Fig. 4-20.   Longitudinal section of the apex of a lateral bud of a terete *Vanda*.

the apex of many *Paphiopedilum* hybrids. The apex, excised and cultivated on Thomale medium, develops in a perfectly normal manner producing new leaves and quickly forming a plantlet. To induce callus formation we added to the medium a strong auxin at high concentration
(2,4-D at 1 ppm). In these conditions, some of the explants made calluses (Fig. 4-25) and formed protocorms. The callus has been subcul-
tured on the same medium several times and when transferred into a
medium deprived of 2,4-D it formed plantlets. These preliminary experiments show that the techniques of meristem culture can also be applied
to *Paphiopedilum*.

CULTURE TECHNIQUES

Mineral Nutrition.   As we saw earlier, the apical meristems of most
higher plants have proved to be very difficult to grow *in vitro*. By contrast, we found that most of the orchid meristems are much less specific
and very tolerant as far as mineral nutrition is concerned.

When one looks through the literature on mineral nutrition of orchid

Fig. 4-21. Leafy shoot growing from the apex of a lateral bud of a terete *Vanda* after two months of cultivation. (Photograph by J. F. Muller.)

Fig. 4-22. *Vanda* callus differentiating protocorms. (Photograph by J. F. Muller.)

seedlings, one must come to the conclusion that they are very adaptable on a wide combination of various inorganic salts. Like the embryos, the orchid meristems seem to thrive on a great variety of media.

The old Knudson C solution gives very good results with many genera, such as *Cymbidium, Miltonia, Odontoglossum.* Nevertheless, the rate of success in growing *Cattleya, Dendrobium,* or *Vanda* on straight Knudson C is low. For these genera, especially the *Cattleya* and their hybrids, several different mineral solutions have been proposed.

Fig. 4-23. Protocorms of *Phalaenopsis* newly formed from the meristem of a lateral bud of the inflorescence. (Photograph by J. F. Muller.)

Major Elements. Raghavan and Torrey (1964a) found that the *Cattleya* embryos are unable to utilize the ion NO_3^- during germination and early stages of growth. During that period they require a source of ammonium nitrogen. However, after two months of culture on ammonium ion, they become progressively able to utilize NO_3^-. At that time, they have the first one or two well-formed leaves; earlier, however, as long as they are in the protocorm stage, the growth on NO_3^- is very poor.

Raghavan and Torrey (1964b) showed that this ability to use the NO_3^- ion is parallel to the formation of nitrate reductase in the seedling tissues. At a certain stage of morphological development, with the ap-

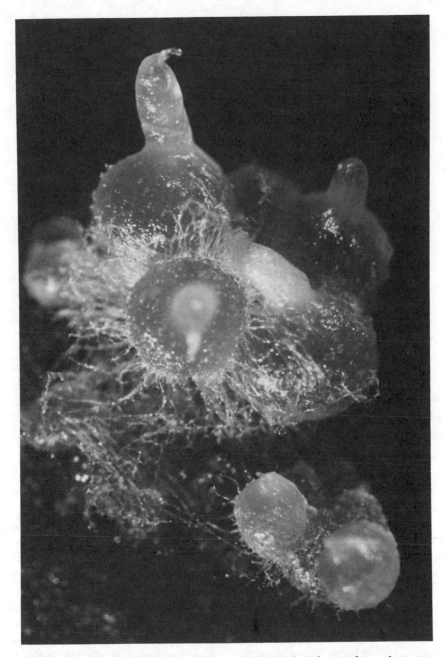

Fig. 4-24.   Protocorms formed on a piece of tuber of *Ophrys* cultivated *in vitro*. (Photograph by J. F. Muller.)

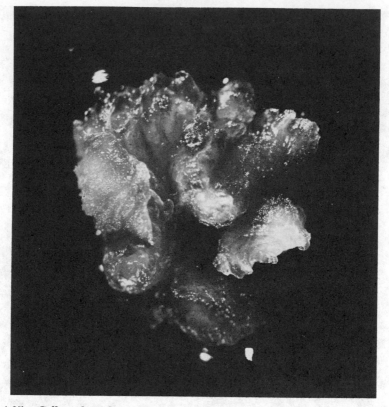

Fig. 4-25. Callus of *Paphiopedilum* produced by the cultivation of the apex on a medium with high auxin content. (Photograph by J. F. Muller.)

pearance of the first leaves, the seedlings acquire the ability to synthesize nitrate reductase in the presence of nitrate as a substrate, but they are unable to do so when younger.

We did not investigate the nitrate reductase activity of the *Cattleya* meristems or of the protocorms of regeneration, but very early in these studies we noticed that the growth is much better when ammonium is used as a nitrogen source.

Our studies on the cultivation of the apical meristem of various plants—potato, dahlia, and chrysanthemum—indicate a very high potassium requirement for these organs. In the case of *Cattleya* we also noticed a much better growth on media with high potassium content than on Knudson C. Finally, using Knudson C medium as a starting point, we had fairly good results by lowering the calcium nitrate to 500 mg/l instead of using 1 g, adding 500 mg/l of ammonium sulfate or ammonium nitrate and 500 mg/l of potassium chloride.

Other media giving equally good results have been proposed by other workers. They also provide the nitrogen in reduced forms, such as NH_4^+ or urea and have a fairly high content in K^+ (Reinert and Mohr, 1967; Lindemann, 1967; Scully, 1967).

One of the difficulties encountered by the early workers in the field of mineral nutrition was a very rapid drop in pH when an ammonium salt was used, the NH_4 ion being absorbed and metabolized much faster than the corresponding cation. Urea does not present this inconvenience and has been shown to be a good source of nitrogen. Another way to avoid this inconvenience is to buffer the solution. Burgeff (1936) proposed a Sörensen mixture of KH_2PO_4, or a McIlvaine combination of KH_2PO_4 and citrate.

We have found that citrate is highly beneficial, but its action might also be explained if one remembers that it may function as a chelating agent and provide a better use of iron.

Minor Elements. Following Knudson, most orchid growers paid very little attention to the minor elements other than iron and manganese. In most cases, the other ions such as zinc, boron, and molybdenum are brought with the inoculum (seeds) and as impurities, mostly from the agar. The seedlings are never kept aseptically long enough to show a deficiency in these minor elements.

The situation is quite different when one propagates the same clone for very long periods, and we noticed, after two years of culture of *Cymbidium* protocorms transferred every month into a new medium, some necrosis and aberrant development of the primary leaves. These symptoms disappeared when the explants were transferred into a medium containing all the minor elements. Several solutions have been prepared for tissue culture. We have always used the one devised by Heller (1953) (see recipes).

Vitamins and Growth Regulators. It is the discovery of growth regulators and vitamins which made possible the success of plant tissue culture.

Vitamins. Isolated roots of many species, such as tomato, do not grow without added thiamine. In many cases thiamine is the only essential growth factor, but other species require pyridoxine. Inositol is essential for the growth of many plant tissues, such as tobacco.

It has become a common practice to add to the medium used for tissue culture natural products or a vitamin mixture. Various solutions have been proposed (Gautheret, 1959) and are routinely added to the medium.

Auxins. Auxins are essential factors for cell division. Most plant tissues do not grow *in vitro* without auxin. Beside their effects on cell division, they induce root formation and inhibit development or interfere with the morphogenesis, producing abnormal organs, calluses, or teratomas. The most commonly used are indole acetic acid (IAA), indole butyric acid (IBA), naphtalene acetic acid (NAA), and 2,4-dichlorophenoxyacetic acid (2,4-D).

In 1945, Meyer and in 1962 Boesmann again pointed out the beneficial effect of IAA on *Cattleya* seedlings. At the beginning of our work on *Cattleya* meristems, we also noticed the stimulating effect of auxin: IAA, NAA, and IBA seem equally active. When used at a concentration of 1 or 2 ppm, they improve the rate of growth, and the percentage of surviving meristems is much higher. They do not seem to be essential growth factors, since it is possible to grow meristems without exogenous auxin in the media, but in that case the amount of explants developing protocorms might drop below 5%. Thus it has been a common practice to use auxin for the excision of *Cattleya*. IAA or IBA, more effective on root formation, or NAA is added to the media at a concentration of about 1 ppm. Reinert and Mohr recommend a mixture of 1.75 ppm of NAA and 1.75 ppm of IBA; Lindemann: 1 μM NAA (approximately 2 ppm). We commonly use a mixture of NAA and IBA at .5 ppm.

Cytokinins. Before the discovery of cytokinins by Miller, Skoog, et al. (1955), the cultivation of the tissues of many species, like tobacco and soybean, was possible only by adding to the media complex factors of unknown composition, such as coconut milk, yeast extract, and corn endosperm. The main compound active on cell division in these natural products is a cytokinin, namely zeatin or isopentenyladenine (IPA), but in addition they contain many substances beneficial to growth, including cyclitols (as shown by Steward), vitamins, and amino acids.

The natural cytokinins, zeatin or IPA, are difficult to obtain, so synthetic analogs are generally used. These include kinetin (6-furfurylaminopurine) and benzylaminopurine (BAP), which is about 10 times more active.

Beside being an essential factor for cell division, cytokinins have a strong effect on differentiation, that is, bud formation. Skoog and Miller (1971) discovered that in tobacco, in order to promote cell division or cell differentiation, cytokinins have to be associated with an auxin. Skoog showed that each phenomenon can be induced only with a proper balance of each regulator. When a higher concentration is used, the cytokinin concentration must also be higher. In tobacco, for example, a certain ratio of auxin/kinetin induces only cell division. With a higher ratio, buds are formed. These results have been confirmed by many authors on very different species.

In the case of orchids, the juvenile tissues of the buds of many species such as *Cymbidium* and *Miltonia* will form meristems and then regenerate plants without added cytokinins. However, even in that case, the effects of cytokinins are well marked. On *Cymbidium*, Fonnesbech (1972) found that kinetin at 100 μM induced growth of many small shoots and promoted callus formation and fresh weight increase. We have ourselves noticed the same effect on *Odontonia* meristems, the best concentration being 1 mg/l kinetin along with 1 mg/l IAA. On *Cattleya*, Lindemann (1967) found that the best concentration of kinetin was 5 μM (about 1 ppm) associated with 0.5 μM NAA (1.75 ppm). In our experience we had better results with IPA than with kinetin. The best concentration being 1 ppm with 1 ppm of NAA.

Pierik and Steegmans (1972) investigated the effects of 6-benzylaminopurine (BA) on the protocorms of *Cattleya aurantiaca*. At low concentrations of 10^{-8} and 10^{-7} the seedlings do not produce adventive plantlets or protocorms, and the dry and fresh weight decrease. Higher BA concentrations of 10^{-6} and 10^{-5} have an opposite effect. The number of plantlets and protocorms from a single explant increases strongly as do the fresh and dry weight, but the root formation is inhibited.

On the other hand, Reinert and Mohr (1967) observed that in their conditions the addition of kinetin at 1 ppm to the medium used for the excision of *Cattleya* meristems resulted in the browning and poor growth of the explants. It is only when new protocorms were formed that kinetin could be used.

It is indeed very difficult to compare the results obtained by different authors, since the *Cattleya* alliance is very complex. Some hybrids like *Potinara* combine as many as four different genera: *Cattleya*, *Laelia*, *Brassavola*, and *Sophronitis*. It is not surprising that big differences are encountered in the growth of the primary explant and, further, in the reaction of the protocorm to various growth regulators.

Gibberellins. Gibberellins (GA) are a third group of regulators. Their effects are mainly on stem elongation. The effect on the growth of the meristems of various dicots is very striking. Meristems of plants like potato or chrysanthemum will not develop without GA.

On the other hand, on orchid meristems an exogenous supply of GA seems to have a deleterious effect, resulting in very thin, threadlike, and chlorotic stems. Nevertheless, Lindemann (1967) used it at a concentration of 0.5 μM associated with 1 μM NAA and also kinetin for *Cattleya*.

Other Additives

Miscellaneous Additives.   Orchid growers add to the orchid media an extraordinary variety of ingredients either for seed germination or for the further growth of the plantlets. Since Noël Bernard introduced salep—an extract of *Ophrys* bulbs—compounds as different as peptone, banana, apple, tomato or pineapple juice, coconut milk, and fish emulsion have been employed (see Vitamins above, see also Chapter 3 of this book).

Some have a very striking effect, inducing growth at a fantastic rate. Each grower seems to have his own mixture, generally kept secret. Banana, pineapple juice, and fish emulsion seem to be the most popular.

The effect is real but most difficult to study, due to the chemical complexity of these ingredients, and we shall examine only those that have been mentioned in the scientific literature.

Peptone.   This compound should really be studied under nitrogen nutrition, since it has been used mainly as a nitrogen source for the culture of microorganisms. It is the dried product of pancreatic digestion of proteins, usually meat, but is not defined. Beside amino acids, it is a mixture of various peptides which, in addition to their effect as a nitrogen source, might have a growth-promoting action.

It has been occasionally used in plant tissue culture. Kandler (1950) found it necessary for the growth of sunflower tissue. Following Lami (1927), who found that peptone increases the germination of *Vanda* and *Phalaenopsis*, many investigators have used peptone to grow orchid embryos. Mariat (1952), Curtis (1947), and Withner (1955) reported beneficial results from peptone on various species.

In our experiments, we found a definite stimulation of the growth of *Cattleya, Dendrobium,* and *Vanda* by adding .1–.2% of peptone to the medium. We made a determination of the free amino acid in the sample used, but we could not duplicate the effect of peptone by a mixture of the amino acids at the same concentration. Raghavan and Torrey (1964a, 1964b), who investigated the effect of various amino acids on the growth of *Cattleya* seedlings, showed that only those related to the ornithine cycle, such as arginine, ornithine, and also urea, were able to induce rapid growth of the seedlings and to replace NH_4NO_3. These two observations lead us to conclude that the stimulating effect of peptone must be due to some peptide or other unknown constituent.

Coconut Milk.   Coconut milk was introduced in tissue culture by Van Overbeek et al. in 1941. Trying to grow *in vitro* very young embryos of

Datura still at the globular stage, they had the idea of giving them a natural fluid which normally feeds the embryo. They were able, after excising *Datura* embroys only a few days old, to grow them on a medium with coconut milk and raise the plants to maturity. Without coconut milk, the embryos gave only calluses, stopped growing, and died.

Caplin and Steward (1948) discovered then that this product exhibits a fantastic effect on the promotion of cell division of carrot. In 15 days the increase of the explant was 350%, whereas during this time the controls on IAA increased only 20%. Morel and Wetmore (1951) showed that coconut milk is necessary for the growth of the tissues of certain monocotyledons, like *Amorphophallus*.

Several investigators observed stimulating effects of coconut milk on the germination of various orchid seedlings, but it is difficult to evaluate properly the effects obtained in comparing results when various species are used.

The effect of coconut milk on the growth of apical meristems of *Cattleya, Dendrobium,* or *Vanda* is very striking. It definitely promotes the division of epidermal cells, leading to the formation of the protocorms. These are regenerated much faster, and the percentage of explants which do not grow and become necrotic is much lower.

The optimum concentration is between 10 and 15%. It may be autoclaved without loss of activity (Morel, 1965b; Lindemann, 1967; Kunisaki et al., 1972; Kim et al., 1970; Intuwong and Sagawa, 1973). The milk is taken out of green nuts that have reached their full size. It is filtered through paper and stored frozen.

Among the many other ingredients proposed for embryo cultures, two have been used for growing the orchid apices: green banana and pineapple juice. Although these two products have a very striking effect on the speed of growth of seedlings, we found them rather toxic on the early development of the apex. Even coconut milk can be toxic at this stage (Reinert and Mohr, 1967). These materials should not be used before the first transfer of a protocorm until it is big enough to be sectioned into three or four pieces. Then their effect is very beneficial, and on such media the speed of growth of the plantlet might be five to ten times faster than on Knudson C.

Media

We recommend the use of a different medium for the excision and for the subsequent growth of the protocorm. Also, different media should be used for the genera that are easy to grow, such as *Cymbidium, Miltonia,* and *Odontoglossum,* and for the more delicate, such as *Cattleya, Dendrobium,* and *Vanda.*

A.   For the excision: Knudson C medium.

B.   For the propagation of the protocorm: modified potato meristem medium (Morel and Muller, 1964).

Double distilled water	1 liter
Ammonium sulfate $(NH_4)_2SO_4$	1 g
Calcium nitrate $Ca(NO_3)_2 \cdot 4H_2O$	0.5 g
Potassium chloride KCl	1 g
Magnesium sulfate $MgSO_4 \cdot 7H_2O$	0.125 g
Potassium dihydrogen phosphate KH_2PO_4	0.125 g
Sucrose	20 g
Heller solution for minor elements	1 ml
Agar	6 g
Green banana	ca. 40 g
Homogenize in a blender	

C.   For the excision: use liquid medium on a culture rotator.

Double distilled water	1 liter
Ammonium sulfate $(NH_4)_2SO_4$	1 g
Potassium chloride KCl	1 g
Magnesium sulfate $MgSO_4 \cdot 7H_2O$	0.125 g
Calcium nitrate $Ca(NO_3)_2 \cdot 4H_2O$	0.5 g
Potassium dihydrogen phosphate KH_2PO_4	0.125 g
Urea	0.5 g
Citric acid	0.125 g
Naphthalene acetic acid NAA	1 mg
Sucrose	20 g
Coconut milk	100 ml
Heller solution for minor elements	1 ml

D.   For protocorm multiplication: the same as the liquid medium recipe above, but solidified by 8 g of agar.

E.   For raising the plantlets: one can use the same solidified medium, but better results are obtained when the coconut milk is replaced with 40 g green banana, 200 ml pineapple juice (unsweetened), and 0.5 g fish emulsion.

In all media the pH is adjusted to between 5 and 5.5 with concentrated ammonium hydroxide.

Heller solution for micronutrients consists of two solutions added at the rate of 1 ml each per liter of nutrient media.

Solution A:
Double distilled water	1 liter

Zinc sulfate $ZnSO_4$ 1 g
Boric acid H_3BO_3 1 g
Manganese sulfate $MnSO_4 \cdot 4H_2O$ 0.1 g
  (or with $\cdot 1H_2O$, .075 g)
Copper sulfate $CuSO_4 \cdot 5H_2O$ 0.03 g
Aluminum chloride $AlCl_3$ 0.03 g
Nickel chloride $NiCl_2 \cdot 6H_2O$ 0.03 g
Potassium iodide KI   0.01 g
Solution B:
  Double distilled water   1 g
  Ferric chloride $FeCl_3 \cdot 6H_2O$ 1 liter

The Excision of the Apical Meristem

The technique of excision varies with each genus, so after considering basic practices we shall discuss three different examples: *Cymbidium, Cattleya, Vanda.*

Sterilization. Although the sterilization technique is precise, it should not present any major difficulty for the orchid grower used to handling aseptic seedlings. For the excision of the meristem, one has to work under a dissecting microscope, and this can be done in any clean room. The explants are so small that the chances of contamination by airborne spores are very small. An aseptic room or boxes are necessary only for the subsequent transfers.

It is a good practice to wash first the table and the stage of the microscope with ethanol; then one places near the microscope a sheet of sterile paper towel, a beaker of ethanol, and several beakers of sterile water covered with aluminum foil. The dissecting tools are sterilized by dipping in ethanol, then rinsed with sterile water, and kept under a paper towel. During the course of the excision, each time that they are used they are sterilized and dried in this way, since it is impossible to flame the razor blades that we use without taking the temper out.

For the buds, we always use as a sterilizing agent a solution of calcium hypochlorite at 60 g/l. The suspension has to be stirred 10–15 min, then filtered. One can also use commercial sodium hypochlorite (Clorox) in the same way as for the seeds. It is a good practice to dip the buds first into ethanol to wet the waxy surface entirely before putting them into hypochlorite. Usually, 20 minutes of immersion are enough. They are rinsed with sterile water and dried under the sterile paper towel.

Tools.   We found watchmakers' tools very handy for the excision. We use the very thin Swiss watchmaker tweezers and holders for thin needles. We make microscalpels by breaking small pieces of razor blades with large tweezers and holding them with the tools called pin vises.

The Plant Material.   Generally, it is better to start from an actively growing shoot. The risks of contamination are reduced and the growth is better. One should remember that in sympodial plants the terminal meristem stops dividing in the early stage of shoot formation and is no longer functional; it can be used in very young shoots only. In most cases, the results are better with lateral buds still dormant.

With *Cattleya, Miltonia,* and *Odontoglossum* (Fig. 4-26) it is best to use the lateral bud of a new leaf, 4–5 in. long. This is not necessary with *Cymbidium.* This plant has such power of regeneration that the dormant buds of backbulbs, even several years old, may be used with success. Of course, it is easier and safer to start with a green bulb, but when you do not want to disturb your plant, just take a backbulb. In that case, if the eyes are too small and little developed, it is a good practice to keep the bulb some time in damp peat moss in a polyethylene bag to let it swell before the excision. According to the variety, one finds from two to four good-sized eyes in a *Cattleya* bulb and from two to six in a *Cymbidium* bulb.

With *Vanda,* beside the terminal bud, one can use the meristematic areas found on the stem in the leaf axils. Normally, these meristematic areas will, according to the condition, give a flower stem, or even a new shoot in some cases when the main bud is removed.

Cymbidium.   First all the leaves, as well as the roots and the necrotic areas, are carefully separated from the bulb with a sharp knife. Then all the tissues situated around and between the eyes are removed to permit a free access to the sterilizing agent. It is more convenient not to remove the eyes because once free they are more difficult to hold under the dissecting microscope. After the whole bulb is dipped in ethanol, it is put in a calcium hypochlorite solution at 60 g/1 20 min and rinsed with sterile water. It is then ready for the dissection. One starts to take out the upper part by sectioning it at midlength (Fig. 4-27); the outer scales are split with a scalpel by longitudinal cuts made on each side of the bud, taking care not to damage the apex. These scales are thus more easily removed. For that, we use the thin watchmaker tweezers. One by one, all the scales and most of the leaf primordia are taken out. For the smallest leaf pri-

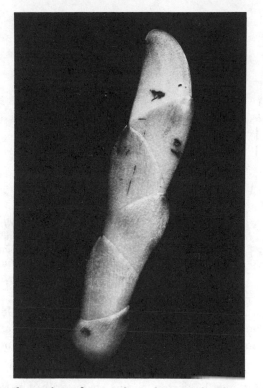

Fig. 4-26. Young shoot of *Cattleya* at the right stage for the excision of the meristems of lateral buds. (Photograph by J. F. Muller.)

mordia, a needle is very convenient. In early experiments we exposed the apex by removing all leaf primordia, but this is not necessary; doing so, one might damage the apical dome. The very small explant thus obtained is delicate and has to establish itself, so now we usually leave the two first primordia.

The excision of the explant is done by making, with the razor blade, four cuts at right angles around the meristem and one just below the insertion of the first leaf primordium, about at the procambium level. These operations are easier to do than to describe, and with a little practice on valueless plants, anyone can learn to perform them quickly.

The explant is then immediately planted on the nutrient medium in a Kahn test tube (7 mm × 7 cm). These small tubes are very convenient for such tiny pieces. Not only do they take less space, but when working

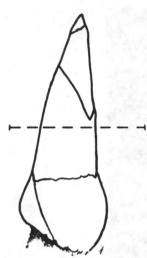

Fig. 4-27.   Lateral bud of *Cymbidium* ready for the excision of the apical meristem. The upper part will be cut away at the level of the dashed line, then the scales will be split longitudinally and removed.

in an open room the chances of contamination are much lower than with large vials. In the case of *Cymbidium,* the growth is just as good on agar medium as on liquid medium, and we prefer the solid medium because it is much easier to handle.

Cattleya.   Cattleyas have four or five scales and one or two leaves. It is a very simple matter to remove them from the leader to expose the eyes. This is done by splitting the scales opposite the eye with a scalpel. The eyes are so well protected that they are generally germ-free. Nevertheless, the whole shoot is sterilized in the usual way before the excision. *Cattleya* buds are usually bigger than those of *Cymbidium.* The scales are more succulent and much thicker; accordingly, they are more difficult to remove without injury. The excision is done essentially in the same way as in *Cymbidium.* First, the upper part of the scales is cut out (Fig. 4-28), then each one is removed with tweezers. It might be necessary to cut the base of each scale to remove it properly. Since in the case of *Cattleya* the mortality rate of the small explants is very high, we never remove all the scales. The explant is not made from the meristem only but is really a whole bud from .5 to 1.5 mm long. As we pointed out earlier, the leaf primordia are able to grow and to regenerate protocorms. This is another reason for not removing them all. One should section the explant just below the leaf primordia, since the rib meristem and pith tissue do not proliferate and may become necrotic.

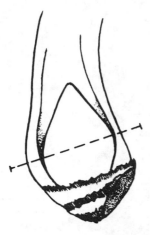

Fig. 4-28.   Bud of *Cattleya* exposed after removing the scales. The explant will be taken inside after sectioning the upper part of the scales along the dashed line. Their basal parts will be split longitudinally to make them easier to remove.

In the case of *Cattleya*, it is possible to grow the explant on solid medium, but the results are much better when one uses a liquid medium. With that method, the explants are put into 16-mm test tubes with a few milliliters of medium so that the solution is spread on a very large surface (5–8 mm long) and gets plenty of aeration. The tubes are put on a roller such as those used for animal tissue cultures (Fig. 4-29). The rotation speed should not exceed 2 rpm.

The tissues of many varieties get brown very quickly on the cut surfaces. This browning is due to the oxidation of phenolic material by the tyrosinase, which is very abundant in these varieties. To avoid oxidation, it is a good practice to do the excision in a Petri dish under sterile water. Coconut milk seems to inhibit this browning to some extent, so it has also been recommended that the excision be done in the same manner but under sterilized coconut milk (Lindemann, 1967). The oxidation products seem to be toxic; they diffuse into the medium and discolor it. The only practical way to get around this inconvenience is to transfer the explant every two or three days, as long as the cut surfaces are not healed.

Vanda.   This plant, being monopodial, has only one active meristem, but in the leaf axils there are undifferentiated meristematic areas that normally produce the inflorescences (Fig. 4-30).

The anatomical structure of these lateral meristems has not yet been studied, but they can be used for propagation purposes. They are not protected from contamination by the leaves, and the area is always dirty. When the leaves are removed, the lateral meristems are exposed, and it is

Fig. 4-29. Roller used for the culture of the meristems on liquid media.

Fig. 4-30. Meristematic area at the axis of a *Vanda* bud. It will be removed by making a circular incision (dashed circle).

210

necessary to wash the stem very gently with dilute solution of detergent before sterilizing. For that purpose, we usually use a solution of calcium hypochlorite at 30 g/l instead of 60 g/l, and we watch the tissue carefully in order to remove it from the hypochlorite before bleaching takes place. Usually 10–12 min are enough.

Then, with a razor blade, we make a circular incision around the meristem and detach it. It will grow just as well either on solid or liquid medium. It is easy to adapt one of these techniques described for other genera.

The Multiplication of the Protocorm *in vitro*

The rate of growth of the initial explant varies a great deal, according to the size of the inoculum and the nature of the species. We shall examine separately the rates of *Cymbidium* and *Cattleya*.

Cymbidium. Very small inocula of *Cymbidium*, comprising the apical dome only, do not show any sign of growth before one or even two months. On larger pieces, with two or three leaves (the size we recommend), the swelling is noticeable after two or three weeks. At that time, the tissues become green, but it is never before one month that individual protocorms start to differentiate and that rhizoids are formed. Some varieties like the famous Alexanderi 'Westonbirt' are much slower. Others like Babylon 'Castlehill' are a little faster.

The same wide range of variation is found in the number of protocorms regenerated from a single explant. Some like Rosanna 'Pinkie' may regenerate as many as a dozen. Usually the number is much lower (Fig. 4-31). Here again, Alexanderi 'Westonbirt' never produces much more than one or two. The optimum size for sectioning of the new protocorms seems to be 2–3 mm, just before the differentiation of the terminal bud. If this bud is well organized, it must be removed and may be grown separately. It will soon produce a small plant.

It is best to do the transferring in a sterile transfer room or sterile box. The protocorms are taken out of the test tubes, put into a sterile Petri dish, and sectioned under the cover of the dish to avoid contamination. During the first transfer each one might be cut into four pieces by two sections at right angles passing through the central axis.

Later, when one has hundreds of explants to section, it becomes impossible to take such care. We then slice them with the use of the small gadget seen in Fig. 4-32. It is made of a piece of glass tubing 6 mm in diameter, 6 cm long, stuck to a small plate of glass, in the middle of which a hole, of the outside diameter of the tube, has been drilled.

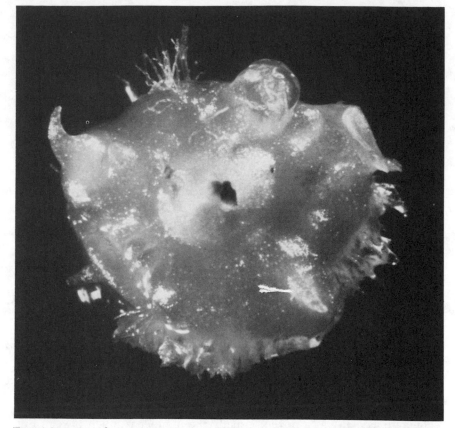

F'g. 4-31. Apical meristem of *Cymbidium* apex after six weeks of culture. The view is taken from above so that the axis of the stem is perpendicular to the picture. Notice the beginning of the formation of some new protocorms on the equatorial plane. (Photograph by J. F. Muller.)

The tube is filled with the protocorms and then pushed out with a glass rod exactly fitting the tube. As they come out, these protocorms are sliced by a razor blade running against the glass plate and fall into the Petri dish. They are then spread on the surface of the agar in Erlenmeyer flasks filled with the transfer medium. There the growth is very fast. Most of the varieties can be sectioned again in about three weeks.

Cattleya.   As we saw earlier, in order to have good results, it is necessary to start in the case of *Cattleya* with a much bigger explant than with *Cymbidium*. This has to be from two to three times larger. At first, the

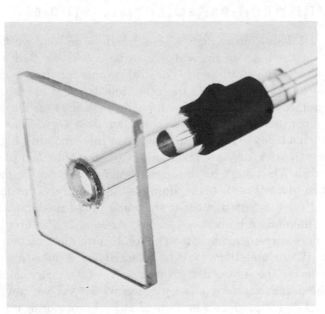

Fig. 4-32. Device used to section large amount of protocorms. The tubing is filled with protocorms. They are pushed out with a glass rod and sliced, as they come out, with a razor blade running along the glass plate.

Fig. 4-33. Clump of protocorms of *Cattleya* ready to transfer. (Photograph by J. F. Muller.)

growth is much lower; no change is seen for three to four weeks. Then the swelling of the bases of the leaf primordia cracks the dark sheet of material made by the dead cells. The leaf primordia then become so thick that they separate one from the other and the bud opens. In most cases new protocorms are formed in the axils as well as on the bases of these leaf primordia. They are always smaller than *Cymbidium* protocorms and are not ready to transfer before two or three months (Fig. 4-33).

For the first transfer, they are taken out of the test tube and put into a sterile Petri dish. Then one tries to separate each of the new-formed protocorms with a piece of razor blade, doing as little injury as possible. The medium used for this first transfer is the same liquid medium. *Cattleya* protocorms differentiate buds very quickly. As soon as the first leaf appears, it is necessary to remove it. The protocorms are sectioned into two or four pieces and then may be planted on a solid medium in Erlenmeyer flasks. The subsequent growth is very fast; many clones have to be sectioned and transferred every three weeks. As we saw earlier, many cultivars do not produce well-individualized protocorms but mostly

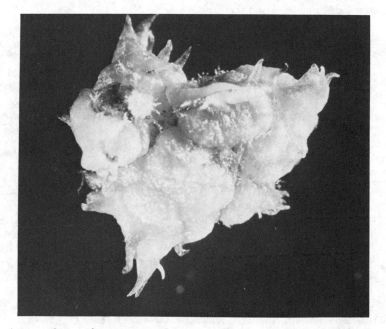

Fig. 4-34.   Clump of portocorms of *Lycaste*. (Photograph by J. F. Muller.)

Fig. 4-35.   Short rhizomes produced by meristems of *Phaius* grown *in vitro*. (Photograph by J. F. Muller.)

buds and plantlets. These buds and plantlets are sectioned at regular intervals, and the propagation in that case depends on the budding of their leaf sections, according to the process already described earlier.

Other Genera.   The growth of the apical meristem of other genera follows a similar pattern. In *Lycaste* the protocorms are less individualized. The explant forms a mass with small conical protuberances (Fig. 4-34). This phenomenon is accentuated even more in *Phaius,* where a callus first appears, from which a great many buds evolve and elongate into short rhizomes, each one forming a new plant (Fig. 4-35).

In some cases, without apparent reason, a single protocorm instead of differentiating a terminal bud and forming a plantlet keeps dividing into hundreds of new ones. This phenomenon is rather frequent with *Odontoglossum* or *Miltonia*. Figure 4-36 shows a test tube containing several hundred *Odontoglossum* protocorms, all coming from a single one. Figure 4-37 shows a very great number of miltonias coming from a single meristem. In *Cattleya* this process is greatly increased by cytokinins, as shown by Pierik and Steegmans (1972).

Fig. 4-36. Protocorms of *Odonto-glossum* produced by the spontaneous division of a single meristem. (Photograph by J. F. Muller.)

Conclusions

Our first studies show that among the species that were tried a very large number can be propagated by meristem culture. Of course, a very small fraction of the cultivated orchids has been investigated. So far as I know, entire groups with a very special morphology, such as terrestrial orchids with bulbs (*Orchis, Ophrys*) or fibrous roots (*Habenaria, Listera*), have barely been examined. The multiplication of *Phalaenopsis* is always difficult, that of *Paphiopedilum* exceptionally so.

Phalaenopsis tissues seem to have a very high content of phenolic compounds. The excised meristem turns black quickly, oxidative products of these phenols diffuse into the medium, and most of the explants die. Thus far we have not been able to get around this difficulty.

The apex of *Paphiopedilum* has very little protection from the leaves. It is always contaminated by bacteria, and it is very difficult to get aseptic cultures. About 90% of the cultures appear to be contaminated. Few

Fig. 4-37.   Several dozen young *Miltonia,* coming from a single meristem, by spontaneous division. (Photograph by J. F. Muller.)

Fig. 4-38. New protocorms of *Paphiopedilum* produced by sectioning a single protocorm coming from a seed. (Photograph by J. F. Muller.)

of them grow, and then they differeniate a plantlet almost immediately. It is only by inhibiting this plantlet formation, by a high 2,4-D concentration of 1 ppm, that it has been possible to get callus formation. But one must remember that 2, 4-D is a weed killer and few cultivars can stand it. This callus will produce protocorms, and we have noticed for a long time that the *Paphiopedilum* protocorms coming from the seed can be sectioned and propagated exactly like the protocorms of other species (Fig. 4-38). Thus it appears to be possible that in the near future all usual cultivated orchids will be propagated by meristem culture.

Bibliography

BALL, E. 1946. Development in sterile culture of stem tip and subadjacent regions of *Tropaeolum majus* L. and of *Lupinus albus* L. *Am. J. Bot.* **33**:301–318.

BALL, E. A., ARDITTI, J., and CHURCHILL, M. E. 1971. Clonal propagation of orchids from leaf tips. *Orchid Rev.* **79**:281–288.

BLOWERS, J. W. 1964. Meristem propagation and possible effects. *Orchid Rev.* **72**: 407–408.

BOESMANN, G. 1962. Problemes concernant le semis et l'amélioration des orchidées. *Adv. Hort. Sci.* **2**:368–372.

BURGEFF, H. 1936. *Samenkeimung der Orchideen.* G. Fischer, Jena.

BUVAT, R. 1955. Le méristème apical de la tige. *Ann. Biol.* 31:595–656.

CAPLIN, S. M., and STEWARD, F. C. 1948. Effect of coconut milk on the growth of explants from carrot roots. *Science* 108:655–657.

CHAMPAGNAT, M. 1971. Recherches sur la multiplication végétative de *Neottia nidus-avis* Rich. *Ann. Sc. Nat. Bot. et Biol. Vég. 12è sér.*, 12:209–247.

CHAMPAGNAT, M., and MOREL, G. 1969. Multiplication végétative des *Cattleya* à partir de bourgeons cultivés *in vitro*. *Soc. Bot. France Mémoires* 116:111–132.

———. 1972. La culture *in vitro* des tissues de tubercules d'Ophrys. *Compt. Rend. Acad. Sci. Paris* 274:3379–3380.

CHAMPAGNAT, M., MOREL, G., CHABUT, P., and COGNET, A. M. 1966. Recherches morphologiques et histologiques sur la multiplication végétative de quelques orchidées du genre *Cymbidium*. *Rev. Gen. Bot.* 73:706–746.

CHAMPAGNAT, M., MOREL, G. and GAMBADE, G. 1968. Particularités morphologiques et pourvoir de régénération du *Cymbidium virescens* cultivé *in vitro*. *Soc. Bot. France Mémoires* 115:236–249.

CHURCHILL, M. E., ARDITTI, J., and BALL, E. A. 1971. Clonal propagation of orchids from leaf tips. *Am. Orchid Soc. Bull.* 40:109–113.

CHURCHILL, M. E., BALL, E. A., and ARDITTI, J. 1970. Production of orchids from seedling leaf tips. *Orchid Digest* 34:271–273.

———. 1973. Tissue culture of orchids, I. Methods for leaf tips. *New Phytol.* 72:161–166.

CURTIS, J. T. 1947. Studies on the nitrogen nutrition of orchid embryos, I. Complex nitrogen sources. *Am. Orchid Soc. Bull.* 16:654–660.

DERMEN, H. 1947. Periclinal cytochimeras and histogenesis in cranberry. *Am. J. Bot.* 34:32–43.

FONNESBECH, M. 1972. Growth hormones and propagation of *Cymbidium in vitro*. *Physiol. Plant.* 27:310–316.

FOSTER, A. S. 1941. Comparative studies on the structure of the shoot apex in seed plants. *Bull. Torrey Bot. Club* 68:339–350.

GAUTHERET, R. J. 1942. Le bourgeonnement des tissus végétaux en culture. *Sciences* 40:96–128.

———. 1944. Recherches sur la polarite des tissus végétaux. *Rev. Cytol. Cytophysiol. Vég.* 7:45–297.

———. 1969. *La culture des tissus végétaux.* Masson Ed., Paris.

HALPERIN, W., and WETHEREL, D. F. 1964. Adventive embryony in tissue culture of *Daucus Carotta. Am. J. Bot.* 51:274–283.

———. 1965. Ontogeny of adventive embryos of wild carrot. *Science* 147:756–758.

———. 1966. Alternative morphongenetic events in cell suspensions. *Am. J. Bot.* 53:443–453.

HELLER, R. 1953. Recherches sur la nutrition minérale des tissus végétaux cultivés *in vitro. Ann. Sci. Nat. Bot. Biol. Veg.* 14:1–223.

IRMISCH, T. 1853. *Beiträge zur Biologie und Morphologie der Orchideen.* Ambrosius Abel, Leipzig.

INTUWONG, O., and SAGAWA, Y. 1973. Clonal propagation of sarcanthine orchids by aseptic culture of inflorescences. *Am. Orchid Soc. Bull.* 42:209–215.

KANDLER, O. 1955. Versuche zur Kultur isolierten Pflanzengewebes *in vitro. Planta* 38:564–585.

KELLER, G. U., and SCHLECHTER, R. 1930–1940. Monographie und Iconographie der Orchideen Europas und des Mittelmeergebietes. II: 328; V: 485.

KIM K.-K., KUNISAKI, J. T., and SAGAWA, Y. 1970. Shoot tip culture of *Dendrobium. Am. Orchid Soc. Bull.* 39:1077–1080.

KOCK, W. E. 1972. Surprising success with a little known fertilizer. *Am. Orchid Soc. Bull.* 41:338–340.

KOHLENBACH, H. W. 1965. Über organisierte Bildungen aus *Macleaya cordata* Kallus. *Planta* 64:37–40.

KONAR, R. V., and NATARAJA, K. 1965. Production of embryoids in tissue cultures of floral buds of *Ranunculus sceleratus* L. *Naturwiss.* 42:140–141.

KUNISAKI, J. T., KIM, K.-K., and SAGAWA, Y. 1972. Shoot tip culture of *Vanda. Am. Orchid Soc. Bull.* 41:435–439.

KÜSTER, E. 1925. *Pathologische Pflanzenanatomie.* G. Fischer, Jena.

LAMI, R. 1927. Influence d'une peptone sur la germination de quelques vandées. *Compt. Rend. Acad. Sci. Paris* 184:1579–1581.

LINDEMANN, E. G. 1967. Growth requirements for meristem culture of *Cattleya.* University Microfilm, Ann Arbor, Mich. Order No. 67–14, 733.

MARIAT, F. 1952. Recherches sur la physiologie des embryons d'orchidées. *Rev. Gén. Bot.* 59:324–377.

MEYER, J. R. 1945. Ação de una heteroauxina sobre o crescimento de "seedlings" de orchideas. *O Biologico* 11:151–153.

MILLER, C. O., SKOOG, F., VON SALTZA, M. H., and STRONG, F. M. 1955. Kinetin—A cell division factor from deoxyribonucleic acid. *J. Am. Chem. Soc.* 17:1392.

MOREL, G. M. 1960. Producing virus-free cymbidiums. *Am. Orchid Soc. Bull.* 29: 495–497.

——. 1963a. La culture *in vitro* du méristème apical de certaines orchidées. *Compt. Rend. Acad. Sci. Paris* 256:4955–4957.

——. 1963b. La culture du méristème apical. *Rev. Cytol. Biol. Vég.* 27:307–314.

——. 1964. A new means of clonal propagation of orchids. *Am. Orchid Soc. Bull.* 31:473–477.

——. 1965a. Eine neue Methode erbgleicher Vermehrung: Die Kultur von Triebspitzen Meristemen. *Die Orchidee* 16:165–176.

——. 1965b. Clonal propagation of orchids by meristem culture. *Cymb. Soc. News* 20:3–16.

——. 1970. Neues auf dem Gebiet der Meristem Forschung. *Die Orchidee* 21:433–443.

MOREL, G. M., and CHAMPAGNAT, M. 1969. Divers modes d'évolution de l'apex au cours de la formation des protocormes, in *l'Orchidée—Proceedings of the Second European Orchid Congress.* Paris, 1964, pp. 20–26.

MOREL, G., and MULLER, J. F. 1964. La culture *in vitro* du méristème apical de la pomme de terre. *Compt. Rend. Acad. Sci. Paris* 258:5250–5252.

MOREL, G. M., and WETMORE, R. H. 1951. Tissue culture of monocotyledons. *Am. J. Bot.* 38:38–140.

MURASHIGE, T., and SKOOG, F. 1962. A revised medium of rapid growth and bio-assays with tobacco tissue cultures. *Physiol. Plant.* 15:473–497.

NAGL, W. 1972. Evidence of DNA amplification in the orchid *Cymbidium in vitro*. *Cytobios.* 5:195–234.

NAGL, W., and RUCKER, W. 1972. Beziehungen zwischen Morphogenese und nuklearem DNS Gehalt bei aseptischen Kulturen von *Cymbidium* nach Wuchsstoff Behandlung. *Z. Pflanzenphysiol.* 67:120–134.

PIERIK, R. L. M., and STEEGMANS, H. H. M. 1972. The effect of 6-benzylaminopurin on growth and development of *Cattleya* seedlings grown from unripe seeds. *Z. Pflanzenphysiol.* 68:228–234.

PLANTEFOL, L. 1946. Fondements d'une théorie phyllotaxique nouvelle. La théorie des hélices foliaires multiples. *Ann. Sci. Nat. Bot.* IIè sér. 7:158–222; 8:1–66.

POPHAM, R. A. 1951. Principle types of vegetative shoot apex organization in vascular plants. *Ohio J. Sci.* 51:249–270.

PREVOT, P. C. 1939. La néoformation des bourgeons chez les végétaux. *Mém. Soc. Roy. Sci. Liège*, nè sér. 3:173–340.

RAGHAVAN, W., and TORREY, J. B. 1964a. Organic nitrogen nutrition of the seedlings of the orchid *Cattleya*. *Am. J. Bot.* 51:264–274.

———. 1964b. Effect of certain organic nitrogen compounds on growth *in vitro* of seedlings of *Cattleya*. *Bot. Gaz.* 125:260–267.

REINERT, A., and MOHR, H. C. 1907. Propagation of *Cattleya* by tissue culture of lateral bud meristem. *Proc. Am. Soc. Hort. Sci.* 91:664–671.

ROBBINS, W. J. 1922. Cultivation of excised root tip and stem tip under sterile conditions. *Bot. Gaz.* 73:376–390.

RUTKOWSKI, E. 1971. How meristems multiply. *Am. Orchid Soc. Bull.* 40:616–622.

SAGAWA, Y., and SHOJI, T. 1967. Clonal propagation of dendrobiums through shoot meristem culture. *Am. Orchid Soc. Bull.* 36:856–859.

SATINA, S., BLAKESLEE, A. F., and AVERY, G. A. 1940. Demonstration of the germ layers in the shoot apex of *Datura* by means of induced polyploidy in periclinal chimeras. *Am. J. Bot.* 27:895–905.

SCHMIDT, A. 1924. Histologische Studien an Panerogamen Vegetationspunkten. *Bot. Arch* 9:345–404.

SCULLY, R. 1967. Aspects of meristem culture in the *Cattleya* alliance. *Am. Orchid Soc. Bull.* 36:103–108.

SKOOG, F. 1950. Chemical control of growth and organ formation in plant tissue. *Ann. Biol.* 26:545–562.

———. 1964. Growth and organ formation in tobacco tissue culture. *Am. J. Bot.* 31:19–24.

SKOOG, F., and MILLER, C. O. 1957. Chemical regulation of growth and organ formation in plant tissue cultured *in vitro*. *Symp. Soc. Exp. Biol.* 11:118–131.

STEWARD, F. C., BLAKELY, L. M., KENT, A. C., and MAPES, M. O. 1963. Growth and organization in free cell cultures. *Brookhaven Symp. Biol.* **16**:73–88.

STEWARD, F. C., and MAPES, M. O. 1963. The totipotency of cultured carrot cells: Evidence and interpretation from successive cycles of growth from phloem cells. *J. Indian Bot. Soc., Maheshwari Comm. Issue,* **42a**:237–247.

STEWARD, F. C., MAPES, M. O., and MEARS, K. 1958. Growth and organized development of cultured cells, II. Organization in cultures from freely suspended cells; III. Interpretations of the growth from free cell to carrot plant. *Am. J. Bot.* **45**:705–713.

TEO, C. K. H., KUNISAKI, J. T., and SAGAWA, Y. 1973. Clonal propagation of strap leaf *Vanda* by shoot tip culture. *Am. Orchid Soc. Bull.* **42**:402–405.

THOMALE, H. 1957. *Die Orchideen.* Stuttgart.

UEDA, H., and TORIKATA, H. 1972. Effects of light and culture medium on adventitious root formation by cymbidiums in aseptic culture. *Am. Orchid Soc. Bull.* **41**:322–327.

VAJRABHAYA, M., and VAJRABHAYA, T. 1970. The culture of *Rhynchostylis gigantea,* a monopodial orchid. *Am. Orchid Soc. Bull.* **40**:907–910.

VAN OVERBEEK, J., CONKLIN, M. E., and BLAKESLEE, A. F. 1941. Factors in coconut milk essential for growth and development of very young *Datura* embryo. *Science* **94**:350–351.

WETMORE, R. H. 1954. The use of *"in vitro"* cultures in the investigation of growth and differentiation in vascular plants. *Brookhaven Symp. Biol.* **6**:22–40.

WETMORE, R. H., and MOREL, G. 1949. Growth and development of *Adiantum pedatum* L. on nutrient agar. *Am. J. Bot.* (*Suppl.*) **36**:805–806.

WHITE, P. R. 1933. Plant tissue cultures. Results of preliminary experiments on the culturing of isolated stem-tips of *Stellaria media. Protoplasma* **19**:97–116.

WIMBER, D. D. 1963. Clonal multiplication of the *Cymbidium* through tissue culture of the shoot meristem. *Am. Orchid Soc. Bull.* **32**:105–107.

———. 1965. Additional observations on clonal multiplication of *Cymbidium* through culture of shoot meristems. *Cymb. Soc. News* **20**:7–10.

WITHNER, C. L. 1955. Ovule culture and growth of *Vanilla* seedlings. *Am. Orchid Soc. Bull.* **17**:662–663.

YAMADA, T., NAKAGAWA, H., and SINTO, Y. 1967. Studies on the differentiation in cultured cells, I. Embryogenesis in three stains of *Solanum melongena. Bot. Mag. Tokyo* **80**:68–74.

5

Development of the Embryo and the Young Seedling Stages of Orchids

YVONNE VEYRET

Until 1804, when Salisbury found that orchid seeds could germinate, the seeds were considered sterile. Much later, in 1889, Bernard's discovery of the special conditions necessary for their growth and development explained the lack of success in previous attempts to obtain plantules, and the success, although mediocre, when sowing of seed took place at the foot of the mother plant. Knowing the rudimentary state of minute orchid embryos, one can better understand that an exterior agent, usually a *Rhizoctonia*, can be useful in helping them through their first stages of development (Bernard, 1904).

Orchid embryos, despite their rudimentary condition, present diverse patterns of development, the most apparent of which are concerned with the character of the suspensor; there are other basic patterns that are revealed in the course of the formation of the embryonic body. These characteristics have been used differently in the classification of orchid embryos. This will be discussed in the first part of this chapter, to which will be added our knowledge of other phenomena concerning the embryo, in particular polyembryonic and apomictic seed formation. The second part of this chapter is devoted to the development of the young embryo, a study necessary for understanding the evolution of the different zones of the embryonic mass. We will also examine the characteristics of the embryo whose morphology, biology, and development are specialized when compared to other plants.

The Embryo

Embryological Classifications.   The orchid embryo is still little known, since only about 60 kinds have been the object of embryological investigations, while the family numbers more than 20,000 species. In spite of that, a comprehensive view on the development of the embryo in the different groups is beginning to become clear. It is still necessary to pursue these investigations, specifically among the POLYCHONDREAE and the KEROSPHEREAE.

The first important works related to orchid embryology were due to the Dutch scientist Treub, in 1879, whose work offered a suggestive glimpse of the appearance, and sometimes the segmentation, of the embryo in diverse orchid groups. The first attempt at the classification of orchid embryos came a long time later, in 1949. At this time, Swamy distinguished two groups, A and B, following the absence or not of the suspensor. In these two groups, the first division of the egg is transverse and the proembryo after the second cellular generation shows two juxtaposed cells coming from the terminal cell; the basal cell, following a horizontal division, has yielded a middle cell and a lower cell of the suspensor. The role taken by each of these cells in the further stages of development is taken account of in the classification system. In the A group the embryo is formed, following in the order of the importance of their contribution, by the terminal cells, then by the middle cell, and finally by the initial cell of the suspensor. In the B group the early embryo is formed in large part from the cells coming from the terminal cell, and in a lesser degree by those coming from the median cell. The A group embryos correspond to "Asterad type" and those of the B group to the "Onagrad type," as Johansen (1950) understands them.

Some rare types cannot be classified in either of these groups because they do not present the same proembryonic form in the second cellular generation. The first divisions are made without definite order and end in the formation of an undifferentiated mass of six to eight cells. The majority of these cells elongate to form a suspensor complex while one of them, located near the chalazal end, becomes the mother cell of the embryo proper. This generates a proembryo armed with filamentous suspensors so that the embryo will in the end be provided with two. This way of development is defined as being the "*Cymbidium* form." In this classification, Swamy (1949) makes an attempt to give ontogenetic interpretation to these characteristics, but his study ends at only the second cellular generation, the author thinking generally that "only the first 2 or 3 cell generations in the zygote are consistent: The subsequent cell divisions take place with no definite sequence or pattern." Swamy acknowledges, on the other hand, five kinds of suspensors that are diagramed and described on pages 174–175 of *The Orchids* and keyed on p. 179 (Wirth and Withner, 1959).

Johansen's classification (1950) is done uniquely according to the suspensor characteristics, notably characteristics relative to the presence or absence of embryonal tubes, to the number and arrangement of these tubes, and to the appearance of these cells in the case of the filamentous suspensors. Very often it is certain that the appearance of the suspensor gives a recognizable distinctiveness to the embryos of different genera (Fig. 5-1). Thirteen plans are described in his scheme and you may find that these are also keyed and described in *The Orchids* (Wirth and Withner, pp. 177–179).

If one closely follows the continuing development of the embryo proper, one ascertains that they are not without distinction and that the young forms do follow well-definable patterns. It is in this way that Veyret could determine a certain number of embryonic types among orchids that would fit into the system devised by Souèges (1936–1939). The embryonic types are characterized by the four great embryogenic laws as Souèges defines them: laws of origin, of number, of disposition or arrangement, and of destinies of the first four cellular generations. Only the law of destinies cannot be determined in the development of the orchids since the embryo at time of fruit maturity is, in general, morphologically undifferentiated. In Souèges' classification, the embryos are grouped into various megarchetypes, these being defined by the sum of the constructive potentials of the apical and basal cells, resulting from the first transverse segmentation of the zygote.

The embryos among orchids are of three kinds:

1. Fundamental types that develop following fixed rules.
2. Irregular types presenting forms which are constructed following clear rules but varying with individual species.
3. Secondary types with the early embryo differing from the fundamental types by differences in
   a. the speed of the segmentation of the homologous blastomeres; or
   b. the direction of certain segmentation walls leading to an earlier differentiation; or
   c. the very variable differentiation of the most inferior component parts of the embryo which have no constructive role to play.

The different types are classified according to Souèges' system into the first and second periods of the periodic embryogenic classification or

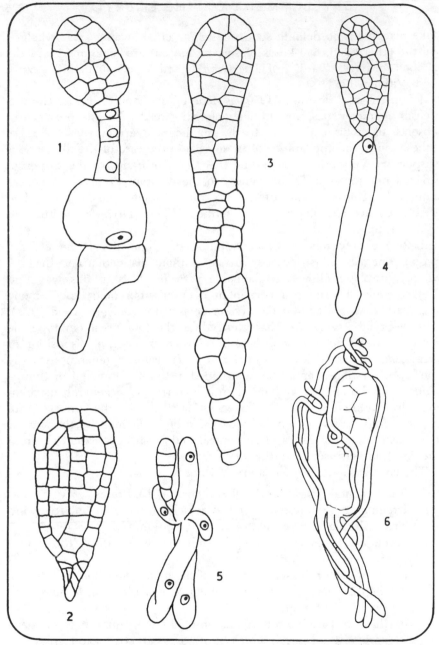

Fig. 5-1. Embryos presenting different types of suspensors. (1) *Herminium monorchis*, according to Treub. (2) *Vanilla fragrans*, according to Veyret. (3) *Epidendrum ciliare*, according to Treub. (4) *Spathoglottis plicata* and (5) *Geodorum densiflorum* (proembryo), according to B. G. L. Swamy. (6) *Phalaenopsis grandiflora*, according to Treub.

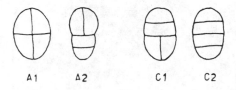

A1   A2   C1   C2

Fig. 5-2. The different forms of proembryonic tetrads among orchids.

grand divisions of the system. In the first period, the first two cells coming from the zygote, both the apical and basal, take part in the construction of the embryo, and the laws of development enter the process as
soon as the first segmentation wall is established in the fertilized egg. In
the second period the embryo develops only from the apical cell and the
laws apply only to the apical cell. This classification and the sequence of
divisions is established on presumed phylogenetic criteria, and the relative evolutionary position of the types is based mainly on the orientation
of the segmentation walls and the tendency of constructive potentials to
be localized in the apical cell. Thus the tetrads in A configuration are
phylogenetically earlier than the tetrads in B, these being earlier than
the tetrads in C; the 1 forms of these tetrads are more primitive than the
2 forms (Fig. 5-2). There are no tetrads of the B configuration in the
orchids. The megarchetypes form a phylogenetic series as the role of the
basal cell is reduced to only generating the suspensor.

The conventionally accepted abbreviations used in the description of
the different parts of the embryo are the following:

ca: apical cell of the two-celled proembryo
cb: basal cell of the two-celled proembryo
cc: the upper daughter cell of ca
cd: the lower daughter cell of ca
m: the upper daughter cell of cb
ci: the lower daughter cell of cb
d: the upper daughter cell of m
f: the lower daughter cell of m
n: the upper daughter cell of ci
n': the lower daughter cell of ci
h: the upper daughter cell of n
k: the lower daughter cell of n
o: the upper daughter cell of n'
p: the lower daughter cell of n'
h': the upper daughter cell of h
h'': the lower daughter cell of h
q: the quadrant stage
l: the stage of the upper octants
l': the stage of the lower octants

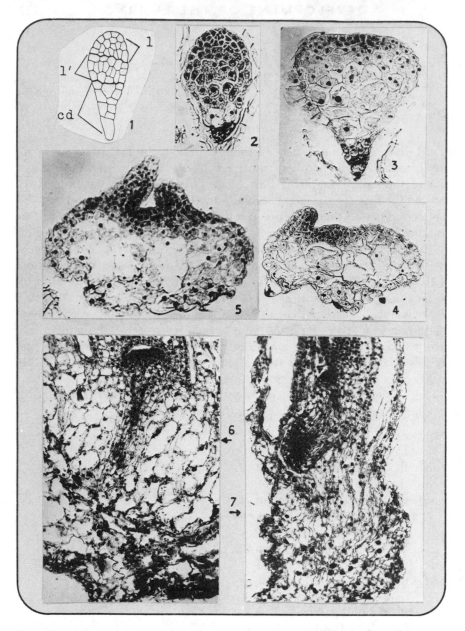

Fig. 5-3. Germination of a species without cotyledon, *Stanhopea costaricensis*. (1) Longitudinal section of a mature embryo. (2–5) Longitudinal section of a young protocorm showing the formation of the apical meristem and the first foliar member. *Bulbophyllum bufo*. (6) Differentiation of the stele in the protocorm. (7) Formation of the first root.

Embryonic Types

Before going further with the embryogenic classification of the orchids, it is necessary to anticipate and examine what is going to happen at the moment of germination when the embryo differentiates. Two situations are presented, embryos with and those without cotyledons. In the case of acotyledonous species, which are in the majority, the vegetative point of the stalk is formed by the upper part of the embryo represented by the layer *l* (Figs. 5-3 and 5-4), in the case of the rare species with cotyledons, it is the cotyledon that is formed by the terminal layers while the vege-

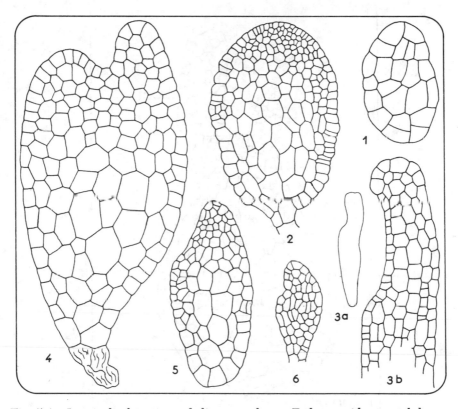

Fig. 5-4.   Longitudinal sections of diverse embryos. Embryos without cotyledons: (1) *Listera ovata,* one of the most rudimentary; and (2) *Epidendrum radicans,* one of the most differentated histologically, according to Veyret. Embryos with cotyledons: (3) *Sobralia macrantha:* (3a) mature embryo, (3b) cotyledon at tip, according to Treub; (4) *Bletilla hyacinthina,* according to Bernard; (5) *Polystachya microbambusa* and (6) *Epidendrum vitellinum,* according to Veyret. All these embroys are mature except for *E. vitellinum.* All 136X except for Fig. 3a, which is 41X.

tative point of the stalk is derived from the layer of more centrally located cells, l', located directly under the terminal cells (Figs. 5-4 and 5-5). The degeneration of the lowermost cell layers of the embryo has been ascertained among some species where it is possible to follow rigorously the pattern of development and to observe the definable boundary in the seed between the suspensor and the embryo proper. With such develop-

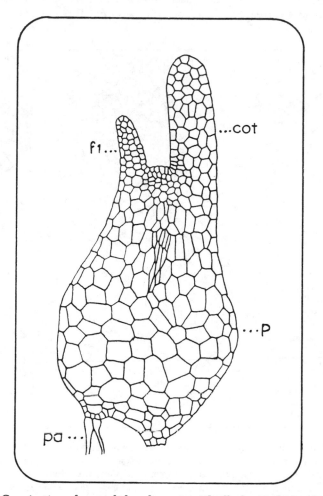

Fig. 5-5. Germination of a cotyledoned species, *Bletilla hyacinthina*, after Bernard. Longitudinal section of a young plantlet, 100X. cot = cotyledon; pvt = apex of stem; p = protocorm; pa = absorbing hairs.

mental information one can establish only a general formula relating to two patterns of the megarchetypes:

1. Among the cotyledonous species, the terminal cells may form the cotyledon.
2. Among the acotyledonous species, the terminal cell may form the vegetative apex of the stalk.

Within these two principal divisions—depending on presence or absence of cotyledon—the types can be grouped according to the form of the tetrad (second cellular generation), and then the subsequent formation of the embryo proper. One can emphasize here the difficulty that exists at times in determining the boundary between the so-called embryo and the suspensor. Classification of any orchid may be temporary—to be revised at the time when additional data relating to the destiny of all the blastomeres are obtained. The further study of orchid embryonic differentiation will thus enable us to establish the megarchetypes more precisely (Figs. 5-6–5-8).

The following is a key for the determination of the different groups of orchidaceous embryonic types:

A. Species without cotyledon
    B. Species of the first period
        C. Tetrad in A1
            Embryo $= ca + cb$ *Neottia nidus-avis* group
        CC. Tetrad in A2
            Embryo $= ca + cb$ *Manniella gustavi* group
            Embryo $= ca + m + n + o$ *Limodorum abortivum* group
            Embryo $= ca + m + n$ *Lecanorchis japonica* group
            Embryo $= ca + m + h'$ *Orchis maculata* group
            Embryo $= ca + m$ *Liparis pulverulenta* group
        CCC. Tetrad in C2
            Embryo $= ca + m + n + o$ *Eulophia oedoplectron* group
    BB. Species of the second period
        D. Second tetrad in A1
            Embryo $= cc + cd$ SARCANTHEAE group

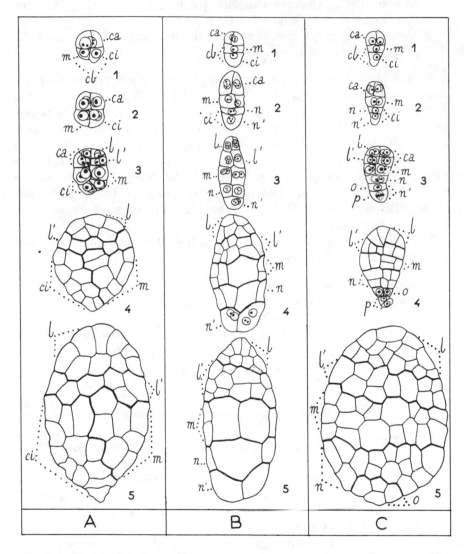

Fig. 5-6. Stages of embryogenesis among regular species. (A) *Epipactis atrorubens,* (B) *Manniella gustavi,* and (C) *Limodorum abortivum,* all according to Veyret. 1 = second cellular generation (tetrad); 2 = third cellular generation (quadrant stage); 3 = fourth cellular generation (octant stage); 4 = intermediary stage; 5 = adult embryo. 136X.

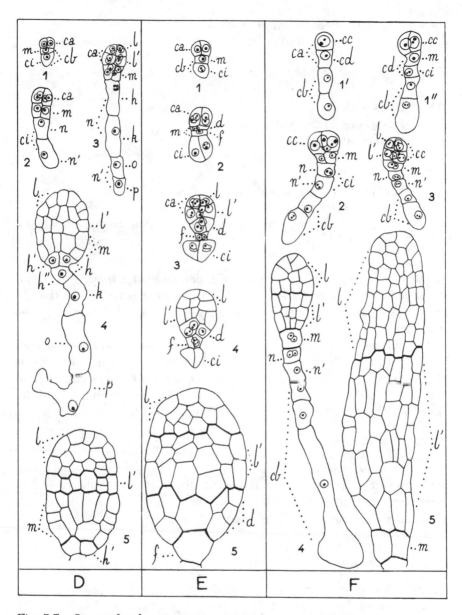

Fig. 5-7. Stages of embryogenesis among regular species. (D) *Serapias longipetala,* (E) *Liparis pulverulenta,* and (F) *Coelogyne parishii,* all according to Veyret. 1′ = first tetrad; 1″ = second tetrad (second period); for the other stages see caption of Fig. 5-6. 136X for Fig. 5-7D; 225X for the others.

DD. Second tetrad in A2
  Embryo $= cc + m + ci$ *Polystachya geraensis*
                                  group
  Embryo $= cc + m$ *Coelogyne parishii*
                                  group

AA. Species with cotyledon
    E. Species of the first period
        Tetrad in A2
  Embryo $= ca + m$ *Epidendrum vitellinum*
                                  group

    EE. Species of the second period
        Second tetrad in A2
  Embryo $= cc + m + ci$ *Polystachya microbam-*
                                  busa group

Another group may be represented by *Eulophia epidendraea;* this would be placed in this classification after the *Coelogyne parishii,* but the observations of Swamy do not provide the developmental details, only the final structure.

The different embryonic types are defined by the developmental laws in the first four cellular generations. Since the first division of the egg is always transverse and the types of each group exhibit the same tetrad formation, these common characteristics will not be repeated in the descriptions of the various types within the groups (Figs. 5-6–5-8).

Neottia nidus-avis Group

Neottia nidus-avis Type.   In the third cellular generation, the quadrants present a circumaxial arrangement, the division of the m and ci cells occurring following a longitudinal division. In the fourth cellular generation the octants are distributed in two layers:

l = layer of the upper octants
l' = layer of the lower octants

The m and ci cells are transversally divided.

Listera ovata Type.   This differs from the preceding by the transverse division of the ci cell in the third cellular generation.

Spiranthes autumnalis Type.   This type is irregular: in addition to some forms identical to those of the *Neottia,* the quadrant cells can arrange themselves in two superimposed layers or in a tetrahedron.

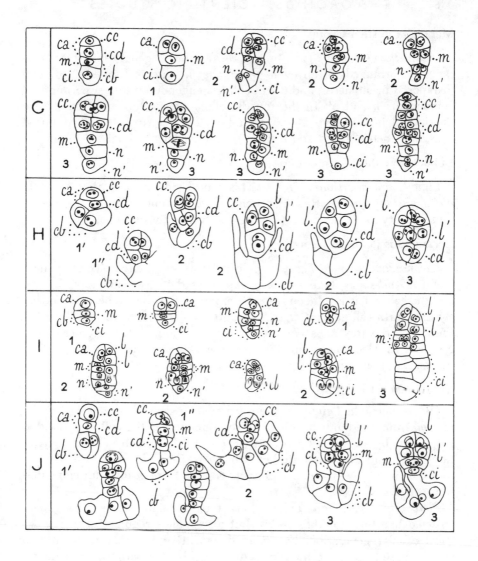

Fig. 5-8. Stages of embryogenesis among irregular species. (G) *Eulophia oedoplectron*, (H) *Angraecum distichum*. (I) *Epidendrum radicans*, (J) *Polystachya microbambusa*, all according to Veyret. Same stages as for Figs. 5-6 and 5-7. 320X for Fig. 5-8H; 225X for the others. In some figures the quadrant stage may be found slightly ahead, or the octants not all formed as yet, but these figures clearly indicate the mode of segmentation of the blastomers issued from *ca*.

Manniella gustavi Group

Manniella gustavi Type. In the third cellular generation the quadrants are arranged on an even plane, the m state is formed from two juxtaposed blastomeres, and the ci cell has engendered two superimposed components n and n'. In the fourth cellular generation the octants are distributed in two layers, the m stage is formed from four juxtaposed cells, and each of the n and n' cells is divided longitudinally.

Limodorum abortivum Group

Limodorum abortivum Type. This type is identical to that of *Manniella gustavi* type in the third cellular generation only. In the octant stage, this type differs from it by the transverse division of n'.

Lecanorchis japonica Group

Lecanorchis japonica Type. The quadrants are in circumaxial arrangement; at the same stage tier m is represented by two juxtaposed cells, and the cell ci has produced the two superimposed blastomeres n and n'. By the fourth cell generation, the octants are distributed in two layers of four cells, the row m is composed of four juxtaposed blastomeres, the cell n is divided longitudinally, and the cell n', which forms a short suspensor, already shows signs of degeneration.

Orchis maculata Group

Orchis maculata Type. In the third cellular generation the ca layer is formed from quadrants arranged on the same plane, the cell m is divided longitudinally and the cell ci transversely. In the fourth cellular generation the octants are distributed in two layers, the m layer is formed from four circumaxial cells; the components coming from ci form an elongated four-celled suspensor.

Platycoryne paludosa Type. This type presents some forms constructed like those of the *Orchis*, but also some irregular forms, in the third cellular generation, by an arrangement of the quadrants in two superimposed dyads, or in a tetrahedron, and by a superimposed arrangement of the two blastomeres of m; and in the fourth cellular generation by a tetrahedral arrangement or in two superimposed dyads of the four components of m.

Liparis pulverulenta Group

Liparis pulverulenta Type. The embryo of the third cellular generation presents some quadrants in circumaxial arrangement, two superimposed cells d and f issued from m, two juxtaposed blastomeres in ci. The octants are normally distributed in two layers and at this stage the cells d and f are divided vertically and the layer ci is formed from four circumaxial cells.

Bulbophyllum oreonastes Type. This differs from the preceding one by the lengthwise division of m and the transverse division of ci in the octant stage, by a new longitudinal division of the two cells of m, and by the joint possession of ci in the octant stage.

Masdevallia veitchiana Type. This differs from that of the *Liparis* by the arrangements of the quadrants in two superimposed dyads and the joint possession of the cells m and ci in the course of the third and fourth cellular generations.

Eulophia oedoplectron Group

Eulophia oedoplectron Type. This type is especially irregular. Some forms, uncommon ones, originate from a tetrad in A2, but the majority of the forms are constructed from a tetrad in C2. The quadrants are represented by the circumaxial cells or arranged in two superimposed dyads or in a tetrahedron. The octants can distribute themselves normally in two superimposed layers, or they can present a normal lower layer while the upper layer will be provided with homologous octant cells; they can still arrange themselves in two superimposed tetrahedrons or else the l' layer only will have its blastomeres arranged in tetrahedron, while the l layer will be made of two juxtaposed lower cells and of two other superimposed cells. In the third generation the m cell segments vertically, and the ci cell segments horizontally in n and n'. These new blastomeres remain undivided in the fourth generation.

Sarcantheae Group. The types in this group differ little among themselves; they are all irregular.

Angraecum distichum Type. The quadrants are arranged either in two superimposed dyads, or in a tetrahedron, or following a circumaxial arrangement. The octants distribute themselves normally in two layers. The two lower cells of the tetrad, which constitute the cd layers, remain undivided in the third cellular generation, sometimes still undivided in the octant stage; their division is generally transverse, more rarely periclinal.

Acampe renschiana Type. One finds again the same kinds of quadrants as those in the *Angraecum distichum* except for those whose arrangement is circumaxial; in the third cellular generation the two cells of the cd layer can divide and arrange themselves in a tetrahedron.

Cyrtorchis sedeni Type. In this species there has only been observed an arrangement of the quadrants, that of two superimposed dyads; the two cells of the *cd* layer divide in a periclinal manner.

Polystachya geraensis Group

Polystachya geraensis Type. This type is especially irregular since some forms originating from a tetrad in A1 add themselves to those derived from the tetrad in A2. The quadrants can arrange themselves in a tetrahedron or in two superimposed dyads; in this stage the cell *m* has engendered two juxtaposed blastomeres, and the cell *ci* has remained undivided. In the fourth generation the octants constitute the two layers *l* and *l'*; the two cells of the *m* layer and the *ci* cell are not segmented.

Coelogyne parishii Group

Coelogyne parishii Type. The third generation is represented by a layer of quadrants, two juxtaposed cells in *m*, and two superimposed blastomeres *n* and *n'* issued from the transverse division of *ci*. In the fourth generation there is the formation of two layers of octants, the cells of the *m* layer. The blastomeres *n* and *n'* remain undivided.

Epidendrum vitellinum Group

Epidendrum vitellinum Type. *Epidendrum vitellinum* embryos present two sorts of irregularities: the first comes from the presence of a supplementary tetrad in A1, and the second from a variable distribution of the blastomeres; in the forms derived from the tetrad in A2, the second are the most numerous. It is thus that in the third cellular generation the quadrants are found grouped on one layer only, or they are distributed in a tetrahedron with the two cells of *m* separated by a longitudinal transverse division, and that in the fourth cellular generation. The four blastomeres from *m* form a tetrahedron, or two superimposed dyads, while the octants are normally distributed in two layers.

Polystachya microbambusa Group

Polystachya microbambusa Type. This type is irregular because some proembryonic stages proceed from a tetrad in A2, while others, less numerous, derive from a tetrad in C2. The quadrants present two arrangements, either circumaxial or in a tetrahedron. At this same stage the cell *m* is found vertically divided, and the blastomere *ci* remains undivided. It will still be so in the fourth cellular generation while the *m* layer will be formed from four circumaxial cells and the quadrants will have engendered two layers of octants.

In Table 5-1 we classify all the species whose embryology is known, although often in a very imperfect manner so that it is necessary to distribute some of them in several columns. The species that are irregular are noted with an asterisk. They are destined to be replaced by species of regular type as they are discovered and described. In the third column the species with secondary differences are listed for each type.

Each species is followed by a letter in parentheses which indicates the tribe or one of its divisions according to Schlechter (1926):

C: CYPRIPEDILOIDEAE
O: OPHRYDOIDEAE
P: POLYCHONDREAE
A: KEROSPHAEREAE-ACRANTHAE
S: KEROSPHAEREAE-PLEURANTHAE-SYMPODIALES
M: KEROSPHAEREAE-PLEURANTHAE-MONOPODIALES

The number in brackets corresponds to a reference in the bibliography.

The Embryo in the Seed

Embryogenesis is a relatively short stage, in comparison to other concurrent phenomena, in the formation of the fruit. It starts toward the middle of the period, occurring between the pollination and the dehiscence of the ovary, and lasts an average of about two weeks. When the pod is ready to dehisce, the internal integument of the ovule as well as the deepest layers of the external integument are generally found to be degenerated. In the majority of the POLYCHONDREAE, however, and in many European OPHRYDOIDEAE, the cuticle of the epidermis of the inner integument of the ovule persists, and this seems to have the effect of impeding the hydration of seeds and thus hindering their germination *in vitro* (Veyret, 1969). The outer cell layer of the external integument constitutes by itself the cover of the seed. The embryo in the absence of *Rhizoctonia* may be in limited contact with the seedcoat or may be more or less isolated in the center according to the importance of the development of the external integument in the course of the embryogenesis. The cells of the coat are dead, empty, and generally transparent. Their walls can be simple or diversely ornamented, rarely very thick as among seeds of some species of *Apostasia, Neuwiedia, Selenipedium,* and the VANILLEAE—*Vanilla, Galeola, Epistephium.* In the VANILLEAE, the seed of the latter two is also winged.

By the time the fruit has matured, the suspensor of the embryo has degenerated, and the cotyledon is not at all or scarcely developed. There is only slight cotyledonary formation in embryos of *Platyclinis glumacea*

TABLE 5-1

Classification of orchids by embryological type†

Embryological Group	Embryological Type
Neottia nidus-avis	*Neottia nidus-avis* (P)(1)
	Spiranthes autumnalis (P)(1)
	Listera ovata (P)(1)
Maniella gustavi	*Maniella gustavi* (P)(1)
Limodorum abortivum	*Limodorum abortivum* (P)(1)
Lecanorchis japonica	*Lecanorchis japonica* (P)(2)
Orchis maculata	*Orchis maculata* (O)(3)
	Platycoryne paludosa (O)(1)
Liparis pulverulenta	*Liparis pulverulenta* (A)(1)
	Bulbophyllum oreonastes (S)(1)
	Masdevallia veitchiana (A)(1)
Eulophia oedoplectron	*Eulophia oedoplectron* (S)(1)
SARCANTHEAE	*Angraecum distichum* (M)(1)
	Acampe renschiana (M)(1)
	Cyrtorchis sedeni (M)(1)
Polystachya geraensis	*Polystachya geraensis* (A)(5)
Coelogyne parishii	*Coelogyne parishii* (A)(1)
Epidendrum vitellinum	*Epidendrum vitellinum* (A)(1)
Polystachya microbambusa	*Polystachya microbambusa* (A)(1)

†See text, page 239, for details.

TABLE 5-1 (continued)

Species Identical to or Differing by Secondary Characteristics from the Type	Species Definitely Belonging to Group
Epipactis atrorubens (P)(1)	*Epipogium aphyllum* (P)(10,11)
Goodyera repens (P)(1)	*Epipactis palustris* (P) (12) *E. latifolia* (12) *Spiranthes australis* (P)(13, 14) *S. cernua* (27)(15)
Cephalanthera ensifolia (P)(1)	*Hetaeria shikokiana* (P)(16)
Gymnadenia conopsea (O)(6) *Habenaria platyphylla* (O)(7) *Loroglossum hircinum* (O)(8) *Ophrys lutea* (O)(1) *Orchis aristata* (O)(9) *O. laxiflora* (1) *O. longibracteata* (1) *Phyllomphax helleborine* (O)(1) *Platanthera bifolia* (O)(1) *Serapias longipetala* (O)(1)	*Habenaria decipiens* (O)(7) *H. heyenecina* (7) *H. longicalcarata* (7) *H. longicornu* (7) *H. marginata* (7) *H. plantagenea* (7) *H. rariflora* (7) *H. viridiflora* (7)
	Dendrobium microbulbon (A)(7) *Epidendrum ciliare* (A)(12)
Vanilla fragrans (P)(1) *Corallorhiza innata* (S)(1)	*E. lacerum* (17) *E. prismatocarpum* (18) *E. radicans* (1)
Maxillaria variabilis (S)(1)	*Laelia brysiana* (A)(12)
Bulbophyllum muscicola (M)(1)	

*See text, pages 234–238, for details.

241

TABLE 5-1 (continued)

Classification of orchids by embryolygical type†

Species Appearing to Belong to the Group	Species Little Known Embryologically or Known Only from General Characteristics	
	Cypripedium parviflorum (C)(25) *C. reginae* (26) *C. spectabile* (12) *Goodyera tesselata* (P)(27) *Prescottia micrantha* (P)(1) *Spiranthes annua* (P)(28) *Zeuxine sulcata* (P)(29, 30)	No Suspensor
Epipactis pubescens (P)(19) *Galeola septentrionalis* (P)(20)		
Hetaeria nitida (P)(21)		
Anacamptis pyramidalis (O)(12) *Herminium monorchis* (O)(12) *Orchis latifolia* (O)(12) *Peristylus grandis* (O)(4) *P. spiralis* (17) *Serapias pseudocordigera* (O)(23)	*Cypripedium barbatum* (C)(12) *C. venustum* (12) *Gastrodia elata* (P)(31) *Haemaria discolor* (P)(12) *Chamaeorchis alpina* (O)(32) *Cynorkis ampullacea* (O)(33) *C. lilacina* (34) *C. ridleyi* (33) *Habenaria blephariglottis* (O)(27) *H. ciliaris* (35)	Suspensor Present
	H. integra (35) *H. rariflora* (36) *H. tridentata* (27)	
Spathoglottis plicata (S)(17)	*Nigritella angustifolia* (O)(32) *N. nigra* (37) *Ophrys myodes* (O)(38) *Platanthera montana* (O)(32) *Pterygodium newdigatae* (O)(39) *Satyrium nepalense* (O)(40) *Serapias lingua* (O)(12) *Coelogyne breviscapa* (A) (17) *Dendrobium anosum* (A) (41)	
Aerides sp. (M)(17) *Cottonia* sp. (M)(17) *Diplocentrum conjestrum* (M)(17) *Luisia* sp. (M)(17) *Phalaenopsis schillerana* (M)(12) *P. grandiflora* (12) *Rhyncostylis* sp. (M)(17) *Saccolabium* sp. (M)(17) *Vanda parviflora* (M)(24) *V. roxburgii* (24) *V. spathulata* (24) *V. tricolor* (12)	*D. barbatulum* (17) *D. graminifolium* (17) *D. nobile* (42, 43) *Epidendrum cochleatum* (A)(44) *E. variegatum* (44) *E. verrucosum* (44) *Cattleya* sp. (45) *Arundina bambusifolia* (A)(46) *Pleurothallis clausa* (A)(47) *P. racemiflora* (47) *P. procumbens* (A)(48) *Restrepia vittata* (A)(47) *Trichosma suavis* (A) (49) *Aplectrum hiemale* (S)(27)	
	Bletia shepherdii (S)(44) *Bulbophyllum mysorense* (S)(17) *Calanthe madagascariensis* (S)(1) *C. veitchii* (50) *Catasetum* sp. (S)(43) *Corallorhiza multiflora* (S)(27) *Maxillaria crassifolia* (S)(51) *M. punctulata* (S)(47) *Odontoglossum crispum* × *O. adrianae* (S)(52) *Peristeria elata* (S)(17) *Phajus grandiflorus* (S)(44) *P. wallichii* (12) *Zygopetalum mackayi* (S)(24) *Cymbidium bicolor* (S)(24, 53) *C.* sp. (52) *Eulophia epidendraea* (S)(54) *E. cucculata* (1) *Geodorum densiflorum* (S)(17) *Stanhopea costaricensis* (S)(1) *S. oculata* (12)	No Cotyledon Present
	Bletilla hyacinthina (P)(52) *B. striata* (55) *Sobralia macrantha* (P)(12)	Cotyledon Present

References:
1. Veyret, 1965.
2. Tohda, 1971b.
3. Montéverdé, 1880.
4. Treub, 1883.
5. Chaiyasut, unpublished.
6. Ward, 1880.
7. Swamy, 1946c.
8. Heusser, 1915.
9. Tohda, 1971a.
10. Afzélius, 1954.
11. Geitler, 1956.
12. Treub, 1879.
13. Maheswari and Narayanaswami, 1950.
14. Maheswari and Narayanaswami, 1952.
15. Swamy, 1948a.
16. Tohda, 1967.
17. Swamy, 1949.
18. Swamy, 1948b.
19. Brown and Sharp, 1911.
20. Kimura, 1971.
21. Olsson, 1967.
22. Baranov, 1924.
23. Baranov, 1915.
24. Swamy, 1942a.
25. Carlson, 1940.
26. Schnarf, 1931.
27. Leavitt, 1901.
28. Leavitt, 1900.
29. Seshagiriah, 1941.
30. Swamy, 1946b.
31. Kusano, 1915.
32. Dumée, 1910.
33. Veyret, 1972.
34. Veyret, 1967.
35. Brown, 1909.
36. Swamy, 1943a.
37. Afzélius, 1928.
38. Senianinova, 1924.
39. Duthie, 1915.
40. Swamy, 1944.
41. Pastrana and Santos, 1931.
42. Poddubnaja-Arnoldi, 1959.
43. Poddubnaja-Arnoldi, 1960a.
44. Sharp, 1912.
45. Knudson, 1935.
46. Mitra, 1971.
47. Prillieux, 1860.
48. Afzélius, 1966.
49. Baranov, 1917.
50. Poddubnaja-Arnoldi, 1960b.
51. Fleischer, 1874.
52. Bernard, 1909.
53. Swamy, 1946a.
54. Swamy, 1943b.
55. Tohda, 1968.

(Beer, 1863), *Sobralia macrantha* (Treub, 1879), *Bletilla hyacinthina* (Bernard, 1909), *B. striata* (Tohda, 1968), *Epidendrum vitellinum,* and *Polystachya microbambusa* (Veyret, 1965). In a rather larger number of species, it is only possible to distinguish an apical zone of small cells, the rest of the embryo being formed from bulky cells, stuffed with reserve substances, which are all similar morphologically. There is, however, histological and physiological differentiation along the axis.

The orchid embryo has developed in a sac without benefit of an endosperm, and this has generally been interpreted as the cause of the rudimentary state of the embryo. The endosperm does not ordinarily form in the orchids, either from a lack of fusion of the second sperm nucleus with the endosperm nuclei, or from an immediate degeneracy of the nucleus of the endosperm if the double fertilization does take place. Among a few species, however, the segmentation of the nucleus of the endosperm does take place, but it never leads to the production of a normal endosperm. In ovules of *Chamaeorchis alpina* and *Paphiopedilum insigne* (Afzelius, 1916), and in *Limodorum abortivum* (Veyret, 1965) and *Pogonia japonica* (Abe, 1968), there form only two nuclei; *Cephalanthera damasonium* and *C. longifolia* (Hagerup, 1947) form a few; *Cyprypedium guttatum* (Prosina, 1930), *C. parviflorum, C. spectabile,* (Pace, 1907), *Polystachya geraensis,* (Chaiyasut, unpublished), and *Lecanorchis japonica* (Tohda, 1971b) form four; *Bletilla striata* (Abe, 1971) forms eight; and *Vanilla planifolia* (Swamy, 1947), forms twelve. There are sixteen in *Galeola septentrionalis* (Kimura, 1971). This endosperm is entirely reabsorbed before the end of embryogenesis. The embryo in certain species may be more developed than among species which have endosperm. One cannot therefore attribute a cause and effect relationship to this phenomenon.

In the absence of the endosperm, Treub (1879) attributed the role of nourishment to the suspensor; or, in the absence of the suspensor, to the slight contact of the embryonic sac with the embryo. But the influence of such assimilatory structures, nuclei of the sac, or the suspensor, or relations of proximity between the embryo and the walls of the embryonic cavity, do not seem related to a better development of morphological differentiation of the embryo. In addition, in the cases of polyembryony, the embryos of the same sac reach an equal or nearly equal size in development, and, particularly, a histological differentiation comparable to that of single embryos. These facts all favor the unique qualities of orchid embryos.

Polyembryony and Apomixis

Usually the embryo in the seed is single, but in numerous species, the phenomenon of polyembryony occurs, true or false, facultative or predominant. Multiple embryos in many instances are related to apomixis (Fig. 5-9), rather than to true polyembryony.

Cases of pseudopolyembryony are rare in orchids. There are double embryo sacs in the same nucellus in *Orchis morio* (Schacht, 1850; Braun, 1859), *Gymnadenia conopsea* (Strasburger, 1878), *Lecanorchis japonica* (Tohda, 1971b), and *Listera nipponica* (Abe, 1972), and both are functional. In *Pterygodium newdigatae* (Duthie, 1915) there may also be the formation of two archesporial cells in a single ovule that would be partly responsible for the polyembryony in the seed. Often there are ovules enclosing two separate nucelli in *Bletilla striata* (Abe, 1971). One embryo sac seems to develop normally in each of these.

True polyembryony is more frequent, and has been reviewed by Wirth and Withner (1959) with illustrations from the literature. In addition to the examples cited there, Duthie (1915) has also shown that some of the embryos of *Pterygodium newdigatae* are the result of segmentation of the embryo proper.

When they have been formed, the development of a synergid into an embryo, in addition to a normal embryo, is found in a facultative manner in *Orchis maculata* and *Listera ovata* (Hagerup, 1944, 1947). The synergids begin segmentation without there being any fusion with a sperm nucleus, so that, in this case, we are dealing with haploid apogamy. The same process can be responsible for the twin embryos in *Vanilla planifolia* (Swamy, 1947) and *Platanthera sachalinensis* (Abe, 1972).

Other forms of apomixis involve haploid parthenogenesis, somatic apospory, megasporic embryony, omnisaccate diploid embryony, and probably diplospory. Haploid parthenogenetic development has been discovered in a certain number of orchids, particularly those where fertilization is late in occurring (Hagerup, 1945, 1947; Maheswari and Narayanaswamy, 1950), and has been described for *Gastrodia elata* by Kusano (1915). In the F_1 generation of a cross between *Bletilla striata* var. *gebina* and *Eleorchis japonica*, Miduno (1940) ascertained that a certain number of the plants are haploid; he thinks that they must have a parthenogenetic origin. Somatic apospory (Swamy, 1948) and related examples are cited by Wirth and Withner (1959).

Megasporic embryony has been recognized by Seshagiriah (1941) in *Zeuxine sulcata*. Gametogenesis does not occur, and the four meiotically

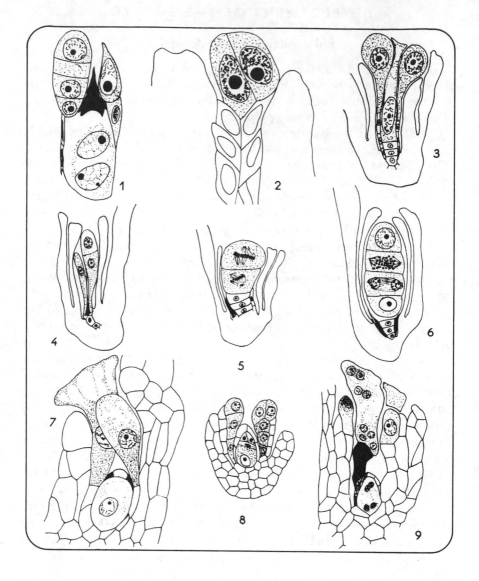

Fig. 5-9. Different forms of apomixis. Parts 1 through 4 show nucellar embryony. (1, 2) *Nigritella nigra*, according to Afzelius. (3, 4) *Zeuxine sulcata*, according to Seshagiriah. (5, 6) Formation of the embryo from the tetrads of macrospores in *Zeuxine sulcata*, according to Seshagiriah. (7, 8) Integumentary embryony in *Spiranthes cernua*, according to Swamy. (9) Somatic apospory in *Spiranthes cernua*, according to Swamy.

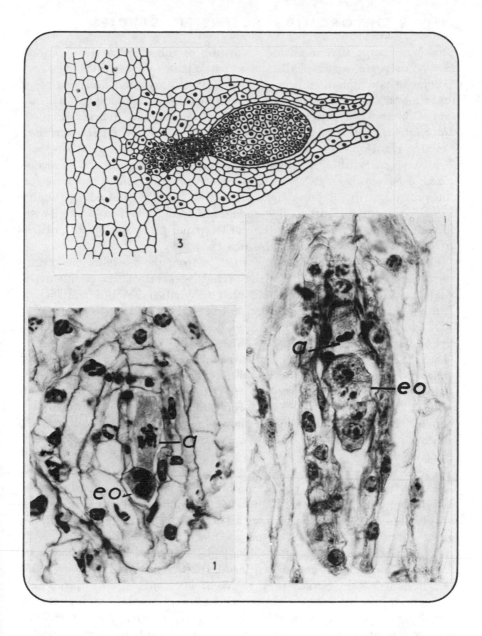

Fig. 5-10. Parts 1 and 2 present omnisaccate diploid embryony. (1) In the presumed hybrid of *Cynorkis lilacina* × *C. ridleyi*. (2) In *C. ridleyi*, *a* = archesporial cell; *eo* = omnisaccate diploid embryo, after Veyret. 565X. (3) Foliar embryo of *Malaxis paludosa*, after Taylor. 100X.

produced spores all join in the formation of the embryo. In *Cynorkis lilacina* only one or two of the spores participate (Veyret, 1965).

Omnisaccate diploid embryony (Fig. 5-10) is known in *Cynorkis ridleyi* and *C. ampullacea,* and the hybrid presumed to be *C. lilacina* × *C. ridleyi.* Diplospory is supposed to occur among triploid hybrids such as *Dactylorchis fuchsii* × *D. purpurella* and *D. fuchsii* × *D. praetermissa* (Heslop-Harrison, 1959).

In some cases, although apomictic formation of seed has been recognized, it has not been determined by which process. This is the case for the probable triploid population of *Listera borealis* studied by Simon (1968). Apomixis occurs in *Zygopetalum mackayi* when pollinated with foreign pollen. In addition to the earlier paper of Suessenguth (1923), Afzélius (1959) has confirmed this process.

Cases of adventitious embryo formation from the nucellus or integuments have been described for the orchids, generally after problems in gametogenesis. Earlier cases are detailed by Wirth and Withner (1959).

A new example of nucellar origin has been described by Veyret (1967, 1972) for *Cynorkis lilacina, C. ridleyi, C. ampullacea,* and the presumed hybrid *C. lilacina* × *C. ridleyi.* It is also noted in *Zeuxine sulcata* (Seshagiriah, 1932), though in a different manner of apomixis. The adventive embryony is integumentary in *Spiranthes cernua* (Leavitt, 1900; Swamy, 1948). In these cases the embryos form from cells of the inner integument (Fig. 5-9). In *Nigritella nigra* it is only the races of high chromosome number ($2n=ca$ 64) known in the mountains of Scandinavia that are adventive (Afzélius, 1932). They are apparently survivors from ancient glacial periods.

The anomalies in the development of the gametes, male or female, are abnormal meiosis in the pollen mother cells in *Zeuxine* (Seshagiriah, 1934); degeneration of the pollen *in situ* in *Spiranthes*; frequent formation of abberrant tetrads with more than four nuclei in *Cynorkis ampullacea*; abortion of the pollen in the presumed hybrid of *C. lilacina* × *C. ridleyi* and also certain races of *C. ridleyi.* The pollen forms normally in *Nigritella* and *Spiranthes,* but its germination has never been observed. On the female side, tetrads form in *Nigritella, Zeuxine,* and *Spiranthes,* but the embryo sac mother cell produces an incomplete and nonfunctional gametophyte. In *Cynorkis* the archesporial cell is the embryo sac mother cell, and it degenerates after an abnormal meiosis and gives rise to a gametophyte consisting generally of two cells. The gametophyte is nonfunctional, with certain exceptions in *C. lilacina.*

Finally, some species are known that produce polyembryonic seeds without the process being known. They are *Orchis latifolia* (Schleiden, cited by Strasburger, 1878), *Orchis morio* (Müller, 1847) *Goodyera tesselata, G. pubescens, Aplectrum hiemale, Habenaria tridentata, H. blephariglottis* (Leavitt, 1901), *Satyrium nepalense* (Swamy, 1944), and *Cypripedium calceolus* (Strasburger, 1877).

Vegetative Apomixis

An unusual means of apomixis was confirmed for the first time by Dickie in 1875. In *Malaxis paludosa* small protuberances or bulbils form at the apical margins of adult leaves. Taylor (1967) showed that these structures simulate an ovule provided with an embryo, and they may properly be called foliar embryos. These embryos are capable of giving rise to secondary foliar embryos by division of their integuments (Fig. 5-10).

The Different Types of Young Seedling

Bernard (1889) found that the seeds of orchids germinate in nature through the symbiosis established between them and the mycelium of the fungal genus *Rhizoctonia*. In the laboratory the simplest way to obtain germination is to use an appropriate culture medium. The germination *in vitro* or under natural conditions is accompanied by a general development pattern common to all species: formation of a tubercle, called the protocorm, which is covered with rhizoids on about two-thirds of its basal part, a lack of formation of the radicle, generally late development of a first root, while several leaves begin developing from the apex.

The form of the young protocorm can be characteristic of certain groups, although there are several exceptions. Among the POLYCHONDREAE the young protocorm is generally elongated and much less thick than in the other taxa and its tip is often curved (Fig. 5-11). Among the KEROSPHAERAE-PLEURANTHAE-MONOPODIALES the protocorm is clearly dorsiventral. It is even provided with a dorsal crest in *Phalenopsis* and *Vanda* (Bernard, 1909) and *Taeniophyllum* (Goebel, 1889) (Fig. 5-12). *Aerides minimum* (Raciborski, 1898) has an appearance resembling *Taeniophyllum*. In asymbiotic germination *in vitro* the dorsiventrality is often less clear. In contrast among the KEROSPHAEREAE-ACRANTHAE and the KEROSPHAEREAE-PLEURANTHAE-SYMPODIALES (Fig. 5-13) the protocorm is a "cattleya" type generally in the form of a top. The protocorm may in a certain species even become discoid. Among several species of these taxa, the protocorm is narrow, as Fuchs and Ziegenspeck (1927a) showed for *Microstylis* (*Acroanthus*), *Corallorhiza innata,* and *Liparis loeselii.* It swells afterward into a club-shaped mass in this last species. The protocorm of a species of *Phymatidium* seems to resemble that of *Taenio-*

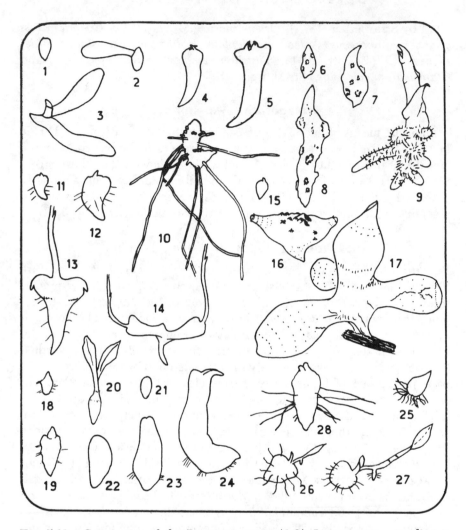

Fig. 5-11. Germination of the POLYCHONDREAE. (1-3) *Listera ovata,* according to Fuchs and Ziegenspeck. (4, 5) *Neottia nidus-avis;* according to Bernard, 5X. (6-8) *Vanilla madagascariensis,* according to Tonnier. (9) *Vanilla fragrans,* according to Bouriquet. (10) *Nervilla crispata,* according to Burgeff, 20X. (11-14) *Epipogium aphyllum,* according to Irmisch. (16, 17) *Didymoplexis minor,* according to Burgeff. (18-20) *Spiranthes spiralis,* according to Fuchs and Ziegenspeck. (21-24) *Manniella gustavi,* according to Veyret 15X. (25-27) *Goodyera repens,* according to Beer. (28) *Goodyera pubescens,* according to Knudson, 10X.

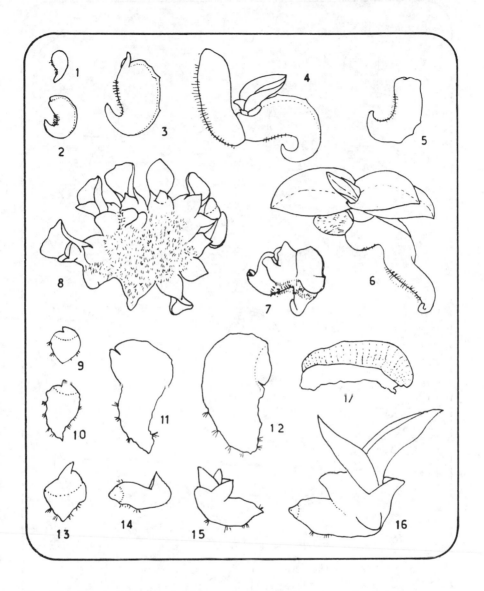

Fig. 12. Germination of the PLEURANTHAE-MONOPODIALES. (1-4) *Phalaenopsis amabilis* × *P. rosea.* 5X. (5-8) *Vanda tricolor,* after Bernard. 5X. (9-12) *Angraecum distichum,* after Veyret. 15X. (13-16) *Acampe renschiana,* after Veyret. 15X. (17) *Taeniophyllum tjibodasanum,* after Burgeff, 5X.

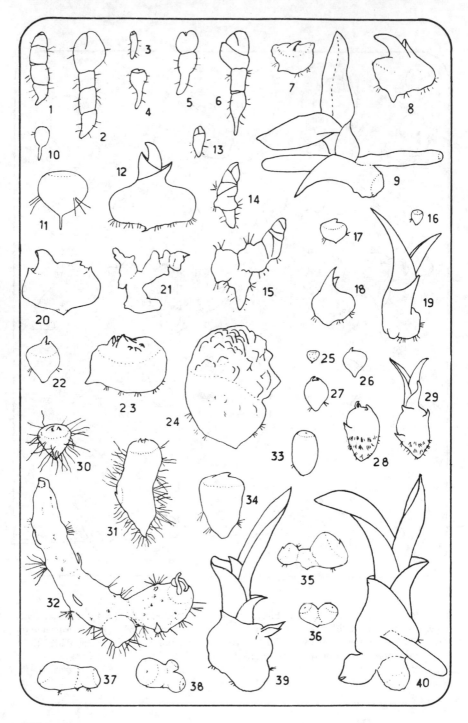

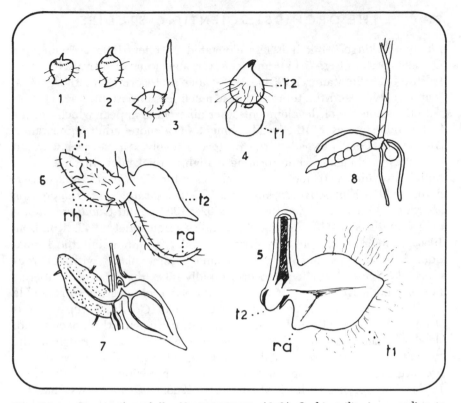

Fig. 5-14. Germination of the Ophrydoideae. (1–3) *Orchis militaris*, according to Irmisch. (4, 5) *Ophrys apifera*, according to Fabre. (4) One of the first forms under which the plantule of the second year is shown. The diameter of the lower tubercule is from 1 to 3 mm. (5) Longitudinal section showing the formation of the tubercule of the second year. (6, 7) *Platanthera montana*, according to Bernard. (6) Plantule of the second year. (7) Section of the same plantule (the infected zone is indicated by the dotted area). 2X. t1 = tubercule of the first year; t2 = tubercule of the second year; ra = root; rh = rhizome.

Fig. 5-13. Germination of the Acranthae (1-12) and the Pleuranthae-Sympodiales. (1, 2) *Microstylis* (*Acroanthus*), and (3-6) *Liparis loeselii*, after Fuchs and Ziegenspeck. (7-9) *Epidendrum nocturnum*, after Veyret. 15X. (10-12) *Cattleya*, after Bernard. 20X. (13-15) *Corallorhiza innata*, after Fuchs and Ziegenspeck. (16-19) *Bulbophyllum bufo*, after Veyret. 10X. (20, 21) *Eulophia maculata* (*Angraecum maculatum*), according to Prillieux and Rivière, after Bernard, and (22-24) after Veyret. (25-29) *Cymbidium giganteum*, after Bernard, 10X. (30-32) *Cymbidium*, after Burgeff. 10X. (33-38) *Odontoglossum pulchellum*, 15X, and (39-40) 10X, after Veyret.

253

phyllum zollingeri, but it is less elongated (Müller, 1895). Among the
OPHRYDOIDEAE (Fig. 5-14) and the CYPRIPEDIOIDEAE (Irmisch, 1853;
Curtis, 1943; Stoutamire, 1964a) the young protocorm is of the general
"cattleya" type, and the tuberization is much less important.

In the sequence of development other differences appear which are the
first manifestations of the establishment of a more adult appearance.
Among the POLYCHONDREAE, the protocorm evolves in several ways: it
continues to elongate while forming a slightly thickened rhizome and is
rapidly provided with roots. This is the case for *Neottia nidus-avis* (Bernard, 1889), *Epipactis atropurpurea, Listera ovata,* and *Limodorum
abortivum* (Fuchs and Ziegenspeck, 1962a, 1962b) and *Goodyera pubescens* (Knudson, 1941). Among the vanillas (Bouriquet, 1947; Knudson,
1950; Tonnier, 1952) one can verify a similar evolution but the formation of the roots is much later; and among some plants with tuberous
roots, like *Spiranthes,* the protocorm rapidly takes the form of a tubercle
(Fuchs and Ziegenspeck, 1927b). *Sobralia macrantha* seems to resemble
it (Hofmeister, 1861). As for the protocorm of *Goodyera repens,* it is
strongly tuberized (Beer, 1863), which is surprising for this species with
slender, branching roots. Finally, among some kinds of *Epipogium,* the
summit of the protocorm enlarges and several primordia appear which
produce the characteristic coralloid roots and aphyllous stems (Irmisch,
1853). The recent work of Stoutamire (1963, 1964b) shows that the
protocorm and young seedlings of plants in the genera *Caladenia, Microtis, Stenostylis, Thelymitra,* and *Diuris* show the classic development of
the POLYCHONDREAE. In the ACRANTHAE and the PLEURANTHAE-SYMPO-
DIALES a cluster of three to five leaves generally forms before the appearance of the first root, which is thick-set and generally similar to that of
the adult. Such young seedlings are found in the following genera:
Cattleya (Bernard, 1903; Knudson, 1935), *Epidendrum* (Pfitzer, 1877;
Veyret, 1965); *Laelia* (Bernard, 1903); *Polystachya, Bulbophyllum, Stan-
hopea,* and *Maxillaria* (Veyret, 1965); *Odontoglossum,* (Bernard, 1909;
Burgeff, 1936; Veyret 1965); *Miltonia* (Prillieux, 1860); and *Arundina*
(Mitra, 1971).

In contrast, Bernard (1909) showed in seedlings of a *Cymbidium* hybrid that after a normal differentiation of the terminal bud at the tip of
the protocorm, tuberization is initiated so that the greatly reduced first
leaves are spread apart. The protocorm thus appears to have several
internodes like the pseudobulb of an adult plant. The protocorm of
Corallorhiza innata (Irmisch, 1853) rapidly evolves into a coralloid root.
The protocorm of *Eulophia maculata* (known then by Prillieux and
Riviere, 1856, as *Angraecum maculatum* or *Eulophidium maculatum;*

Veyret, 1956) also takes on a branched appearance after a conventional germination. The protocorms of *Liparis loeselii* and of one species of *Microstylis* (Fuchs and Ziegenspeck, 1927b) offer a succession of several internodes, somewhat tuberized, before the terminal bud grows, and pseudobulbs in these species are generally scanty.

Among the species of the PLEURANTHAE-MONOPODIALES the elongation of the protocorm in natural conditions is important. The protocorm seems to behave as an aerial root adhering intimately to its support, as do the mature roots of these epiphytes. In an asymbiotic milieu the protocorm is generally shorter and thicker. If in these different groups the habit of the adult is acquired directly at the onset, in several rare exceptions is it different. Among the OPHRYDOIDEAE the sympodial form is slow in establishing itself. This is what Irmisch (1853) showed in *Orchis militaris*, Fabre (1856) in *Ophrys apifera*, Bernard (1900) in *Platanthera montana*, and Fuchs and Ziegenspeck (1927b) among various species of OPHRYDOIDEAE (Fig. 14). In effect the terminal bud develops at most a single small green leaf. The first tubercle of the plant then forms from this bud. It is isolated by the end of the year but is still joined to the original protocorm by a short rhizome that bears a few roots. In the course of the second year the terminal bud with a new tubercle is found drawn away. It is only from this time on that the new growth will be formed from an axillary bud. In *Orchis ustulata*, according to Fuchs and Ziegenspeck (1927a), the young plantule thus produces a succession of tubercles, up to eight, of which the ensemble may reach 20–30 mm. Bernard (1902) thought that this mode of development was closely tied to the mode of infection of the plantule.

In certain cases, as we just noted, the budding of the protocorm is normal and characteristic; it contributes to the building of the plantule. The budding of the protocorm is a general phenomenon among orchids and is able to end in the formation of multiple plantules. They are produced with variable frequency according to the species and the medium used for multiplication.

One or several protocorms can differentiate from the epidermal cells of the initial protocorm. Sometimes, however, the adventitious buds can be directly fixed on the initial protocorm, as has been seen in some plantules of *Bulbophyllum bufo* and *Odontoglossum pulchellum* (Veyret, 1965). This phenomenon can be provoked, augmented, or intensified by provoking a growth factor disorder by the action of a nonspecific species of *Rhizoctonia*, as in experiments by Bernard (1909) on *Vanda tricolor* and *Cymbidium*; or this can be accomplished artificially, as in experiments on *Cattleya* by Mariat (1952), who added some derivatives of barbituric acid to the culture medium. Curtis and Nichol (1948) obtained masses of tissues resembling callus on *Vanda, Cymbidium,* and *Cattleya,* but the exact causes of these proliferations are not yet clearly known.

The Differentiation of the Protocorms

The embryo in mature seeds can be in two different forms, the most frequent by far being the acotyledonous forms. We know of the germination of the cotyledonous species by Bernard's figures (1909) of *Bletilla hyacinthina.* The cotyledon represents the terminal or upper part of the embryo (Figs. 5-4 and 5-5) when its origin can be determined with precision. From the start of germination (Fig. 5-5) what would be the second foliaceous member appears opposite the cotyledon. In the continuation of development, the cotyledon and the first leaf take the form of small, enfolded leaves.

When the embryo presents no morphological differentiation, as in *Stanhopea costaricensis* (Fig. 5-3), germination is accompanied by a general cellular growth except in the upper cells of the embryo. This zone is distinguished increasingly clearly by its meristematic activity, while the greater growth of the cells of the central regions, accompanied by some longitudinal division, ends in the formation of a conical protocorm. The upper part of the protocorm then forms the vegetative apex of the stem. A leaf primordium is not slow in forming from this zone, and the protocorm continues to enlarge. A second foliaceous primordium appears afterward, opposite and a certain distance from the first, thus delimiting the vegetative point of the stem.

The differentiation of the radicle is still incomplete. In the central part of the protocorm axial files of cells organize themselves into conductive elements and form a stele. It is difficult to assign an exact lower limit to this stele because it never entirely differentiates. The stele does not extend to the base of the protocorm even following the degeneracy of some of the lowest cells of the embryo in the course of the germination as is the case with *Polystachya geraensis* or of *Stanhopea costaricensis.* The first root is produced well after the first leaves among the KEROSPHAEREAE and the CYPRIPEDILOIDEAE. Among these taxa, the endogenous origin of the root has been noted in *Bulbophyllum bufo* (Veyret, 1965), *Polystachya geraensis* (Chaiyasut), *Arundina bambusifolia* (Mitra, 1971), *Paphiopedilum* (Bernard, 1909), and *Cypripedium calceolus* (Irmisch, 1853). Among the POLYCHONDREAE, the first root forms while the bud is still little developed. It has an exogenous origin in *Neottia nidus-avis* (Bernard, 1889), and probably in *Vanilla fragrans* (Bouriquet, 1947), also for *C. calceolus* in the CYPRIPEDILOIDEAE (Fuchs and Ziegenspeck, 1926a).

The phenomenon of terminality of the vegetative point seems characteristic for most orchids as compared to most other monocotyledons. Some years ago Souèges (1954) showed in a definitive manner that the vegetative tip of the stem of certain monocots such as *Potomageton natans* or various members of the COMMELINACEAE was also lateral, even though other investigators had thought the origin was terminal. Souèges thought that there might nevertheless be transitional types between the two major groups of angiosperms, as is the case for *Zannichelia palustris* of the NAIADACEAE (Campbell, 1897) where the cotyledon and the vegetative point of the stem are juxtaposed in a terminal position. Among the majority of the orchids this stage is surpassed since the tip of the stem is strictly terminal. One cannot therefore generalize on the opinion of Pfitzer (1877), who, according to his research on the germination of *Dendrochilum glumaceum*, considered the apical part of the embryo as being a cotyledon.

Concluding Remarks

Strongly homogeneous taxa from the morphological point of view are equally so from an embryological viewpoint. The embryos in a broad sense are essentially similar, and the embryogenesis is essentially the same. This applies particularly among the OPHRYDOIDEAE and the KEROSPHAEREAE-PLEURANTHAE-MONOPODIALES. When the various groups are heterogeneous, as among the POLYCHONDREAE or the KEROSPHAEREAE-ACRANTHAE and the PLEURANTHAE-SYMPODIALES, the heterogeneity is manifest in the embryo as well. The POLYCHONDREAE, nevertheless, apart from *Vanilla*, are older than other MONANDRAE (Pfitzer, 1906) and one finds generally among the nonmonopodial KEROSPHAEREAE an embryo belonging to the *Liparis* group, although the species of that group possess different suspensors according to their species or type.

Embryonic characteristics determined by the great laws of embryogeny appear to constitute more certain criteria for classification and for understanding family interrelationships than those relative to the appearance of the suspensor. Resemblances are revealed in the course of embryogenesis that permit one to grasp some of the real affinities by the sequence of the appearance of characteristics in common even before the separation into various taxa. In addition, even in the same genus the suspensor can vary according to the species. For instance, Swamy indicated that the suspensor of *Dendrobium barbatulum* consisted of an enormous haustorial vesicle, while that of *D. microbulbon* was represented by a single short cell.

One could argue that certain principles of an embryological classification are in default when a single genus, such as *Epidendrum* or *Polystachya*, includes both cotyledonous and noncotyledonous species. This phenomenon is explainable since these are distinctly irregular species and possess different types of tetrads as well, revealing probably an old hybrid origin among different species or genera. It is interesting to note that these two taxa are reunited by Dressler and Dodson (1960) in the subtribe EPIDENDRINAE, and that they represent an important crossroad as suggested in their diagram of relationships within the subtribe.

Johansen's classification loses much of its interest because of too detailed emphasis on the character of the suspensor. In Swamy's classification of the plants in the A group, the proembryo of the second cellular generation does not always offer a similar arrangement of blastomeres in a T formation, the wall of the division being vertical in the terminal cell as well as in the basal cell of the two-celled proembryo of *Listera* and *Neottia*. This is important since among certain POLYCHONDREAE where the proembryo is formed by the whole egg cell, the species having a proembryonic tetrad of A1 type (Fig. 5-2) are embryologically less involved than others in the A2 category.

The DIANDRAE are little known embryologically and the APOSTASIEAE are hardly known at all. In the MONANDRAE one passes from the POLYCHONDREAE, the oldest of the MONANDRAE, to the OPHRYDOIDEAE, then to the KEROSPHAEREAE, sympodial and, finally, monopodial. This means that the evolution of the orchid is from a terrestrial habit to an epiphytic one, from the granular to a solid state for the pollinia, and from having only a caudicle to having a stipe. Among the plants that disturb this scheme embryologically are those in the genus *Vanilla*, in the POLYCHONDREAE, placed embryologically in the *Liparis* group closer to the KEROSPHAEREAE. This may be explained since the vanillas are different from other POLYCHONDREAE by their vining growth and by their pollinia with simple grains united by viscous material and not by elastic threads. There are genera as well, *Eulophia* in part, *Cymbidium, Geodorum, Stanhopea,* and *Polystachya,* that by their belonging to a category of embryological classification other than the first period, show even higher level than that occupied by the PLEURANTHAE-MONOPODIALES. But these genera may be considered vandaceous in the first sense of the term as introduced by Lindley (1830–1840)—"with a distinct caudicula united to a deciduous stigmatic gland." The caudicle of these plants is actually properly called a stipe, since the origin of the structure is different from the caudicle of the BASITONAE. These plants are therefore among the most evolved.

Among the plants where the pollen apparatus presents these characters, one also finds some genera, such as *Maxillaria*, belonging to the first period. This would coordinate better with Garay's (1972) recent discussion of evolution in the orchids where he suggests two major tribes in the previously known KEROSPHAEREAE or ACRANTHAE-PLEURANTHAE, the EPIDENDREAE and VANDEAE, based on anther and column structure, rather than the older systems based on the inflorescence and sympodial or monopodial growth habits. It is certain that much research must still be done to explain the heterogeneous mixtures of embryological types in those taxa.

It is interesting to note that among some species of *Eulophia*, such as *E. oedoplectron,* the tetrad is in C2 form apart from the characteristics that bring about the passage from the first to the second period of embryological classification. This species is in addition irregular and does not follow the general mode of embryological development of other *Eulophia*. It is necessary here to underline that the species of *Eulophia* constitute a strongly variable genus with some species terrestrial, others cpiphytic, although more rarely so. The stems may be leafed to the base, or twining, or more or less thickened into rhizomes, or into aerial pseudobulbs. As for the species of *Coelogyne,* their place in the second period cannot yet be explained. We have few data on the embryo of the CYPRIPEDILOIDEAE, and it is a little surprising to find such heterogeneity within the single genus *Cypripedium* where the embryo may or not be provided with a suspensor, and the root has a different origin.

The relative morphological characteristics of the young embryos are very general, but they confirm the reality of the large taxonomic groups and allow one to separate more clearly the MONOPODIALES (or VANDEAE following Garay's usage) in the midst of the KEROSPHAEREAE, and to bring closer the ensemble of the ACRANTHAE with the sympodial PLEURANTHAE. These last two taxa could then be subdivided, first taking into consideration the characteristics of the pollinarium and no longer emphasizing the terminal or lateral position of the inflorescence. Garay, in effect, has done this. The characteristics of the embryo relative to the origin of the vegetative apex of the stem, and the corresponding precocious histological differentiation of the proembryo, as well as those characteristics of subsequent development, confirm the highly evolved state of the orchids.

Bibliography

ABE, K. 1968. Contributions to the embryology of the family Orchidaceae, II. Development of the embryo sac in *Pogonia japonica* Reich. f. *Sci. Rep. Tôhoku Univ. ser. 4, Biol.* 34:59–65.

———. 1971. Contributions to the embryology of the family Orchidaceae, IV. Development of the embryo sac in *Bletilla striata*. *Sci. Rep. Tôhoku Univ. ser. 4, Biol.* 35:213–218.

——. 1972. Contributions to the embryology of the family Orchidaceae, VI. Development of the embryo sac in 15 species of orchids. *Sci. Rep. Tôhoku Univ. ser. 4, Biol.* 36:135–178.

AFZÉLIUS, K. 1916. Zur Embryosackentwicklung der Orchideen. *Svensk. Bot. Tidskr.* 10:183–227.

——. 1928. Die Embryobildung bei *Nigritella nigra*. *Svensk Bot. Tidskr.* 22:82–91.

——. 1932. Zur Kenntnis der Fortpflanzungsverhältnisse und chromosomenzahlen bei *Nigritella nigra*. *Svensk Bot. Tidskr.* 26:365–369.

——. 1954. Embryo sac development in *Epipogium aphyllum*. *Svensk Bot. Tidskr.* 48:513–520.

——. 1959. Apomixis and polyembryony in *Zygopetalum mackayi* Hook. *Acta Horti Bergiami* 19:2.

——. 1966. Cleistogamy in *Pleurothallis procumbens* Lindl. *Acta Horti Bergiami* 20:313–317.

BARANOV, P. 1915. Recherches sur le développement du sac embryonnaire chez les *Spiranthes australis* Lindl. et *Serapias pseudocordigera* Moric. *Bull. Soc. Imp. Nat. Moscou. N.S.* 29:74–92.

——. 1917. Contributions à l'étude de l'embryologie des Orchidées. (French summ.) *J. Soc. Bot. Russ.* 2:20–29.

——. 1924. Contributions à l'étude de l'embryologie des Orchidées, II. *Herminium monorchis* R. Br. *J. Soc. Bot. Russ.* 9:5–9.

BEER, J. B. 1863. *Beiträge zur Morphologie und Biologie der Familie der Orchideen*. Wien.

BERNARD, N. 1889. Sur la germination du *Neottia nidus-avis*. *Compt. Rend. Acad. Sci. Fr.* 128:1253–1255.

——. 1900. Sur quelques germinations difficiles. *Rev. Gén. Bot.* 12:108–120.

——. 1902. Études sur la tubérisation. *Rev. Gén. Bot.* 14:5–101.

——. 1903. La germination des Orchidées. *Compt. Rend. Acad. Sci. Fr.* 137:483–485.

——. 1904. Recherches expérimentales sur les Orchidées. *Rev. Gén. Bot.* 16:405–451; 458–476.

——. 1909. L'evolution dans la symbiose. Les Orchidées et leurs champignons commensaux. *Ann. Sci. Nat. Bot. 9e sér.* 9:1–196.

BOURIQUET, G. 1947. Sur la germination des graines de Vanillier (*Vanilla planifolia* And.) *Agron. Trop.* 2:150–164.

BRAUN, A. 1859. Über Polyembryonie und Keimung von *Coelebogyne*, ein Nachtrag zur der Abhandlung über Parthenogenesis bei Pflanzen. *Abhand. Königl. Akad. Wiss. Berlin* 1859:109–263.

BROWN, W. H. 1909 The embryo sac of *Habenaria*. *Bot. Gaz.* 48:241–250.

BROWN, W. H., and SHARP, L. W. 1911. The embryo sac of *Epipactis*. *Bot. Gaz.* 52:439–452. Pl. X.

BURGEFF, H. 1936. *Samenkeimung der Orchideen und Entwicklung ihre Keimpflazen*. Gustave Fisher, Jena.

CAMPBELL, D. H. 1897. A morphological study of *Naias* and *Zannichelia*. *Proc. Cal. Acad. Sci. 3rd ser. Bot.* 1:1–67.

CARLSON, M. C. 1940. Formation of the seed of *Cypripedium parviflorum*. *Bot. Gaz.* 102:295–301.

CHAIYASUT, K. (Unpublished.) Embryologie et blastologie du *Polystachya geraensis* Rodrig.

CURTIS, J. T. 1943. Germination and seedling development in five species of *Cypripedium* L. *Am. J. Bot.* 30:199–206.

CURTIS, J. T., and NICHOL, M. A. 1948. Culture of proliferating orchid embryos *in vitro*. *Bull. Torrey Bot. Club* 75:358–373.

DICKIE, G. 1875. Note on the bud development of *Malaxis*. *J. Linn. Soc. London, Botany* 14:1–3.

DRESSLER, R. L., and DODSON, C. H. 1960. Classification and phylogeny in the Orchidaceae. *Ann. Mo. Bot. Gard.* 47:25–68.

DUMÉE, P. 1910. Quelques observations sur l'embryon des Orchidées. *Bull. Soc. Bot. Fr.* 47:83–87.

DUTHIE, A. V. 1915. Note on apparent apogamy in *Pterigodium newdigatae*. *Trans. Royal Soc. S. Africa* 5:593–598.

FABRE, J. H. 1856. De la germination des Ophrydées et de la nature de leurs tubercules. *Ann. Sci. Nat. Bot. 4e sér.* 5:163–186.

FLEISCHER, E. 1874. Beiträge zur Embryologie der Monokotylen und Dikotylen. *Flora* 57:419–423. Pl. VIII.

FUCHS, A., and ZIEGENSPECK, H. 1926a. Die Entwicklungsgeschichte der Axen der einheimischen Orchideen und ihre Physiologie und Biologie, I. *Cypripedium, Hellehorine, Limodorum, Cephalanthera. Bot. Arch.* 14:165–260.

———. 1926b. II. *Listera, Neottia, Goodyera. Bot. Arch.* 16:360–413.

———. 1927a. Die Entwicklungsgeschichte der einheimischen Orchideen und der Bau ihre Axen, II. *Bot. Arch.* 18:378–475.

———. 1927b. Entwicklung, Axen und Blätter einheimischen Orchideen, IV. *Bot. Arch.* 20:275–422.

GARAY, L. A. 1972. On the origin of the Orchidaceae, II. *J. Arnold Arboretum* 53:202–215.

GEITLER, L. 1956. Zur Fortpflanzungsbiologie, Embryologie und mechanistichen Deutung der Embryogenese von *Epipogium aphyllum*. *Oester. Bot. Zeit.* 103:312–325.

GOEBEL, K. 1889. Pflanzcn biologische Schilderungen. I.

HAGERUP, O. 1944. On fertilization, polyploidy and haploidy in *Orchis maculatus* L. sensu lato. *Dansk. Bot. Arkiv.* 2:1–25.

———. 1945. Facultative parthenogenesis and haploidy in *Epipactis latifolia*. *Det. Kgl. Danske Vidensk. Selsk. Biol. Medd.* 19:1–14.

———. 1947. The spontaneous formation of haploid, polyploid and aneuploid embryos in some orchids. *Det. Kgl. Danske Vidensk. Selsk. Biol. Medd.* 20:1–22.

HESLOP-HARRISON, J. 1959. Apomictic potentialities in *Dactylorchis*. *Proc. Linn. Soc. London.* 170:174–178.

HEUSSER, K. 1915. Die Entwicklung der generativen Organe von *Himantoglossum hircinum* Spr. (= *Loroglossum hircinum* Rich.) *Beih. Bot. Centralbl. Abt.* 12: 218–277.

HOFMEISTER, W. 1861. Neue Beiträge zur Kenntnis der Embryobildung der Phanerogamen, II. Monocotyledonen. *Abh. Sächs. Ges. Wiss.* 7:629–760.

IRMISCH, T. 1853. *Beiträge zur Morphologie und Biologie der Orchideen.* Leipzig.

JOHANSEN, D. A. 1950. *Plant embryology. Embryology of the Spermatophyta.* Chron. Bot. Waltham, Mass.

KIMURA, C. 1971. Embryological studies of *Galeola septentrionalis* Reich. f. (Orchidaceae). *Sci. Rep. Tôhoku Univ. ser. 4* 35:253–258.

KNUDSON, L. 1935. Germination of orchid seed. *Am. Orchid Soc. Bull.* 3:65–66.

———. 1941. Germination of seed of *Goodyera pubescens. Am. Orchid Soc. Bull.* 10: 199–201.

———. 1950. Germination of seeds of *Vanilla. Am. J. Bot.* 37:341–347.

KUSANO, S. 1915. Experimental studies on the embryological development in an angiosperm. *J. Coll. Agric. Univ. Tokyo* 6:7–120.

LEAVITT, R. G. 1900. Polyembryony in *Spiranthes cernua. Rhodora* 2:227–228.

———. 1901. Notes on the embryology of some New England orchids. *Rhodora* 3:61–63; 202–205.

LINDLEY, J. 1830–1840. *The Genera and Species of Orchidaceous Plants.* London.

MAHESWARI, P., and NARAYANASWAMI, S. 1950. Parthenogenetic development of the egg in *Spiranthes australis. Curr. Sci. India* 19:249–250.

———. 1952. Embryological studies on *Spiranthes australis* Lindl. *Trans. Linn. Soc. London* 53:474–486.

MARIAT, P. 1952. Recherches sur la physiologie des embryons d'Orchidées. *Rev. Gén. Bot.* 59:324–377.

MIDUNO, T. 1940. Chromosomen Studien an Orchidaceen, III. Über das Vorkommen von haploiden Pflanzen bei *Bletilla striata* Reichb. f. var. *gebina* Reichb. f. *Cytologia* 2:156–177.

MITRA, G. C. 1971. Studies on seeds, shoot-tips and stem-discs of an orchid grown in aseptic cultures. *Indian J. Exp. Biol.* 9:79–85.

MONTEVÉRDÉ, M. 1880. Recherches embryologiques sur l'*Orchis maculata. Bull. Acad. Sci. St-Pétersbourg* 26:326–335.

MÜLLER, F. 1895. Orchideen von unsicherer Stellung. *Ber. Dtsch. Bot. Gesell.* 13: 199–210.

MÜLLER, K. 1847. Beiträge zur Entwicklungsgeschichte des Pflanzen Embryos. *Bot. Z.* 5:73–78.

OLSSON, O. 1967. Embryological studies in the Orchidaceae. The genus *Hetaeria. Svensk. Bot. Tidskr.* 61:33–42.

PACE, L. 1907. Fertilization in *Cypripedium. Bot. Gaz.* 44:353–374.

PASTRANA, M. D., and SANTOS, J. K. 1931. A contribution to the life history of *Dendrobium anosmum* Lindl. *Nat. Appl. Sci. Bull. Philippines Univ.* 1:133–134.

PFITZER, E. 1877. Beobachtungen über Bau und Entwicklung der Orchideen. *Nat. Med. Verhandl. Heidelberg* 2:23–30.

——. 1906. On the phylogeny of orchids. Report of the third international conference on genetics. *Royal Hort. Soc. London* **1906**:476–481.

PODDUBNAJA-ARNOLDI, V. A. 1959. (In Russ.) Ovule culture of some orchids on synthetical media. *Dokl. Akad. Nauk. SSSR* **1959**:223–226.

——. 1960a. Study on fertilization in the living material of some Angiosperms. *Phytomorphology* **10**:185–198.

——. 1960b. (In Russ.) Polyembryony in orchids. *Bjull. Glavn. Bot. Sada.* **36**: 56–61.

PRILLIEUX, E. 1860. Observations sur la germination du *Miltonia spectabilis* et de diverses autres orchidées. *Ann. Sci. Nat. Bot. 4e sér.* **13**:288–296. Pl. 14.

PRILLIEUX, E., and RIVIÈRE, A. 1856. Observations sur la germination et le développement d'une orchidée (*Angraecum maculatum*). *Ann. Sci. Nat. Bot. 4e sér.* **5**: 119–136. Pl. 5 to 7.

PROSINA, M. N. 1930. Über die von *Cypripedium*-typus abweichende Embryosackentwicklung von *Cypripedium guttatum*. Sw. Z. Wiss. *Biol. Planta* **12**:532–544.

RACIBORSKI, M. 1898. Biologische Mittheilungen aus Java. *Flora* **75**:325–361.

SALISBURY, R. A. 1804. On the germination of the seeds of Orchideae. *Trans. Linn. Soc.* **7**:29–32.

SCHACHT, H. 1850. *Entwicklungsgeschichte der Pflanzen embryon.* Verhand. Eerste Klasse Koninkl. Nederland Inst., Amsterdam.

SCHLECHTER, R. 1926. Das System der Orchidaceen. *Nat. Bot. Gart. Mus. Berlin-Dahleim* **9**:563–591.

SCHNARF, R. 1931. *Vergleichende Embryologie der Angiospermen.* Berlin.

SENIANINOVA, M. 1924. (In Russ. with French summ.) Étude embryologique de l'*Ophrys myodes. J. Soc. Bot. Russ.* **9**:10–13.

SESHAGIRIAH, K. N. 1932. Development of the female gametophyte and embryo in *Spiranthes australis* Lindl. *Curr. Sci.* **1**:102.

——. 1934. Pollen sterility in *Zeuxine sulcata. Curr. Sci.* **3**:205–206.

——. 1941. Morphological studies in Orchidaceae, I. *Zeuxine sulcata* Lindl. *J. Indian Bot.* **20**:357–365.

SHARP, L. 1912. The orchid embryo sac. *Bot. Gaz.* **54**:372–384.

SIMON, W. 1968. Chromosome numbers and B chromosomes in *Listera. Caryologia* **21**:181–189.

SOUÈGES, R. 1936–1939. *Exposés d'embryologie et de morphologie végétale,* Vol. IV. *La segmentation. Les blastomères;* Vol. VIII. *Les lois du développement;* Vol. IX. *Embryogénie et classification. Essai d'un système embryogénique.* Hermann et Cie., Paris.

——. 1954. L'origine du cône végétatif de la tige et la question de la "terminalité" du cotylédon des Monocotylédones. *Ann. Sci. Nat. Bot. IIe sér.* **72**:1–20.

STOUTAMIRE, W. P. 1963. Terrestrial orchid seedlings. *Australian Plants, Orchidaceae* **2**:119–122.

——. 1964a. Seeds and seedlings of native orchids. *Mich. Bot.* **3**:107–119.

——. 1964b. Terrestrial orchid seedlings, II. *Australian Plants, Orchidaceae* **2**:264–266.

STRASBURGER, E. 1877. Über Befruchtung und Zelltheilung. *Jenaische Z.* 2:435–536.

———. 1878. Über Polyembryonie. *Jenaische Z.* 12:647–669.

SUESSENGUTH, K. 1923. Über Pseudogamie bei *Zygopetalum mackayi* Hook. *Ber. Dtsch. Bot. Gesell.* 41:16–23.

SWAMY, B. G. L. 1942a. Female gametophyte and embryogeny in *Cymbidium bicolor* Lindl. *Proc. Indian Acad. Sci. B.* 15:194–201.

———. 1942b. Morphological studies in three species of *Vanda*. *Curr. Sci.* 2:285.

———. 1943a. Embryology of Orchidaceae. *Curr. Sci.* 12:13–17.

———. 1943b. Gametogenesis and embryogeny of *Eulophia epidendracea* Fisher. *Proc. Nat. Inst. Sci. India* 9:59–65.

———. 1944. The embryo sac and the embryo of *Satyrium nepalense* Don. *J. Indian Bot. Soc.* 23:56–70.

———. 1946a. Some notes on the embryo of *Cymbidium bicolor* Lindl. *Curr. Sci.* 15:139–140.

———. 1946b. The embryology of *Zeuxine sulcata* Lindl. *J. Indian Bot. Soc.* 20:357–365.

———. 1946c. Embryology of *Habenaria*. *Proc. Nat. Inst. Sci. India* 12:413–426.

———. 1947. On the life history of *Vanilla planifolia*. *Bot. Gaz.* 108:449–456.

———. 1948a. Agamospermy in *Spiranthes cernua*. *Lloydia* 2:149–162.

———. 1948b. The embryology of *Epidendrum prismatocarpum*. *Bull. Torrey Bot. Club* 75:245–249.

———. 1949. Embryological studies in the Orchidaceae, II. Embryogeny. *Am. Midl. Nat.* 41:202–232.

TAYLOR, R. L. 1967. The foliar embryos of *Malaxis paludosa*. *Can. J. Bot.* 45:1553–1556.

TOHDA, H. 1967. An embryological study of *Hetaeria shikokiana*, a saprophytic orchid in Japan. *Sci. Rep. Tôhoku Univ. ser. 4, Biol.* 33:83–95.

———. 1968. Development of the embryo of *Bletilla striata*. *Sci. Rep. Tôhoku Univ. ser. 4, Biol.* 34:125–131.

———. 1971a. Development of the embryo of *Orchis aristata*. *Sci. Rep. Tôhoku Univ. ser. 4, Biol.* 36:239–243.

———. 1971b. Development of the embryo of *Lecanorchis japonica*. *Sci. Rep. Tôhoku Univ. ser. 4,* 35:245–251.

TONNIER, J. P. 1952. Le Vanillier. *Bull. Rech. Agron. Madagascar* 1952:55–59.

TREUB, M. 1879. Notes sur l'embryogénie de quelques orchidées. *Naturk. Verh. Koninkl. Akad. Amsterdam* 19:1–50. Pl. I to VIII.

———. 1883. Notes sur l'embryon, le sac embryonnaire et l'ovule. *Ann. Jard. Bot. Buitenzorg* 3:76–87. Pl. XII.

VEYRET, Y. 1965. Embryogénie comparée et blastogénie chez les Orchidaceae-Monandrae. *Mémoires ORSTOM* 12.

———. 1967. L'apomixie chez le *Cynosorchis lilacina* Ridley (Orchidaceae). *Compt. Rend. Acad. Sci. Fr. sér. D*, 265:1713–1716.

———. 1969. La structure des semences des Orchidaceae et leur aptitude a la germination *in vitro* en cultures pures. *Mus. Hist. Nat. Paris, Trav. Lab. "La Jaysinia"* 3e fasc. 89–98. Pl. III and IV.

———. 1972. Études embryologiques dans le genre *Cynorkis* (Orchidaceae). *Adansonia, sér. 2* 12:389–402.

WARD, H. M. 1880. On the embryo sac and development of *Gymnadenia conopsea*. *Q. J. Microsc. Soc.* 10:1–18.

WIRTH, M. W., and WITHNER, C. L. 1959. Embryology and development in the Orchidaceae, in *The Orchids: A Scientific Survey*. Ronald Press, New York, pp. 155–188.

6

The Anatomy of Orchids

CARL L. WITHNER, PETER K. NELSON, AND PETER J. WEJKSNORA

To write a review on the anatomy of orchids is a lengthy and painstaking endeavor, and it becomes difficult to produce an essay that is not just an annotated bibliography. We have chosen, therefore, two themes with which to associate our coverage: evolution and ecology. We hope thereby to avoid the impossible tedium of Solereder and Meyer (1930). This classic source of information on orchid anatomy, though complete in detail and replete with unalphabetized references to even obscure early authors, gives no feeling for the living, growing, evolving orchid plant. We hope this chapter can remedy that situation as well as collate the more current research on orchid anatomy with a perspective that might lead to more complete information on many interesting points. The anatomy of orchids is no less fascinating than their variety or their flowers. As Vermeulen (1966) has said, "The ORCHIDACEAE excel in an overwhelming polymorphy."

Floral Structure and Vasculature

The structure of the orchid flower was first described by Robert Brown (1833). His survey compared its structure with that of closely allied monocots. He concluded that there were three stigmas present in the flower and that the rostellum was a modification of a part of the third stigma. The designation of a numbered stigma was derived from its position relative to the dorsal sepal. Thus the stigma opposite the dorsal sepal was considered number one, with the remaining two numbered counterclockwise (Fig. 6-1). Brown stated that the stamens were arranged in two whorls numbered in like manner to the stigmas, with the outer whorl elements corresponding to sepal arrangements and the inner whorl

A

B

C

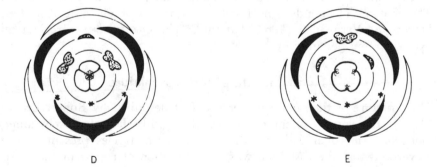

D

E

Fig. 6-1. Diagrams of flower parts. (A) Typical monocot with radially arranged sepals and petals, stamens, and carpels, and with all stamens functional. (B) Idealized orchid as interpreted by Brown and Darwin with two stamens of the outer whorl contributing to the structure of the lip. (C) *Neuwiedia*, a primitive orchid with three functional stamens. (D) Modern diandrous orchid. (E) Modern monandrous orchid. (C, D, E, after Garay, 1960.)

elements corresponding to petal arrangements. The labellum was presumably a compound structure resulting from the fusion of staminal elements with the lip petal. Brown further suggested that the protrusion on the labellum of *Glossodia* corresponded to a staminal trace, though later research described below tends to refute this. He noted that the ovary consisted of three carpels.

Darwin (1884), in studying the paths of vascular bundles in flowers representing various orchid tribes, verified many of Brown's assumptions. He concluded that the labellum was a compound structure made of the median petal and the two lateral stamens of the outer whorl. This notion was also held by Lindley (1853) (Fig. 6-1). Though subsequent investigations have shown this assumption about lip structure to be erroneous, many of Darwin's other observations remain valid.

The relation of floral vasculature to the evolutionary position of orchids has been considered by Garay (1960). His studies and those of Swamy (1948) of the details of vascular structure in orchid flowers (Fig. 6-1) have been successfully used in understanding the nature of the orchid flower and have proved some earlier notions false. Garay's (1972) recent paper seems to resolve finally the various questions of floral anatomy in relation to orchid taxonomy and along with the 1960 paper will be cited many times for the relevance of its comparative anatomical data. Placed finally in the proper perspective, the genera *Apostasia* and *Neuwiedia* no longer seem questionable for inclusion in the ORCHIDACEAE.

Vascular Anatomy. In examining the vasculature of the orchid flower, we are concerned with two points: first, the origin of the six vascular bundles (called traces) which run along the ovary; and second, the pattern of splitting of these traces to form the traces to the various parts of the flower (Fig. 6-2). Swamy, in studying both monandrous and diandrous flowers, found considerable variation in the origins of the six ovarian traces and in the actual number of bundles that enter the flower to form these six traces. In diandrous *Cypripedium* and *Paphiopedilum* (CYPRIPEDIOIDEAE) six bundles enter the flower independently (Fig. 6-2A) and make up the main traces in the ovary. A seventh separate bundle supplies the bract. Garay (1960) reports that investigations into the floral vasculature of APOSTASIOIDEAE (Fig. 6-3) reveal the same pattern of six independent bundles reported in *Cypripedium* by Swamy. The monandrous orchids generally display a pattern of only three bundles entering the floral axis (Fig. 6-2B, C, D, E), which then split to form the six ovarian traces. The bract is supplied by an offshoot from one of the three.

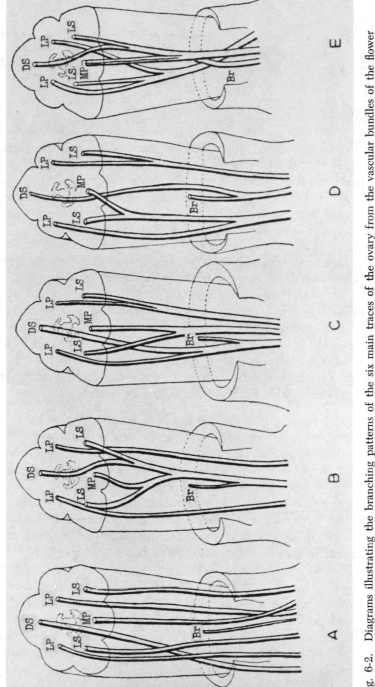

Fig. 6-2. Diagrams illustrating the branching patterns of the six main traces of the ovary from the vascular bundles of the flower stalk. (After Swamy, 1948.)

Among different genera there is variation in both the pattern of divergence and the position at which the divergence takes place. In *Habenaria* (ORCHIDOIDEAE), for instance, a center bundle gives rise to the bract trace (Fig. 6-2B) then splits into four traces, which supply the dorsal and two lateral sepals (DS, LS) and the modified petal, or labellum (MP). The two side traces continue intact to form the two lateral petal traces (LP).

In *Vanilla* (NEOTTIOIDEAE), *Rhynchostylis*, some species of *Dendrobium*, and a few others (EPIDENDROIDEAE) the three vascular bundles each form two ovarian traces (Fig. 6-2C), one to the lateral petal and lateral sepal (LP, LS), and the other to the lateral petal and dorsal sepal (LP, DS), respectively. The bundle that supplies the lateral petal and dorsal sepal branches off before diverging and provides the bundle for the bract.

Other species of *Dendrobium*, *Bulbophyllum*, and *Pholidota* show a pattern (Fig. 6-2D) in which one bundle forms three traces supplying the lateral and modified petal and the lateral sepal (LP, MP, LS). The second bundle supplies only the bract and dorsal sepal (DS). The third bundle forms the traces to the other lateral petal and lateral sepal (LP, LS).

In the final type (Fig. 6-2E), described by Swamy and demonstrated by *Cymbidium, Eulophia, Spathoglottis*, and certain members of the tribe SARCANTHINEAE, one bundle supplies the bract and the other two trifurcate to form the traces to the modified petal, the dorsal and lateral sepal (MP, DS, LS), and to the two lateral petals and lateral sepal (LP, LS), respectively.

The six traces, however formed, travel through the ovary and, except for their positions relative to the axis of the ovary, are indistinguishable from each other. Though uniformity of size is generally maintained in all six, there are exceptions. In certain orchids, such as *Cymbidium, Eulophia*, and *Spathoglottis*, Swamy (1948) observed a double nature in the trace supplying the labellum. He further noted the fact that *Cypripedium*, with a larger labellum than many orchids, shows only a single trace. This would tend to negate any explanation based on the relationship between the size of the organ and the trace size supplying it.

At no place is there any evidence of traces branching off to supply the placentae; thus the ovules are deprived directly of any vascular system, which would be unusual in other flowering plants. The evolutionary trend toward formation of quantities of ovules, so typical of orchids, would almost preclude individual vascular supplies to each one as a simple matter of available space in the placental ridges and their ovule-bearing projections. This makes one wonder whether orchid seeds are so minute because they were never afforded an ample food supply via developed vascular traces. It is also likely that the vasculature was once present in an ancestral form and subsequently degenerated and disappeared.

Ovary.   The nature of the ovary is still subject to debate. The classical view of Brown states that the ovary is unilocular and tricarpellate. Lindley (1853) advanced the view that the ovary was of six basic parts, though it was a view generally ignored in favor of the tricarpellary argument. Saunders (1923, 1937) contended that each of the six traces in the ovary wall represented a separate carpel and that the carpels of the traces continuing to the outer perianth whorl (LS, DS) are sterile and solid. Swamy (1948) questioned the validity of this classical assumption, pointing out a supposed lack of supportive anatomical evidence and inconsistencies inherent in the hypothesis.

Vermeulen (1966), citing numerous workers (Eichler, 1875; Pfitzer, 1889; Pulle, 1952) and his own observations of both prepared materials and the drawings of other investigators (Camus, 1929; Poddubnaya-Arnoldi, 1960; Garay, 1960), concludes that the ovary is, in fact, made up of six parts: three sterile elements alternating with three fertile elements, each of the latter bearing double placentae, and each element supplied by a single trace. Garay's plate (Fig. 6-3) illustrates how the placentation and carpellary arrangement probably evolved. He visualizes it as proceeding from a three carpellate condition with axile placentation, as in some primitive orchids as *Apostasia*, to a unilocular ovary with parietal placentation in more advanced forms. Garay feels the unilocular condition may have come about "by a longitudinal division of the septa which eliminated the torus, followed by a gradual shortening of the lamella, until merely traces are found along the inner wall."

Vasculature of Stamens, Stigmas, and Spurs.   The second level of divergence, that where the ovarian traces break up into the traces of the individual floral parts, occurs just below the level of insertion of the perianth. Three rather distinct, though similar, patterns of splitting to the staminal and stigmatic whorls are described by Swamy (1948).

In the first type, present in CYPRIPEDILINAE (Fig. 6-4A, B, C), the first traces to be separated are those to the stigmas (G1–G3). These arise from the dorsal and lateral sepal traces (DS, LS). Higher up, both trifurcate to become the marginal and median traces of the perianth. The main dorsal sepal trace (DS) gives rise to a smaller trace (A1), which leads into the staminode. The lateral petal traces (LP) form the staminal traces (a1, a2) and then continue intact to form the median vein in their respective petals. Masters (1887), in studying *Cypripedium*, observed that

1
 ́ *Apostasia nuda*

2
Selenipedium Chica

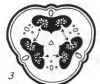

3
*Phragmipedium
       longifolium*

4
*Lecanorchis
       javanica*

5
*Limodorum
    abortivum*

6
*Cephalanthera
       alba*

Fig. 6-3. The evolution of placentation and carpellary arrangement as
interpreted by Garay (1960). See text
for details.

the trace underlying the labellum trifurcates, forming marginal and
median traces to the labellum. The subsequent studies of Swamy, however, do not bear this out.

In the second type, characterized by members of the OPHRYDINAE
(Fig. 6-4D, E, F), the main dorsal sepal trace (DS) forms only the traces
which supply the median stigma (G1), the stamen (A1), and the median
trace underlying the corresponding sepal. The marginal traces of that
sepal are from the traces of the lateral petals (LP). The labellum trace
(MP) forms only the median trace of the labellum and its spur, a downward, pouchlike extension of the labellum. When the spur is single, the
median trace of the labellum (MP) runs through its entire length and
then upward into the labellum. When the spur is double, a marginal
trace from the one lateral sepal enters one spur; a similar marginal trace
from the other lateral sepal enters the second spur. Each runs the length
of its spur before entering the labellum (Fig. 6-5). In this case, the
labellar trace immediately enters the labellum without a detour into
either spur. The lateral stigmal traces (G2, G3) arise from the lateral
sepal trace (LS). Swamy also notes that marginal traces to the lateral

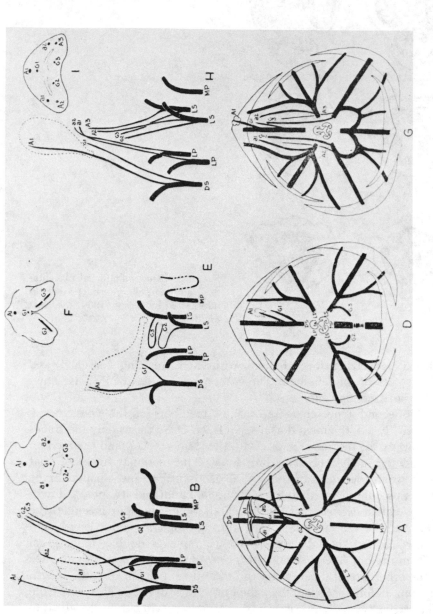

Fig. 6-4. The origin of the vasculature in the various flower parts from the six traces of the ovary.(A, B, and C) *Paphiopedilum.* (D, E, and F) *Habenaria.* (G, H, and I) Typical monandrous orchid. (After Swamy, 1948.) Details in text.

274

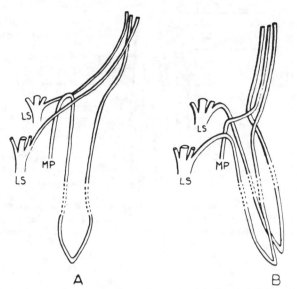

Fig. 6-5.   The nature of the vascular supply to the labellum with a spur. (A) When the spur is single.(B) When the spur is double. (After Swamy, 1948.)

petals are lacking in *Habenaria* and *Platanthera* and poorly developed or absent in some species of *Euhabenaria*.

The majority of the monandrous orchids show the third type of organization (Fig. 6-4G, H, I), in which the lateral petal traces (LP) supply the lateral petals and the marginal traces of the dorsal sepal along with the dorsal marginal traces of the lateral sepals. The two lateral sepal traces (LS) fuse with the modified petal trace (MP), and these fused traces then divide to supply the ventral marginal traces of the modified petal. Finck (1954), in studying typical monandrous *Epidendrum cochleatum* and then its variety *triandrum,* found vasculature similar to that described by Swamy for this type. It was noted, however, that the marginal traces of the lateral petals were branches of the traces of the lateral sepals (LS) (Figs. 6-6 and 6-7), rather than being offshoots of the lateral petal trace (LP), as Swamy found. In this third type, the three stigma traces (G1–G3) are separated first at the level of the perianth insertion. These originate from the main traces of the three sepals (LS, DS).

In the monandrous orchids, the median stamen of the outer whorl (A1) is the one functional stamen and is supplied by the dorsal sepal trace (DS). In some genera, the traces (a1, a2) in the gynostemium, representing lateral stamens of the inner whorl, are given off from the main traces

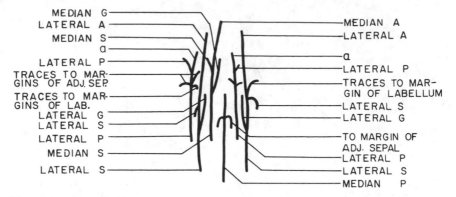

Fig. 6-6. Generalized diagram showing traces to all flower parts of *Epidendrum cochleatum* var. *triandrum*. P = trace to petal. S = trace to sepal. G = trace to carpel. A = trace to stamen of outer whorl. a = to stamen of inner whorl. (After Finck, 1954.)

underlying the lateral petal (LP). Here the stamens of the outer whorl (A2, A3) are represented only by vestigial traces. *Eulophia epidendraea* demonstrates this arrangement, according to Swamy. In others, such as *Eulophia nuda,* it is the main traces of the lateral sepals (LS) which form traces in the gynostemium (A2, A3), representing the lateral stamens of the outer whorl. Here the stamens of the inner whorl are represented only by vestigial traces. Of the two arrangements, the former condition is the more common. Certain genera, *Cymbidium, Dendrobium,* and the like, may exhibit both sets of traces for lateral stamens (A2, A3, a1, a2) in varying degrees. The matter of staminal expression, as represented by the vein pattern, is variable. Swamy (1948) considered this variation to result from the degree of adnation and cohesion of evident parts.

In CYPRIPEDILINAE both staminal whorls contribute members. The inner whorl gives the two functional lateral stamens. The outer whorl contributes the median stamen which manifests itself as the staminode (Fig. 6-1), while the two lateral stamens may be represented by internal traces or staminodal outgrowths.

In monandrous orchids with a high degree of adnation, the origins of the traces may become highly obscured. The median stamen of the outer whorl (A1) may be accompanied by any combination of the inner and outer whorl stamens, any of which may be represented by an intact distinct trace, or may fuse with another to form a compound trace. The A3 and A2 traces may travel intact, or they may fuse as the point of origin moves above the trifurcation of the two traces. This variability in the degree of adnation may occur even in the same species. Swamy (1948) notes this phenomenon in *Geodorum densiflorum*. In *Geodorum* it serves to show the possible conformations of gynostemium vasculature (Fig. 6-8). In the .normal arrangement two compound stamen traces (a1 × A2), (a2 × A3) arise from the junction points of the respective traces of the petal and sepal. The first deviation shows the staminal traces of the outer whorl (A2, A3) as the only ones expressed. In the second deviation an inner whorl stamen (a2) and an outer whorl stamen (A2) are represented as separate, distinct traces. In the third variation an outer whorl trace (A2) and an inner whorl trace (a1) are represented by a single trace arising from the junction of a lateral petal (LP) and a lateral sepal (LS). Also present in this form are the a3 trace and the trace from the labellum (MP). Both these latter traces are separate and distinct, bringing to four the total number of staminal traces in the gynostemium of this form. The gynostemium, having reduced or fused traces, cannot be considered an extension of the floral axis as indicated by Oliver (1895), Rendle (1930), or Willis (1936). Swamy (1948) notes, rather, that the gynostemium is to be considered a "floral tube" (Wilson and Just, 1939).

In considering the expression of staminal structure, it would appear that the ancestral orchid possessed all six functional stamens. The median stamen of the inner whorl (a3) probably disappeared early, as the median petal developed into the labellum. The remaining stamens (A2, A3, a1, a2) then either were reduced or fused into compound traces, and A1 continued its original function. Pfitzer's (1889) diagrams of *Neuwiedia* indicate the early beginnings of the reduction of the three staminal traces (A2, A3, a2). Specialization can be seen to have proceeded along two lines: suppression of the median stamen of the outer whorl (A1) leading to the diandrous condition and suppression of the two lateral stamens of the inner whorl (a1, a2) leading to the monandrous type. This opinon was shared by Rolfe (1909–1910) and Swamy (1946, 1948). The reduction or fusing of these staminal elements indicates the degree to which the present form has advanced from the ancestral.

Triandry. The vascular supply to the stamens may take one of several forms. In certain cases either the inner or outer whorl of stamens is functional, with the nonfunctional organs represented only by traces or vestiges of traces. Finck (1954), investigating *Epidendrum cochleatum* and its variety *triandrum,* found that in the *triandrum* flowers the lateral staminal traces of the inner whorl (a1, a2) disappeared, and the three staminal traces of the outer whorl (A1, A2, A3) passed through the gynostemium and supplied the stamens (Figs. 6-6 and 6-7). The same

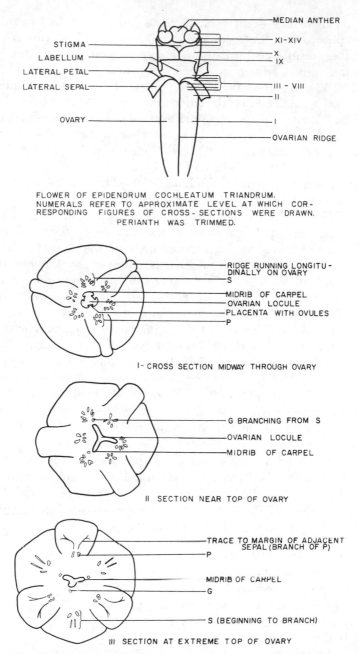

FLOWER OF EPIDENDRUM COCHLEATUM TRIANDRUM.
NUMERALS REFER TO APPROXIMATE LEVEL AT WHICH COR-
RESPONDING FIGURES OF CROSS-SECTIONS WERE DRAWN.
PERIANTH WAS TRIMMED.

I- CROSS SECTION MIDWAY THROUGH OVARY

II SECTION NEAR TOP OF OVARY

III SECTION AT EXTREME TOP OF OVARY

Fig. 6-7. Diagrams showing vasculature at various levels in the flower of *Epidendrum cochleatum* var. *triandrum*. See details in text. (After Finck, 1954.)

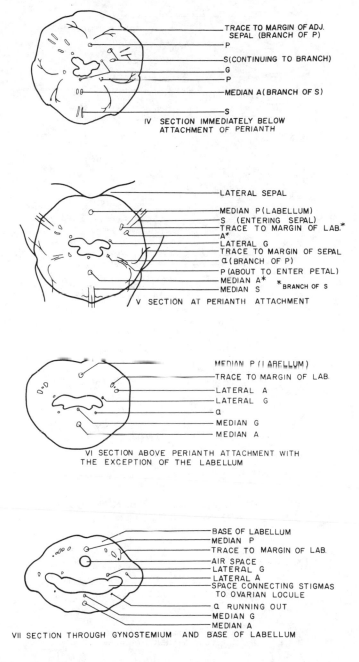

IV SECTION IMMEDIATELY BELOW ATTACHMENT OF PERIANTH

- TRACE TO MARGIN OF ADJ. SEPAL (BRANCH OF P)
- P
- S (CONTINUING TO BRANCH)
- G
- P
- MEDIAN A (BRANCH OF S)
- S

V SECTION AT PERIANTH ATTACHMENT

- LATERAL SEPAL
- MEDIAN P (LABELLUM)
- S (ENTERING SEPAL)
- TRACE TO MARGIN OF LAB.*
- A*
- LATERAL G
- TRACE TO MARGIN OF SEPAL
- α (BRANCH OF P)
- P (ABOUT TO ENTER PETAL)
- MEDIAN A*
- MEDIAN S   *BRANCH OF S

VI SECTION ABOVE PERIANTH ATTACHMENT WITH THE EXCEPTION OF THE LABELLUM

- MEDIAN P (LABELLUM)
- TRACE TO MARGIN OF LAB.
- LATERAL A
- LATERAL G
- α
- MEDIAN G
- MEDIAN A

VII SECTION THROUGH GYNOSTEMIUM AND BASE OF LABELLUM

- BASE OF LABELLUM
- MEDIAN P
- TRACE TO MARGIN OF LAB.
- AIR SPACE
- LATERAL G
- LATERAL A
- SPACE CONNECTING STIGMAS TO OVARIAN LOCULE
- α RUNNING OUT
- MEDIAN G
- MEDIAN A

Fig. 6-7 (continued)

279

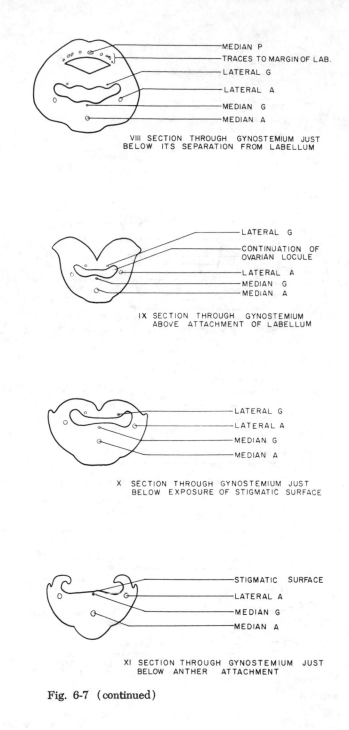

MEDIAN P
TRACES TO MARGIN OF LAB.
LATERAL G
LATERAL A
MEDIAN G
MEDIAN A

VIII SECTION THROUGH GYNOSTEMIUM JUST
BELOW ITS SEPARATION FROM LABELLUM

LATERAL G
CONTINUATION OF
OVARIAN LOCULE
LATERAL A
MEDIAN G
MEDIAN A

IX SECTION THROUGH GYNOSTEMIUM
ABOVE ATTACHMENT OF LABELLUM

LATERAL G
LATERAL A
MEDIAN G
MEDIAN A

X SECTION THROUGH GYNOSTEMIUM JUST
BELOW EXPOSURE OF STIGMATIC SURFACE

STIGMATIC SURFACE
LATERAL A
MEDIAN G
MEDIAN A

XI SECTION THROUGH GYNOSTEMIUM JUST
BELOW ANTHER ATTACHMENT

Fig. 6-7 (continued)

280

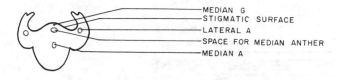

XII SECTION THROUGH GYNOSTEMIUM AT LEVEL OF
BASES OF ANTHERS (NOT ANTHER ATTACHMENT)

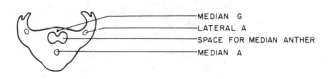

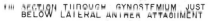

XIII SECTION THROUGH GYNOSTEMIUM JUST
BELOW LATERAL ANTHER ATTACHMENT

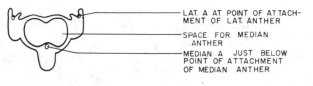

XIV SECTION THROUGH GYNOSTEMIUM JUST
BELOW MEDIAN ANTHER ATTACHMENT

Fig. 6-7 (continued)

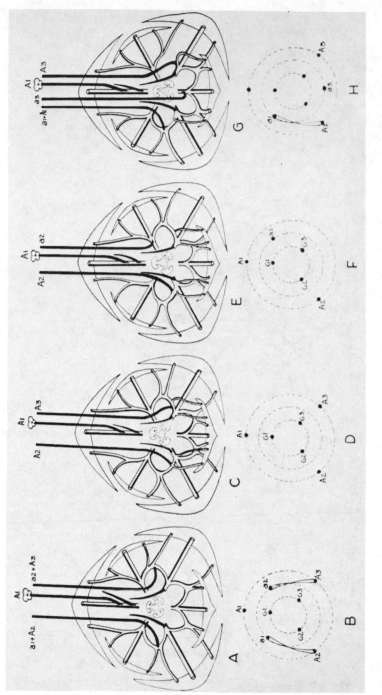

Fig. 6-8. Vascular diagrams of the normal and three deviation patterns found in flowers of *Geodorum densiflorum*. Traces to the stamens and stigmas are in solid black. (After Swamy, 1948.) Details in text.

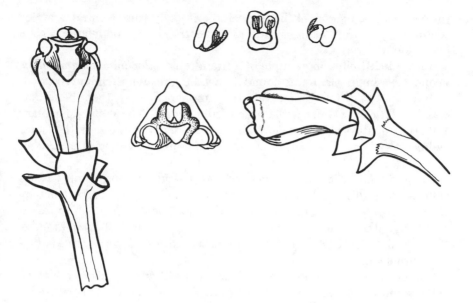

Fig. 6-9.   Triandrous column of *Lc*. Joseph Hampton.

anatomy has been found (Withner, unpublished) in a dissection of a deformed flower of a modern polyploid hybrid, *Lc*. Joseph Hampton, which consistently produced columns with three stamens (Fig. 6-9).

The columns of the monadrous type of *E. cochleatum* differ in that only one trace of the inner whorl appears and travels farther up the gynostemium (Fig. 6-7) than either of the traces in the *triandrum* variety before disappearing. Though only one of the outer whorl staminal traces is functional (A1), all three are present. The two nonfunctional traces (A2 and A3) simply end at a position in the gynostemium where functional anthers would normally be attached. A pattern similar to that of usual *E. cochleatum* exists in many of the other monandrous orchids, as discussed earlier.

Withner (unpublished) hybridized the triandrous and monandrous types of *E. cochleatum* together and found that all of the F_1's that flowered had only one stamen. Apparently the triandrous condition is recessive and must be present in pure form from both parents for the progeny to remain triandrous. Luer (1972) remarks indirectly on this in describing *Encyclia boothiana* var. *erythronioides*, another triandrous species from Florida. He says Dodson suggested that these forms are ordinarily self-fertilizing, since they lack a functional rostellum. Also, since normal pollinators of these species, originally from Central America centuries ago, are lacking in Florida, the triandrous populations have been able to survive.

Ames (1922) cites an example of a triandrous *Psilochilus macrophyllus*, saying, "A single plant was found in which the gynostemium was triandrous in a very interesting way, the supplementary anthers being lateral, one on each side near the base of the stigma. These supplementary anthers probably represent normally suppressed members of the outer androecial whorl, which often occur as staminodia in species of *Orchis*, *Habenaria*, etc."

Bauer (1830–1838) also remarked on triandrous forms of *Cephalanthera grandiflora*. Though no anatomical studies were carried out by Cruger (1865), at least one odd flower of an *Isochilus* species appeared with five anthers and a filament representing the sixth stamen (Fig. 6-10D). Maige (1909) listed *Orchis militaris* and also *Ophrys tenthredinifera* with six anthers.

Godfery (1933) described an *Ophrys arachnitiformis* flower with three anthers and three rostellum pouches (Fig. 6-10E) and another with four. He makes a point that this fourth anther was directly face to face with the usual fertile anther so that their beaks overlapped. He discusses the evolutionary advantage of the elimination of this anther as the lip evolved in the orchid family and became the landing stage for the pollinator. If it had remained, it would have obstructed access to the rostellum and thus to the stigma and pollen of the usually fertile anther.

Perianth and Labellum. The overall vasculature of the perianth varies in complexity. In *Vanilla* and *Coelogyne* the traces of sepals and petals lie close together, whereas in *Dendrobium, Bulbophyllum, Pholidota,* and *Spathoglottis* the adjacent traces have a tendency to join or fuse, forming an arch. In these genera the arch then separates, and the emerging traces have a distinct and independent nature. In *Cymbidium,* the fusing continues and the traces remain joined for some distance. Some of these doubled traces enter their respective member in a doubled state. This greater complexity and joining of the perianth system seems to coincide with the advanced status of the flower. The fusing of traces and the position at which stamen traces separate increase the confusion surrounding the staminal bundles in orchid flowers.

The presence of the lateral staminal traces of the outer whorl, either in fused or intact form in the gynostemium, serves to disprove the notion of Darwin that the labellum is a fused structure. Arguments against the compound nature of the labellum are not new. Cruger (1865) argued

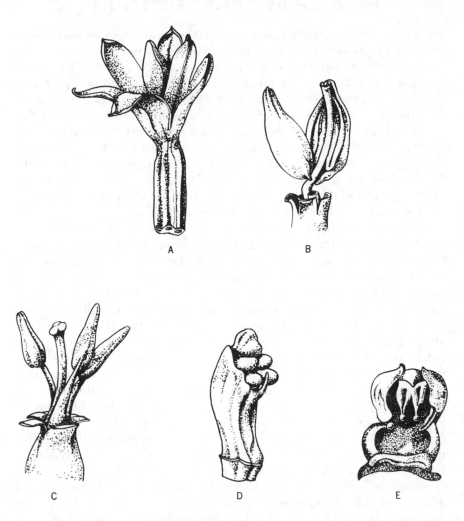

Fig. 6-10. (A, B) *Apostasia nuda.* (A) Bud opened to show parts of perianth. (B) Style with three stigmas and the two anthers attached to it, one pulled back to show the style *in situ.* (C) Flower of *Neuwiedia* to show the three stigmas and the three anthers attached to the style. (D, E) Unusual orchid flowers showing reversion to primitive condition. (A) Column of *Isochilus* with five anthers and a filament representing the sixth. (E) Column of *Ophrys arachnitiformis* showing three anthers. (A, B, D, E after Godfrey, 1933; C after Garay, 1972.)

against this notion, although his views were generally ignored until Worsdell (1916) presented accumulated evidence for the homology of the labellum. The extensive studies of Swamy, accounting for all staminal traces, leave little doubt that the labellum is not composed of a petal and two lateral stamens. Vermeulen (1966), considering developmental and evolutionary aspects, along with anatomical studies, also concluded that the orchid labellum is a single organ, homologous with the median petal. Nelson (1965, 1967), on the other hand, contends that the labellum is composed of three stamens (A2, A3, a3). Citing *Ophrys* as an example, he identified the three lobes of the labellum as the three "missing" stamens, the petal having disappeared. This does not seem likely in view of the data presented by the aforementioned studies.

One line of evidence showing that the usual three lobes of the lip cannot result from the lower petal and two obsolete anthers fused alongside it is cited by Salisbury (1804) in a discussion of peloric flowers. Several orchid species have been recorded with the lateral petals showing the same lip configuration as the lip itself—even with the presence of spurs. In such cases the explanation could not hold. He cites, furthermore, flowers with two lips side by side, or with two lips immediately above and to the sides of a normal lip, both in flowers with normal petals. He explains them as abnormal formations from the vestigial tissues of the obsolete anthers. Since all flower parts, whatever their fusion and final specialized nature, are basically modified leaves attached to an axis (the receptacle), they all seem to maintain an apparent plasticity or ability to revert to other forms. Reychler (1928) presents many peloric or abnormal flowers and illustrates some of the variations that can be expected. The causes of such deformities are open to question, though it is generally believed that unusual environmental conditions, aneuploid chromosome numbers, or other genetic imbalances in hybrids might account for them in some way. In the latter cases, they would reoccur from year to year, as indicated for the *Lc.* Joseph Hampton columns described above.

Porsch (1908) summarized the role of the labellum in bringing about cross-pollination. He noted, in addition to lip color and form—attractive and often bizarre—that the lip can provide nectar, especially in spurred flowers, or imitation or decoy pollen, as described by Janse (1886) for *Maxillaria*. Also, flower wax is produced on the labellum by special epidermal cells of some orchids. Some form nutrient hairs, rich in lipids and other foodstuffs, with thin, pure cellulose walls and preformed regions at their bases for easy separation from the lip. Finally, others have what is called nutrient tissue, superficial layers rich in protein, lipid, and starch, to be eaten or collected by the insect. Such tissues often are scented to help attract the insects (Vogel, 1966a, 1966b).

Nierenberg (1972) described in detail, with comparative photographs made in bright light and ultraviolet, how bees may visit equitant *Oncidium* flowers because of food mimicry. The orchids imitate the flowers of certain malpighs, but they provide no food as the insect expects. This syndrome involves the female *Centris* bees. The males are attracted to other types of orchid flowers because of pseudoantagonism—aggressive actions against the flowers since the bees perceive them as possible intruders into their territories. The UV reflecting central regions of such flowers mimic an enemy insect. Additional research in this area will give us a fuller understanding of the intimate relationships between orchid flowers, speciation, and insects. There is, of course, already a great amount of work on the genus *Ophrys* in relation to the phenomenon of pseudocopulation (Ames, 1937), as the work of Kullenberg (1961) illustrates.

Resupination. Ames (1938, 1946b) reviewed the literature on resupination and made his own experimental observations. He noted that it is mostly the pedicel, not the ovary, that twists. He defined resupination as occurring when "the labellum is visually the lowermost segment of the orchid flower." As most orchid flowers develop, the lip is adaxial, that is, adjacent to the axis of the inflorescence; but by the time the flowers open, many twist through 180° so that the lip is in an opposite position. Darwin (1884) studied this process in *Malaxis paludosa*, which has the lip in adaxial position in the open flower, only to observe that the twist was 360°. He interpreted this behavior of going back to the primitive position as an adaptive mechanism that had enabled the species to survive better in relation to its particular pollinator.

Ames tried a few experiments with *Calopogon, Goodyera,* and *Habenaria,* tying the inflorescences so that their tips were bent toward the ground. The flowers still twisted and opened with the lips in "normal" position for the species. Fuchs and Ziegenspeck (1936) cite klinostat experiments that prevented resupination by the continuous rotating of the developing flowers of European terrestrial species. Whatever the physiological causes of the twisting may be, it is still not explained. In this connection, Ames studied the example of *Catasetum barbatum,* noted first by Schomburgk (1837), which produces both male and female flowers on the same flower stalk. The male flowers resupinate; the female flowers do not. Some selective force seems to be at work.

Sometimes the stalk of the inflorescence may also be twisted, the classic example being plants of the genus *Spiranthes*. The twisting may be in either direction.

Rostellum, Stigma, and Pollination. In the ORCHIDACEAE, a grouping that includes the APOSTASIOIDEAE, CYPRIPEDIOIDEAE, ORCHIDOIDEAE, NEOTTIOIDEAE, and EPIDENDROIDEAE, in the Garay system (1960, 1972) of classification (Fig. 6-11) there appears another modified structure, the rostellum. The term was coined by the French botanist Richards (1818), who defined it as "the part extending above the shining stigma, generally with a narrow top."

In its most highly evolved form, the rostellum consists of a flap or projection of tissue near the end of the column, separating functional stigmas from the anther, and bearing a viscidium or adhesive mass upon its tip, from which a strip of epidermal and adjacent tissue, the stipes or stipe, extends to connect with the pollinia. Various ideas have been offered about the nature of the rostellum:

1. It is a structure derived from the median stigma, which in the process has ceased to function as a fertile stigma.
2. It is derived from the two stamens of the outer whorl (A2, A3).
3. It is a new structure and may not be derived from the median stigma, which is still present and functions as a fertile stigma, though it may be obscured or altered.

The first, or classical, interpretation was favored by Brown (1833), who considered that the rostellum might be a sterile third stigma lobe. Darwin (1884) stated that the median stigma lobe had disappeared and given rise to the rostellum. He did not mention orchid flowers where all three stigma lobes exist.

Hagerup (1952), studying *Herminium,* deduced that the viscidia, the sticky disks to which the pollen attaches, are derived from the two stamens of the outer whorl (A2, A3) and are completely distinct from the rostellum. Subsequent investigation by Vermeulen (1955) indicates that the structure considered to be the viscidia by Hagerup was in fact a part of the rostellum. Wolfe (1865–1866) actually was the first to study the anatomy of European native orchids (OPHRYDEAE) in relation to their pollination and to make observations on the anther, viscidia, and column structure. Vermeulen (1955), working with several genera of OPHRYDEAE, concludes that for that tribe the stigma is composed of three fertile lobes and that there may be a tendency for the reduction of the third lobe. He further states that the rostellum is a separate organ and not homologous with the third stigmatal lobe. Vermeulen finds that in this tribe the viscidia are part of the rostellum, which is a new organ in the OPHRYDEAE, but he is unable to deduce its exact origin. He notes that in most genera of the OPHRYDEAE the rostellum is supplied by a vascular cord, thus differentiating it from the other stigma lobes which he finds to be traceless, or supplied with short traces from the column. Vermeulen (1959) states that in NEOTTIEAE the rostellum is a simple gland; in fact, it may even be lacking, or represented by a secreting stigma. Certain exceptions are noted in the group, however, as flowers of species of the genus *Prasophyllum* bear a long slender rostellum.

Garay believes (1960, 1972) that the matter of rostellum origin and structure is largely settled and that it can be considered to be a manifestation of the third stigma lobe, which has been altered as part of the reorganization the orchid flower has undergone in adapting to insect pollination. He notes the existence of flowers with all three stigmas present, such as *Cephalanthera*, and of flowers with no apparent rostellum at all. Garay does not consider the rostellum described by Vermeulen in OPHRYDEAE to be a true rostellum and calls it instead "connective tissue

AUTHOR	MAIN DIVISIONS OF SYSTEM						
LINDLEY	CYPRIPEDEAE	OPHRYDEAE	ARE-THUSEAE	NEOT-TEAE	MALAXIDEAE	EPIDENDREAE	VANDEAE
BENTHAM	CYPRIPEDIEAE	OPHRYDEAE	NEOTTIEAE		EPIDENDREAE		VANDEAE
REICHENBACH	CYPRIPEDIEAE	OPHRYDEAE	OPERCULATAE				
			NEOTTIEAE		EUOPERCULATAE		
PFITZER	PLEONANDRAE	MONANDRAE					
		ACROTONAE					
		BASITONAE	Acranthae – Convolutae	Acranthae { Duplicatae / Articulatae }		Pleuranthae – Sympodiales Monopodiales	
ROLFE	DIANDRAE	MONANDRAE					
		OPHRYDEAE	NEOTTIEAE		EPIDENDREAE		VANDEAE
SCHLECHTER	DIAN-DRAE	MONANDRAE					
		ACROTONAE					
		BASITONAE	Polychondreae		Kerosphaereae		
MANSFELD	DIANDRAE	THRAUOSPHAEREAE			KEROSPHAEREAE		
		OPHRYDEAE	NEOTTIEAE		EPIDENDREAE		VANDEAE
HATCH	CYPRIPEDIOIDEAE	ORCHIOIDEAE					
		BASITONEAE	ACROTONEAE				
DRESSLER AND DODSON	CYPRIPEDIOIDEAE	ORCHIDOIDEAE					
		ORCHIDEAE	NEOTTIEAE		EPIDENDREAE		
GARAY	APOSTASIOIDEAE	CYPRIPEDIOIDEAE	ORCHIDOIDEAE	NEOTTIOIDEAE	EPIDENDROIDEAE		
					EPIDENDREAE		VANDEAE
			ORCHIDEAE DISEAE DISPERIDEAE	EPIDOGONEAE NEOTTIEAE CRANICHIDEAE	Phaiinae Dendrobiinae Epidendriinae Pleurothallidinae		Cyrtopodiinae Zygopetaliinae Oncidiinae Vandiinae

Fig. 6-11. Summary of various systems of classification of ORCHIDACEAE. (After Garay, 1972.)

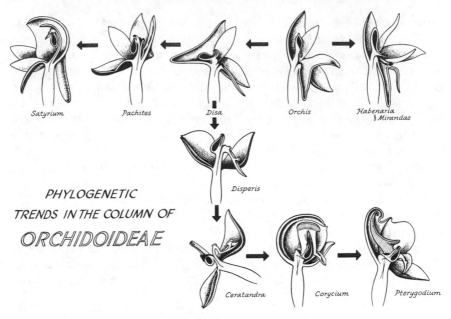

Satyrium   *Pachites*   *Disa*   *Orchis*   *Habenaria § Mirandae*

Disperis

PHYLOGENETIC
TRENDS IN THE COLUMN OF
ORCHIDOIDEAE

Ceratandra   *Corycium*   *Pterygodium*

Fig. 6-12. Phylogenetic trends in the column of subfamily ORCHIDOIDEAE. (After Garay, 1972.)

between the two thecae of the anthers." If it is a new structure and not a developed third stigma lobe, then it cannot be a rostellum, since a rostellum is construed by most botanists to be a developed third stigma lobe. This connective tissue was called a bursicula by Wolfe (1865–1866) and thought by him to be a modified rostellum. Wolfe uses the term retinaculum to describe the adhesive masses, as does Vermeulen (1966).

Garay (1972) published again on his studies of the comparative anatomy of the orchids in relation to their taxonomic groupings (Fig. 6-11). He emphasizes that there is no rostellum in the APOSTASIOIDEAE, the CYPRIPEDIOIDEAE, or the ORCHIDOIDEAE (OPHRYDEAE) subfamilies (Fig. 6-12), and this, naturally, resolves the Vermeulen difficulties over a new structure. Garay concludes by applying various phylogenetic criteria consistently and thoroughly to the family as a whole, that the more advanced species of the subfamily EPIDENDROIDEAE have a column structure most like that of the tribe CRANICHIDEAE of the subfamily NEOTTIOIDEAE. They both have the same angle of insertion of the anther and an ascending rostellum. In the EPIDENDROIDEAE there is finally a complete fusion of the various reproductive parts and a fully developed rostellum. Correlating is

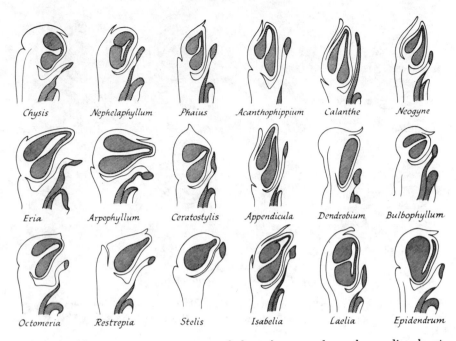

Chysis *Nephelaphyllum* *Phaius* *Acanthophippium* *Calanthe* *Neogyne*

Eria *Arpophyllum* *Ceratostylis* *Appendicula* *Dendrobium* *Bulbophyllum*

Octomeria *Restrepia* *Stelis* *Isabelia* *Laelia* *Epidendrum*

Fig. 6-13. The comparative structure of the column as shown by median longi-tudinal sections in developing buds of members of the tribe EPIDENDREAE. (After Garay, 1972.)

the position of the anther in relation to the rostellum and the position of the rostellum in relation to the evolution of a stipe for the pollinia. These relationships are shown clearly in Garay's two plates reproduced here (Figs. 6-13 and 6-14). Further, the EPIDENDROIDEAE are subdivided by Garay, no longer paying attention to sympody or monopody as such, into two natural units:

1. The EPIDENDREAE with erect anther and ascending or arrect rostel-lum to which the pollinia are attached without appendages.
2. The VANDEAE with an incumbent anther and a horizontally project-ing or porrect rostellum to which the pollinia are attached by a stipe.

These and other evolutionary specializations within the subfamily EPIDENDROIDEAE have enabled the plants of its various species to attain an epiphytic existence and to adapt so variably to insect cross-pollination. The pollinia can be firmly attached to the pollinator and be carried over

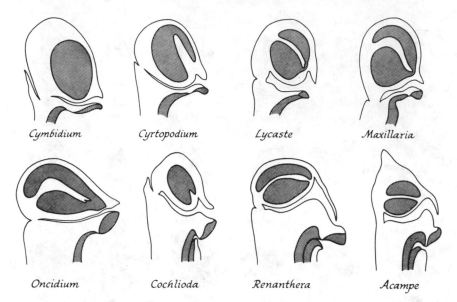

Cymbidium　　　*Cyrtopodium*　　　*Lycaste*　　　*Maxillaria*

Oncidium　　　*Cochlioda*　　　*Renanthera*　　　*Acampe*

Fig. 6-14. The comparative structure of the column as shown by median longitudinal sections in developing buds of members of the tribe VANDEAE. (After Garay,
1972.)

great distances, thus eventually affecting and regulating the size and
spread of the breeding population.

Ames (1946a) makes a point of noting that the pollinator "carried a
structure of bisexual constitution, the male pollen borne on an adhesive
base derived from female tissues." He states also, "my respect for the
rostellum has never suffered through familiarity with orchid flowers."

The rostellum also functions to prevent self-pollination in most orchids
by forming a barrier between the pollen masses and the stigma. A few
orchids, however, are regularly cleistogamous and self-pollinating (Withner, 1970), and others may be if not pollinated by their regular pollinator.
This process occurs because the rostellum in such species is only temporarily effective, soon drying up or shriveling so that the pollen and stigmatic fluid quickly make contact.

Hirmer (1920) and Gellert (1923) studied the various details of rostellar anatomy and stipe formation. Hirmer distinguishes three evolutionary
stages in their phylogenetic specialization. In the primitive monandrous
orchids the three stigma lobes are all functional as stigmas, and the pollinia are attached to insects by the stigmatic fluid that is contacted along
with the pollinia in withdrawing from a flower. As a rostellum evolves,

the median stigma lobe becomes gradually nonfunctional as such; it enlarges and forms a rooflike arch over the entrance to the stylar canal centered between the two lateral stigmatic lobes and itself. At the tip of the primitive rostellum an adhesive mass is developed that is separated from stigmatic secretions proper that may still be produced by its basal part in the area of the stigmatic cavity and stylar canal. The adhesive is segmented and is not removed *in toto* by a single insect but a little at a time. These masses do not have epidermal connections (stipes) on the side of the rostellum toward the anther with the pollinia. The pollinator, pushing into the flower for food or nectar and then withdrawing, presses the adhesive against the adjacent anther and onto itself at the same time. As it departs, the stickiness attaches the massulae of mealy or sectile pollinia to the vector for dispersal.

The next stage involves the epidermal tissue of only the tip of the rostellum on the side toward the anther as well as the underlying parenchymatous cells. This tissue becomes adhesive and is now called the glandula. When disturbed by the pollinator it departs *en masse*, bringing with it the pollinia and their caudicles. The glandula is more or less firmly attached to the pollinia by their own stickiness, according to the species. The stickiness depends on the cells of the tapetum that surround and connect the tetrads, and the caudicles are composed primarily of the same material with a few tetrads mixed in. This tapetum increases tremendously in size on the side of the anther toward the rostellum. The rostellum no longer produces any stigmatic secretions.

The final stage involves stipe formation from the epidermis of the

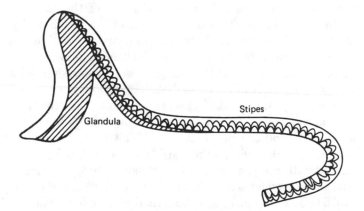

Fig. 6-15.   Median longitudinal section through the stipes of *Vandopsis lissochiloides* showing glandula with adhesive disk. (After Gellert, 1923.) See text for detail.

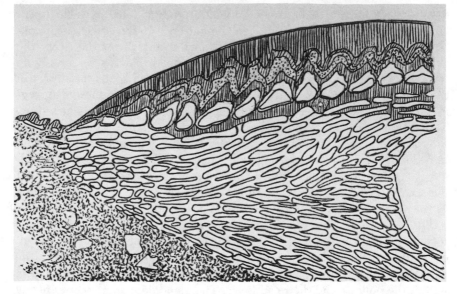

Fig. 6-16.   Longitudinal section through column tip of *Vandopsis lissochiloides* showing transition from stipe to adhesive disk of glandula. (After Gellert, 1923.)

rostellum and a few cell layers beneath it extending between the glandula, or viscidium, as it may now be termed, and the pollinia (Fig. 6-15). Species having this development were grouped together as the VANDEAE by early orchid taxonomists (Lindley, 1853); this unit is retained with additional supporting evidence as a natural group by Garay (1972).

When the stipe is forming, its cells and those of the viscidium (Fig. 6-16) continue to enlarge after the other cells of the rostellum have completed their growth. A tension between these tissues is thus generated, and it may be enhanced by irregular wall thickenings and "mucus" formation in or between the cell walls of the stipe and underlying tissues (Fig. 6-17). The tension leads to an immediate separation of the whole apparatus when lightly pushed or disturbed from below. The stipe tissue, with thickened walls or cuticle toward the epidermal side, is then exposed for dehydration on the other side (Fig. 6-18). This brings about a bending, as much as 90°, so that the pollinia are better oriented by the stipe for direct contact with the stigma of the next flower to be visited, a critical matter that helps prevent self-pollination and promotes outcrossing (Nierenberg, 1972). Gellert (1923) has concerned herself particularly with the formation of the stipe and describes it in detail for *Cymbidium, Maxillaria, Odontoglossum, Vanda, Angraecum,* and still other species. Haberlandt

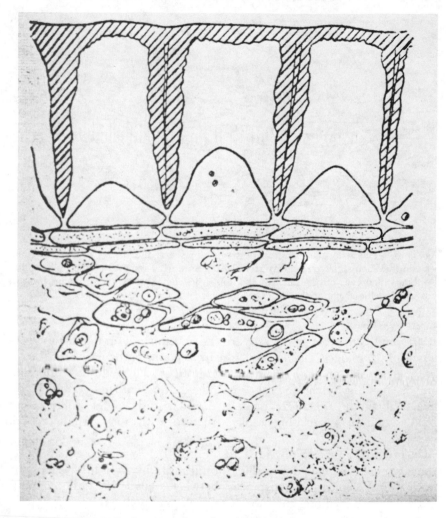

Fig. 6-17. Longitudinal section through a segment of a young stipes of *Vandopsis lissochiloides* showing thickening of epidermal walls on the outer layer and the beginning lysis of underlying cells from which it will be separated. (After Gellert, 1923.)

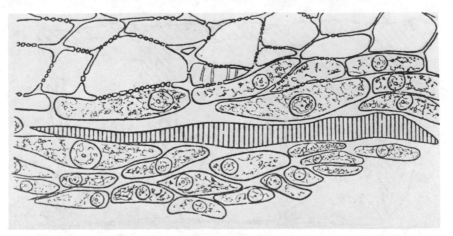

Fig. 6-18. The zone of separation between stipes and the tissues of the rostellum beneath it. When stipes pulls away from rostellum during removal of the pollinarium, the thin-walled cells on the back of the stipes will be exposed for drying in the air, thus altering the shape of the stipes. More details in text. (After Gellert, 1923.)

(1901) and Guttenberg (1915) were the first to study in detail the pollination of *Mormodes, Cycnoches,* and *Catasetum,* particularly the explosive mechanisms of the latter genus with relation to adhesive formation and the bending of the forcibly discharged stipe with its attached pollinia. Williams (1970, 1972) made a detailed study of the pollinaria of the ONCIDIINAE, correlating their structure with the various genera involved.

Gellert noted that stigmatic fluid (Fig. 6-19) is composed of loose, elongated cells imbedded in a thick liquid resulting from the lysis of the pectinaceous cementing substances between cells. The fluid contains lipids and carbohydrates that nourish developing pollen tubes.

Anther and Pollen. The anthers and pollen of orchids have been studied from the beginning of early orchid classification systems, and they are no less specialized than other flower parts. The basic number of stamens in the hypothetical ancestral orchid flower is six (Fig. 6-1), of which two or three survive in members of the APOSTASIOIDEAE (Fig. 6-10A, B, C), two in the CYPRIPEDIOIDEAE, and one for the balance of the family. The anthers are attached at their bases in the subfamilies APOSTASIODEAE, CYPRIPEDIOIDEAE, and ORCHIDOIDEAE (Fig. 6-12), while the attachment is acrotonic in the NEOTTIOIDEAE with the anther erect and still free (Garay, 1960, 1972). In the EPIDENDROIDEAE, derived through the tribe

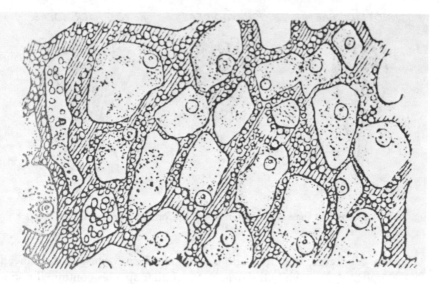

Fig. 6-19. Section through stigmatic surface of *Vandopsis lissochiloides* showing dissolution of parenchyma tissue to form stigmatic fluid. (After Gellert, 1923.)

CRANICHIDEAE of the subfamily NEOTTIOIDEAE, the anther has become imbedded in rostellar tissue in either an erect (tribe EPIDENDREAE) or incumbent (tribe VANDEAE) position (Figs. 6-13 and 6-14). Anthers are basically two-celled containing the usual pollen or with two or more pollinia in higher forms.

The pollen produced in these various types of stamens has undergone similar phylogenetic specializations. In the apostasias it is free and granular, the tetrads separating into individually walled pollen grains, as is typical for most plants. The anthers each have four locules. In the lady's slippers the pollen grains are still individual, but they are viscid and stick together as a transportable mass. In the ORCHIDOIDEAE and NEOTTIOIDEAE most of the pollen (Fig. 6-20) occurs in tetrad formation, without exine development and with the tetrads loosely united by elastic threads from the tapetum (Dressler and Dodson, 1960). The ORCHIDOIDEAE are distinguished by their granular pollen masses, the contents of each locule being organized into many interconnected packets or massulae. In the NEOTTIOIDEAE the pollinia can easily separate into mealy or powdery masses as a result of their sectile nature. When visited by an insect only a portion of the pollen, a few segments or massulae, become detached, the balance remaining for the next visitor.

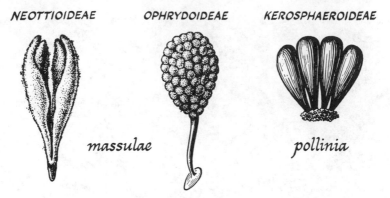

Fig. 6-20.   The pollen masses characteristic of the orchids. (After Garay, 1960.)

Darwin estimated that a single flower of *Orchis mascula* produced 120,000 pollen grains, while the ovary of a similar species contained about 6200 ovules, a safety margin of 20 to 1, if the sectile masses were to be transferred *in toto*. As Garay (1972) noted, the kind of pollinia and their degree of aggregation can determine population size and spatial distribution of a species. Pollinia proper in the higher orchids are associated usually with spatially extensive populations, since insects can carry them intact for great distances. The orchids, as Ames (1944) aptly described, have managed to decoy pollinivorous insects, especially bees, by nectar, and at the same time have made the pollen transportable *en masse* and not suitable for food. They have also produced decoy pollen, as described previously in the discussion of the labellum.

In the ORCHIDOIDEAE the basal portions of the pollinia are modified to form slender, taillike elastic strands called caudicles. These are not developed in the NEOTTIOIDEAE (Fig. 6-20). In the higher orchids the pollen is in hard, compact masses, and the two pollinia from one anther cell may be connected by the caudicle. Since these pollinia often are subdivided, four or eight pollinia per anther is possible. Dressler and Dodson (1960) have attempted to show the evolution of different types of pollinia (Fig. 6-21), coordinating changes in number and shape with caudicle formation and stipe development. In the EPIDENDREAE they consider the primitive number of pollinia as eight, with reduction to six, four, or two showing an advancement. We know that six occur sometimes in hybrids between parents with eight and four pollinia, such as *Laelia* and *Cattleya*; it is not possible to judge what effect ancient natural hybridizations may have had on the number of pollinia in any given taxon.

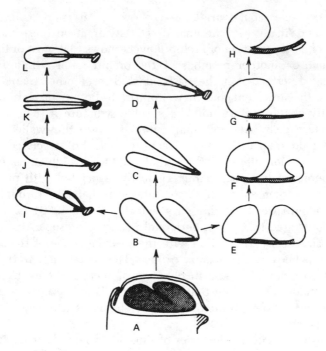

Fig. 6-21. Diagram showing some of the major patterns of evolution in pollinia within the EPIDENDREAE. Mealy pollen is stippled; the rostellum, and viscidium when present, are cross-hatched. (After Dressler and Dodson, 1960.)

Nectaries. Though there is little question about the role of floral nectaries in attracting orchid pollinators (van der Pijl and Dodson, 1966), and although accounts of pollination procedures (Müller, 1873) have been available for a long time, there has been almost no research on the anatomy of these structures. The same applies to extrafloral nectaries, although Jeffrey, Arditti, and Koopowitz (1970) and others have observed or analyzed the sugar contents of their nectar and have supposed that such secretions attract ants, which in turn deter larger insects or herbivores from consuming the plant or flower. It would also seem possible that extrafloral nectar might function to decoy the ants away from flowers, thus reserving floral nectar for proper pollinators.

Floral nectaries are generally classified (Fahn, 1952) according to their location on the flower parts: (1) torus type, on the receptacle; (2) marginal type, between bases of sepals and petals; (3) annular, if forming a ring on the surface of the receptacles; (4) tubular, if lining a spur in the receptacle; (5) ovarial, on the ovary; (6) perigonal, if on the perianth; and finally (7) staminal or (8) stylar, if on these parts of the flower. Simple observation of orchid flowers shows that most of these types exist in one or another member of the orchid family, though few are· specifically described. On the basis of studies of other plants, one can only make the statement that there appears to be no correlation between the locality of the nectary and the specific structure it may have. Typically, nectary cells have granular cytoplasm, are thin-walled, and lack cuticle if epidermal in location. Nectaries may be associated with vascular tissue or not, although those that produce a high sugar concentration are more likely to be characteristically associated with phloem. The nectar may be secreted through stomata, from hairs, or through the surface of epidermal cells. Preliminary work (Rubino, 1966) in our laboratory demonstrated that green perigonal nectaries on sepal tips of various cattleyas show close association with the vascular tissue of the sepal midrib, which penetrates the nectary proper. The nectary apparatus is amply provided with stomata. The mechanism of secretion remains unknown, but one cannot help feeling that photosynthesis in the nectary itself may be significant in the production of the sugars. The nectar is secreted through the stomata.

The illustrations of the nectaries of dorsal and lateral sepals of *Cattleya harrisoniana* (Fig. 6-22) show a generally similar configuration with a green, conelike structure at the sepal tip that extends onto the outside surface. Two green labia flank the cone on either side and follow the sepal edge until they taper away. Except that the labium along the lower edge of the asymmetrical lateral sepals is longer and wider than the labium along the upper edge, the structure is the same. A section through the sepal tip is shown in Fig. 6-23.

Laychock (1971) worked on the extrafloral nectaries associated with the floral bracts of a *Cymbidium* flower stalk. The nectar-producing area is at the base of the bract on its abaxial (lower) surface (Fig. 6-24). No particular vascular tissue is specifically associated with the area, but a distinct change in epidermal cell and stomatal patterns is easily noted. There is an abscission layer at the base of the ovary (Fig. 6-25). Stomata are rounder and more numerous on the bract and protrude like small papillae above the surface, and the nectar is secreted through them. Zimmerman (1932) described similar patterns on bracts and leaves of *Epidendrum cochleatum* and *Catasetum inornatum*, where secretions occur through slitlike crevices. He noted further that nectaries of some species show no specialized subepidermal structure, whereas in others the cells are small and dense and isodiametric.

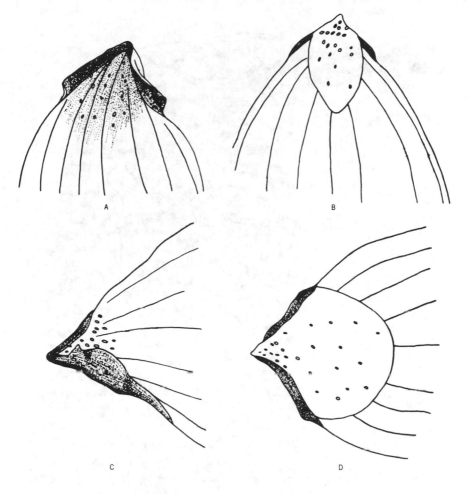

Fig. 6-22. Front and back views of the dorsal and lateral sepal tip nectaries of *Cattleya harrisoniana*. (A, B) Dorsal sepal. (C, D) Lateral sepal. (After Rubino, 1966.)

Fahn (1952) determined that the amount of nectar produced is proportional to the amount of nectariferous tissue present. The most common constituents of nectar are sucrose, glucose, fructose, mucilage, some protein, and organic cells. *Cymbidium* exudates may also contain stachyose (Jeffrey et al., 1970). Frey-Wyssling with colleagues (1954) showed invertase also present, though absent in phloem tissues, demonstrating that nectaries have their own metabolic processes.

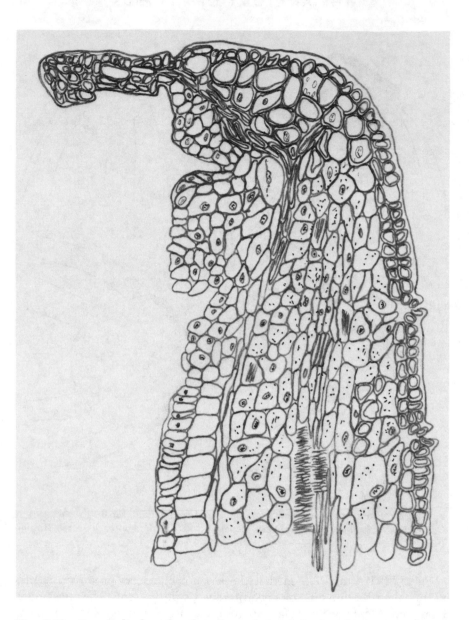

Fig. 6-23. Longitudinal section through sepal tip of *Cattleya harrisoniana* showing termination of midvein in nectary, prominent stomata on back side of sepal, and the lack of a cuticle on the front. The flap of tissue at the top of the section is a portion of the labium. (After Rubino, 1966.)

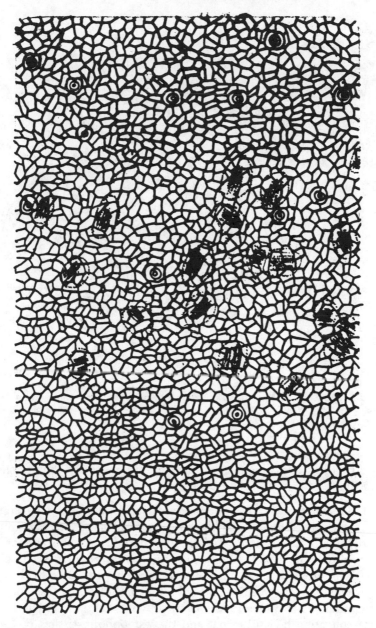

Fig. 6-24.   Change in epidermal cell and stomatal pattern over nectary area at base of bract subtending a *Cymbidium* flower. Stomata become more numerous, and mesophyll cells with raphide crystal bundles may be observed. (After Laychock, 1971.)

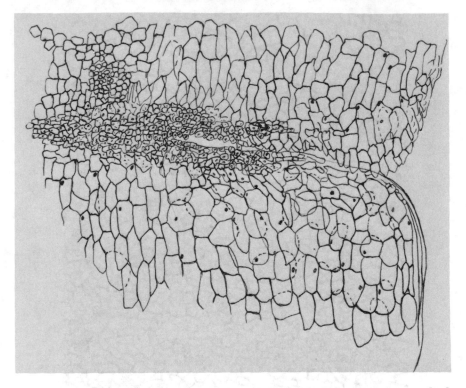

Fig. 6-25. The cells of the abscission zone at the base of a *Cymbidium* pedicel where it joins the flower stalk. Stalk tissue is below, pedicel-bract tissue above. (After Laychock, 1971.)

The structures and functions of most orchid nectaries thus remain to be investigated.

Scent Tissues. Scent or osmophoric tissues were little investigated until the work of Stefan Vogel, who has long been interested in the native orchids of South Africa and the pollination mechanisms that have evolved with them (1954, 1959). Until his work, genuine scent glands were known previously only in animals, and he has described typical scent-producing cells (1962) not only for orchids but also for other plant families. This research is concerned with bat pollination (1968), perfume-collecting bees (1966b), and the osmophoric tissues of *Catasetum* and *Stanhopea* species (1963). The latter tissue (*Futtergewebe*) was originally described by Darwin (1884) in *Catasetum saccatum* and later studied by Haberlandt (1901) in *Catasetum darwinianum*. Vogel's work (1963, 1966a) shows that bees do not actually eat the tissue but only collect the oils produced by it. The richness of the osmophoric cells in reserve materials relates to their production of fragrant terpene oils, which are secreted in small droplets through the outer walls. Since only male bees collect the scent, their perfume loads may be a way of attracting the females during swarming. Nevertheless, the flowers are pollinated in the process, providing no real food value to the bee, as was originally hypothesized. This scent formation is predominantly on the lips of flowers without nectaries in the CATASETINAE, GONGORINAE, LYCASTINAE, HUNTLEYINAE, and ZYGOPETALINAE.

The scent-producing tissue is usually diffuse, a whole surface, such as a portion of the labellum, exuding the scent rather than a specific gland being formed. In fact, the whole surface of the flower may be scented in some cases, or only the lip in others. Scent production may occur only on vegetative parts of some plants, this condition being a forerunner of scent concentration on floral parts alone—a derived adaptation ecologically uniting pollinator and flower.

From an evolutionary point of view, the scent tissue is ultimately organized into scent glands, or osmophores. The cells contain massive protoplasts, and the nucleus is often enlarged by endomitosis. The cells are filled with starch or fatty oils which otherwise are characteristically found only in seeds or bulbs. The gland tissues "suggest a division of labor, being differentiated into storage, production, and epidermal emission layers."

Vegetative Structures

The primitive orchid plant, following the basic sympodial habits of other monocotyledons (Holttum, 1955), is presumed to display the following features: terrestrial fleshy roots without velamen; slender, subterranean rhizomes; elongate stems; corms, or possibly pseudobulbs, of many internodes; and several spirally arranged plicate leaves. Advanced orchids, by contrast, show epiphytic aerial roots with velamen; rhizomes fleshy or absent altogether; a monopodial habit; modified stems with pseudobulbs of single internodes; reduced numbers of leaves which are conduplicate, clustered, or distichous (Dressler and Dodson, 1960). The inflorescence may progress from a terminal to a lateral or to an axillary position. Dressler and Dodson believe that pseudobulbs may be derived phyletically from either of two organs: from terrestrial cormlike forerunners involving several nodes at the base of a stem, such as those found in *Bletia* or *Phaius;* or from the thickening of the stem of the whole aerial shoot, resulting eventually in pseudobulbs consisting of a single internode (Fig. 6-26).

Fig. 6-26. Schematic representation of the possible evolution of the vegetative habits of the orchids. The primitive sympodial orchid plant with terminal inflorescence has become modified by forming fleshy perennial underground roots, rhizomes, tubers, or

corms; the stem has become vinelike, monopodial, almost absent, reedlike, canelike, or
modified into homoblastic or heteroblastic pseudobulbs. Details of explanation in text.

Leaf Anatomy

The leaves of orchids from over 150 genera are listed in Solereder and Meyer (1930) as having been described. The investigations of Chatin (1857), Krüger (1883), Möbius (1887), and Tominski (1905) are particularly cited. The work of Cyge (1930) has since appeared. Möbius attempted to work out a correlation between leaf anatomy and an orchid classification scheme. More recently, Ayensu and Williams (1972) correlated leaf anatomy within the ONCIDIINAE with other generic traits as an aid in taxonomic study.

For purposes of our discussion here, the leaves of orchids may be divided into two major types: (1) ribbed or plicate; and (2) leathery, soft or hard, often fleshy. These divisions are by no means absolute, and a plant exhibiting a combination of these two characteristics may well be the rule rather than the exception. Nevertheless, the creation of these two categories allows for the discussion of orchid leaf characteristics with a certain perspective, rather than just listing species with a description of their anatomy. Dressler and Dodson (1960) point out that the plicate type is ordinarily convolute and the leathery type conduplicate in development. These authors emphasize that the conduplicate leaf probably evolved independently in several evolutionary lines and is strongly associated with the epiphytic habit. *Vanilla* seems to be intermediate, showing convolute formation during development and, finally, fleshy leaves with conduplicate appearance when mature. *Cymbidium*, they also note, is transitional in the same way, having a conduplicate development but showing many veins characteristic of the plicate type.

General Structure of Plicate Leaves. The first type, plicate, represented by such species as *Coelogyne barbata, Calanthe furcata,* and most species of *Catasetum,* is generally elongate, and may be narrow or broad. The plicate leaf is thin and membranous and is ribbed parallel to the course of the fibrovascular bundles. Plicate leaves are often deciduous, lasting only a single season (e.g., *Lycaste*). The primary function of such leaves is photosynthetic. The retention of water, so important to the epiphytic varieties of orchids, is taken over almost entirely by pseudobulbs, fleshy roots, rhizomes, or various other underground structures.

The veins, and hence the ribs, run parallel along the length of the leaf and may form an internal pattern indicative of the species (Fig. 6-27). The leaf of *Epidendrum piniferum* has a large midrib, with closely spaced groupings of two or three ribs on either side. *Catasteum* has a large midrib with another large rib on either side; equally spaced between them are several smaller ribs. *Coelogyne barbata* shows a midrib with two large ribs on either side, and *Cymbidium* Flirtation has five large ribs on either side of the midrib. The latter two species have several smaller ribs spaced between the larger ones. In the plicate leaf, each rib projects on the underside of the leaf, with the upper side either indented at the site of the rib, as in *C. barbata,* or bulging as in *Cymbidium* Flirtation.

Epidermis. Both sides of the plicate leaf have a surface covered with cutinous layers. The upper surface is generally covered with a thicker cutin layer, as in *Calanthe furcata*; or the cutin layer may be about equal on both sides, as demonstrated by *Epidendrum piniferum.* The former condition is by far the commoner one and is present in varying degrees. Cyge (1930) notes that the thickness of the cutinous covering is determined by the degree of exposure to the sun, and that leaves which are more exposed on both sides (i.e., upright) tend to exhibit a more nearly equal distribution of cutin than do leaves which have only their upper side exposed to the sunlight.

Stomata are present and are generally restricted to the lower surface of the leaf. Stebbins and Khush (1961) find that the stomatal complex in APOSTASIOIDEAE, and the ORCHIDALES in general, are anomocytic, with two guard cells and no subsidiary cells. They are generally found running parallel to the veins and may be distributed on both surfaces, though this is common only in certain terrestrial varieties and leaves which are exposed to sun on both sides. When stomata are present on both sides, however, Cyge (1930) notes that distribution is unequal, with a greater concentration on the underside. In upright leaves, stomatal distribution varies, with the greater number of stomata being at the base of the leaf.

The stomata are oval, with the long dimension running parallel to the length of the leaf. They are found singly and may be sunken below the surface of the epidermis. Cyge reports the presence of such stomata in certain terrestrial orchids, and Solereder and Meyer (1930) consider them rare. They state that such stomata have prominent outer ledges where the drop below the epidermal surface occurs. Such stomata also show a thickening of the guard cells and have small air pores. This type is common to plants in dry environments.

A second type of stoma may be observed in leaves that are devoid of strongly developed epidermal cuticle or outer walls. This type features ledges that are neither prominent nor pronounced and air pores that are larger and more similar to the stomata of *Tradescantia.* They are present in epiphytes and many terrestrial types from relatively humid niches. Substomatal chambers are present and vary in size. Subsidiary cells, as noted previously, are not generally present, though modified epidermal cell types may occur adjacent to the guard cells.

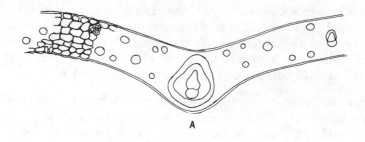

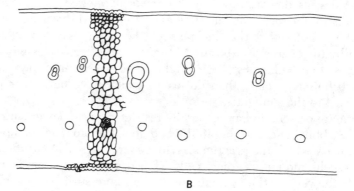

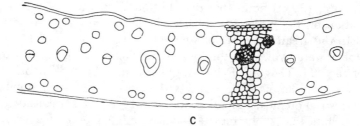

Fig. 6-27. Diagrammatic representation of typical leaf cross sections. (A) *Catasetum* species. (B) *Schomburgkia lueddemanniana*. (C) *Epidendrum prismatocarpum*. (D) *Vanda coerulea* (E) *Rhyncholaelia* (*Brassavola*) *digbyana*. (F) *Phalaenopsis lueddemanniana*.

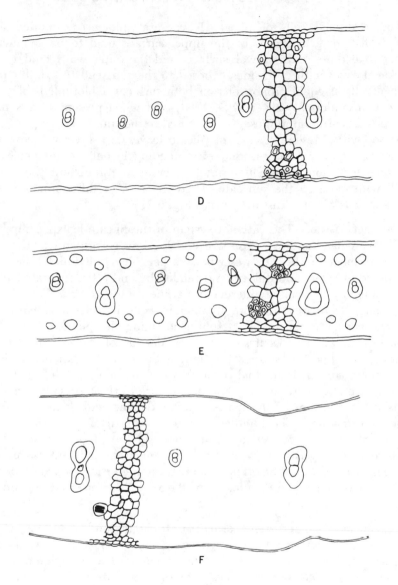

Fig. 6-27 (continued)

The epidermal cells of plicate leaves are oblong or rounded and slightly flattened. The cells of the upper surface tend to be somewhat larger than those of the lower surface, and the outer walls tend to be thicker. Leaves with both sides exposed to the sun tend to exihibit epidermal cells of equal proportions on both surfaces. Chlorophyll may be present in epidermal cells (Cyge, 1930) and when present occurs in a greater concentration in those cells exposed to the sun.

Mesophyll. Mesophyll cells in plicate leaves are directly below the epidermis and are mostly isodiametric, tending in only some species to be somewhat elongated. Chlorophyll is scattered throughout the mesophyll with more on the sun side. There is no clear differentiation into palisade cells or other distinct types (Fig. 6-27).

Vascular Tissue. The vascular system of the plicate leaf is composed of fibrovascular bundles whose size and location correspond to the size and placement of the ribs described earlier. The individual bundle is composed of xylem and phloem surrounded by and often separated from each other by a layer of sclerenchyma fibers (Fig. 6-27). The sheath may range from two to more than eight cells in thickness, with the greatest concentration of fibers found adjoining the phloem cells. The phloem cap, when it occurs, faces the underside of the leaf and forms continuous supportive strands throughout the entire length of the leaf. The structure of the individual constituents of the bundle closely parallels those found elsewhere in the plant. The nature of these specific elements will be covered more fully in the sections dealing with leathery leaves and stem anatomy. The bundles in the leaves are always oriented, in usual fashion, with the xylem uppermost and the phloem below. This arrangement is consistent with the arrangement of the bundles in the stem; thus a leaf would be oriented so that the underside (abaxial) of the leaf is homologous with the outside of the axis. The upper side would be adaxial.

Trichomes. The occurrence of hairs or trichomes on leaf surfaces is not common, and they are often restricted to the lower surface. Hairs are similar, whether from plicate or other types of leaves. Solereder and Meyer (1930) note that trichomes, when present, are often secretory in nature, the secretion being slimy or oily. Such hairs are often sunken with the bases set in depressions, and they are arranged singly or in groups. The emerging tip is either simply pointed or more complicated with branched structure. Trichomes may be either unicellular or shaggy multicellular structures composed of bundles of single-celled stalks. *Dendrobium cariniferum,* we note, has hairs on both sides of the leaf, though the underside is more thickly covered than the upper.

Leathery Leaves. The second leaf type, leathery, is a more heterogeneous grouping, including all those leaves not classed as plicate. The designation "leathery" refers mostly to the feel of the leaf surface. The leaves of plants in this grouping are similar in that protruding veins, except for the midrib, are not generally present. The cuticle is heavier and more waxy, and the leaf is thicker than in the plicate type, and may even be fleshy. A characteristic which the harder or tougher members of this group share is a uniform rigidity, and the supportive tissue is not confined to the fibrovascular bundles. Such leaves will usually snap or crack when bent beyond a point. We will discuss three subclasses of leathery leaves; soft, hard, and fleshy.

Soft Leathery Leaves. The thin, soft, sometimes almost fleshy leathery leaf is present in terrestrial orchids, including *Paphiopedilum,* and also in epiphytes such as *Phalaenopsis.* Leaves of this type are pliable and are without firmness except for the epidermal walls and cutin. The leaves are rounded or elongated and do not generally grow in an upright fashion. Characteristically, such leaves bear a cutinous coating, generally thicker on the upper epidermis than the lower. The epidermal layer is composed of elongate, flattened cells, which tend to be thin-walled, though a slight thickening of the wall can sometimes be detected on the outer side (Fig. 6-27). The epidermal cells themselves tend to be larger on the upper surface of the leaf and are arranged in a more orderly fashion than those on the lower. Chloroplasts, when present in epidermal cells, show greater concentration on the upper surface of the leaf, as Cyge (1930) reports for *Orchis maculatus* and *Cephalanthera rubra,* or may be present exclusively on the upper surface as in *Listera ovata.* Stomata are generally present as single structures either in slight depressions or on the surface. They appear to be of the previously discussed *Tradescantia* type, with poorly developed ledges and large air pores. Subsidiary cells are not generally present, though modified cells are present in some plants such as *Gymnadenia conopsea.* Though single stomata are prevalent, Raunkiaer (1899) notes the presence of double stomata in *Sturmia, Liparis loeseli,* and *Epipactis palustris.* Stomata are generally present on the underside of the leaf, though Cyge reports the presence of stomata on both sides of *Epipactis rubiginosa* and *Orchis globosa.* When present on both sides, the upper side has the greater number.

Substomatal chambers are present. Hairs may be present on either or both sides or only on the lower side, as noted before.

Mesophyll cells are present and are generally rounded. Although there is a tendency for the upper cells to be larger, there is no specific palisade layer. Chloroplasts are present in greater concentration toward the upper side of the leaf, but the mesophyll cells in general contain relatively fewer chloroplastids than those of the plicate leaves. Conversely, meso- phyll cells of soft leathery leaves are more involved with water storage than similar layers in plicate leaves. There may occasionally be enlarged intercellular spaces.

Fibrovascular bundles are present, sometimes in layers, and travel parallel through the length of the leaf. Often an alternation of bundle size is observed, generally taking the pattern of one large bundle being followed by one to three smaller bundles of different or similar size. As in plicate leaves, the bundles are composed of xylem, phloem, and a sheath of sclerenchyma fibers surrounding and separating them. As before, the phloem is capped by a concentration of fibers; but in these soft leathery leaves, the phloem cap tends to be larger than in the plicate.

In addition, fiber strands may be present that are not associated with any conductive tissue. These strands, which appear supportive in nature, may be found either in layers above and below the conductive tissue or scattered among the veins. The occurrence of such fiber bundles is closely correlated with the stiffness of the leaf, and it should be noted that the gradation from soft and thin to hard and rigid, or fleshy and terete, is dis- tinctly related to the number of such extravascular strands and their loca- tion in relation to the leaf surface.

Hard Leathery Leaves. The subgrouping of hard leathery leaves in- cludes such plants as *Rhyncholaelia* (*Brassavola*) *digbyana, Oncidium splendidum, Epidendrum prismatocarpum,* and *Angraecum eburneum.* In thickness, there is considerable variation, ranging from the rather thin *E. prismatocarpum* to the very thick *O. splendidum* (Fig. 6-27). In shape the leaves are generally elongate, show conduplicate form with a midrib protruding, and are from plants predominantly epiphytic in habit. Their V-shape appears to decrease the flat surface area directly exposed to the sun and also adds support to the leaf. Increased folding, with subsequent fusion or modification, may be nearly complete, as in certain equitants such as *Oncidium variegatum* (Fig. 6-28).

The hard leathery leaves have a thick, heavy covering of cutin that is either thicker on the upper leaf surface, or equally thick on both surfaces, depending on leaf orientation. The epidermal cells on the upper surface are usually larger than those of the lower. The upper epidermal cells of *E. prismatocarpum* are more elongated than those of the lower epidermis; *O. splendidum* shows similar cell shapes on both surfaces. The epidermal cells are slightly flattened with distinctly thickened outer walls. The surface of many hard leathery leaves is uneven or rugose to the touch from the bulging of these walls.

Stomata are present on the lower side of the leaf, though leaves of certain plants such as *O. splendidum* show stomata on both sides. In such cases, they are present in greater concentration on the lower surface. Stomata are generally located in slight depressions; however, some may project as small papillae above the leaf surface. The guard cells are large, and the air spores are small. Thick, distinct ledges can readily be distinguished. Substomatal chambers are present. Such stomata are arranged in rows parallel to the veins of the leaf. Distinct subsidiary cells are lacking, though some modified adjacent cells may occur. Hairs are not generally present on hard leathery leaves, though *Epidendrum conopseum* shows moderate trichation on the lower surface. These hairs are unicellular and not seated in depressions.

Mesophyll cells may be isodiametric, as in most leaves of this type, with the upper mesophyll cells being slightly larger; or there may be differentiation into elongated, almost palisadelike cells, as we have observed in *R. (B.) digbyana* (Figs. 6-27 and 6-28). Secondary thickenings in bands on the walls of mesophyll cells occur in *O. splendidum* and are oriented mostly parallel to the direction of the veins. These thickenings apparently strengthen the cells and hence lend support against collapse due to dehydration. Solereder and Meyer (1939) note that the occurrence of such markings in epiphytic orchid leaf mesophylls is not uncommon and occurs in several species. They are illustrated in the work by Curtis (1917) on epiphytic orchids of New Zealand. Occasional mesophyll cells of *R. digbyana* may have typical Casparian strips banding their walls.

Fibrovascular bundles travel parallel to the leaf for its entire length. There may be more than one layer of bundles, as in *Schomburgkia lueddemanniana,* or only one layer, as in *Angraecum eburneum.* The midrib is the largest bundle and may be double. The anatomy of the individual bundles conforms to the general plan, though the sclerenchyma, both that capping the phloem and that of the entire sheath, tends to be considerably thicker than in previously discussed leaves. Small bundles may be present in great numbers. In leaves of *E. prismatocarpum* the bundles are practically adjacent to each other; larger bundles may be spread out evenly as in *R. digbyana.*

The feature that seems to differentiate the hard leathery from the soft leathery leaves, besides thicker epidermal walls and greater cutinous covering, is the presence of a series of sclerenchyma fiber bundles running through the mesophyll and/or distinctly sclerified cells at or near the surface of the leaf. These fiber bundles are composed of from four to more than 20 sclerenchyma fibers, which appear similar to the fibers found surrounding the vascular bundles. They strengthen the leaf and

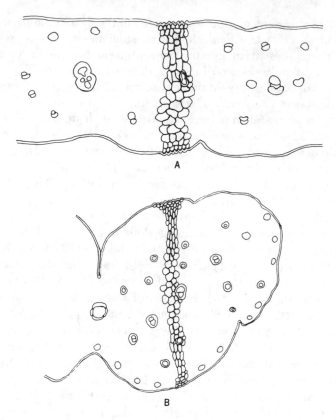

Fig. 6-28.   Diagrammatic representation of various leaf cross sections. (A) *Epidendrum parkinsonianum.* (B) *Brassavola nodosa.* (C) *Oncidium variegatum.* (D, E) *Eria pannea.* (D) Section toward base of leaf showing channel. (E) Section toward tip of leaf.

seemingly provide for greater protection for the inner mesophyll from the outside environment. These bundles were mentioned earlier in conjunction with the soft leathery leaves; it must be emphasized that an increase in the quantity of these extravascular fibers corresponds to the degree of toughness characteristic of the leaf.

E. prismatocarpum shows a great number of such fiber bundles, packed closely together and situated in two layers, one above and one below the vascular bundles proper. Occasionally there is a connection between adjacent fiber bundles. The bundles vary in size, but no particular size pattern can be detected. The bundles may give the epidermis a decidedly ribbed appearance, reminiscent of the plicate leaf, but the ribbing is not

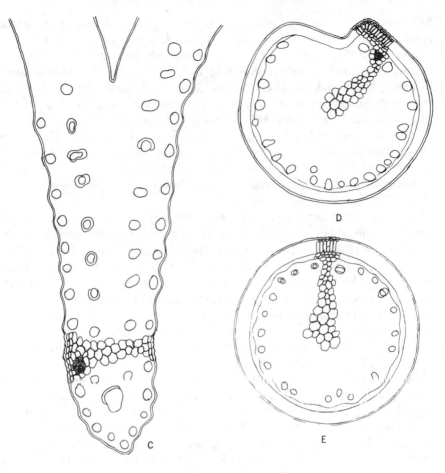

Fig. 6-28 (continued)

solely derived from fibrovascular bundles as it is in the plicate leaf. The occurrence of so many fiber bundles and fibrovascular bundles in the leaves of *E. prismatocarpum* severely limits the amount of mesophyll. Leaf area not occupied by bundles is crammed with the slightly elongated mesophyll cells that are all equally spaced and equally supplied with chloroplasts.

Leaves which are certainly tough, hard, and stiff but, surprisingly, practically devoid of fiber bundles are from plants of *R. digbyana*. Instead of fibers they have an extremely thick cuticle and a very thick-walled epidermal cell layer to help provide the firmness. Although extravascular fiber bundles are lacking as such, the sclerenchyma sheath surrounding the fibrovascular bundles is greatly expanded so that there is also a thick cap or layer of sheath fibers on the xylem side resembling the phloem cap in all typical vascular bundles of the other plants. In this instance, however, the individual fibers on the xylem side more closely resemble the extravascular fibers present in *E. prismatocarpum* than they do the usual vascular fibers concentrated around the phloem. That is, they are thick walled but with a larger lumen than is usually present. The xylem cap is situated directly subepidermally, whereas the smaller phloem cap is separated from the epidermis by a mesophyll layer. Smaller vascular bundles among the larger main bundles have typical phloem caps and show no xylem cap fibers at all.

The mesophyll layer near the upper epidermis is composed of large, elongate cells about five cells deep. They are heavily sclerified and virtually chloroplast-free. The epidermis itself is also heavily sclerified. This almost colorless layer of cells, backed by fibers and surfaced with thick cutin, gives the leaf a glaucous or whitish appearance on the upper surface and affords protection from the intense light, heat, and dryness to which the plant is normally exposed. In comparison, the superficially similar leaves of *O. splendidum* show many fiber bundles. The epidermal cells are also heavily sclerified, and the mesophyll cells are marked with sclerified thickenings across the cell walls.

Leaves of *Vanda coerulea,* though classed as leathery, are noticeably more pliable than the other hard leathery types. The rigidity of *V. coerulea* is due not to the occurrence of sclerenchyma fibers, either incorporated or scattered, but to a heavily sclerified epidermis and occasionally sclerified cells in the mesophyll (Fig. 6-27). This provides a leathery feel to the leaf and protection to the mesophyll. The lack of stiffness is due to the small number of fibers within the leaf.

It thus appears that several factors can contribute to the leatheriness or toughness of a leaf. Thick cuticle and sclerified epidermal cells, extravascular fiber bundles, xylem or phloem cap fibers, and sclerified mesophyll cells are all features that occur separately or in combination. The quality of a given leaf is determined by the number of such characteristics present and the relative number of such cells compared with other cell types.

Fleshy Leaves.   The subgroup "fleshy" is more of a compromise than anything else. It includes leaves that are almost completely terete, such as those of *Eria pannea* or *Brassavola nodosa,* and leaves that are thick and somewhat broadened, such as those of *Epidendrum parkinsonianum* (Fig. 6-28). Though some fleshy leaves exhibit a certain toughness, they are considerably softer than any of the hard leathery leaves. One should not imagine them as being succulent, however, for in fact the softest of this type would be considerably tougher than a fleshy leaf of some *Crassula* or related species. These leaves possess thickened cuticles and thick-walled epidermal cells that provide the toughness. A combination of very thick mesophyll layers and the lack of large numbers of extra-vascular fibers imparts the fleshy character. A number of these leaves and other leaves have been illustrated by Möbius (1887).

The terete type of leaf, such as that of *Brassavola nodosa* or *B.* David Sander, grows in a "rolled up" habit, actually a completely folded and fused, basically conduplicate structure (Fig. 6-28), with a channel de-tectable where the leaf margins meet. The consequence of this folding or encirclement is to remove the upper surface from contact with the outside environment, and it provides a cylindrical area, rather than two flat planes, in contact with the atmosphere. The outer (lower) surface is tough and covered with a thick cutin. The inner (upper) surface (the channel only) does not evidence the same toughness, and the cutin cover is thinner. The outer surface is comparable functionally to the upper surface of hard leathery leaves, whereas the inner surface shows charac-teristics leading to the development of fleshiness. The fibrovascular bundles are sheathed thickly with sclerenchyma, and a layer of sclerified fiber bundles exists both above and below the vascular bundles. The con-centration of bundles is greater toward the outer (lower) surface. The mesophyll cells between the inner (upper) fiber bundle level and the lower epidermis are isodiametric. From the level of the inner bundles toward the upper epidermis, the cells have an elongate character, being three times as high as wide. These cells contain few or no chloroplasts and appear entirely involved with water storage. Thus by rounding, *B. nodosa* and *B.* David Sander retain the leathery characteristic on their outer surfaces and are afforded ample protection for the fleshy water storage tissue toward the center of the leaf.

This encirclement has proceeded in *E. pannea* to a point where the lower (outer) epidermal surface extends almost completely around the inner mesophyll mass composed of isodiametric water-filled cells (Fig. 6-28). A small depression only toward the base of the leaf, approximately 20 cells across at its widest point, forms the channel that represents the upper surface of the leaf at the joining point of the two edges. The en-circling lower surface is composed of slightly thickened epidermal cells backed up by often sclerified, elongate hypodermal cells. A layer of fibro-vascular bundles follows this hypodermal layer, and extravascular bundles are lacking. The channel surface is covered by similar epidermal cells but lacks correlating vascular components and is backed by several elongated palisadelike sclerified mesophyll cells. A layer of mesophyll cells separates these from the vascular bundles that are arranged, as described, in a ring completely around the leaf. The cells between the epidermis and the bundles are filled with chloroplasts, whereas the interior mesophyll cells are devoid of chloroplasts and apparently only concerned with water storage. Thus one way in which a leaf can achieve fleshiness is by the encirclement of the mesophyll by the lower surface and the elimination of the upper surface, resulting in a mesophyll water-storage core surrounded by chlorenchyma and an epidermal layer.

The second type of fleshy leaf includes those that result simply by expansion of and increase in the number of mesophyll cells. Such fleshiness is demonstrated by *E. parkinsonianum*. The leaf structure is similar to that of the hard leathery leaves, with thick-walled epidermal cells, heavy cutin, sclerenchyma sheaths around bundles, and also extravascular bundles, but the proportion of fiber constituent of the vascular bundles is considerably smaller (Fig. 6-28). The many large, isodiametric mesophyll cells present show no special differentiation patterns and have almost uniform distribution of chloroplasts.

Ecological Adaptation. It must be realized that orchid leaf types, whether plicate, soft leathery, hard leathery, or fleshy, are primarily related to the environment in which the plant lives. Modifications such as leatheriness and fleshiness are related directly to the ecology of the individual plant and not to a genus determinant. One would thus expect several orchids sharing the same ecological niche to possess similar leaf types. That is often the case, in fact, whether or not the various species are closely related. Orchid leaf types are a direct result of modifications enabling the plants to cope with a specific microhabitat and should be viewed as such.

Three general directions of modification are thus observed. These are plicateness, leatheriness, and fleshiness, which correspond to the types described in our discussion. The plicateness provides for a large surface area and exposure to light. It is questionable whether plicateness can truly be considered as a direction of modification, for it is felt by some investigators that original ancestral orchids, likely in southeastern Asian jungles, were already of the plicate terrestrial type. Accepting this, one can consider that orchids evolved vegetatively by developing pseudobulbous epiphytic habits along with the plicate leaves, thus enabling them to inhabit new niches; further, the two other characteristics represent further adaptations for survival in other still different niches. The main function of plicate leaves, as previously stated, is to allow for maximum exposure to light for photosynthesis, particularly in shady, moist situations. Dessication is not a major problem for plants with such leaves, and food and water storage is accomplished by the presence of fleshy roots, corms, or rhizomes and later by the fleshy stems. This is indicated in a general fashion by the illustrations in Fig. 6-26.

The evolution and proliferation of orchid species have been accompanied by responses to environments where rainfall and temperatures are seasonal and sun and heat are considerably more direct than in shady, moist forests. These more xerophytic conditions necessitate ample provisions for water storage and retention as a prerequisite for survival. The fleshy or soft leathery leaves provide splendid modifications for water storage, but they depend on a somewhat shaded or cool environment; otherwise strong sun and heat cause them to dehydrate and lose the precious water upon which survival depends. Finally, the hard leathery leaves provide the most effective water retention. The thick epidermal cells and, more important, the layers of sclerenchyma fibers provide excellent protection against dehydration, as well as affording support. Anyone growing or handling orchids can attest to the efficiency of such devices as reflected in cultural procedures.

Stem Structure

Orchid stems are of two main types, sympodial and monopodial. The sympodial habit is ancestral and follows the general habit of many monocots (Holttum, 1955). The sympodial habit is characterized by successive growths, each originating from the base of the preceding one. The initial portion of the new growth, with its shortened internodes, may form a rhizome or corm at the base of the new growth, or it may develop directly into an upright shoot. These upright shoots are ordinarily leafy and determinate, and they bear the flowers, though occasionally there may be a specialization into separate vegetative and flowering branches. It is in these stems, each producing a new supply of roots, that the great variety of vegetative adaptations typical of epiphytic orchids is apparent (Fig. 6-26). They may be pseudobulbous or reedy, one- to many-leaved or bracted, and flowers may be produced either terminally or in axillary fashion. Solereder and Meyer (1930) cite investigations of the axis in over 100 genera. Though the life of a single stem is limited, the growth of the plant as a whole is unlimited.

The monopodial types are characterized by a uniform axis producing leaves and stem at the apex in indeterminate fashion. Flowers are always axillary in origin, and the roots may be produced at all nodes successively. The vandaceous orchids and their relatives have stems characteristic of this type. Such orchids are never deciduous and terminal leaves are always present for water storage and to protect the growing apex. There are monopodial types, however, that are "leafless," the buds being protected by minute scales and with the roots fleshy for water storage and green for photosynthesis.

Intermediate between these two major groups are the plants such as *Dichaea* or *Lockhartia* with pseudo-monopodial habit. Individual stems may continue to produce leaves for a long period of time, but when new stems finally develop they are formed in sympodial fashion from the base of the previously formed growth. The monopodial habit is thus a specialized pattern that has evolved from sympodial ancestry.

Vanilla, although ostensibly fitting the description of a monopodial orchid, must be considered in a class by itself, with its vine habit and different internal anatomy. It will be described separately.

Sympodial Habit: Terrestrial.   Three general types of organization are demonstrated within the sympodial group. One of the most primitive arrangements, typical of many terrestrial species with basal leaves, is found in the group called the lady's slipper orchids, such as plants of *Paphiopedilum* or *Cypripedium.* The other two habits are found among epiphytic orchids: those with fleshy pseudobulbous stems and those with reedlike, woody stems.

The slipper orchids grow from short underground rhizomes terminating in short leafy stems (Rosso, 1966) bearing flowers on stalks from the axils. A new shoot or growth arises annually from an axillary bud at the base of the previous growth, according to the growth cycle of the plant. The rhizome produces adventitious roots and some scale leaves, though the latter are often small and inconspicuous.

Rhizome.   Two types of rhizome are evidenced by the slipper orchids. The conduplicate genera, such as *Paphiopedilum,* feature a condensed axis, whereas in the plicate-leaved forms of *Cypripedium* the axis is generally elongated. In external appearance the transition from shoot to rhizome occurs quickly. In cypripediums a sharp area of demarcation can readily be observed; the other genera display a more gradual transition without any sharp demarcation between the two.

An epidermal layer and cutin covering may or may not be present. Hairs or trichation of any form are absent. When an epidermis is present, as may occur in very new rhizome growth, the epidermal cells are squat, thick-walled, and cutinized or suberized. An explanation for the occasional lack of epidermal tissue seems to be that in growth, with constant substrate abrasion, that layer is often rubbed away (Rosso, 1966).

Below the epidermis, or epidermal region, if a proper epidermis is lacking, the arrangement of tissues varies. The cortex in *Cypripedium* constitutes a major portion of the rhizome, while in the conduplicate varieties the cortical region is reduced. *Selenipedium* shows a cortex reduced to only a thin cylinder of cells. The demarcation between cortex and central tissues is indicated by an endodermis and a pericycle with a sclerenchymatous band.

Endodermis and pericycle are not limited to roots in the orchids. According to Solereder and Meyer (1930), the orchids often show an endodermis in stems, or a pericycle with a sclerenchymatous ring delimiting the vascular cylinder from the cortex. Such pericycle in stems may be multiseriate; it is usually uniseriate in orchid roots.

Rosso (1966) notes that two types of endodermal cells occur in the slipper orchids, based on suberization patterns of the walls, O or U types. For those conduplicate species examined, including *Paphiopedilum insigne, P. niveum,* and *P. venustum,* and *Phragmipedium caudatum* and *P. schroederae,* the endodermis is multiseriate. In all other slipper orchids only a uniseriate endodermis is detected. In work discussed in Solereder and Meyer (1930) and Rosso (1966), several authors note that the cell types present, whether O or U, are related to the individual species and are not general throughout a given genus. Directly within the endodermis of all genera a sclerenchyma band is found. The long, narrow, thick-walled fibers composing this ring differ in number according to species; some *Cypripedium* species show reduced pericycle; the conduplicate genera feature a multiseriate one. Vascular tissue in all species remains restricted within the circle formed by the cortex, but a strict, definite pattern is lacking.

The vasculature is composed of individual fibrovascular bundles consisting of phloem and xylem with a sclerenchyma sheath surrounding the conductive members and occasionally separating them. The sheath tends to be considerably thicker in the region of the phloem and forms a phloem cap over that area. The sheath may or may not completely encircle the entire bundle. Each bundle is oriented so that the phloem, and hence the phloem cap, is facing the exterior of the rhizome.

The phloem of the bundle consists of sieve tube elements with simple transverse end plates; companion cells occur too. All xylem conductive elements are tracheidlike in the rhizome, though vessel elements may be found in root xylem (see Cheadle, 1942). The individual elements are long, with spiral, annular, or scalariform thickenings, along with pits with reduced borders and highly extended apertures.

The cells of the ground parenchyma, and also those within the cortex-pericycle layer, are of medium size and roughly isodiametric. The walls are of medium to heavy thickness. Rosso notes the appearance of parenchyma cells, particularly in the conduplicate forms, with a cross-reticulate pitting. Cells with simple pitting and no reticulations occur in many other genera.

Aerial stems. The leafy above-ground stem of the lady's slipper, although externally distinguishable from the rhizome, is similar in internal anatomy. It is generally short, often being overlooked beneath the leaves in such genera as *Paphiopedilum* or *Cypripedium*. A trichated epidermis is present in the plicate genera, giving such species a hairy appearance. Some people are allergic to the hairs. Rosso (1966) notes the presence of two types of trichomes: a nonglandular type, with attenuated end cells; and a second type with bulbous end cells. When present, hairs are more densely concentrated on the upper portion of the leafy stem.

The epidermis is covered with a cutin of varying thickness. As with orchid leaves, the thickness tends to be directly related to the ecology rather than to any phylogenetic determinant. The epidermal cells are straight-sided, thick-walled, and suberized. Epidermal cell measurements show considerable interspecific variability. The presence of stomates on stems has been noted by various workers, including Solereder and Meyer (1930). Rosso (1966) found variations in the stomate size which corresponded to epidermal cell size. As with stomata in orchid leaves, subsidiary cells are absent, and the stomata are subtended by substomatal chambers. Chloroplasts are generally present in epidermal cells of the slipper orchids, and the concentrations appear higher in young cells than in more mature ones. The concentration of chloroplasts in the epidermis, as can be expected, is somewhat less than in cells from the interior of the stem. Below the epidermis layer of the stem is found a thin cortex, composed of cylindrical cells with walls of medium thickness, though walls in plants of the conduplicate genera may show secondary thickening.

Within the cylinder bounded by the pericycle the vascular bundles and ground parenchyma are distributed. The bundles, like those of the rhizome, are composed of phloem and xylem, surrounded by a sclerified sheath tending to be thicker about the region of the phloem. The conducting elements of the xylem are tracheids, with the same types of wall architecture found in the rhizome. The phloem is as described before, though Rosso notes that plicate genera show more elongated cells. The bundles are oriented with the phloem facing outward, tending toward a circular arrangement in the nodal areas but showing a scattered arrangement in the internodal regions. Leaf traces are located outside the pericycle ring, when it is present, or in the area corresponding to that region when a clearly defined ring is absent.

The ground parenchyma is composed of cells with walls of varying thickness. In *Cypripedium* thick-walled cells are restricted to the basal portion of the stem. *Paphiopedilum* has a gradual thickening approaching the center of the stem. *Paphiopedilum* has sclerified cell walls with considerable coarse-reticulate pitting. The parenchyma of *Selenipedium* shows cell walls with an increase in thickness relative to their nearness to the center of the stem.

Many details for other terrestrial species might be reviewed here, but they are catalogued systematically with reference to the species in Solereder and Meyer (1930), particularly for European ground orchids, and in Camus (1929) and Fuchs and Ziegenspeck (1936); thus there is no point in repeating them. The lady's slipper plants covered in our descriptions are not unique morphologically.

Sympodial Habit: Epiphytic. The pseudobulbous orchids are largely epiphytic plants possessing a horizontally growing rhizome and a vertically oriented pseudobulb. The pseudobulb functions as a water-storage organ and is often large and spherical in form. A new shoot sprouts from the rhizome annually, bearing pseudobulb, bracts, leaves, and flower spikes. The pseudobulbs can be one of two types, according to Pfitzer's system (1889), regardless of shape: heteroblastic or homoblastic. Heteroblastic pseudobulbs are composed of only one internode; homoblastic of two or more internodes by varying or similar length (Fig. 6-26). The latter is considered the more primitive of the two.

Data obtained by examination of 35 species and hybrids by Wejksnora (1971) indicate that the pseudobulb and rhizome anatomy of such plants is essentially the same for all species studied. The evidence indicates that homogeneity of structure can be expected for pseudobulbous species not yet examined. Thus, with a few minor exceptions that will be noted, the generalized anatomy of orchid rhizome and pseudobulb can be described by a careful examination of one selected species such as *Oncidium sphacelatum*. Those plants that differ discernibly from this species can represent extremes, with *Oncidium sphacelatum* representing a generalized type. Though most specimens studied were greenhouse grown, subsequent examinations of the same, as well as other species obtained in the wild, revealed little difference worth noting, and it was not expected that the cultivated plant specimens would differ anatomically from those that grew wild.

The Rhizome. The horizontally growing rhizome of *O. sphacelatum* produces both adventitious roots and pseudobulbs. The length between

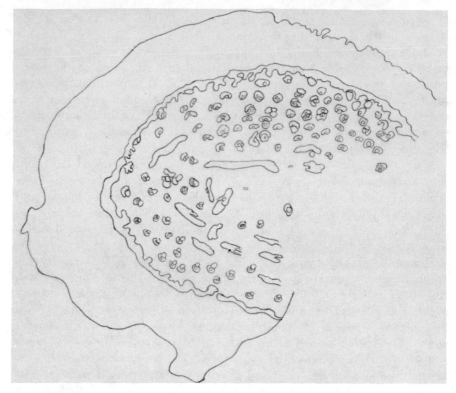

Fig. 6-29. Section through rhizome of *Oncidium sphacelatum* showing vascular bundles running in various planes. (After Wejksnora, 1971.)

pseudobulbs varies; it ranged on the plants examined from 2 to 5 cm from pseudobulb to pseudobulb; it was about 7–13 mm wide. The rhizome was covered with a cutinous layer which varied in thickness, and below this was a layer of sclerified cells, ranging in diameter from 50 μ at the periphery to about 80 μ farther in. Directly below these cells is the most conspicuous internal feature of the rhizome—one which is visisble to the unaided eye—a ring of vascular bundles. This ring surrounds the ground parenchyma with its scattered bundles and separates them from the cortex (Fig. 6-29), similar to the pattern in rye and certain other grasses which possess two concentric bundle rings and a ring of sclerenchyma (Arber, 1925). In *O. sphacelatum* the ring is not merely a sclerenchyma band with occasional bundles, but rather a circle of connected bundles. This ring may be broken, and in those areas in which no bundle occurs there is generally no sclerenchyma bridging the gap. Occasionally the ring is made of groups of connected bundles, which though not continuous maintain the ring shape. The presence of a gap in the ring structure ordinarily correlates with a root outgrowth from the rhizome. The bundles scattered in the ground parenchyma are generally only single.

The bundles forming the ring are similar in structure to those scattered about (Fig. 6-30), except in the amount of phloem present. The ring bundles contain less phloem than their scattered counterparts. Though exceptions occur, on the average the difference in phloem quantity prevails. Within the ring the parenchyma is composed of isodiametric cells, averaging 79 μ near the circle, decreasing to about 50 μ nearer the center. The individual cell walls are smooth and of medium thickness. The bundles are numerous, randomly scattered, and do not run parallel to the axis of the rhizome or to each other. In any section bundles can be found at a variety of angles to the axis. Bundles at right angles are not uncommon (Fig. 6-29). A great deal of bending takes place, and it is doubtful that any bundle travels in a straight line through the rhizome.

Vascular Tissue. Examination of an individual bundle reveals the structure common to most monocots, that is, phloem and xylem completely surrounded by a sheath of thick-walled sclerenchyma fibers and sometimes sclerified parenchyma (Fig. 6-30). The ring bundle is oriented so that the phloem and its sclerenchyma cap is facing the surface of the

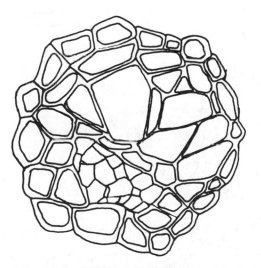

Fig. 6-30. Transverse section of vascular bundle from rhizome of *Oncidium sphace-latum*. See text for explanation. (After Wejksnora, 1971.)

rhizome, and the xylem is toward the center. There are no ruptured cells or conspicuous air spaces associated with bundles.

Bundles are either round or oblong in cross section. When circular, the diameter ranges from 1000 to 1500 μ. The larger oblong bundles swell out at the phloem cap and also on the xylem side. Such bundles have dimensions of up to 2500 μ in the long direction and in width up to 1500 μ. Bundles of intermediate dimensions are most common.

Two types of fiber make up the sheath. The first type may surround the entire bundle or be somewhat restricted to the phloem side. It is always present as a cap around the outside of the phloem and may separate the phloem and xylem (Fig. 6-30). These fibers will be referred to as phloem cap fibers. When occurring around the phloem, the sheath may be up to

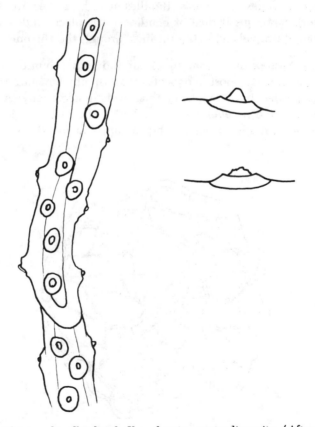

Fig. 6-31.   Rhizome bundle sheath fiber showing protruding pits. (After Wejksnora, 1971.)

six cells thick, though two to three cells are the most common. The band between the phloem and xylem is generally only one or two cells thick. The cells have blunt ends and are rather long. They range in diameter from 20 to 50 μ, and their walls are thick and heavily lignified. The walls show secondary thickenings with as many as five distinct layers totaling 3.5–5 μ in thickness. Numerous pits dot the fiber walls. These pits (Fig. 6-31) are prominent and large with diameters of 30–35 μ. They are bordered, and the apertures are 12–19 μ. The pit border forms a mound or nipple that extends upward about 7–10 μ. This mound is somewhat pointed or rounded, or has the appearance of being broken off. The central point is possibly an enlargement or thickening of the pit membrane. This may be pushed out by internal or external pressure through the aperture. Numerous other species show such pits in varying degrees, and their presence seems normal for orchid vasculature in general.

The second type of fiber is found mainly around the xylem. It occurs, joining the phloem cap fibers or sclerified parenchyma, or may be the only sheath member. These fibers, referred to here as xylem cap fibers, are 70–110 μ wide and rather long. They have lignified walls, tapered ends, simple pits, and thinner walls, about 1.3 μ, than the phloem cap fibers (Fig. 6-32) and lack the peculiar pits formed on the phloem cap fibers.

The third type of cell forming the bundle sheath is that of the sclerified parenchyma. These are parenchyma cells that have become lignified and appear to be midway between sclerenchyma fibers and parenchyma in structure. They are oblong and flat (Fig. 6-32) about $250 \times 800 \mu$ to $400 \times 1200 \mu$. Some cells, particularly those actually touching the xylem, are elongated and thinner, reaching lengths of up to 2000 μ. The usual size is about $300 \times 1450 \mu$. The walls are thin and lignified and are profusely dotted with simple oval pits. These pits are rather large and distributed evenly over the walls. These sclerified parenchyma cells are generally found to be restricted to the xylem area of the bundle and may not be in direct contact with the conducting cells.

These three cell types make up the sclerenchyma sheath. Around the phloem the cells are generally homogeneous, only cap fibers, while the xylem side may show three types of cells with little or no pattern or consistency in either distribution or organization.

The xylem in each bundle generally consists of five to ten cells. In cross section, these cells are angular, having three to five sides (Fig. 6-30), the diameter of each cell ranging from 250 to 450 μ.

In longitudinal view, two separate and different cell types can be distinguished, differentiated by the markings on the cell walls. One xylem

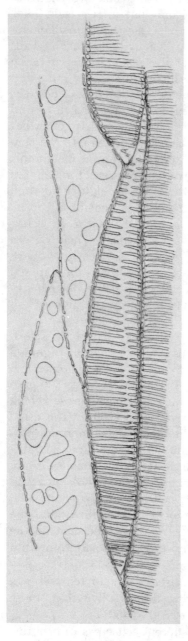

Fig. 6-32. Xylem parenchyma and adjacent scalariform and spiral tracheids from rhizome vascular bundle of *Oncidium sphacelatum*. (After Wejksnora, 1971.)

type is distinguished by the presence of scalariform pitted walls. These cells are long and have tapering tips (Fig. 6-32). There does not appear to be any distinct end wall nor any area that could be considered a differentiated end wall region. In fact, the ends of each cell do not connect with other ends; they often approximate the middle of the adjacent cells. Examination of macerations confirms that the cells are tracheids.

The other type of xylem cell is identical to the first in size and shape, differing only in that they have helical thickenings on the walls (Fig. 6-32). Like the scalariform tracheids, they have tapered ends with no clear end plates, so they may also be referred to as tracheids.

These two types are found together, forming no particular pattern and entwining with each other. Within the bundle the tracheids repeatedly bend, twist, and cross. A xylem fiber is occasionally found included in the twisting mass of tracheids, though the fibers are generally restricted to the outer edge of the xylem.

The phloem, separated from the xylem by a sclerenchyma band, is seen in cross section to be composed of 8 to 15 sieve-tube elements and companion cells (Fig. 6-30). The sieve-tube elements are angular, having a diameter of about 120–180 μ. The companion cells are smaller, 50–75 μ. Elements are about 500–750 μ in length, and the companion cells are somewhat shorter. The elements have distinct sieve plates with large pores. The slime bodies are spheroidal and scattered.

Pseudobulb Structure. The pseudobulb is the enlarged vertical stem from which leaves and flower stalks arise. *Oncidium sphacelatum* represents the heteroblastic type, having a pseudobulb consisting of only one internode and functioning as a storage organ. It is 10–14 cm long, 2–4 cm wide, and flattened, being about 1.3 cm thick at the widest point. The surface shows three to six ridges.

The surface of the pseudobulb is covered with a layer of cutin about 15 μ in thickness, and trichomes are completely lacking. Below the cutin is a layer of squat, thick-walled, sclerified epidermal cells, and below this is a layer two to five cells thick of large sclerified parenchyma cells, about 100 μ in diameter, that extend unevenly into the ground parenchyma (Fig. 6-33). The ground parenchyma consists of irregularly shaped cells ranging from 60 to 140 μ in diameter with large air spaces adjacent to the bundles. The fibrovascular bundles are scattered. Sclerified parenchyma cells in the ground parenchyma are found infrequently.

The vascular bundles of the pseudobulb lack any orientation of xylem and phloem with respect to the outer surface. There is no ring of bundles, nor any apparent grouping, and the bundles are scattered rather far apart

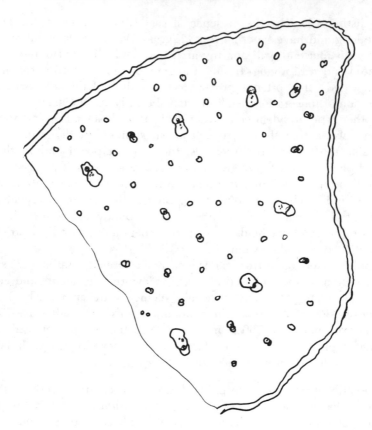

Fig. 6-33.   Segment of pseudobulb of *Oncidium sphacelatum* shown in diagrammatic cross section. (After Wejksnora, 1971.)

(Fig. 6-33), though they run more or less directly through the pseudo-bulb without the bending and twisting characteristic of rhizome bundles. The bundles themselves extend into intercellular spaces, which are found throughout the pseudobulb. These spaces are probably the result of early differentiation and/or differential growth of the bundles with respect to the surrounding parenchyma cells.

The vascular bundles are about the same size as those in the rhizome, with a sclerenchyma sheath bounding the vascular cylinder (Fig. 6-34). The sheath is composed of sclerenchyma fibers of the phloem cap type described earlier and sclerified parenchyma cells. Xylem fibers, as de-scribed before, are generally not found in the pseudobulb. Around the

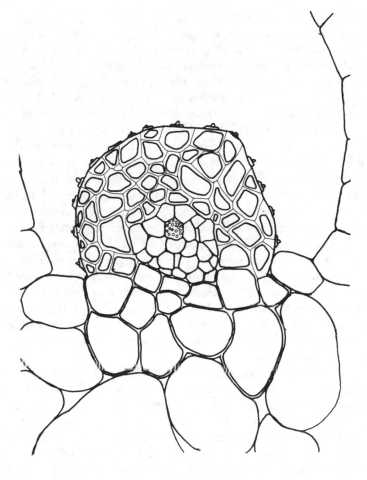

Fig. 6-34. Cross section of pseudobulb vascular bundle showing adjacent air space and phloem cap fibers protruding into it. (After Wejksnora, 1971.)

phloem, the cap of fibers is heavy, in places reaching a thickness of eight cells, though four to six is more common. The phloem cap fibers do not generally extend around the bundle, so that on the xylem side the main component of the sheath is sclerified parenchyma. The parenchyma cells found in the bundle are similar to those of the rhizome. The pits found on the fibers are reminiscent of those in the rhizomes and protrude into the intercellular spaces.

The xylem in each bundle, as it is seen in transverse sections made at random, consists of about five tubes. These are tracheids, roundish in shape, and ranging to 300 μ in diameter. They are straight, long, and like the bundles themselves extend parellel to the axis of the pseudobulb. Their walls are dotted with scalariform pits. Smaller tracheids with helical thickenings also occur. There is generally no sclerenchyma separating the phloem from the xylem. The phloem is composed of sieve-tube elements and companion cells. These cells are identical to those found in the rhizome, and the description for that phloem applies also in the pseudobulb. The rhizome contains about four times as many bundles as does the pseudobulb. This probably accounts in part for the woodiness of the rhizome compared with the relative softness of the pseudobulb.

In general, the pattern shown by *O. sphacelatum* is found throughout the other sympodial species examined. Vascular tissue remains unchanged in structure; arrangements and quantity, however, may vary from species to species. The pseudobulbs of *Lycaste cruenta*, for instance, show large bundles with more xylem present, whereas plants such as *Epidendrum adenocarpon, E. belizense,* and *Odontoglossum crispum* have bundles significantly smaller than *O. sphacelatum*. The pseudobulbs of *Brassia keiliana* have a series of large bundles arranged in a ring along with smaller scattered bundles (Fig. 6-35). There is no connection between these arranged bundles, as exists in the rhizome; otherwise, they match the previous descriptions.

More variation of this kind is noted in the pseudobulbs examined than in the rhizomes. But both organs may vary tremendously in size, and the pseudobulbs may display a wide variety of shapes from flattened to spherical. Pseudobulbs may be smooth, ridged, or angled. Both homoblastic and heteroblastic varieties were examined, and no pattern of internal differences could be distinguished. Epidermal cell size and the amount of cutin varied as with the leaf types described earlier. Again, the differences noted were related to ecology rather than genealogy.

Parenchyma cell sizes in rhizomes of different species appeared generally constant, whereas variation was noted in both the size of pseudobulb parenchyma cells and their intercellular spaces. *Odontoglossum crispum, Epidendrum belizense,* and *Bulbophyllum ornatissimum* have parenchyma cells of larger dimensions than those of *O. sphacelatum,* while those of such plants as *Cattleya bowringiana* and *Dendrochilum glumaceum* show smaller parenchyma. Although considerable variety can exist in external features of the pseudobulbs, little variation occurs in the internal arrangements.

As mentioned previously, many additional details of rhizome and stem anatomy may be found in Solereder and Meyer (1930). Dannecker

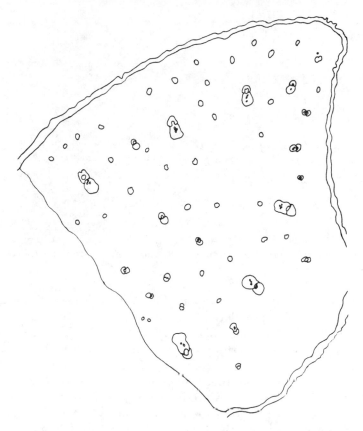

Fig. 6-35. Segment of pseudobulb of *Brassia keiliana* in cross section. (After Wejksnora, 1971.)

(1898) and Ross (1909) both studied the ant-inhabited pseudobulbs of various *Schomburgkia* and *Diacrium* plants. The hollows in the pseudo-bulbs are the result of growth and do not occur from the activity of the ants. There may be a single main cavity or a main cavity with a series of side cavities as well. Openings to the outside occur naturally at the base of the pseudobulbs.

Orchids in general do not form wound tissue. The only response to wounding, or the abscission of leaves, appears to be a suberization of the walls of exposed cells. Rao (1963) observed that even protocorm callus cells *in vitro* may turn brown from suberization.

Sympodial Habit: Reed-Stemmed Orchids. The reed-stemmed or-

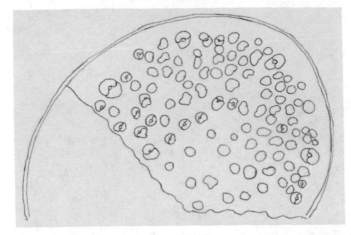

Fig. 6-36.   Diagrammatic cross section of the stem of *Epidendrum nocturnum*.

chids, characterized by such species as *Epidendrum ibaguense* or *Dichaea glauca*, resemble the pseudobulbed orchids in that they are largely epiphytic and exhibit a sympodial growth pattern. In this type, the rhizome gives rise not to a fleshy storage organ, like the pseudobulb, but rather to a thin, tough stem. Examination of the rhizome of such plants reveals a structure similar to that evidenced by the pseudobulbous varieties. Often reed-stemmed plants produce large, thick roots that offset the lack of pseudobulbs. The roots function to store water or food. Size of the rhizome may vary, but the general arrangement of tissues and cell types remains largely the same. The vascular bundles are similar to those in pseudobulbous species, and the individual elements comprising the xylem, phloem, and phloem caps are virtually the same as those present in pseudobulbed, rhizomatous types.

As can be expected from the external appearance of the structures, the reedy secondary stem is somewhat different from the fleshy pseudobulb. The sheathing bases of the leaves and also the bracts may almost completely cover the stems. The external surface of the stems is largely devoid of hairs of any type, and the epidermis is covered by a cutinous coat which varies in thickness from species to species. In *D. glauca* the coating is thin; a thicker coat is present in *E. nocturnum*. The epidermal cells are flattened with walls varying in thickness according to the species. In all plants examined, however, the epidermal walls were thickest on their exposed sides. A sclerified subdermal layer is present in *E. nocturnum*, although *D. glauca* lacks such a layer (Figs. 6-36 and 6-37). *Dichaea*

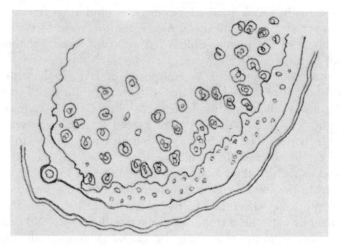

Fig. 6-37. Diagrammatic cross section of the stem of *Dichaea glauca.*

shows instead a ring composed of sclerified cells and fibrovascular bundles that *E. nocturnum* lacks, while in the rest of the stem approximately the same arrangement of bundles is present in both species. The parenchyma of both species, and indeed of all reed stems examined, is virtually identical: roughly isodiametric, regular in size, and about 70 μ in diameter. The walls are of medium thickness, and large intercellular spaces rarely occur.

The fibrovascular bundles display a similarity in all members of this group. Variations in the size, amount of xylem, and extent of the bundle sheath covering do exist, but no serious deviation in form occurs, and the bundles resemble those described for *O. sphacelatum*. A great number of bundles occurs within each stem, and the concentration of bundles more closely resembles that of the rhizome of *O. sphacelatum* than that of the pseudobulb. It is this dense concentration of bundles, with their sclerenchyma, that gives the reedy stems their characteristic toughness.

Monopodial Stemmed Orchids. Sympodial varieties of orchids form new rhizomes and shoots yearly, whereas monopodial varieties continue to grow on the same shoot by renewed apical development. The monopodial growth habit (excluding *Vanilla*) is represented by such plants as *Vanda suavis* or *Phalaenopsis lueddemanniana*. In *Phalaenopsis* or related forms such as *Aerangis,* the stem may be so short as to be almost invisible at the base of the leaves, while in *Vanda* it may reach several feet in length. In forms such as *Vanda*, the epidermis is covered by a cutinous layer of varying thickness. Again, the thickness is related directly to the ecology of the plants. Epidermal cells are relatively thick-walled and squat in form.

In *Vanda suavis* the cortex is composed of large cells with medium thick walls and is separated from the epidermis by a layer of parenchyma with sclerified walls of medium thickness (Fig. 6-38). A pericycle layer about eight cells thick with regularly spaced air spaces bounds the inside of the cortex. The fibrovascular bundles are similar to those described for other orchid stems, although the sclerenchyma sheath is enlarged and especially thick over the phloem. Great numbers of bundles are present in the stem, making it appear more like a sympodial rhizome than anything else. Of course in the monopodial habit the stem would be homologous with the rhizome of the sympodial types. No definite pattern of scattering arrangement can be detected, though all bundles are oriented with phloem facing outward. Parenchyma cells are small, isodiametric, and fairly thick walled with simple pits. *Phalaenopsis* is similar, but a clearly defined pericycle layer is absent, and the outermost bundles

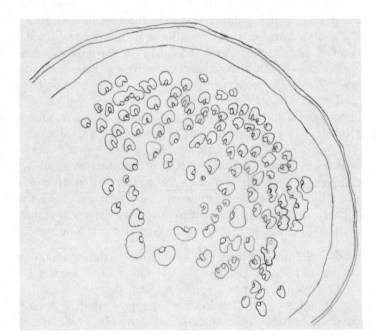

Fig. 6-38.   Diagrammatic representation of the stem of *Vanda suavis* in cross section.

in the ground parenchyma seem to describe a fairly regular circle. No sclerenchyma connections, such as those found in pseudobulbed rhizomes, are present. Little variation is observed between these two and other monopodial plants examined.

Vanilla Habit. The stem arrangement of *Vanilla planifolia* is essentially monopodial in habit, in that the growth of the plant is accomplished by an elongation from the apical meristem, rather than by growth from yearly shoots. Such growth is indeterminate. With its vining habit and a root emerging from each node to attach the plant as it grows, *Vanilla* can climb many feet to the tops of shrubs and trees. It is almost the only orchid that behaves like a vine; and though its seedling stages may pass on the ground, it eventually achieves epiphytic status as the bases of the stems die and terminal growth and axillary branching occur. The internal anatomy of the *Vanilla* stem is unique and deserves special mention because of the xylem (Fig. 6-39).

The xylem elements of *Vanilla* are larger in diameter than is usual for orchids. Unlike those of other orchids we have studied, the conducting cells are vessel elements, with gently sloping end walls and scalariform perforation plates, instead of the usual tracheids (Fig. 6-40). The walls

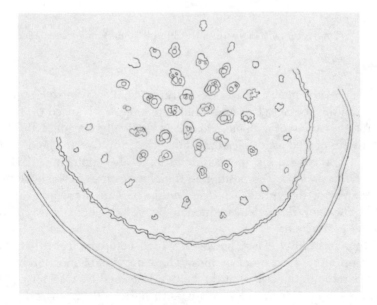

Fig. 6-39. Diagrammatic representation of the stem of *Vanilla planifolia* in cross section.

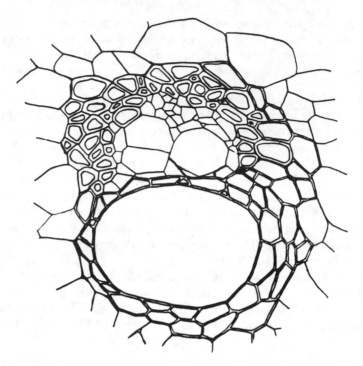

Fig. 6-40.   Transverse section of vascular bundle from *Vanilla* stem. See text for details.

of the elements show scalariform pitting; smaller protoxylar elements have spiral thickenings. This might indicate a more advanced evolutionary state for the stems of *Vanilla*, since it is generally accepted that vessel elements developed as a specialization of the primitive tracheid conducting cell. This would be in accord with Garay's opinion (1960) that *Vanilla* holds an advanced position in the NEOTTIOIDEAE. The evolutionary development of the vining habit of *Vanilla* perhaps depended upon the greater efficiency for water conduction by vessel elements as compared to the basic tracheids.

Another possibility exists, however, which could indicate that *Vanilla* is more primitive than other orchids. The tracheids of most orchids may be comparable to the "vascular tracheids" described by Carlquist (1961). These are degenerate vessel elements with narrowed lumen and small or no real end walls, and occurrence is in specialized members of advanced families such as CACTACEAE. Cheadle (1942) found in 20 species of orchids, including *Vanilla*, a considerable variation in occurrence of vessel elements in stems. Tracheids are considerably more common. The whole story is not yet understood.

A thin cutin covers the epidermis, and trichomes appear to be lacking. The epidermal cells are thick walled and sclerified. A uniseriate suberized hypodermal layer underlies the epidermis. A multiseriate cortical layer is present with the cells roughly isodiametric, thin walled, and bearing no indication of being suberized. A distinct uniseriate pericycle layer composed of suberized or sclerified cells encircles the ground parenchyma and vascular bundles. The ground parenchyma is composed of uneven isodiametric cells with thin walls. The vascular bundles are scattered about the parenchyma with little orientation. These are composed of xylem, phloem, a sheath of sclerenchyma fibers, and sclerified parenchyma. The sheath is multiseriate but considerably thinner than other sheaths examined, being only three to four cells thick at the thickest point of the phloem cap.

Root Anatomy

Orchid roots are described in Shushan's chapter (1959) in *The Orchids* and are detailed by Solereder and Meyer (1930), so that only a little information need be added here.

Garay (1972) made a point of the distinction between terrestrial and epiphytic types of root—a velamen or multilayered epidermis being characteristic of the latter, and root hairs being a possibility for both. Garay notes that some species of terrestrial orchids may have as many as three layers in the epidermis, and Holm (1904) and Moss (1923) both described terrestrial orchid root forms, some with velamen. Garay points out that the polystele structure is found only in the root tubers of the OR-CHIDOIDEAE (White, 1907).

But according to Rosso (1966) the supposition that the velamen is coincidental with the epiphytic habit must be examined further. He describes a uniseriate velamen for *Cypripedium irapeanum* roots and also *Selenipedium*, and notes that Meinecke (1894) also cites similar anatomy for roots of *Pleione, Liparis*, and some other orchids that are at least semiterrestrial. Rosso further states that a multiseriate velamen occurs in *Paphiopedilum* and is also typical of most epiphytic monandrous orchids. The velamen may be partly sloughed so that the thicker-walled exodermis cells are exposed to the environment. It would seem that the distinction between having only an epidermis, even multilayered, or having a velamen, even with a single layer, would lie in the presence or absence of an exodermal layer delimiting the cortex proper. However, we cannot find this explicitly stated in any of this information and are therefore at a loss to interpret all of the descriptions critically. Linsbauer (1930) says the velamen is a dermal tissue arising from the dermatogen by tangential divisions, and it is bounded proximally by an exodermis with passage cells. He further notes that it is seldom single layered, has no intercellular spaces, and it has dead cells filled with air in a dry condition. Goebel (1922) described terrestrial orchid roots with usual velamens, also uniseriate velamens, and, in addition, velamens on roots of other monocots such as lilies and amaryllids. Mulay et al. (1958) observed similar occurrences. Others have considered that only the ARACEAE and the ORCHIDACEAE had velamenous roots. We agree with Rosso that the matter could be reexamined with profit.

The velamen is often interrupted by pneumathodes which are more readily observed when roots are moistened. They, of course, provide for gas exchange into the central regions of the root. The exodermis, the inner layer of the velamen delimiting it from the root cortex, is often composed of alternating short and long cells. Some of the long cells may be polyploid. Geitler (1940) demonstrated 16-32 ploidy in such cells of *Epidendrum ciliare*. The polyploid cells apparently occur by endomitoses after a cell has matured and no longer will divide.

Solereder and Meyer (1930) noted that root hairs may be branched in *Chysis bractescens* or *Haemaria discolor*. Other species have spiral or reticulate thickening or helical banding, similar to that which occurs sometimes on parenchyma cells of cortex, mesophyll, or pith. Root hairs may be found on roots of some epiphytic species, especially at their points of contact with a substrate surface. Several of the saprophytic orchids seem to lack root hairs, but others have them, so it is difficult to make any correlation with their specialized nutritional status. Mycorrhizal invasion can destroy the epidermis and some of the underlying cells.

Cheadle (1942) found scalariform and simple perforation plates in xylem elements of orchid roots, particularly late-formed metaxylem, showing that true vessel elements may be present in root xylem. Rosso (1966) pictures them from the roots of *Selenipedium* and *Phragmipedium*, and describes them from *Cypripedium*. Spiral or annular tracheids are more common. Apparently vessels arose first in the later xylem of orchid roots and then appeared successively higher in the plant, as in *Vanilla*, until they were found in stem tissue as well. Further research is necessary to establish the extent of this variation, that is, how often and in which groups of orchids it may occur. It is hardly possible at this time to make any phylogenetic predictions based on the occurrence of these cells.

Bibliography

AMES, O. 1922. A triandrous form of *Psilochilus macrophyllus*. Orchidaceae 7:45.

———. 1937. Pollination of orchids through pseudocopulation. *Bot. Museum Leaflets, Harvard Univ.* 51:1–30.

———. 1938. Resupination as a diagnostic character in the Orchidaceae with special reference to *Malaxis monophytos*. *Bot. Museum Leaflets, Harvard Univ.* 6:145–183.

———. 1944. The pollinia of orchids. *Am. Orchid. Soc. Bull.* 13:190–194.

———. 1946a. The evolution of the orchid flower. *Am. Orchid. Soc. Bull.* 14:355–360.

———. 1946b. Notes on resupination in the Orchidaceae. *Am. Orchid Soc. Bull.* 15:18–19.

ARBER, A. 1925. *Monocotyledons, a Morphological Study.* Cambridge University Press.

AYENSU, E., and WILLIAMS, N. 1972. Leaf anatomy of *Palumbina* and *Odontoglossum* subgenus *Osmoglossum*. *Am. Orchid Soc. Bull.* 41:687–696.

BAUER, F. A. 1830–1838. Illustrations of orchidaceous plants. London.

BROWN, R. 1833. On the organs and mode of fecundation in Orchideae and Asclepiadeae. *Trans. Linn. Soc., London* 16:685–745.

CAMUS, M. A. 1929. *Iconographie des Orchidées d'Europe*, 1921; text, 1928.

CARLQUIST, S. 1961. *Comparative Plant Anatomy.* Holt, Rinehart and Winston, New York.

CHATIN, . 1857. Anatomie des plantes aériennes de l'ordre des Orchidées. 2. Mem. rhizome tige, feuilles. *Mem. Soc. Sc. Nat. de Cherbourg* 5:33–69. (Cited in Solereder and Meyer.)

CHEADLE, V. 1942. The occurrence and type of vessels in the various organs of the plant in the Monocotyledoneae. *Am. J. Bot.* 29:441–450.

CRUGER, H. 1865. A few notes on the fecundation of the orchids and their morphology. *Linn. Soc. J. Bot.* 8:127–135.

CURTIS, K. M. 1917. The anatomy of the six epiphytic species of the New Zealand Orchidaceae. *Ann. Bot.* 31:133–149.

CYGE, T. 1930. Etudes anatomiques et ecologiques sur les feuilles des Orchidées indigenes. *Extr. Mem. Acad. Polon. Sci. Lett., ser. B, Sci. Nat.*

DANNECKER, 1898. Bau und Entwicklung hohler ameisenbewohnter Orchideenknollen nebst Beitrag zur Anatomie der Orchideen Blätter. Thesis, Freiburg, Switzerland. (Cited in Solereder and Meyer.)

DARWIN, C. 1884. *The Various Contrivances by which Orchids are Fertilized by Insects*, 2nd ed. London.

DRESSLER, R., and DODSON, C. 1960. Classification and phylogeny in the Orchidaceae. *Ann. Mo. Bot. Gard.* 47:25–68.

EICHLER, A. W. 1875. *Blüthendiagramme.* Leipzig.

FAHN, A. 1952. On the structure of floral nectaries. *Bot. Gaz.* 113:464–470.

FINCK, G. 1954. Report on the floral vascular anatomy of *Epidendrum cochleatum* and *E. cocheatum triandrum*, with particular attention to staminal traces. Senior thesis, Brooklyn College.

FREY-WYSSLING, A., ZIMMERMAN, M., and MAURIZIO, A. 1954. Über den enzymatischen Zuckerumbau in Nektarien. *Experientia* 10:490.

FUCHS, A., and ZIEGENSPECK, H. 1936. Lebensgeschichte der Blütenpflanzen Mitteleuropas. p. 81.

GARAY, L. A. 1960. On the origin of the Orchidaceae. *Bot. Museum Leaflets, Harvard Univ.* 19:57–96.

———. 1972. On the origin of the Orchidaceae, II. *J. Arnold Arboretum* 53:202–215.

GEITLER, L. 1940. Die Polyploidie der Dauergewebe höherer Pflanzen. *Ber. Dtsch. Bot. Gesel.* 58:131–142.

GELLERT, M. 1923. Anatomische Studien über den Bau der Orchideenblüte. *Feddes Repert. Beih.* 25:1–65.

GODFREY, M. 1933. *Monograph and Iconograph of Native British Orchidaceae.* Cambridge University Press.

GOEBEL, K. 1922. Erdwurzeln mit velamen. *Flora* 15:1–26.

GUTTENBERG, H. VON. 1915. Anatomisch-physiologische Studien an den Blüten der Orchideengattungen *Catasetum* Rch. and *Cycnoches* Lindl.

HABERLANDT, G. 1901. *Sinnesorgane im Pflanzenreich, zur Perception mechanischer Reize.* Engelmann, Leipzig.

HAGERUP, O. 1952. The morphology and biology of some primitive orchid flowers. *Phytomorphology* 2:134–138.

HIRMER, M. 1920. Beitrage zur Organographie der Orchideenblüte. *Flora* 113:213–309.

HOLM, T. 1904. The root structure of North American terrestrial Orchideae. *Am. J. Sci. Ser. IV* 168:197–212.

HOLTTUM, R. 1955. Growth habits of monocotyledons—Variations on a theme. *Phytomorphology* 5:399–413.

JANSE, J. M. 1886. Imitierte Pollenkorner bei *Maxillaria* sp. *Bericht. Dtsch. Bot. Ges.* 4:277–283.

JEFFREY, D., ARDITTI, J., and KOOPOWITZ, H. 1970. Sugar content in floral and extrafloral exudates of orchids: Pollination, myrmecology and chemotaxonomy implications. *New Phytol.* 69:187–195.

KRUGER, P. 1883. Die oberirdischen Vegetationsorgane der Orchideen in ihren Beziehungen zu Klima und Standort. *Flora* 66:435, 451, 467, 499, 515.

KULLENBERG, B. 1961. *Studies in* Ophrys *pollination.* Almquist, Uppsala.

LAYCHOCK, S. 1971. The extra-floral nectaries of a *Cymbidium* hybrid. Senior thesis, Brooklyn College.

LINDLEY, J. 1853. *The Vegetable Kingdom,* 3rd ed. London.

LINSBAUER, K. 1930. Die Epidermis, in K. Linsbauer, *Handbuch der Pflanzenanatomie,* Vol. 4. Gebruder Borntraeger, Berlin.

LUER, C. 1972. *The Native Orchids of Florida.* N. Y. Botanical Garden, New York.

MAIGE, M. 1909. Sur quelques cas teratologiques. *Rev. gen. Bot.* 21:312–322.

MASTERS, M. J. 1887. On the floral conformation of the genus *Cypripedium. Linn. Soc. J. Bot.* 22:403–422.

MEINECKE, E. 1894. Beitrage zur Anatomie der Luftwurzeln der Orchideen. Dissertation. Heidelberg. Also in *Flora* (1894) **78**:133–203. (Cited in Solereder and Meyer.)

Möbius, M. 1887. Über den anatomischen Bau der Orchideenblätter und dessen Bedeutung fur das System dieser Familie. *Jahrb. Wiss. Bot.* **18**:530–607.

Moss, C. E. 1923. On the presence of velamenous roots in terrestrial orchids. *Proc. Linn. Soc.* **135**:47.

Mulay, B. N., Deshpande, B. D., and Williams, H. P. 1958. Study of velamen in some epiphytic and terrestrial orchids. *J. Indian Bot.* **37**:123–127.

Müller, H. 1873. *Die Befruchtung der Blumen durch Insekten.* Engelmann, Leipzig.

Nelson, E. 1965. Zur organophyletischen Natur des Orchideenlabellums. *Bot. Jahrb.* **84**:175–214.

———. 1967. Das Orchideenlabellum ein Homologon des einfachen medianen Petalums der Apostasiaceen oder ein zusammengesetztes Organ? *Bot. Jahrb.* **84**:22–35.

Nierenberg, L. 1972. The mechanism for the maintenance of species integrity in sympatrically occurring equitant oncidiums in the Caribbean. *Am. Orchid Soc. Bull.* **41**:873–882.

Oliver, F. W. 1895. *Natural History of Plants.* New York.

Pfitzer, E. 1889. Orchidaceae (in "Die natürlichen Pflanzenfamilien") by Engler and Prantl **2**:52–218.

Pijl, L. van der, and Dodson, C. 1966. *Orchid Flowers, their Pollination and Evolution.* University of Miami Press.

Poddubnaya-Arnoldi, V. A. 1960. Study of fertilization in the living materials of some angiosperms. *Phytomorphology* **10**:185–198.

Porsch, O. 1908. Die Honigersatzmittel der Orchideen Blüten. *Knys. Bot. Wandtafeln XII.* (Cited in Solereder and Meyer.)

Pulle, A. A. 1952. *Compendium van de Terminologie, Nomenclatuur en Systematiek der Zaadplanten*, 3d ed. A. Osthoek, Utrecht.

Rao, A. N. 1963. Organogenesis in callus cultures of orchid seeds, in P. Maheshwari and N. S. Ranga Swamy, Eds., *Plant Tissue and Organ Culture.* New Delhi.

Raunkiaer, C. 1899. *De Danske Blomsterplanters Naturhistorie.* Copenhagen, pp. 305–382.

Rendle, A. B. 1930. *The Classification of Flowering Plants*, Vol. I. *Monocotyledons.* London.

Reychler, L. 1928. *Mutation with Orchids.* Brussels.

Richards, L. C. 1818. *Des Orchideis Europaeis Annotationes. Mem. Mus. Hist. Nat. Paris* **4**:23–61.

Rolfe, R. A. 1909–1910. The evolution of the Orchidaceae. *Orchid Rev.* **17**:129–132, 193–196, 289–292, 353–356; **18**:33ff.

Ross, . 1909. Pflanzen und Ameisen. *Naturwiss. Wochenschr.*, 13–15. (Cited in Solereder and Meyer.)

Rosso, S. W. 1966. The vegetative anatomy of the Cypripedioideae (Orchidaceae). *Linn. Soc. J. Bot.* **59**:309–341.

Rubino, R. 1966. Preliminary investigation of the floral nectaries in orchids. Senior thesis, Brooklyn College.

SALISBURY, R. A. 1804. Germination of seeds of the Orchidaceae. *Trans. Linn. Soc.* 7:29–32.

SAUNDERS, E. R. 1923. A reversionary character in the stock (*Matthiola incana*) and its significance in regard to the structure and evolution of the gynoecium in the Rhoedales, the Orchidaceae and other families. *Ann. Bot.* 37:451–482.

———. 1937. *Floral Morphology.* Cambridge University Press.

SCHOMBURGK, R. 1837. On the identity of three supposed genera of orchideous epiphytes. *Trans. Linn. Soc.* 17:551–552. Pl. 29.

SHUSHAN, S. 1959. Developmental anatomy of an orchid *Cattleya* × Trimos, in C. Withner, Ed., *The Orchids.* Ronald Press, New York.

SOLEREDER, H., and MEYER, F. 1930. *Systematische Anatomie der Monokotyledonen.* Vol. 6, pp. 92–242.

STEBBINS, G., and KHUSH, C. 1961. Variation in the organization of the stomatal complex in the leaf epidermis of the monocotyledons and its bearing on their phylogeny. *Am. J. Bot.* 48:51–59.

SWAMY, B. G. L. 1946. Embryology of *Habenaria. Proc. Nat. Acad. Sci. India* 12: 413–426.

———. 1948. Vascular anatomy of orchid flowers. *Bot. Museum Leaflets, Harvard Univ.* 13:61–95.

TOMINSKI, P. 1905. Die Anatomie des Orchidenblattes in ihrer Abhängigkeit von Klima und Standort. Thesis, Berlin. (Cited in Solereder and Meyer.)

VERMUELEN, P. 1955. The rostellum of the Ophrydeae. *Am. Orchid Soc. Bull.* 24: 239–245.

———. 1959. The different structure of the rostellum in Ophrydeae and Neottieae. *Acta Bot. Neer.* 8:338–355.

———. 1966. The system of the Orchidales. *Acta Bot. Neer.* 15:224–253.

VOGEL, S. 1954. *Blütenbiologische Typen als Elemente der Sippengliederung. Bot. Studien 1.* Gustave Fischer, Jena.

———. 1959. Organographie der Blüten kapländischer Ophrydeen, I and II. *Akad. Wiss. Lit. Mainz* 6; 7.

———. 1962. Duftdrusen im Dienste der Bestaubung. *Akad. Wiss. Lit. Mainz* 10.

———. 1963. Das sexuelle Anlockungsprinzip der Catasetinen- und Stanhopeen-Blüten und die wahre Funktion ihres sogenannten Futtergewebes. *Ost. Bot. Zeitschr.* 110:308–335.

———. 1966a. Scent organs of orchid flowers and their relation to insect pollination. In *Proceedings of the Fifth World Orchid Conference,* pp. 253–259.

———. 1966b. Parfümsammelnde Bienen als Bestäuber von Orchidaceen und *Gloxinia. Oster. Bot. Zeitschr.* 113:302–361.

———. 1968. Chiropterophilie in der neotropischen Flora. *Flora* 157:562–602.

WEJKSNORA, P. 1971. Anatomy of orchid leaves and stems, with special reference to *Oncidium sphacelatum.* Senior thesis. Brooklyn College.

WHITE, J. H. 1907. On polystely in roots of Orchidaceae. *Univ. Toronto Studies, Biol. Ser.* 6:165–179.

WILLIAMS, N. 1970. Some observations on pollinaria in the Oncidiinae. *Am. Orchid Soc. Bull.* 39:32–43, 207–220.

——. 1972. Additional studies on pollinaria in the Oncidiinae. *Am. Orchid Soc. Bull.* 41:222–230.

WILLIS, J. C. 1936. *A Dictionary of Plants and Ferns.* Cambridge University Press.

WILSON, C. S., and JUST, T. 1939. The morphology of the flower. *Bot. Rev.* 4:97–131.

WITHNER, C. L. 1970. Die kleistogamen *Epidendrum*-arten der Sektion Encyclia. *Die Orchidee* 21:4–7.

WOLFE, T. 1865–1866. Beiträg zur Entwicklungsgeschichte der Orchideen-Blüthe. *Jahrb. Wiss. Bot.* 4:261–304.

WORSDELL, W. C. 1916. *Principles of Plant Teratology.* London.

ZIMMERMAN, M. 1932. Über die extrafloralen Nectarien der Angiospermen. *Beih. Bot. Centralbl.* 49:99–196.

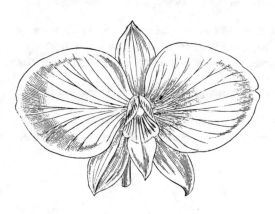

7

Alkaloids of the Orchidaceae

BJÖRN LÜNING

In the early days of natural product chemistry, scientists succeeded in isolating from natural sources only the products easiest to obtain, usually crystalline solids with good separating properties from other compounds simultaneously occurring in the biological material. Investigators usually used only vegetable material and worked mostly with species growing in their home countries unless the material was well known for a special pharmacological action such as opium, whose alkaloids (papaverine, morphine) were isolated and characterized during the first half of the nineteenth century. Most of the chemistry of that time was concentrated in Europe and later in North America. Few orchids, if any, with proven pharmacological action grow in these areas, so the investigation of the natural products of the ORCHIDACEAE, although it is the largest of all plant families, was, except for vanillin, deferred to recent times.

Natural product chemistry deals with the investigation of compounds not generally occurring in all plants or animals but confined to one species. The so-called natural products are usually not involved in the main life processes of the cells (primary metabolism), but they are formed as peculiar offshoots along *specific* biogenetic pathways and referred to as *secondary metabolites*. There are several classes of secondary metabolites, the best known one, covering the terpenes and steriods, is built up from a branched C_5-unit (isoprenoid type). The biologically active precursor to this group of compounds is mevalonic acid.

Another important group of secondary metabolites is made up of the aromatic compounds of which vanillin is an example. There are two main biogenetic routes to the aromates, one proceeds via shikimic acid, leading in the primary metabolism to the well-known amino acids phenylalanine and tyrosine, but with many offshoots into various groups of aromatic

349

compounds. The other route rests on the cyclization of polyacetate chains, which usually, in the primary metabolism, do not cyclize but end up in lipids.

The alkaloids constitute a heterogeneous group of nitrogen-containing compounds. Many suggestions have been made to define an alkaloid. One definition that rests solely on chemical facts requires that an alkaloid contain at least one heterocyclically bound basic nitrogen atom. By this definition all quaternary salts would be classified as nonalkaloids. In a group of related compounds produced by a botanic species, those with heterocyclic as well as open chain nitrogen are often found together, and the latter would then fall outside the definition, which would be impractical.

Hegnauer proposed a biogenic definition of the alkaloids, so that all secondary metabolites containing basic or quaternary nitrogen being of amino acid origin should be classified as *true alkaloids*, and compounds of the same classes biosynthesized from other sources should be called *pseudoalkaloids*. The pharmacologists usually add the criterion that alkaloids should have more or less pronounced pharmacological action in animals. As is easily seen, these diverse definitions do not make the delineations of the scientific field too distinct but, of course, do not alter the chemical facts being elucidated therein. It is, however, of great interest for the taxonomist hoping to use chemical criteria in classification to know the biogenetical background of the chemical compounds under consideration.

History

Early investigations on the alkaloid content of ORCHIDACEAE were made on material in European orchid collections by de Wildemann (1892) and de Droog (1896), the latter's comprising no less than 104 species in 78 genera. de Wildemann also noticed the high alkaloid content in *Dendrobium nobile* Lindl. and *Phalaenopsis lueddemanniana*, Rchb.f. and, besides these findings, de Droog proved the occurrence of alkaloids in a number of *Catasetum* species. Around 1900 Boorsma (1899, 1902) investigated a few orchids of the Buitenzorg Botanic Garden (Java), and he found additional weak positive alkaloid reactions in *Paphiopedilum javanicum* Pfitz. and *Liparis parviflora* Lindl. All these findings were more or less hidden in inaccessible literature, and in 1921 Wester published a new screening of 30 species without, however, finding any new alkaloid-containing species except *Chysis bractescens*. Lindl.

Suzuki et al. (1932, 1934) and Chen and Chen (1935) published works on attempts to determine the structure of the main alkaloid, dendrobine, from *Dendrobium nobile* Lindl., but they failed to present a complete structure. Since 1963 many workers have published investigations on ORCHIDACEAE alkaloids from many aspects. These include screening for alkaloids in different species, separation, purification, and structural elucidation of individual alkaloids. Some of the biogenetic pathways leading to certain of the key alkaloids have also been investigated. These works will be reviewed below.

Methods and Results of Orchid Alkaloid Screening

Methodology.   Different methods for the evaluation of the alkaloid content of plants have been used from time to time. de Wildemann (1892), de Droog (1896), and Wester (1921) used a staining technique and microscopic visualization of alkaloids. By this method it is possible to get information not only of the occurrence but also on the site of the alkaloids in the plant. The method has the disadvantage of being insensitive and is not specific.

Modern investigators (Lüning, 1964, 1967; Slaytor, 1969, 1971) have used a purely chemical procedure of alkaloid identification in extracts of homogenized living material. In accordance with the recommendation of the International Union of Pure and Applied Chemistry Committee on Chemotaxonomy, living reference specimens are now being grown of those species tested.

In the laboratory method, the living plant material is extracted, the alkaloids are enriched, and the concentrated solution is tested by a set of analytical reactions. In a simplified field method only a Dragendorff spot test is made on the plant juice. This is usually sufficient to tell whether the alkaloids in a species occur to such an extent that a further chemical investigation is possible.

Results.   The alkaloid tests of 2044 species from 281 genera should constitute a gold mine for phytochemists. Only a few of the alkaloid rich species so far found, however, have not already been investigated. The main reason for lack of further investigation lies in the rarity or inaccessibility of the alkaloid-rich species. Structural elucidation of any alkaloids from plant material usually requires the collection of several kilograms of living plants, and this is often a difficult matter.

It is also inferred that the mere presence of nitrogen-containing material, classed as alkaloids, could be used as a taxonomic tool. The more or less doubtful use of the alkaloids in this respect will be treated in the discussion below. However, since much work has been devoted to the search for alkaloid-accumulating species, and this work may be of some taxonomical interest, I have chosen to present the screening results at a generic level, listing as positive only those species containing .1% or more alkaloid. Some 214 species in 64 genera came up to this alkaloid content. The genera are presented in alphabetical order, since work leading to a phylogenetically more natural system of the ORCHIDACEAE down to generic level is under way, and any attempt in classifying alkaloid-rich genera according to older systems gives unexplainable results. The data in Table 7-1 are compiled from Lüning (1964, 1967, 1973) and Lawler and Slaytor (1969, 1971).

Dendrobine Alkaloids

Dendrobium nobile Lindl. Suzuki et al. (1932) isolated from the drug "Chin Shih Hu" made of dried stems of *D. nobile* a crystalline alkaloid which they called dendrobine. This was the first successful attempt to isolate an alkaloid from an orchid. The drug "Chin Shih Hu" had been used for long time in China as a tonic and antipyretic. Dried stems of different dendrobiums are sold under the name of Shih-Hu (see below under *D. lohohense*), and various other *Dendrobium* species except *D. nobile* have been shown to contain dendrobine, as does *D. linawianum* Rchb.f. Suzuki (1934) could reveal a few details of the structure. At the IUPAC Symposium on Natural Products in Kyoto 1963, Inubushi et al., Yamamura and Hirata, and Okamoto et al. independently presented a complete structure of dendrobine. In fact, they initially suggested different structures for the alkaloid, which can be learned from the proceedings of the symposium, but they finally agreed that one of them was right. The structure was gained only after the use of ingenious degradations and the highest standard of spectroscopic methods (see fig. 7-1). In 1972 Inubushi et al. and Yamada et al. independently published two synthetic routes to *dl*-dendrobine, thereby finally establishing its structure.

A further 14 alkaloids closely related to dendrobine have been isolated from *Dendrobium* species. Thus Inubushi and Nakano (1965), Inubushi, Tsuda and Katarao (1966), and Okamoto et al. (1966a) have worked on *D. nobile* and found dendramine (6-hydroxydendrobine), and dendrine (Fig. 7-2); dendroxine (Okamoto et al., 1966b), as well as 4-hydroxydendroxine (Okamoto et al., 1972) and 6-hydroxydendroxine (Fig. 7-3), (Okamoto et al., 1966b.) In *D. findlayanum* Par. et Rchb.f., Granelli et al. (1970) found 2-hydroxydendrobine (Fig. 7-3).

In *D. nobile* Lindl. there is also an open chain alkaloid related to dendrobine, earlier called nobiline (Fig. 7-4) (Yamamura and Hirata, 1964). To exclude possible confusion with a nonalkaloidal molecule called nobi-

TABLE 7-1

Alkaloid-Accumulating Species Arranged Generically

Genus	Species with an Alkaloid Content $\geqslant 0.1\%$	Total Number of Species Tested
Acacallis	0	1
Acampe	0	1
Acanthephippium	0	4
Acianthus	0	2
Acineta	0	1
Acriopsis	0	6
Acrochoene	0	1
Aerangis	3	4
Aerides	0	14
Agrostophyllum	0	23
Amblostoma	0	1
Amitostigma	0	1
Anacamptis	0	1
Ancistrorhyncus	0	2
Angraecum	3	12
Anoectochilus	1	11
Anota	0	1
Ansellia	0	1
Anthogonium	0	1
Apostasia	0	1
Appendicula	0	5
Arachnanthe	1	2
Arachnis	0	1
Arethusa	1	1
Armodorum	0	1
Arpophyllum	0	1
Arundina	0	1
Ascocentrum	0	1
Ascoglossum	0	1
Aspasia	0	1
Barbosella	0	2
Batemannia	0	1
Bifrenaria	0	7
Bletia	0	3
Bletilla	0	1
Botriochilus	0	1
Bolusiella	0	1
Brassavola	1	6

TABLE 7-1   (continued)

Genus	Species with an Alkaloid Content $\geqslant 0.1\%$	Total Number of Species Tested
Brassia	0	8
Bromheadia	0	2
Bulbophyllum	9	134
Cadetia	0	5
Caladenia	0	3
Calanthe	2	26
Caleana	0	2
Calochilus	1	4
Calypso	0	1
Calyptrochilus	0	1
Camaridium	0	1
Camarotis	1	6
Campylocentrum	0	3
Capanemia	0	2
Catasetum	2	15
Cattleya	0	34
Cephalanthera	0	1
Ceratostylis	1	8
Chamaeangis	0	1
Chamaeorchis	0	1
Cheirostylis	0	2
Chiloglottis	0	2
Chiloschista	1	3
Chondrorhynca	0	1
Chysis	1	3
Cirrhaea	0	3
Cirrhopetalum	0	4
Cleisostoma	0	3
Coelogyne	2	26
Colax	0	1
Comparettia	0	1
Corallorhiza	0	1
Coryanthes	0	3
Corybas	0	3
Corymborkis	1	4
Cremastra	0	1
Cryptocentrum	0	1
Cryptochilus	0	2
Cryptophoranthus	0	1
Cryptostylis	3	3

TABLE 7-1 (continued)

Genus	Species with an Alkaloid Content $\geqslant 0.1\%$	Total Number of Species Tested
Cyclopogon	0	1
Cycnoches	0	4
Cymbidium	2	37
Cypripedium	0	4
Cyrtopodium	0	3
Cyrtorchis	1	4
Dactylorrhiza	0	3
Dendrobium	32	384
Dendrochilum	0	3
Diacrium	0	1
Diaphananthe	1	8
Dichaea	0	7
Dimerandra	0	1
Diplocaulobium	0	2
Diploprora	0	1
Dipodium	0	4
Disa	0	2
Diuris	0	2
Doritis	1	2
Eggelingia	1	1
Elleanthus	1	8
Epiblastus	0	2
Epidendrum	8	86
Epipactis	0	3
Eria	14	77
Erythrodes	1	3
Eulophia	2	13
Eulophidium	0	1
Eurochoene	0	1
Guleundra	1	3
Galeola	0	2
Gastrochilus	1	2
Gastrodia	1	2
Geodorum	0	3
Giulianetta	0	1
Glomera	0	7
Glossodia	0	2
Glossorrhynca	0	2
Gomeza	0	4
Gongora	0	7

TABLE 7-1 (continued)

Genus	Species with an Alkaloid Content $\geqslant 0.1\%$	Total Number of Species Tested
Goodyera	1	11
Govenia	0	1
Grammatophyllum	0	5
Grobya	0	2
Gymnadenia	0	1
Habenaria	2	12
Hammarbya	1	1
Herminium	0	2
Hetaeria	0	5
Hexisea	0	1
Hippeophyllum	1	1
Hormidium	0	1
Houlletia	0	3
Huntleya	1	4
Ione	0	4
Ionopsis	0	1
Isabelia	0	1
Isochilus	0	1
Jacquiniella	0	1
Jumellea	0	2
Kingiella	1	1
Laelia	0	26
Lanium	0	1
Lankesterella	0	1
Leochilus	0	2
Lepanthes	1	4
Lepidogyne	0	1
Leptotes	0	1
Liparis	28	67
Listera	0	1
Lockhartia	0	2
Luisia	0	8
Lycaste	0	6
Lyperanthus	1	1
Macodes	0	1
Malaxis	18	49
Malleola	1	1
Masdevallia	0	3
Maxillaria	0	23
Mediocalcar	0	4

TABLE 7-1 (continued)

Genus	Species with an Alkaloid Content $\geqslant 0.1\%$	Total Number of Species Tested
Mendoncella	0	1
Mesospinidum	0	1
Microtis	0	2
Miltonia	0	8
Mischobulbon	0	1
Mobilabium	0	1
Monomeria	0	1
Mormodes	0	4
Nageliella	1	1
Neolauchia	0	1
Neomoorea	0	1
Neottia	0	1
Nephelaphyllum	0	3
Nervilia	4	12
Notylia	0	1
Oberonia	5	29
Octomeria	0	7
Odontoglossum	0	12
Oncidium	0	42
Ophrys	0	1
Orchis	0	4
Ornithocephalus	0	1
Ornithochilus	0	2
Otochilus	0	3
Pachyphyllum	0	1
Panisea	0	1
Paphiopedilum	1	23
Paradisianthus	0	1
Parasarcochilus	0	2
Pelatantheria	0	2
Pelexia	0	1
Peristeranthus	0	1
Peristeria	0	1
Pescatorea	0	2
Phaius	0	13
Phalaenopsis	17	35
Phloeophila	0	1
Pholidota	0	14
Phragmipedium	0	2
Phreatia	1	22

TABLE 7-1 (continued)

Genus	Species with an Alkaloid Content $\geqslant 0.1\%$	Total Number of Species Tested
Phymatidium	0	1
Physosiphon	0	1
Platanthera	0	1
Platyclinis	0	2
Platyglottis	0	2
Plectorrhiza	0	1
Plectrophora	0	1
Pleione	0	4
Pleurobotrium	0	1
Pleurothallis	0	40
Plocoglottis	2	7
Podangis	1	1
Podochilus	0	6
Polystachya	0	15
Pomatocalpa	1	8
Ponera	0	1
Ponthieva	0	1
Porpax	0	1
Porphyrodesme	0	1
Prasophyllum	0	5
Prescottia	0	1
Promenaea	0	2
Pseudolaelia	0	3
Pteroceras	0	1
Pterostylis	0	12
Rangaeris	1	1
Reichenbachanthus	0	1
Renanthera	1	5
Rhinerrhiza	1	1
Rhyncostylis	0	2
Robiquetia	0	3
Rodriguezia	0	5
Rodriguezopsis	0	1
Rudolphiella	0	1
Saccolabiopsis	0	1
Saccolabium	1	12
Sarcanthus	2	8
Sarcochilus	2	21
Sarcopodium	0	3
Sarcorhyncus	0	1
Satyrium	0	2
Scaphyglottis	0	5

TABLE 7-1 (continued)

Genus	Species with an Alkaloid Content $\geqslant 0.1\%$	Total Number of Species Tested
Schistostylus	0	1
Schoenorchis	0	2
Schomburgkia	0	2
Scuticaria	0	2
Selenipedium	0	1
Sigmatostalix	0	5
Sobralia	0	5
Sophronitis	0	5
Spathoglottis	0	8
Spiculaea	0	1
Spiranthes	0	7
Stanhopea	0	8
Stauropsis	0	1
Stelis	0	7
Stenocoryne	0	2
Stenorrhyncus	1	3
Stereochilus	0	1
Taeniophyllum	3	10
Tainia	0	4
Tetragamestus	0	1
Teuscheria	0	1
Thelymitra	0	3
Theodorea	0	1
Thunia	0	2
Trichocentrum	0	4
Trichoglottis	2	7
Trichopilia	1	8
Tridactyle	0	2
Trigonidium	0	3
Thrixspermum	0	6
Trizouxis	0	1
Tylostylis	0	1
Vanda	3	22
Vandopsis	5	13
Vanilla	0	5
Warrea	0	1
Warscewiczella	0	2
Xylobium	0	4
Zeuxine	1	5
Zygopetalum	0	4
Zygostates	0	1

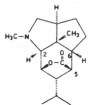

Fig. 7-1.   Dendrobine.

lin, the alkaloid should have the name of second preference, nobilonine (Onaka et al., 1965). In *D. hildebrandii* Rolfe Elander and Leander (1971) found 6-hydroxynobilonine (Fig. 7-4). In this species and in *D. friedricksianum* Lindl. the isopentenyl derivatives of dendroxine and 6-hydroxydendroxine were found to be major alkaloidal constituents (Fig. 7-5a). (Hedman, Leander, and Lüning, 1971).

After the original report on the occurrence of N-methyldendrobinium and N-isopentenyldendrobinium ion in *D. nobile* Lindl. by Inubushi (1967) and Inubushi, Sasaki, et al. (1964) these compounds have been extensively investigated. The relative and absolute stereochemistry of the N-isopentenyl derivatives of dendrobine, dendroxine, and 6-hydroxydendroxine was elucidated by Hedman and Leander (1972). The N-oxide of dendrobine was also found in *D. nobile* by the same authors (Fig. 7-5b).

In *D. wardianum* a somewhat different quaternary alkaloid, dendrowardine, was found. Two of the rings of the dendrobine skeleton are retained, but the ring containing the nitrogen atom has been opened and an oxygen appears in place of the nitrogen atom. The nitrogen atom instead forms part of a choline molecule and is attached to the system with one bond as in nobilonine (Blomqvist et al., 1973) (Fig. 7-6).

Inubushi et al. (1964a) and Huang (1965) suggested an absolute configuration for dendrobine (Fig. 7-1) and a number of related alksdaloids,

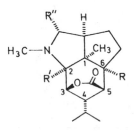

Fig. 7.2.   Tertiary derivatives of dendrobine. R = R′′ = H, R′ = OH, 2-hydroxydendrobine. R′ = R′′ = H, R = OH, dendramine. R = R′ = H, R′′ = CH₂COOCH₃, dendrine.

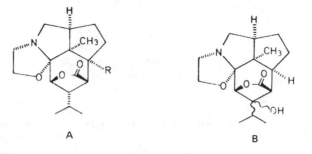

Fig. 7-3. Dendroxine and hydroxylated dendroxines. (A) R = H, dendroxine[24-26], R = OH, 6-hydroxydendroxine. (B) 4-hydroxydendroxine.

based on studies of optical rotation. This configuration was identical with that known for the nonalkaloidal sesquiterpenoid picrotoxinine.

However, one year later Matsuda et al. (1965) published the opposite configuration for nobilonine methiodide based on X-ray diffraction. Nobilonine was suggested to have the same configuration as dendrobine and the related alkaloids.

This confusion was solved by elaborate chemical transformations of nobilonine and pictrotoxinin into compounds with similar circular dichroism curves (Behr and Leander, 1972). The absolute configuration of dendrine was finally established by synthesis from dendrobine (Granelli and Leander, 1969).

In Table 7-2 the dendrobine alkaloids are collected with their empirical formulas, occurrence, and complete bibliography.

Dendrobium Alkaloids not Related to Dendrobine

Dendrobium primulinum Lindl. produces small, volatile alkaloids, one of which is the well-known hygrine (Fig. 7-7a) and the other a previously unknown alkaloid, dendroprimine, the structure of which was elucidated

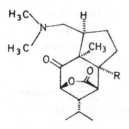

Fig. 7-4.   Nobilonine and 6-hydroxynobilonine. R = H, nobilonine = nobiline. R = OH, 6-hydroxynobilonine.

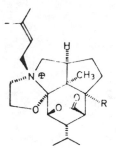

Fig. 7-5a.   Quaternary derivatives of dendroxine. R = H, N-isopentenyldendroxine. R = OH, N-isopentenyl-6-hydroxy-dendroxine.

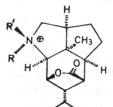

Fig. 7-5b.   Quaternary derivatives of dendrobine. R = R′ = CH₃, N-methyldendrobine. R = CH₃, R′ = CH₂ — CH = C(CH₃)₂, N-isopentenyldendrobine. R = O⁻, R′ = CH₃, dendrobine N-oxide.

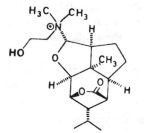

Fig. 7-6.   Dendrowardine.

by spectroscopic means (IR, NMR, and mass spectrometry) and by chemical degradations (see Fig. 7-7b) (Lüning and Leander, 1964).

The relative configuration of dendroprimine was assigned by synthesis and interpretation of NMR and IR spectra of the four possible isomers to the naturally occurring alkaloid (Lüning and Lundin, 1967), and the absolute configuration was established by degradations and comparison with compounds of known absolute configuration (Fig. 7-7b) (Blomqvist et al., 1972).

When *Dendrobium chrysanthum* Wall. was treated by the methods used for *D. primulinum*, it appeared to contain only hygrine. However,

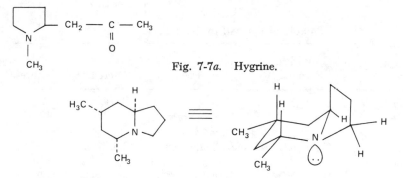

Fig. 7-7a.   Hygrine.

Fig. 7-7b.   Dendroprimine, right illustration showing its preferred conformation.

considerable doubt existed that this was the whole truth. By use of other methods nonvolatile alkaloids were found when hygrine had been removed. Building blocks in the two remaining alkaloids were *cis-* and *trans*-cinnamic acid, and therefore the alkaloids were called *cis-* and *trans*-dendrochrysine. The structures of the two alkaloids were assigned by spectral methods, but chemical degradations were unsuccessful. Finally, the structure of one of them was confirmed by synthesis, and the two alkaloids could also be interrelated (Fig. 7-8) (Ekevåg et al., 1973).

Inubushi, Tsuda, Konita, and Matsumoto (1964, 1968) found in *D. lohohense* a new phtalide pyrrolidine, alkaloid shihunine (Fig. 7-9). This plant was obtained as a yellow, strawlike drug called Shih-Hu at the Hong Kong market.

The alkaloid-rich *D. pierardii* Roxb. was later shown to contain an alkaloid related to shihunine called pierardine (Fig. 7-10), which in contrast to shihunine is optically active (Elander et al., 1969). The absolute configuration of pierardine was shown, and a total synthesis was performed of the optically active alkaloid (Elander et al., 1971). During this work shihunine was found also in *D. pierardii* Roxb., and furthermore it could be shown that shihunine, under conditions prevailing in the plant, exists in the open betain form (Fig. 7-11), which explains its lack of optical activity.

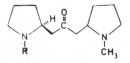

Fig. 7-8.   *Cis-* and *trans*-dendrochrysine. *Cis*-dendrochrysine: R = *cis*-cinnamoyl. *Trans*-dendrochrysine: R = *trans*-cinnamoyl.

TABLE 7-2
Alkaloids Related to Dendrobine

Name	Empirical Formula	Source	Reference[a]	Figure
Dendrobine	$C_{16}H_{25}NO_2$	*D. nobile, linawianum, findlayanum*	A-H	1
2-Hydroxy-dendrobine	$C_{16}H_{25}NO_3$	*D. findlayanum*	I	2
6-Hydroxy-dendrobine (Dendramine)	$C_{16}H_{25}NO_3$	*D. nobile*	J	2
N-Methyldendrobinium ion	$C_{17}H_{28}NO_2$	*D. nobile*	E	5b
Dendrobine *N*-oxide	$C_{16}H_{25}NO_3$	*D. nobile*	K	5b
N-Isopentenyl dendrobinium ion	$C_{21}H_{34}NO_2$	*D. nobile*	KL	5b
Nobilonine	$C_{17}H_{27}NO_3$	*D. nobile, findlayanum, hildebrandii*	C, I, M, N	4
6-Hydroxy-nobilonine	$C_{17}H_{29}NO_4$	*D. hildebrandii, friedricksianum*	N	4
Dendrine	$C_{19}H_{29}NO_4$	*D. nobile*	F, O	2
Dendoxine	$C_{17}H_{25}NO_3$	*D. nobile*	P	3
4-Hydroxy-dendroxine	$C_{17}H_{25}NO_4$	*D. nobile*	Q	3
6-Hydroxy-dendroxine	$C_{17}H_{25}NO_4$	*D. nobile*	P	3
N-Isopentenyl dendroxinium ion	$C_{22}H_{34}NO_3$	*D. friedricksianum, hildebrandii*	R	5a
N-Isopentenyl-6-hydroxydendroxinium ion	$C_{22}H_{34}NO_4$	*D. friedricksianum, hildebrandii*	R	5a
Dendrowardine	$C_{25}H_{32}NO_4$	*D. wardianum*	S	6

[a] References:
A. Suzuki et al., 1932, 1934.
B. Inubushi et al., 1963.
C. Yamamura and Hirata, 1964a, 1964b.
D. Onaka et al., 1964.
E. Inubushi, Ishii, et al., 1964; Inubushi, Katarao, et al., 1964; Inubushi, Tsuda, et al., 1964; Inubushi, Sasaki, et al., 1964.

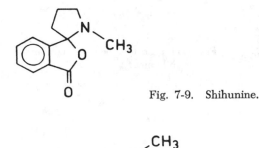

Fig. 7-9.   Shihunine.

Fig. 7-10.   Pierardine.

The two species, *D. parishii* Rchb.f. and *D. anosmum*, Lindl., both with rhubarb-smelling flowers, in spite of the name of one of them, give an unusually strong positive reaction for alkaloid. In fact, these two species contain only one quaternary alkaloid, for which I here propose the name dendroparine (Fig. 7-12). The structure of the optically inactive alkaloid was revealed by spectroscopic means and confirmed by synthesis (Leander and Lüning, 1968a). A further confirmation was given by X-ray crystallography on dendroparine bromide (Söderberg and Kierkegaard, 1970).

Dendrobium crepidatum Lindl. provides the most recently studied group of alkaloids of *Dendrobium* species. Originally a great number of alkaloids were believed to occur in the plant, but later the number was

F.   Inubushi and Nakano, 1965; Inubushi et al., 1965.
G.   Inubushi et al., 1972.
H.   Yamada et al., 1972.
I.   Granelli et al., 1970.
J.   Inubushi et al., 1966.
K.   Hedman and Leander, 1972.
L.   Inubushi, 1967.
M.   Onaka et al., 1965.
N.   Elander and Leander, 1971.
O.   Granelli and Leander, 1970.
P.   Okamoto et al., 1966a, 1966b.
Q.   Okamoto et al., 1972.
R.   Hedman et al., 1971.
S.   Blomqvist et al., 1973.

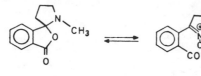

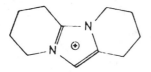

Fig. 7-11. Lactone-betaine tautomerism of shihunine.

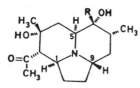

Fig. 7-12. Dendroparine.

reduced to five. Two of these were finally shown to be artifacts produced during the isolation process, so that only three alkaloids are originally present in the plant.

The first obtained in crystalline form is crepidine. Attempts to elucidate its structure by spectroscopic and chemical means appeared so difficult that an X-ray crystallographic structure determination was performed (Kierkegaard et al., 1970). The result of this investigation showed that the alkaloid contained an unknown tricyclic ring system (Fig. 7-13). The chemical and spectroscopic findings were fully consistent with the structure arrived at by X-ray crystallography. On the basis of the structure of crepidine, the structures of the other two alkaloids, dendrocrepine and crepidamine, and their transformation products could be solved by chemical and spectroscopic means (Elander et al., 1973) (Fig. 7-14).

Phenylisoquinoline Alkaloids

Most of the alkaloid-accumulating species are found in the subfamily EPIDENDROIDEAE of the ORCHIDACEAE (Garay, 1972). Thus far, the only genus outside this subfamily from which alkaloids have been isolated and their structures determined is *Cryptostylis*. From *Cryptostylis fulva* Schltr. collected by Lüning in New Guinea, 1967, Leander and Lüning

Fig. 7-13. Crepidine. R = phenyl.

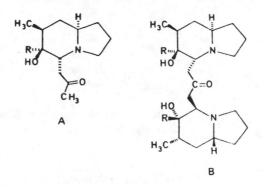

Fig. 7-14. Crepidamine (A) and dendrocrepine (B). R = phenyl.

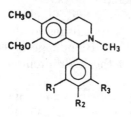

Fig. 7-15. Cryptostylines. Cryptostyline I: R_1 = H, R_2R_3 = O-CH$_2$-O; Cryptostyline II: R_1 = H, R_2 = R_3 = OCH$_3$; Cryptostyline III: R_1 = R_2 = R_3 = OCH$_3$.

(1968b) and Leander et al. (1969) isolated three alkaloids, cryptostylines I, II, and III, belonging to a completely new group of alkaloids, the so called 1-phenylisoquinolines (see Fig. 7-15) related to but unlike the PAPAVERACEAE alkaloids. The alkaloids are optically active and their absolute configuration was revealed by X-ray diffraction of cryptostyline I methioidide (Leander et al., 1973; Westin, 1972; Brossi and Teitel, 1971; Kametani et al., 1971). In the related species *Cryptostylis erythroglossa*. Hay. from Taiwan the enanthiomeric alkaloids were found, and related compounds have been found in *Cryptostylis erecta* R. Br. (Slaytor, 1969).

Eria Alkaloids

Alkaloids of phenylalanine origin are found in *Eria* species. Many erias have shown an elevated alkaloid content during the screening tests, but so far alkaloids have been isolated only from *Eria jarensis* Ames. These alkaloids are simple methylated phenethylamines, some of which, surprisingly, have not been observed previously in nature (Hedman et al., 1969) (Fig. 7-16).

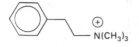

Fig. 7-16.   Phenethylamines from *Eria jarensis*. R₁ and R₂ are H or methyl.

Pyrrolizidine Alkaloids

A widespread group of alkaloids in the ORCHIDACEAE is that containing the bicyclic pyrrolizidine ring system.

The first pyrrolizidine alkaloid isolated from an orchid is chysine (Fig. 7-17). Its structure was first wrongly formulated (Lüning and Tränkner, 1965), but the correct structure was established by transformation to lindelofidine and synthesis (Lüning and Tränkner, 1968; Brandänge and Lundin, 1971), and later also by X-ray diffraction of the hydrobromide of the corresponding acid (Söderberg, 1971).

Simple pyrrolizidine derivatives have been found in many *Vanda* and *Vandopsis* species. Thus *Vanda cristata* Lindl. (Lindström and Lüning, 1969) and *Vanda hindsii* Lindl. contain laburnine acetate (Fig. 7-18) and *Vandopsis lissochiloides* Pfitz. and *Vandopsis gigantea* Pfitz, contain a mixture of laburnine and lindelofidine together with their acetates. *Vanda helvola* Bl. contains laburnine and its acetate. *Vanda luzonica* Loher contains laburnine or lindelofidine but no acetate (Brandänge and Granelli, 1973).

The ease of work with the acetates of the pyrrolizidine carbinols in

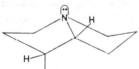

Fig. 7-17.   Chysine.

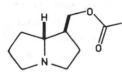

Fig. 7-18.   Laburnine acetate from *Vanda cristata*.

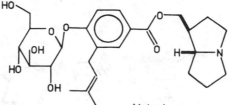

Malaxin

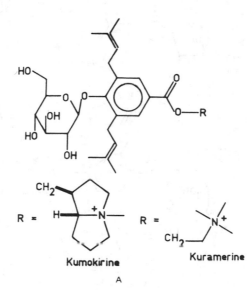

Kumokirine   Kuramerine

A

Fig. 7-19a.   *Liparis-Malaxis* alkaloids.

minute quantities relative to that of the free alcohols has opened up the field of these and related alkaloids for study.

As early as 1902 Boorsma found a trace of alkaloid in *Liparis parviflora* Lindl. Eguchi and Wakasugi (1936) found a bitter alkaloid in *Liparis krameri* Franch et Sav., and Lüning in his survey (1964) published a high alkaloid content in *Liparis loeselii*. Since these are rare orchids, other species were tested. Lüning and Leander (1967) published the structure of malaxin (Fig. 7-19) from *Malaxis congesta* comb. nov. (Schltr.), and Nishikawa et al. (1967) published the structures of nervosine (Fig. 7-19) from *Liparis nervosa* Lindl., kuramerine from *L. krameri* Franch et Sav., and kumokirine (Fig. 7-19) from *L. kumokiri* F. Maekawa.

Nishikawa et al. (1969) published the occurrence of malaxin, free and in complex with guanidine, in *Liparis bicallosa* Schltr. and *Liparis hachijoensis* Nakai, as well as an alkaloid isolated from *L. auriculata* called

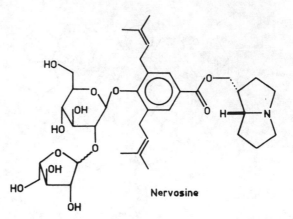

Nervosine

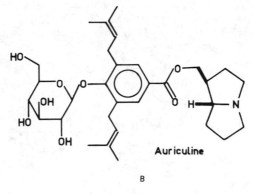

Auriculine

B

Fig. 7-19b. *Liparis-Malaxis* alkaloids.

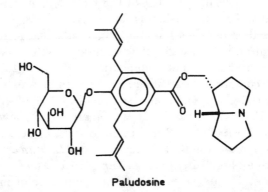

Paludosine

Fig. 7-19c. *Liparis-Malaxis* alkaloids.

370

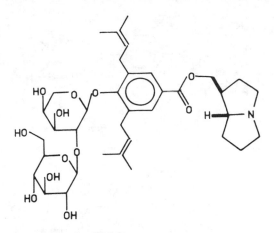

Fig. 7-20. Grandifoline.

auriculine, which was, however, not fully characterized (Fig. 7-19).

After the advent of the possibility to study small amounts of pyrrolizi-dine carbonols by way of their acetates, even the rare species could become the subject of study. Thus Lindström and Lüning (1971) published the structure of an alkaloid from *Liparis loeselii* L. C. Rich which has a structure identical to that postulated for auriculine. The same paper deals with the major alkaloid paludosine from *Hammarbya paludosa* (L.) O.K. (Fig. 7-19). In 1967 Lüning collected *inter alia* a number of *Liparis* and *Malaxis* species in New Guinea, and from one of them, *M. grandifolia* Schltr., a new alkaloid grandifoline was obtained (Lindström et al., 1971). The saccharide part of the glycoside consisted of a previously unknown disaccaride, the identity of which was confirmed by synthesis (Ekborg et al., 1969). In contrast to nervosine, the arabinose moiety in grandifoline is attached directly to the aglucone (Fig. 7-20).

The minor alkaloid hammarbine (Fig. 7-21) from *Hammarbya palu-dosa* was shown to carry a methoxy group instead of the isopentenyl group at position 5 of the aromatic nucleus. This feature also characterizes the alkaloids from the Formosan species *Liparis keitaoensis* Hay. keitaoine (Fig. 7-21) and keitine (Fig. 7-21). The latter is also the first *Liparis-Malaxis* alkaloid so far found which does not contain any sugar (Lindström and Lüning, 1972).

Phalaenopsis and Vandopsis Alkaloids

It has long been known that certain *Phalaenopsis* species contain alkaloids (de Droog, 1896). It was not until 1966 that the first alkaloid in

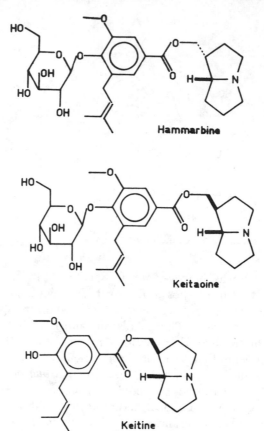

Fig. 7-21. *Liparis-Malaxis* alkaloids with a methoxyl group on the aromatic ring.

this group, phalaenopsin, was isolated from *P. amabilis* Bl. (Lüning et al., 1966). This alkaloid was later called phalaenopsine T (T for trache-lanthamidine, the pyrrolizidine carbinol esterified) (Fig. 7-22) and was soon followed by phalaenopsine La (La for laburnine) (Fig. 7-22) from *P. mannii* Rchb.r. and phalaenopsine Is (Is for isoretronecanol) (Fig. 7-22) from *P. equestris* Rchb.f. (Brandänge and Lüning, 1969; Brandänge et al., 1970, 1971). *P. cornucervi* Rchb.f. contains a somewhat different alkaloid, cornucervine (Fig. 7-23). In many species there are mixtures of phalaenopsines La and T, which can best be demonstrated by the optical rotation and GLC-MS of the mixture. These values for a number of *Phalaenopsis* species and the occurrence of the alkaloids in the genus are summarized in Table 7-3, adapted from Brandänge et al. (1972).

TABLE 7-3

Alkaloid content in 17 Phalaenopsis and related species

	Phalae-nopsine La I		Phalae-nopsine T II	Cornu-cervine III	Phalae-nopsine Is IV	$[\alpha]_D$ of Alkaloid
Ph. amabilis			+			−15°
Ph. amboinensis	+					+9.3°
Ph. aphrodite			+			−13°
Ph. cornu-cervi				+		−4.3°
Ph. equestris					+	−42°
Ph. fimbriata			+			−12°
Ph. gigantea	−		−	−	−	
Ph. hieroglyphica	+	or	+			
Ph. lindenii	−		−	−	−	
Ph. lueddemanniana	+	or	+			
Ph. mannii	+					+10°
Ph. sanderiana	+		+			+2.3°
Ph. schilleriana	+					+10°
Ph. stuartiana	+		+			−7.7°
Ph. sumatrana	+					+6.6°
Ph. violacea	+	or	+			
Kingiella taenialis	+					
Doritis pulcherrima	+	or	+			

I will end this survey of the individual alkaloids isolated from orchids by noting a few odd molecules. Plant juice of *Vandopsis longicaulis* from New Guinea gives a strong positive Dragendorff reaction. This is due to a high content of two simple compounds N-methylpyridinium ion (Fig. 7-24) and N-methylpiperidine N-oxide (Fig. 7-24) (Brandänge and Lüning, 1970). The same compounds have recently been found in *Trichoglottis perezei*. *Vandopsis parishii* Schltr., on the other hand, has given only hygrine, and *Vanda lamellata* Lindl. has shown the presence of a small amount of nicotine.

Biosynthesis of Orchid Alkaloids

Not very long after the elucidation of the structure of dendrobine and its allies, speculations started about its formation in the plant. Yamazaki et al. (1966) showed that 2-^{14}C labeled mevalonate was incorporated in dendrobine formed in *Dendrobium nobile* at a yield of .012%. After a speculation by Conroy (1968) that the C_{15}-skeleton of dendrobine was

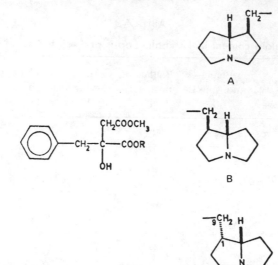

Fig. 7-22.   Phalaenopsines. Phalaenopsine La: R = A. Phalaenopsine T: R = B. Phalaenopsine Is: R = C.

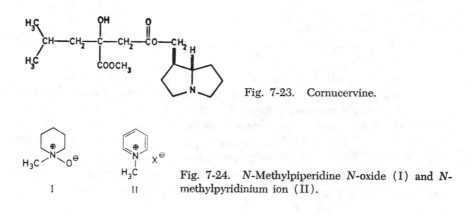

Fig. 7-23.   Cornucervine.

Fig. 7-24.   N-Methylpiperidine N-oxide (I) and N-methylpyridinium ion (II).

formed by oxidative removal of rings A and B of a steroid, Edwards et al. (1970) showed by using 4-^{14}C labeled mevalonic acid that the route starts with *trans-cis* farnesol. This is then cyclized to a cadalane system, which gives the dendrobine skeleton by cleavage and recyclization by one of the two ways, *a* or *b*, seen in Fig. 7-25. The recovery of ^{14}C from relevant positions after degradation of labeled dendrobine clearly showed

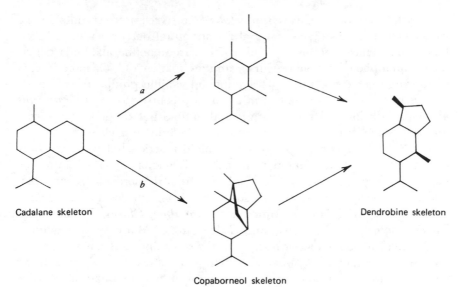

Cadalane skeleton

Dendrobine skeleton

Copaborneol skeleton

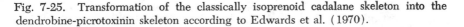

Fig. 7-25. Transformation of the classically isoprenoid cadalane skeleton into the dendrobine-picrotoxinin skeleton according to Edwards et al. (1970).

this pathway. Methionine-^{14}C-methyl gives a fairly strong labeling (.10%) of dendrobine. No degradations were made, but the high yield of label indicates that the methyl carbon is introduced at a late stage of biosynthesis and constitutes the N-methyl group. Dendroparine is labeled by feeding radioactive lysine to *D. parishii*. No labeling of dendroprimine was observed upon feeding radioactive mevalonic acid to *D. primulinum*. Radioactive tracer experiments are going on with *Cryptostylis* species (Lüning et al., unpublished).

Discussion

As more structures of alkaloids have been elucidated, it has become more difficult to draw any but a few very obvious conclusions about their occurrence. Looking on the family as a whole, there are scattered species of high alkaloid content in most taxa. Since it is not known, and not very easy to reveal, whether many species produce alkaloids and further metabolize them and only a few accumulate them, any firm conclusions should not be drawn from the mere presence of alkaloid-positive reactions of a species. In the future when these regulating factors may be better known this feature may again be a useful taxonomic tool.

Thus far only one genus, *Cryptostylis* from NEOTTIOIDEAE outside the EPIDENDROIDEAE, has given material for structural determination of alkaloids. This genus contains a set of 1-phenylisoquinoline alkaloids generated from a phenylalanine precursor and an unknown C_6-C_1 moiety. It is interesting to note that the C_6-C_1 compound vanillin produced by *Vanilla* species of the same subfamily can act as a precursor to the *Cryptostylis* alkaloids. Alkaloids of the same general outline but differently cyclized are well known from the AMARYLLIDACEAE (Wildman, 1968).

Neottia nidus-avis gives a positive alkaloid reaction but produces only choline to any extent. Although all taxa in EPIDENDROIDEAE contain species with a high alkaloid content, as drawn from the screening tests, alkaloids have been successfully isolated only from species in the genera *Dendrobium, Eria, Liparis, Malaxis, Hammarbya, Chysis, Phalaenopsis, Vanda, Vandopsis, Kingiella,* and *Trichoglottis.* Many species with a moderate alkaloid content according to the screening tests have not given any isolable amounts of alkaloids upon reinvestigation, whereas other species with low alkaloid values have proven rich in quaternary salts, which often cannot be extracted by organic solvents thus being lost during screening.

The most intensely studied group, the section *Dendrobium* in the genus *Dendrobium,* consists of about 150 closely related, mostly intercrossable species. Out of 57 species tested in this taxon 18 were found to contain more than .05% alkaloid calculated on the dry weight. Some 14 of these alkaloids have been isolated and characterized. However, unlike the situation in other families, the 28 alkaloids isolated from these species are not related chemically but belong to six different groups, the largest centered around dendrobine with 15 alkaloids.

Although experiments linking the different alkaloids of this group to each other biogenetically still are lacking, one can derive from current biosynthetical laws how the alkaloids are formed from the experimentally established skeleton A (Fig. 7-26). Three processes would account for all alkaloids but dendrine from this skeleton: (1) introduction of nitrogen as ammonia or ethanolamine derived from serine; (2) reduction; (3) methylation. The necessary ring closures would follow automatically. Addition of ammonia to A produces B, which through a one step reduction and monomethylation gives the known 2-hydroxydendrobine. Reduction of this produces dendrobine, which by alkylation or oxidation can be transformed into quaternary salts or the *N*-oxide. If 2-hydroxydendrobine is instead methylated, which might occur directly on its tautomeric form nor-nobilonine, nobilonine is formed and ring closure is made impossible. The somewhat different alkaloid, dendrine, might well be formed through

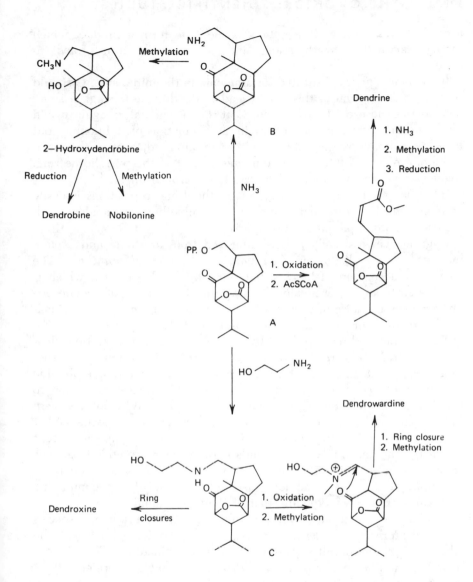

Fig. 7-26. Tentative scheme of biogenetic interrelations of dendrobine akladoids.

377

a condensation of A with acetyl-coenzyme A to form a product which through amination, methylation, and reduction is transformed to dendrine.

If ethanolamine is substituted for ammonia in the initial step, C should be formed. Reduction of this would lead to dendroxine through a spontaneous ring closure. Methylation of C at the nitrogen atom again would block the ordinary ring closure, as in the formation of nobilonine, but here the carbonyl oxygen attacks the carbon atom adjacent to the nitrogen atom to form a tetrahydrofuran ring instead of the pyrrolidine ring. Reduction and further methylation lead to dendrowardine.

Introduction of the hydroxyl groups in the 4 and 6 positions is easily accounted for by hydroxylation of original double bonds in analogy with the corresponding picrotoxinin derivatives.

Hygrine, which usually is biosynthesized from ornithine and acetate units, is a major alkaloid in *D. primulinum* and *D. chrysanthum*. The alkaloid is accompanied in *D. primulinum* by dendroprimine, which is not related to any other alkaloids known and is probably of acetate origin, whereas the accompanying alkaloids in *D. chrysanthum* are cinnamoyl derivatives of norcuskhygrine, closely related to hygrine.

The remaining three types of alkaloids occurring in this section, dendroparine from *D. parishii* and *D. anosmum* of lysine origin, the structurally complicated alkaloids from *D. crepidatum*, and the simple phtalide alkaloids from *D. pierardii* and *D. lohohense* add further confusion to the chemotaxonomical view of the genus *Dendrobium*. Why do these very different alkaloids appear in such a homogeneous group of species? What kind of common denominator can be found between these species except for their common ability to accumulate fairly complicated nitrogenous compounds irrespective of their mode of formation. In other sections of *Dendrobium* there is a random distribution of alkaloid-accumulating species of a frequency equaling that of the whole family.

Another taxon with an elevated number of alkaloid-producing species is that comprising the genera *Liparis*, *Malaxis*, and *Hammarbya*. The alkaloids are almost all glycosides containing glucose and arabinose bound to the phenol group of a *p*-hydroxybenzoic acid esterified with a nitrogenous alcohol. The aromatic ring usually carries one or two isopentenyl residues. Recently one alkaloid was found in which one of these alkyl groups was replaced by a methoxyl group. The alcohols generally are pyrrolizidine carbinols, but in one alkaloid choline has been found. There appear to be no boundaries among the three genera so far studied; from the alkaloid taxonomic point of view all species in the genera merge into a common pool. However, the alkaloids may serve to distinguish otherwise very similar species.

Chysis bractescens, with its simple pyrrolizidine carboxylic ester, stands fairly isolated, although some *Mormodes* and *Catasetum* species are fairly alkaloid-rich during active growth. No successful isolation of alkaloid has, however, been reported from these genera.

Several *Bulbophyllum* species and many *Eria* species are fairly alkaloid-rich, but so far only the simple phenethylamine derivatives isolated from *Eria jarensis* have been isolated and their structures determined.

Among the VANDEAE, the taxon VANDIINAE appears as an invariably alkaloid-rich group. Most species investigated contain detectable amounts of alkaloid, although alkaloids have been isolated from only 28 species.

The 17 *Phalaenopsis, Kingiella,* and *Doritis* species tested all contain one or two of the phalaenopsines or cornucervine. These are all mono-methyl esters of hydroxy-diacids, of which the remaining carboxyl group is esterified with a pyrrolizidine carbinol. The diacids can easily be related biogenetically to phenylalanine or leucine. Two possible ways are inferred: (1) oxidative deamination of amino acids *in situ* to the corresponding β-keto acids, which then are condensed with acetyl coenzyme A; or (2) condensation of acetyl coenzyme A with the β-keto acid prior to its transformation to the amino acid. The alkaloids could then be assumed as products formed by trapping intermediates in the one-carbon extension of a carbon chain. It is noteworthy that in those alkaloids derived from phenylalanine, the amino alcohol is attached to the last-introduced acetate unit, whereas in cornucervine, derived from leucine, the amino alcohol is attached to the original carboxyl group of the amino or keto acid. The alkylmalic acid always has (R)-configuration (Brandänge et al., 1973b), but the configuration of the pyrrolizidine carbinol varies with different species. In some cases the amino alcohol appears partly racemic. Whether this is due to an ancient intercrossing leading to these species is a challenging question. To assure purity of the plant material some of the critical analyses were made on single large plants. In the alkaloid from *P. equestris* the very uncommon isoretronecanol appears as the amino alcohol.

The alkaloids occurring in *Phalaenopsis* and allied species thus seem fairly homogeneous, but that is not the case in the genera *Vanda* and *Vandopsis.* Many of the species contain simple pyrrolizidine carbinols or their acetates, but there are reported findings of N-methylpiperidine N-oxide, N-methylpyridinium ion, hygrine, and even nicotine.

Conclusion

Thus far most of the alkaloid-rich orchids which are fairly easy to obtain in quantity have been investigated in a few specialized laboratories in the world. In the future, work will have to be performed locally, near the habitats of the rare, alkaloid-rich species. With the present knowledge it is a possible task to predict where in the ORCHIDECEAE the alkaloid-rich species should be looked for, and I think this possibility and the chemical structures so far elucidated are the most valuable offspring of the early work on orchid alkaloids.

Bibliography

BEHR, D., and LEANDER, K. 1972. *Acta Chem. Scand.* 26:3196.

BLOMQVIST, L., BRANDANGE, S., GAWELL, L., LEANDER, K., and LUNING, B. 1973. *Acta. Chem. Scand.* 27:1439.

BLOMQVIST, L., LEANDER, K., LUNING, B., and ROSENBLOM, J. 1972. *Acta Chem. Scand.* 26:3203.

BOORSMA, W. 1899. *Meded s'Lands Plantent.* 31:123.

———. 1902. *Bull. Inst. Bot. Buitenzorg* 14:36.

BRANDÄNGE, S., and GRANELLI, I. 1973. *Acta Chem. Scand.* 27:1096.

BRANDÄNGE, S., GRANELLI, I., and LUNING, B. 1970. *Acta Chem. Scand.* 24:354.

BRANDÄNGE, S., and LUNDIN, C. 1971. *Acta Chem. Scand.* 25:2447.

BRANDÄNGE, S., and LUNING, B. 1969. *Acta Chem. Scand.* 23:1151.

———. 1970. *Acta Chem. Scand.* 24:353.

BRANDÄNGE, S., LUNING, B., MOBERG, C., and SJÖSTRAND, E. 1971. *Acta Chem. Scand.* 25:349.

———. 1972. *Acta. Chem. Scand.* 26:2558.

BROSSI, A., and TEITEL, S. 1971. *Helv. Chim. Acta* 54:1564.

CHEN, K., and CHEN, A. LING-. 1935. *J. Biol. Chem.* 111:653.

CONROY, H. 1952. *J. Am. Chem. Soc.* 73:1889.

COSCIA, C. 1969. Picrotoxin. In W. Taylor and A. Battersby, Eds. *Cyclopentanoid Terpene Derivatives,* Marcel Dekker, New York, Chap. 2.

DE DROOG, E. 1896. *Mem. Cour. Autr. Mem., Acad. Roy. Belg.* 55:1: Reprinted in: *Rec. Inst. Botan. Univ. Bruxelles* 2:347 (1906).

EDWARDS, O. 1969. Sesquiterpene Alkaloids. In W. Taylor and A. Battersby, Eds., *Cyclopentanoid Terpene Derivatives,* Marcel Dekker, New York, Chap. 6.

EDWARDS, O., DOUGLAS, J., and MOOTOO, B. 1970. *Can. J. Chem.* 48:2517.

EGUCHI, T., and WAKASUGI, K. 1936. *Nagasaki Igakuzasshi* 14:1324.

EKBORG, G., ERBING, B., LINDBERG, B., and LUNDSTROM, H. 1969. *Acta Chem. Scand.* 23:2914.

EKEVÅG, U., ELANDER, M., GAWELL, L., LEANDER, K., and LUNING, B. 1973. *Acta Chem. Scand.* 27:1982.

ELANDER, M., GAWELL, L., and LEANDER, K. 1971. *Acta Chem. Scand.* 25:721.

ELANDER, M., and LEANDER, K. 1971. *Acta Chem. Scand.* 25:717.

ELANDER, M., LEANDER, K., and LUNING, B. 1969. *Acta Chem. Scand.* 23:2177.

ELANDER, M., LEANDER, K., ROSENBLOM, J., and RUUSA, 1973. *Acta Chem. Scand.* 27:1907.

GARAY, L. 1972. *J. Arnold Arboretum* 53:202.

GRANELLI, I., and LEANDER, K. 1969. *Acta Chem. Scand.* 24:1108.

GRANELLI, I., LEANDER, K., and LUNING, B. 1970. *Acta Chem. Scand.* 24:1209.

HEDMAN, K., and LEANDER, K. 1972. *Acta Chem. Scand.* 26:3177.

HEDMAN, K., LEANDER, K., and LUNING, B. 1969. *Acta Chem. Scand.* 23:3261.

———. 1971. *Acta Chem. Scand.* 25:1142.

HEGNAUER, R. 1963. Taxonomic Significance of Alkaloids, in T. Swain, *Chemical Plant Taxonomy.* Academic Press, London and New York.

HUANG, W.-K. 1965. *Hua Hsueh Hsueh Pao* 31:333.

INUBUSHI, Y. 1967. Personal Communication.

INUBUSHI, Y., ISHII, H., YASUI, B., KONITA, T., and HARAYAMA, T. 1964. *Chem. Pharm. Bull.* 12:1175.

INUBUSHI, Y., KATARAO, E., TSUDA, Y., and YASUI, B. 1964. Chem. Ind. (London): 1689.

INUBUSHI, Y., KIKUCHI, T., IBUKA, T., TANAKA, K., SAJI, I., and TOKANE, K. 1972. *J. Chem. Soc. Chem. Comm.:* 1252.

INUBUSHI, Y., and NAKANO, J. 1965. *Tetrahedron Lett.:* 2723.

INUBUSHI, Y., SASAKI, Y., TSUDA, Y., YASUI, B., KONITA, T., MATSUMOTO, J., KATARAO, E., and NAKANO, J. 1964. *Tetrahedron* 20:2007.

INUBUSHI, Y., SASAKI, Y., YASUI, B., KONITA, T., MATSUMOTO, J., KATARAO, E., and NAKANO, J. 1963. *Yakugaku Zasshi* 83:1184.

INUBUSHI, Y., TSUDA, Y., and KATARAO, E. 1965. *Chem. Pharm. Bull.* 13:1482.

———. 1966. *Chem. Pharm. Bull.* 14:668.

INUBUSHI, Y., TSUDA, Y., KONITA, T., and MATSUMOTO, S. 1964. *Chem. Pharm. Bull.* 12:749.

———. 1968. *Chem. Pharm. Bull.* 16:1014.

KAMETANI, T., SUGI, H., and SHIBUYA, S. 1971. *Tetrahedron* 27:2409.

KIERKEGAARD, P., PILOTTI, A.-M., and LEANDER, K. 1970. *Acta Chem. Scand.* 24:3757.

LAWLER, L., and SLAYTOR, M. 1969. *Phytochemistry* 8:1959.

———. 1970. *Proc. Linn. Soc. New South Wales* 94:237.

LEANDER, K., and LUNING, B. 1967. *Tetrahedron Lett.* 3477.

———. 1968a. *Tetrahedron Lett.* 905.

———. 1968b. *Tetrahedron Lett.:* 1393.

LEANDER, K., LUNING, B., and RUUSA, E. 1969. *Acta Chem. Scand.* 23:244.

LEANDER, K., LUNING, B., and WESTIN, L. 1973. *Acta Chem. Scand.* 27:710.

LINDSTROM, B., and LUNING, B. 1969. *Acta Chem. Scand.* 23:3352.

———. 1971. *Acta Chem. Scand.* 25:895.

———. 1972. *Acta. Chem. Scand.* 26:2963.

LINDSTROM, B., LUNING, B., and SHRALA-HANSEN, K. 1971. *Acta Chem. Scand.* 25:1900.

382   THE ORCHIDS: SCIENTIFIC STUDIES

Lüning, B. 1964. *Acta Chem. Scand.* **18:**1507.

———. 1966. Chemotaxonomy in a *Dendrobium* Complex. *Proceedings of the Fifth World Orchid Conference,* p. 211.

———. 1967. *Phytochemistry* **6:**857.

———. 1973. Unpublished results.

Lüning, B., and Leander, K. 1965. *Acta Chem. Scand.* **19:**1607.

Lüning,B., and Lundin, C. 1967. *Acta Chem. Scand.* **21:**2136.

Lüning, B., and Tränkner, H. 1965. *Tetrahedron Lett.:* 921.

———. 1968. *Acta Chem. Scand.* **22:**2324.

Lüning, B., Tränkner, H., and Brandänge, S. 1966. *Acta Chem. Scand.* **20:**2011.

Matsuda, M., Tomiie, Y., and Nitta, I. 1965. *18th Annual Meeting of Chemical Society of Japan, Osaka, April 2.* Abstract of Papers, p. 215.

Nishikawa, K., and Hirata, H. 1967. *Tetrahedron Lett.:* 2591.

———. 1968. *Tetrahedron Lett.:* 6289.

Nishikawa, K., Miyamura, M., and Hirata, Y. 1967. *Tetrahedron Lett.:* 2597.

———. 1969. *Tetrahedron* **25:**2723.

Okamoto, T., Natsume, M., Onaka, T., Uchimaru, F., and Shimizu, M. 1966a. *Chem. Pharm. Bull.* **14:**676.

———. 1966b. *Chem. Pharm. Bull.* **14:**672.

———. 1972. *Chem. Pharm. Bull.* **20:**418.

Onaka, T., Kamata, S., Maeda, T. Kawazoe, Y., Natsume, M., Okamoto, T., Uchimaru, F., and Shimizu, M. 1964. *Chem. Pharm. Bull.* **12:**506.

———. 1965. *Chem. Pharm. Bull.* **13:**745.

Porter, L. A. 1967. *Chem. Rev.* **67:**441.

Slaytor, M. 1969. Personal communication.

Suzuki, H., and Keimatsu, I. 1932. *J. Pharm. Soc. Japan.* **52:**996 (*Ger. Abstr.* **162**), 1049 (*Ger. Abstr.* **183**).

Suzuki, H., Keimatsu, I., and Ito, K. 1934. *J. Pharm. Soc. Japan* **54:**802 (*Ger. Abstr.* **138**), 820 (*Ger. Abstr.* **146**).

Söderberg, E. 1971. *Acta Chem. Scand.* **25:**615.

Söderberg, E., and Kierkegaard, P. 1970. *Acta Chem. Scand.* **24:**397.

Wester, D.H., 1921a. *Pharm. Weekblad* **28:**1438.

———. 1921b. *Ber. Pharm. Ges.* **31:**170.

Westin, L. 1972. *Acta Chem. Scand.* **26:**2305.

Wildman, W. C. 1968. The Amaryllidaceae Alkaloids, in R. H. F. Manske, Ed., *The Alkaloids,* Vol. 9. Academic Press, New York. p. 307.

de Wildemann, E. 1892. *Bull. Soc. Belge Microscopie* **18:**101. Reprinted in *Rec. Inst. Botan. Univ. Bruxelles* **2:**337 (1906)

Yamada, K., Suzuki, M., Hayakawa, Y., Aoki, K., Nakamura, H., Nagase, H., and Hirata, Y. 1972. *J. Am. Chem. Soc.* **94:**8280.

Yamamura, S., and Hirata, Y. 1964a. *Nippon Kagaku Zasshi* **85:**377.

———. 1964b. *Tetrahedron Lett.:* 79.

Yamazaki, M., Matsuo, M., and Arai, K. 1966. *Chem. Pharm. Bull.* **14:**1058.

8

Cytology and the Study of Orchids*

KEITH JONES

Although the fascination most orchidologists find in the orchids lies in their remarkable array of floral intricacy and in the cool beauty of their flowers, it is equally possible to find satisfaction in understanding them from the scientific point of view. Here there need be no clash between esthetics and science, since our understanding of the reasons for variation and knowledge of the ways in which new changes can be brought about contributes greatly to the production of new forms and to the effectiveness of our methods of classification. Any such inquiry must revolve around characteristics that are not the products of environmental differences but are genetically controlled and predictably inherited.

Those changes that result from modifications of the environment are, of course, of value to us in our culture of the orchids, but if they are not inherited they are of little significance to the taxonomist and less to the breeder. A knowledge of the genetic constitutions of individuals is thus of prime importance, but for the overwhelming majority of plants, including orchids, we know relatively little of their makeup in this respect, and indeed the unraveling of the genetic structure of any species can be a long and laborious process. This is particularly true when, as in the orchids, there is a long life cycle and a slow turnover of generations. Although genes and gene interactions are difficult to observe, we are able to see the chromosomes—the carriers of genes and the controllers of destinies—with relative ease, and these can give information of immediate benefit to us. The study of the chromosomes provides knowledge of the fundamental constitution, of evolution and taxonomic affinities, and of genetic potential, revealing features of the biology of the organism which may not be evident in its other aspects.

* Part of this chapter has been published previously in *L'Orchideé* 2(1970):172–179.

Each chromosome can be regarded as a linear subdivision of the total gene array, and it is the frequency, form, and behavior of chromosomes that characterize a species and determine the manner in which genes are transmitted from one generation to another. At the same time, comparisons of the chromosome organization of related individuals may reveal the extent of their relationships, the variety of ways in which they have evolved, and the adequacy of current taxonomic classifications. The chromosomes can, in other words, tell us things genes cannot, and since they are easily seen, whereas genes are hard to discover, we can use them to our advantage providing that we possess the technical skills and the knowledge with which to extract the information they contain. Let us consider first the main attributes of orchid chromosomes, leaving until later the methods by which we are able to observe them.

The number of chromosomes found in somatic tissues can vary widely between four as in a member of the COMPOSITAE, *Haplopappus gracilis,* to over 1200 in some ferns. Whatever the number, we regard it as characteristic for an individual, usually of a species, and sometimes of a genus or higher taxonomic category. This chromosome number we observe in dividing cells in somatic tissues and in those that give rise to the gametes, and we see it as a characteristic that can reflect the evolutionary past of the individual and also indicate to some extent its current requirements and future potentialities. It is important here to realize that although we may speak of the chromosome constancy of individuals or species, both the number of chromosomes and their form can change during evolution. Spontaneous mutations are the cause of these changes, a common effect of which is to double the chromosome number, giving rise to so-called polyploid individuals. Here perhaps we may pause to consider the apparently simple concept of chromosome number. It is essentially simple, to be sure, but it can be confusing in the way in which it is described.

The number of chromosomes found in a typical dividing cell in a root or in other somatic tissue is the total (or somatic) chromosome number of that individual. If such a plant is a diploid this number represents two sets of chromosomes, one derived from the male gamete and the other from the female. In this case, the number of chromosomes contributed by each gamete is called the basic number, which can be represented by x. To describe a diploid number we can, therefore, preface it by $2x =$, but if polyploidy has occurred and the number of sets increased, then we may use the symbols $3x =$, $4x =$, $5x =$, $6x =$, and so on. It is this basic number which is so widely used in evolutionary and taxonomic studies as a means of assessing relationships and the ways in which chromosome differences have been produced. If we do not know the basic number, and sometimes this is difficult to determine with certainty, we cannot use the terminology, and then it is conventional to indicate the total chromosome number by the symbol $2n =$. To say this, however, is to disguise the fact that chromosome cytologists, evolutionists, and taxonomists have found it impossible to reach an agreement on these terms, some using $n =$ to indicate the basic number, and to describe polyploidy as $3n$, $4n$, $5n$, and so on. This is a situation which causes a great deal of confusion to those who are not fully familiar with the subject, but its discussion is beyond the scope of this chapter.

In the orchids, known basic numbers are relatively high when compared with most flowering plants. The modal frequency is around 19 and 20 and the maximum about 34. It has been suggested, therefore, that the orchids are ancient polyploids, that is, they have been derived from plants with more than two chromosome sets, but during the course of evolution there has been loss or gain of individual chromosomes coupled with structural alteration of sets, which has led to their now being effectively diploid. The somatic chromosome number of $2n = 38$ is, for example, a common one in this family. Here we have only one possible basic number, $x = 19$. Where we find species with $2n = 57$ or $2n = 76$, we can then make the assumption that these are triploid ($3x$) or tetraploid ($4x$) plants, respectively.

But basic numbers are not only useful in providing an indication of existence of polyploidy; indeed they are more valuable in suggesting evolutionary relationships and taxonomic classifications. Similarities in basic number generally indicate that species or genera may be more closely related in evolutionary terms than those which differ in this character, though only when the chromosome parameter is paralleled by other features, particularly external morphology, as is the case in the revision of the VANDEAE which I will describe later. The erection of subtribes in this tribe now more effectively reflects a major divergence of the genera concerned, and no doubt there will be similar divisions set up in other tribes as their cytology is better understood. Chromosome differences are, then, a most useful adjunct to classification, but it should be said that chromosome similarities must be treated with caution, for they often disguise differences that are not observable in species, though they may be when hybrids are made. Each case must be taken on its merits, and each potential adjustment to classification must be viewed in the context of all available evidence.

Somatic chromosome numbers and basic numbers show wide variation in the ORCHIDACEAE, as was indicated by Duncan (1959). At the low end of the scale lies *Oncidium pusillum* with only $2n = 10$ chromosomes ($x = 5$), while at the other are the high-numbered taxa such as *Catasetum planiceps* ($2n = c.$ 162), *Oncidium varicosum* ($2n = 168$), and an unidentified species of *Aerangis* ($2n = c.$ 200). Such numbers as these are a certain indication that polyploidy has played a part in the evolution of orchids, but many would say this is already suggested in the high basic numbers that abound in the family. When much more evidence is available to us, and when we are able to examine chromosome pairing in species and in hybrids, we shall be in a better position to appraise the structure and evolution of the family.

Chromosome Techniques

Although orchids cannot be considered ideal material for chromosome analysis, they will respond well to modern squash methods of a simple and direct sort. Despite the disadvantages of small chromosome size and relatively high chromosome number, this family has two advantages which are not present in most other plant groups. First, the epiphytic species produce aerial roots which are both clean and readily accessible for collection. Second, because of the nature of pollen development in a pollinium the nuclear division known as pollen mitosis (which occurs after meiosis but before pollen is fully mature) is well synchronized and provides excellent opportunities for chromosome counting.

Aerial roots are usually quite bulky and possess on their outsides either mature or immature velamen. Roots in which the first centimeter or so of tip are free of mature velamen are removed from the plant and placed in a reagent which will disturb normal phasing of mitosis and produce a degree of chromosome contraction. The consequences are that the chromosomes are spread throughout the cell and appear compact. The main problem here is to ensure that the reagent is able to penetrate into the interior of the root tip, and to facilitate this, it is our practice at Kew to slit the tip along several longitudinal planes with a very sharp blade and to mop the cut surfaces thoroughly with filter paper. The diagram in Fig. 8-1 shows the types of cuts we employ. Following this procedure, roots are immersed in the reagent, which they readily absorb and pass into their interiors. Several different chemical reagents may be used, but we have found two to be most satisfactory. The first, 8-hydroxyquinoline, is used at a concentration of $0.002M$, and roots are left in this at a temperature around 16°C for 4 hours. Alternatively, monobromonaphthalene can be used as a saturated solution (a single globule shaken up with any volume of water), again for 3-4 hours at room temperature, or alterna-

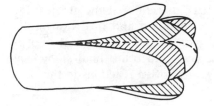

Fig. 8-1. Cutting of the root-tip prior to pretreatment. Cut surfaces are shaded; the broken line indicates the position of the fifth cut.

tively at about 4°C (the usual temperature of a domestic refrigerator) overnight.

Following these particular treatments, roots are fixed in a mixture of absolute alcohol and glacial acetic acid, 3 parts to 1, for at least ½ hour, and are then washed in water, hydrolyzed in normal hydrochloric acid at 60°C for about 8 min, and then placed in Feulgen reagent and left in a dark cupboard. The meristematic tips of the roots become deep purple, usually within an hour, but are better left for two or more, following which small pieces of the tip are taken and squashed, taking care to remove the immature velamen which may cohere and prevent complete squashing. Observations are made from the temporary slide or from the permanent mounts made by removing the cover slip after freezing with a cylinder of CO_2; both slide and cover slip are immersed in absolute alcohol for 15 seconds and then reunited, with Euparal as the mounting medium.

Pollen mitosis takes place in pollinia in buds which are a week or 10 days from flowering. Young pollen grains with only a single nucleus are too young, and those with two nuclei are too old. With practice it becomes very easy to find the correct stage; when it is present it can be spectacular. Usually a large number of tetrads are in mitosis at the same time, and when this is so each cell of each tetrad is at an identical stage of division. Because of the geometrical arrangements of the tetrads, many very clear polar views and metaphases can be seen and are generally highly suitable for chromosome counting. The procedure for making the preparation is very simple. Fresh (or fixed) pieces of pollinia are squashed in a 1% solution of acetic orcein or acetocarmine and are gently heated over a spirit flame to aid differentiation. The slide can be made permanent by the method outlined for root tips. Meiosis can be examined by the same method, but pollen mitosis can be superior if only a chromosome count is required. Further details of the method can be found in Jones (1963, 1966) and in Jones and Daker (1968).

One point has to be stressed: there are in the literature a proportion of reported chromosome counts which are inaccurate. Some are the result of poor techniques and bad preparations. It is, therefore, of the greatest importance to maintain high standards of technique and to discard ruthlessly preparations which are not worthy of analysis. Care must also be taken to differentiate between chromosomes and other bodies, such as bacteria and fungal spores, which may give positive reactions to the stain used.

Variations in Basic Number

It was mentioned earlier that orchid species can show very different basic numbers and that these may have significance in indicating affinities between species or between genera. One example of such use of basic number occurs within the tribe VANDEAE. Summerhayes (1966) recognized two subtribes, ANGRAECINAE and the AERANGIDINAE, which had been separated earlier by Schlechter into two groups on the basis of rostellum characteristic. It had been found at Kew that the ANGRAECINAE consisted almost entirely of species and genera with a basic number of $x = 19$, but that numbers of the AERANGIDINAE were based on $x = 25$ (Jones, 1967). This knowledge showed Summerhayes that rostellum characteristics reflected others of a fundamental nature and confirmed him in his views that these two subtribes should be erected. Thus a combination of external and internal characteristics can produce a classification which probably reflects broads and important evolutionary developments in the VANDEAE. In this case, morphological diagnosis preceded the cytological discovery, but the position could have been reversed, the cytologists pointing out clear chromosome discontinuities and suggesting to the taxonomists reconsideration of current classifications in the light of new information. It seems likely, for example, that new chromosome studies in the SPIRANTHINEAE, where the unusual chromosome number of $2n = 46$ occurs in some genera but not in others, may eventually result in some revision of this subtribe (Martinez and Jones, unpublished).

Such observations as these are of value not only to the evolutionists and taxonomists but also can indicate to the breeder areas where hybridization may be profitable and those where the prospects of fertile hybrids are least. Surveys of chromosome number can thus be of continuing value, though we should always bear in mind the fact that variation in number can occur within the bounds of a taxonomic species. That being so, the chromosome number of any individual can only be known when it has been determined in that individual, although, of course, it is necessary for us to make certain presumptions concerning the probable constitution of a species based on the studies of only very few individuals.

The tribe CATASETINAE is a second example of the significance of basic number. This is a compact tribe with only three genera, *Catasetum, Mormodes,* and *Cychnoches,* all characterized by mechanisms which forcibly discharge their pollinia. Chromosome studies show that they possess chromosome numbers of a relatively high order, $2n = 54$ to $c.$ 162. *Catasetum* and *Mormodes* share the same basic number of $x = 27$, *Cychnoches* differs from them in being $x = 32$ and $x = 34$ (Jones and Daker, 1968). This demonstration that *Cychnoches* is chromosomally distinct from its companion genera is of interest in view of the fact that Dodson has suggested that this genus may have evolved from the fleshy flowering species of the subgenus *Clowsia* of *Catasetum* (Dodson, 1962). Their cytological relationship as now revealed does not necessarily contradict this suggestion, but it may give rise to some reappraisal.

Although we can now see the value of basic numbers, it may be pointed out that some orchids do not easily give up the secret of their own particular basic number. One obvious fact is that ancient polyploidy may obscure completely original basic numbers, but apart from that, there can be real difficulties of interpretation in some plants, even where preparations are of the highest quality. For example, in our own experience at Kew we have found it difficult to be sure of the chromosome numbers within the genus *Stanhopea*. Some chromosomes in a complement can show a secondary constriction, which is the site at which nucleoli are formed. In orchids this constriction can be unusually wide and lacking in staining capacity, so that the two chromosomes segments separated by the constriction can appear as individual chromosomes, thus leading to an incorrect count. That is to say, a number may be determined as $2n = 40$, rather than $2n = 38$, and the basic number consequently deduced to be $x = 20$ and not $x = 19$. Being aware of this sort of error, we can make many observations to determine the true position, but even with this knowledge some plants almost defy a positive determination. In our studies we have concluded that although $x = 20$ seems to be the correct value for *Stanhopea inodora, S. insignis,* and *S. tigrina,* we remain in some doubt concerning the other species we have examined (Daker and Jones, 1970). Being aware of our own failings, we are conscious also of others', and as with other plant groups, we must presume a percentage of error in published chromosome counts of orchids. It is necessary, of course, to be particularly concerned when chromosome number varies with author rather than with taxon. For example, in the genus *Bulbophyllum* those species counted by Pancho are reported to have a basic number $x = 20$ (Pancho, 1965), while at Kew my colleague finds $x = 19$ for this genus (Daker, 1968). Such data are illustrative of what was said earlier. The orchids are not easy cytological material and can contain many pitfalls for the unwary.

Variations at the Polyploid Level

In addition to modifications of basic number, complete replication of chromosome sets above the diploid level (= polyploidy) is a common phenomenon in plants, and it is assumed that about 45% of all species are polyploid. This phenomenon is also widespread in the orchids and is responsible for some interspecific and intraspecific discontinuities (cf. Mehlquist, Chapter 9 in this book). In the genus *Catasetum*, for example, although diploids appear to be most common, tetraploids with $2n = 108$ ($x = 27$) are found in *C. discolor*, *C. fimbriatum*, and *C. pileatum*, while in *C. planiceps* the two plants which I examined were tetraploid $2n = 108$ and hexaploid $2n = 162$. Intraspecific variation also seems likely in *C. atratum* where both diploid and tetraploid counts have been recorded. In the genus *Polystachya* most species are diploid with $2n = 40$, but others are tetraploid $2n = 80$ and one hexaploid $2n = 120$. In this genus the polyploids are not distributed at random, for almost all are segregated according to taxonomic groups or geographical distribution. That is to say, all the species so far examined in section *Affines* are polyploid, but in section *Polystachya* tetraploidy is confined entirely to the American species (Jones, 1966). Such conclusions as these, if supported by more extensive investigations, must be of value both in supporting taxonomic classifications and in pointing to different methods of chromosome evolution in different regions.

In considering chromosome variations in the orchids we need constantly to remind ourselves that investigations are still in the early exploratory stages, and there remains perhaps 80% of orchids for which no chromosome counts are available. This being so, it should not be too surprising that in general very little is known of the extent of possible chromosome variation within a species, since its discovery requires the examination of a substantial number of individuals within its taxonomic limits. Although there is a tacit assumption that a single chromosome count will represent the mode for the species, this need not be the case and, indeed, it is not unusual for more detailed analysis to reveal the extent of numerical variation. As a case in point, we may take *Dendrobium kingianum*, which I examined at Kew (Jones, 1963). Six plants were investigated. One was diploid $2n = 38$, two triploid $2n = 57$, three tetraploid $2n = 76$, and one hexaploid $2n = c.$ 114. This species was subsequently surveyed cytologically in its natural habitats in Australia by Keith Maxwell of New South Wales. As a result of his studies, we are now aware of the relative frequency of these chromosome numbers and their precise location in the wild. He was unable to detect any natural hexaploids but found that the diploids and tetraploids had distinctive distributions, which seemed to be correlated with altitude or degree of exposure, and that triploids occur only where these two are in contact (Maxwell, 1967a, 1967b). We are thus able to see in one rare example the process of evolution as expressed by change of chromosome number, hybridization, and the selection of different types in different habitats. The taxonomist must take this information into account in any future revision of the species. The plant collector also benefits by knowing where the various chromosome aspects of the species are to be found. We must hope that other orchids will be subjected to this type of investigation in future years.

The Analysis of Hybrids

From what has been said so far, it is hoped that reader will appreciate the value of chromosome investigation in the study of the orchids. In conclusion, a few words can be said on the use of chromosomes in the detection of hybridity.

In a hybrid the chromosomes of both parents are brought together, each contributing half of its own total chromosome complement. If the parents differ in their chromosome number, then, of course, it will be possible to detect the hybrids, which will have an intermediate chromosome number. This is the first and simplest use to which the chromosome can be put. However, when the numbers of chromosomes of the parents are the same, a chromosome count will not necessarily detect hybridity. In such cases, it may be essential to examine meiosis to determine whether this division shows irregularity. If meiosis is regular, this is not a certain indication that the plants are not hybrids, for in the orchids we know of several cases of even intergeneric hybridization with regular meiosis and high fertility, but where meiosis is irregular hybridity is presumed. As an example of this use of both chromosome number and pairing, we can take the case of the supposed hybrid *Phaio-cymbidium* Chardwarense.

This plant has been claimed to be hybrid of *Phaius tancurvilliae* and *Cymbidium giganteum*. Others have not been convinced that it is an intergeneric hybrid and have felt that hybridization between two species of *Phaius* is a more likely mode of origin. Chromosome studies of *Phaius* and *Cymbidium* species are incomplete, but at Kew we find *P. tancarvilliae* to have 50 somatic chromosomes, while *C. giganteum* has been counted by others as $2n = 40$. In the supposed hybrid the chromosome number is either 44 or 45, which is more or less what would be expected in the cross of these two species. Its meiosis is highly irregular, resulting in variable chromosome numbers being distributed to the gametes. Both chromosome number and chromosome behavior thus confirm that this plant is a hybrid between chromosomally distinct parents. On the face of it, there is some support for its supposed origin, but it would be far more satisfactory if further information was available on chromosome numbers in other *Phaius* species. We know that species with $2n = 42$ and 46 are present in the genus, in addition to $2n = 50$ for *tancarvilliae*. However, we do not yet have the chromosome number of *P. maculatus*, which some have presumed to have been involved in the parentage of the hybrid. The chromosomes therefore are indicative of a possible situation but cannot be taken in themselves as proving a precise hybrid origin. Indeed chromosome information can be interpreted in a variety of ways. Although I have emphasized their value in both taxonomic breeding and evolutionary studies, their characteristics must be assessed with care and with caution and appraised against the background of other relevant information, whether this be anatomical, physiological, biochemical, or morphological. Finally, we must guard against treating the chromosome as some abstract number, for although our results may be represented by a figure, we must remind ourselves that we are describing bodies which transmit and control patterns of heredity, and their behavior is thus of greatest importance in the control of variation and in the progress of evolution. Indeed some might say that chromosome reproduction is what evolution is all about.

Bibliography

DAKER, M. G. 1970. The chromosomes of orchids, IV. *Bulbophyllum*. *Kew Bull.* 24: 179–185.

DAKER, M. G., and JONES, K. 1970. The chromosomes of orchids, V. STANHOPEINAE. *Kew Bull.* 24:457–459.

DODSON, C. H. 1962. Pollination and variation in the subtribe CATASETINAE. *Ann. Mo. Bot. Gdn.* 59:35–56.

DUNCAN, R. 1959. Orchids and cytology, in C. L. Withner, Ed., *The Orchids*, Ronald Press, New York, Chap. 6.

JONES, K. 1963. The chromosomes of *Dendrobium*. *Orchid Soc. Bull.* 32:634–640.

———. 1966. The chromosomes of orchids, I. *Polystachya*. *Kew Bull.* 20:357–359.

———. 1967. The chromosomes of orchids, II. VANDEAE. *Kew Bull.* 21:31–38.

JONES, K., and DAKER, M. G. 1968. The chromosomes of orchids, III. CATASETINAE. *Kew Bull.* 22:421–427.

MAXWELL, K. 1967a. The *Dendrobium kingianum* Brdw. ex Lindl. Complex I. *Austral. Orchid Rev.* 32:25–30.

———. 1967b. The *Dendrobium kingianum* Brdw. ex Lindl. Complex III. *Austral. Orchid Rev.* 32:139–141.

SUMMERHAYES, V. S. 1965. Notes on African orchids, XXXI. *Kew Bull.* 20:165–199.

9

Some Aspects of Polyploidy in Orchids, With Particular Reference to Cymbidium, Paphiopedilum, and the Cattleya Alliance.

Gustav A. L. Mehlquist

In orchids, as in many other plants, the effect of polyploidy was experienced by breeders long before the existence of polyploidy was conclusively proved. In retrospect it is difficult to ascertain exactly when polyploidy was first encountered, but in at least some of these orchids, the recorded appearance and behavior of certain plants in actual breeding furnish a reasonably accurate basis for establishing this point.

In the genus *Cymbidium* the appearance of *C.* Alexanderi 'Westonbirt' and *C.* Pauwelsii 'Compte d' Hemptinne' in the early 1920s marked the beginning of a new era in hybridizing in this genus. Although some twenty-five years were to pass before these famous parent plants were conclusively proved to be polyploids, their appearance and breeding behavior left little doubt as to their being polyploids (Mehlquist, 1949, 1952). Both of these plants were noticeably slower in their development than their respective sibling seedlings, but they eventually reached greater size and were of much better substance in both flowers and vegetative characteristics. When they were crossed to other *Cymbidium* species and hybrids, both plants showed greater dominance in the offspring than was evident when their siblings and other hybrids of similar parentage were used, and most of the resultant seedlings were rather sterile. As will be evident later in this chapter, all of these features are to be expected when polyploids are bred to diploids.

Before discussing polyploidy any further it might be well to consider a few definitions. The term *polyploidy* refers to the state of an organism having more than two sets of chromosomes (*diploidy*); thus we have *triploidy*, three sets; *tetraploidy*, four; *pentaploidy*, five; *hexaploidy*, six; *octoploidy*, eight; *decaploidy*, ten sets, and so on.

The number of chromosomes an organism transmits to its progeny through the sex cells or gametes, n, is ordinarily one-half of the number contained in the body cells, or the somatic number, of the organism. Thus a diploid would produce gametes with a single set of chromosomes (*haploidy* or *monoploidy*). A tetraploid would produce *diploid* gametes, an octoploid *tetraploid*, and so on. The odd-numbered polyploids such as triploids, pentaploids, and heptaploids usually produce gametes with irregular numbers. These gametes with irregular chromosome numbers, if they survive and function, give rise to irregularly numbered individuals referred to as *aneuploids* in contrast to the *euploid* individuals with numbers in strict multiples of the basic n number usually referred to as x. In diploids the x number of chromosomes is synonymous with the gametic number but since in polyploids the gametic number is usually made up of more than one set of chromosomes, n is used in this discussion to indicate the gametic number, which in most instances equals one-half the somatic number. The basic or lowest n number is referred to as x. Thus in the genus *Cymbidium* where $x = 20$ (Mehlquist, 1949, 1952) the diploids having 40 chromosomes may be described as $2x$ individuals, whereas *C*. Pauwelsii 'Compte d' Hemptinne' being a tetraploid with 80 chromosomes is referred to as a $4x$ plant. The diploids, except in rare circumstances, produce gametes with 20 chromosomes, $n = x$. *C*. Pauwelsii 'Compte d' Hemptinne,' on the other hand, normally produces gametes with 40 chromosomes, that is, here $n = 2x$.

When polyploids began to be extensively studied genetically it soon became apparent that different polyploids often behaved in different ways. It is now customary to distinguish two broad categories of polyploids: *autopolyploids* and *allopolyploids*. Although some polyploids in nature and certainly many produced in culture are in fact intermediate between these categories, an understanding of the behavior expected in each of these groups is a prerequisite to the evaluation and utilization of polyploids in general.

The autopolyploids arise from an increase in chromosome number within the species (intraspecific), whereas an allopolyploid involves two or more species (interspecific). Thus in autopolyploids the same chromosome genome (chromosome set) occurs three or more times, whereas in allopolyploids genomes from two or more species are involved. For instance, *Cymibidum* Pauwelsii 'Compte d'Hemptinne' (*C. lowianum* 'St. Denis' x *C. insigne* 'Bieri') is an allotetraploid having two sets of chromosomes from *C. lowianum* and two sets from *C. insigne.*

Cymbidium Alexanderi 'Westonbirt' too is an allotetraploid but having the parentage (*C. eburneum* × *C. lowianum*) × *C. insigne* it is a trispecific hybrid with two sets of chromosomes from *C. insigne* and two sets composed of a mixture of chromosomes from *C. eburneum* and *C. lowianum.*

An allotetraploid can arise in four different ways:

1. Through hybridization of two diploid species followed by spontaneous doubling of the chromosomes.
2. Through hybridization of two diploid species, each of which contributes unreduced gametes. (Occasionally gametes are formed which contain twice the usual number of chromosomes, that is, they contain the 2n or somatic chromosome number instead of the n or gametic number.)
3. Through hybridization of two autotetraploid forms of the species.
4. Through pollination of an autotetraploid by a diploid or vice versa.

Since autotetraploid forms are very rare in nature and unreduced gametes are normally produced in very low numbers, most allotetraploids probably arise through the first process. For instance, if one unreduced gamete is formed for every 1000 gametes, the probability that two such unreduced gametes would meet to form a zygote (the products of an egg cell and a male gamete) is one in 1 million zygotes. On the other hand, autotriploids would theoretically occur with the same frequency as unreduced gametes are produced, that is, in this instance once in 1000. This difference in probability is probably the main reason autotriploids are relatively frequent in nature, whereas autotetraploids are very rare. Another reason autotetraploids are rare in nature may be their slower growth, which is a disadvantage in competition with normal diploids.

The 'St. Denis' form of *C. lowianum* is known to be a diploid (Mehlquist, 1952) but the 'Bieri' form of *C. insigne* apparently is no longer in culture; hence nothing is known about its chromosome number. Thus the exact mode of origin of *C.* Pauwelsii 'Compte d'Hemptinne' can only be guessed at, but there is no guessing about the importance of this tetraploid hybrid in the development of our modern *Cymbidium* hybrids. The two parents of *Cymbidium* Alexanderi 'Westonbirt' are known to be diploids, *C.* Eburneo-lowianum 'concolor' × *C. insigne* 'Westonbirt,' but whether this tremendously important tetraploid hybrid arose from unreduced gametes or from chromosome doubling of a diploid hybrid remains conjecture. The genetic consequences of the cytological differences between the autotetraploid and the allotraploid are illustrated in Table 9-1.

In gamete formation in an autotetraploid the genes will go to the gametes in two's. If one assumes an autotetraploid to be of the genotype *AAaa*, there are six different ways in which two genes can be drawn from these four, 1*AA*: 4*Aa*: 1*aa*, but only one combination will contain the two recessive genes. Thus one-sixth of the gametes will carry the recessive combination, but the probability that two of these recessive combinations will unite to form a zygote (new individual) is only one in 36. That is, selfing such a hybrid would be expected to yield 35 colored plants for every white, a ratio of 35:1. The corresponding ratio in a diploid would be three colored to one white or 3:1.

However, the number of recessive individuals from selfing such hybrids often greatly exceeds these proportions. In fact, the number of recessives often reaches about one-twentieth of the total. This is explained by the fact that in most cases segregation is not by whole chromosomes but by half-chromosomes or chromatids. Thus in the autotetraploid used as illustration above there would be eight segregating units instead of four thus the genotype might be written *AAAAcccc*. Since only two chromatids will be included in a given gamete, the number of different ways in which two chromatids can be drawn from these eight is much greater than before. Actually there are 28 different ways in which two chromatids can be drawn at random from these eight, but six of them will result in recessive combinations. Thus the gametic ratio will be 22:6 or 11:3 instead of 5:1. The probability that two of these recessive combinations will meet to form a zygote is $(3/14)^2 = 9/196$ or approximately 1 in 22, that is, 21:1.

There is evidence to indicate that the separation of the chromatids in meiosis is not entirely random but that certain restrictive conditions exist which tend to favor the occurrence of recessives resulting in an actual ratio of approximately 19:1.

The 21:1 ratio is based on so-called random chromatid separation (Allard, 1960) whereas the 19:1 ratio is based on so-called equational separation (Burnham, 1962). In practice one does not know which mode of separation of the chromatids is the basis for the observed segregation except by analysis of the segregation ratios, and very large numbers of plants are required to give any reasonably accurate indication of which one actually operates. The equational ratios are somewhat more time consuming to calculate; hence the random chromatid ratio is often used as the basis for comparison. For the purpose of comparison, calculations

TABLE 9-1

Segregation of Recessives in Tetraploids

	Autotetraploids			Allotetraploids	
Genotype	Chromosome Segregation	Random Chromatid Segregation	Equational Chromatid Segregation	Genotype	Amphidiploid Segregation
$AAAA$	0	0	0	$A^aA^aA^bA^b$	0
$AAAa$	0	783:1	575:1	$A^aA^aA^ba^b$	0
$AAaa$	35:1	20.78:1	19.25:1	$A^aA^aa^ba^b$	0
$Aaaa$	3:1	2.48:1	2.40:1	$A^aa^aA^bA^b$	0
$aaaa$	all	all	all	$a^aa^aA^bA^b$	0
				$A^aa^aA^ba^b$	15:1
				$A^aa^aa^ba^b$	3:1
				$a^aa^aa^bA^b$	3:1
				$a^aa^aa^ba^b$	all

based on both modes of segregation are presented in Table 9-1. Actually, the observed segregation is often intermediate between the 21:1 (or 10:1) and the 35:1 ratio.

The reason for this intermediacy is that the mode of segregation a given gene will follow depends on its location on the chromosome, specifically on its distance from the centromere. The centromere is that point on the chromosome which appears to be attached to the spindle and in which separation of the chromatids is initiated.

It is customary to speak of chromatid separation and consequent segregation as if the chromatid was an inviolate unit. This is not so. In the natural course of gametogenesis (gamete formation) the homologous chromosomes, one derived from the egg and one from the male gamete, pair (synapse) in the early stages of meiosis, then divide longitudinally so as to be composed of two chromatids each held together at the centromere. The separation of the two chromatids into separate gametes does not take place until the second miotic division. The chromatids from one chromosome are referred to as sister chromatids. When the chromosomes begin to separate again, it can be seen through the microscope that non-sister chromatids appear to be connected to each other through crosslike connections. These crosslike connections are known as *chiasmata* (sing. *chiasma*) and it is now known that these chiasmata represent points of interchanges of segments between the chromatids. By this mechanism genes are transferred from one chromosome to its homolog and vice versa in the course of meiosis. This interchange, which is known as *genetic crossing over*, is a random phenomenon limited largely by distance. That is, two genes located close together on the chromosome are less likely to effect an interchange than are two genes spaced farther apart. Plants vary greatly as to the number of crossovers that may occur per chromosome but usually the longer the chromosomes, the greater the number of crossovers. It is known, however, that the genes that are located close to the centromere are less likely to be involved in crossovers than those located some distance away; in fact, some may be so close to the centromere that they always stay with the chromosome on which they began synapsis during meiosis, in which case the so-called chromosomal or 35:1 ratio will prevail. On the other hand, if the gene is located so far from the centromere that a crossover always occurs, chromatid segregation will be the result. Thus the same plant may give a 35:1 or 21:1 (or 19:1) ratio of dominants to recessives for different genes, or any combination of these ratios depending on the location of the genes.

These ratios may be further complicated by the fact that autotetraploids often are less fertile than the corresponding diploids because each chromosome has three homologs instead of only one as in diploids. This means that each chromosome in the genome can pair with any one of three other chromosomes, any one of which can satisfy the forces which lead to pairing. Often pairing occurs with one chromosome in one part and with another in another part, so that so-called multivalents of three or four chromosomes are formed. When four chromosomes are thus joined the configuration is known as a quadrivalent. A configuration of three is a trivalent, in which case one chromosome would be left unpaired. If the chromosomes pair only in two's, they are referred to as bivalents.

When the chromosomes separate again, two chromosomes (the chromatids have now become chromosomes) would normally enter each daughter cell, occasionally the separation is three to one daughter cell and one to the other. When separation of a trivalent takes place usually two go to one daughter cell and only one to the other. Furthermore, when a trivalent is formed in a tetraploid plant one chromosome of the four homologs is left without a partner. In such cases the behavior of the unpaired chromosome may vary from cell to cell, from plant to plant, and species to species. It may divide with one half-chromosome (chromatid) going to each daughter cell, the whole chromosome may go to one daughter cell, or it may fail to be included in either daughter cell and thus would be lost altogether.

These variations in distribution of the chromosomes from multivalents often result in a high frequency of gametes with one or two chromosomes more or less than the n number (aneuploidy). Not only are such aneuploid gametes less viable, especially on the male side, but the net result is often a fairly high proportion of aneuploids in the offspring, most of which not only are likely to grow less well than the euploids but are also likely to be less fertile. Thus autopolyploids are almost always less fertile than the corresponding diploids.

In contrast, allopolyploids are almost always more fertile than the corresponding diploid hybrids. Most interspecific hybrids show some degree of sterility even when the species are readily crossed. Often interspecific hybrids are sterile. If this sterility, whether partial or complete, is due to reduced pairing of the chromosomes from the two or more species involved because the chromosomes are so different that they either cannot pair or do so with difficulty, it follows that if chromosome doubling occurs, each chromosome will be provided with a mate which is an exact replica of itself. That is, if the chromosome doubling is effected by a longitudinal split of each chromosome, it will have an exact duplicate to pair with, and this seems nearly always to be the case. Now that each formerly "single" chromosome has acquired a "perfect" mate, pairing takes place readily and fertility is increased. In fact, the greater the sterility of the diploid hybrid, the greater the fertility is likely to be in the doubled hybrid.

If the original diploid hybrid was sterile because of poor pairing of the chromosomes, and chromosome doubling restored the fertility, the resulting allopolyploid will show little or no segregation. In other words, such a hybrid may be almost as true-breeding as either of the parent species. The old saying that a hybrid will betray its parentage by segregation in its progeny thus would not be true. Since orchid hybrids usually show a great deal more fertility than hybrids in other groups, it follows that so-called true-breeding hybrids are less likely to occur in orchids. Generally, fertile interspecific hybrids are so common in orchids that one might wonder if the species in some of the genera in this group are correctly defined.

It might be pointed out that if pairing in a newly formed allotetraploid is between sister chromosomes only, there will be no segregation whatsoever and, because of its double chromosome number, tetraploidy, it might produce nothing but sterile triploids on crossing to either parent species, so we have in fact a new species formed in one step. However, no such true-breeding hybrid has been reported in any orchid, probably because the hybrids at the diploid level generally show various degrees of fertility and consequently will segregate at both diploid and tetraploid levels.

It has already been intimated that chromosome versus chromatid segregations in autopolyploids should be looked on as limiting conditions, rather than as typical ones, since in practice the segregation will more likely be intermediate rather than one or the other. Likewise autotetraploids and allotetraploids should be looked upon as limiting categories since, in orchids at least, the generally high fertility in hybrids at the diploid level is likely to result in polyploids that will show an intermediate behavior (Stebbins, 1950).

The reason for this intermediacy is that for chromatid segregation to occur, two conditions must be fulfilled:

1. Pairing in meiosis must result in quadrivalent formation.
2. The distance between the gene and the centromere must be such that a crossover always takes place between the gene and the centromere.

If the first condition is fulfilled but the second is not, the 35:1 ratio of segregation will prevail, whereas if both conditions are fulfilled a 19:1 or a 21:1 ratio of segregation will result. As indicated earlier, whether the 19:1 or the 21:1 occurs depends on the exact mode of separation, but in most instances this question is of academic interest only.

If, on the other hand, the tetraploid is an allotetraploid which has originated from hybridization of two diploid species a and b whose chromosomes are so dissimilar that little or no pairing takes place between the corresponding chromosomes in the two species, the result is essentially a true-breeding hybrid. In such a hybrid, which is often referred to as an *amphidiploid*, segregation would ordinarily occur only in response to mutations causing heterozygosity, and a mutation in each of the genomes would be required before segregation could occur.

This would mean that the appearance of a recessive trait would be dependent on two genes, that is, the genotype would be $A^a a^a A^b a^b$ resulting in a segregation ratio of 15 dominants to 1 recessive. The a^a and a^b represent the recessive genes which have originated by mutation from $A^a A^b$, respectively.

The exception to this statement is the case where a chromosome from one genome would occasionally pair with a chromosome from the other genome thereby causing heterozygosity that would be chromosomal rather than genetic (Table 9-2). Because of the low compatibility between these nonhomologous chromosomes many gametes being thus exceptionally constituted probably would not be functional. The few that might function, however, could lead to the transfer of genes from one genome to another and result in the segregation of new traits in small numbers. This transfer of genes from one genome to another is of great importance in many breeding projects, especially where only a few genes or a few traits are desired from one species. In practical breeding, transfers of this type are accomplished by repeated backcrossing to the species possessing the greatest number of desirable genes.

It follows from what has been said that backcrossing an allotetraploid to either of the diploid parental species is likely to result in sterile triploids, but by raising as large numbers of seedlings as practical, using the tetraploid hybrid as pollen parent, occasional tetraploid backcross hybrids are likely to occur as a result of the fertilization of an unreduced egg.

If only triploid progeny is produced from such backcrosses, tetraploid hybrids can usually be obtained by pollinating the triploid backcross hybrids with pollen from the original allotetraploid. On the other hand, if it is desired to return to the diploid level, the triploid backcross hybrid should be pollinated by the diploid parent species possessing the greatest number of desired characteristics.

At whatever level the hybridization is carried out, it is, of course, necessary to select in each generation those seedlings that come the closest to one's objectives.

The term amphidiploid was used earlier as a synonym for an allotetraploid composed of only two diploid species. *Cymbidium* Pauwelsii 'Compte d'Hemptinne' fits this specification, whereas *C.* Alexanderi

TABLE 9-2

Types of Heterozygosity in Amphiploids

Heterozygosity through Gene Mutations		
ABCDEFGH	*ABCDEFGH*	No segregations
ABCDEFGH	*ABCDEFGH*	
ABCDEFGH	*ABCDEFGH*	One mutation in one of the *C* chromosomes
ABcDEFCH	*ABCDEFGH*	only, no segregation
ABCDEFGH	*ABCDEFGH*	Same mutation in each of the *C* chromosomes
ABcDEFGH	*ABcDEFGH*	leading to a segregation ratio of 15 dominants to 1 recessive (Table 9-1)
Heterozygosity through Nonhomologous Pairing		
ABCDEFGH	*ABCDEFGH*	Irregular segregation due to reduction in
ABCDEFGH	*ABCDEFGH*	viability of changed chromosomes, leading to reduction in survival of gametes

'Westonbirt' does not. Although it contains two genomes from *C. insigne,* the other chromosomes come from two species *C. eburneum* and *C. lowianum* making it a trispecific allotetraploid but not an amphidiploid.

It should not be assumed that *C. eburneum* and *C. lowianum* have necessarily contributed an equal number of chromosomes in this hybrid, for in gamete formation in the *C.* Eburneo-lowianum parent there might have been produced some gametes with more *eburneum* and fewer *lowianum* chromosomes or the other way around. In fact, the appearance and behavior of *C.* Alexanderi 'Westonbirt' suggest that it has more chromosomes from *C. eburneum* than *C. lowianum.*

Ordinarily interspecific hybrids in which chromosome doubling has not taken place would be expected to show some sterility, but in many orchid hybrids this is not the case. The orchid family seems to be characterized by a large number of cases where the species have undergone considerable morphological differentiation without the corresponding differentiation of the chromosomes. As a result, species that appear to be distinct will often give fully fertile hybrids. For example, the *Cymbidium eburneum, C. grandiflorum, C. insigne, C. lowianum,* and *C. tracyanum,* which represent the main background of our modern *Cymbidium* hybrids, are morphologically distinct yet produce fertile hybrids. On the other hand, where *C. erythrostylum,* which is farther removed taxonomically from the others, was crossed to any one of these five species or their hybrids, the hybrids often showed considerable sterility. Even greater sterility was encountered when the miniature species *C. pumilum* was crossed to the large-flowered species. The first two hybrids involving this species that were registered, *C.* Minuet and *C.* Pumander, were almost completely sterile, and it was not until the introduction of *C.* Sweetheart (*C. pumilum* × *C.* Alexanderi 'Hamilton Smith') that a partially fertile hybrid between *C. pumilum* and the large-flowered hybrids was obtained. From this point hybridization and selection have produced a complex of hybrids linking *C. pumilum* with most, if not all of the large-flowered species at all the chromosome levels usual for this genus.

It is not known exactly where and when polyploidy first occurred in the *Cattleya* alliance. The first published account was by Kamemoto (1950), who found a *C.* Enid to have 84 chromosomes instead of the usual 40. However, from the breeding behavior of notable earlier parent plants it seems reasonable to assume that *Lc.* Lustre (1907), *Lc.* Momus (1916), *C.* Dinah (1919), *Lc.* Areca (1922), *C.* Titrianaei (1923), and *Lc.* Princess Margaret (1930), among others, were tetraploids or at least triploids. Although triploids are notoriously poor breeders, they were often used in the early days of orchid breeding since no cytological information was available and selection of parent plants had to be based entirely on appearance. Discussion with English breeders in 1947 indicated that many of the best looking parent plants often gave relatively few seedlings of variable quality, results which are expected today whenever triploids are used. It is interesting to note that the six English crosses mentioned represented four of England's leading hybridizers, Alexander (Holford Coll.), Charlesworth, Low, and McBean. The first tetraploid *C*. Enids probably came from Harold Patterson and Sons in New Jersey (United States). The name Enid Orchidhaven apparently was applied to the whole group of this parentage. In 1957 Zuck confirmed the tetraploid condition of *C*. Titrianaei.

However, it was the introduction in 1945 of *C*. Bow Bells (Black) and Joyce Hannington (Dane) that really added fuel to polyploid breeding in the *Cattleya* alliance, for they were not only outstanding crosses in their own right but in both crosses various investigators, including this author, found not only both triploids and tetraploids but aneuploids as well, indicating that one of the parents in each cross was tetraploid or near tetraploid. De Tomasi in 1952 reported *C*. Barbara Dane 'Perfection' to be tetraploid, whereas Kamemoto (1952) reported it to be triploid. Both investigators were probably right, for it is likely that different plants were studied. This author has investigated two plants labeled *C*. Barbara Dane 'Perfection'; one was triploid the other tetraploid. The two plants were virtually indistinguishable in growth and flower, a feature that is characteristic of both Bow Bells and Joyce Hannington clones. In fact, when these crosses as a whole are considered they are virtually indistinguishable, raising grave doubts as to the reported parentages. Joyce Hannington behaves as might be expected from its parentage, that is, it is predominantly a fall bloomer, whereas Bow Bells, also a fall bloomer, should according to its reported parentage be a predominantly spring bloomer.

Nomenclatorial confusion aside, through the use of the crosses already mentioned plus many others of lesser influence, polyploid cattleyas and laelio-cattleyas are now produced at will in large numbers. It is perhaps ironic that now that this can be done, interest in the large polyploids has decreased to a great extent in favor of intermediate-sized hybrids obtained by crossing the large-flowered hybrids from the *C*. *labiata* complex with the smaller-flowered species *C*. *claesiana, C*. *intermedia,* and *C*. *loddigesii*. The first hybrid of note in this group was *C*. Henrietta Japhet (*C*. Eucharis × *C*. *loddigesii*). Predominantly a day length controlled fall bloomer, this diploid white-flowered hybrid was easily the most popular in its class until large-flowered fall blooming tetraploid whites were crossed to selected forms of *C. intermedia alba* producing white-flowered hybrids of intermediate size with a tendency to bloom year round. To date all plants of these latter parentages which this author has had opportunity to investigate have proved to be triploids, though it is probably only a matter of time until tetraploids occur in these crosses as well.

One section of the *Cattleya* alliance where a systematic utilization of polyploidy might lead to desired results is in *Sophronitis-Cattleya* breeding. It is now almost a century since the first *Sophro-Cattleya* was registered in 1886, yet the dream of a large-flowered truly red hybrid has not been realized. All the hybrids with truly red flowers tend to approach the *Sophronitis grandiflora* in size. Only in backcrosses to the larger-flowered parents are good-sized flowers obtained, and these tend to approach the colors of the other parents with some reddening from the *Sophronitis* parent. This apparent dominance of smallness from the *Sophronitis* can be explained in terms of intermediacy, which appears to be the usual inheritance pattern in interspecific hybrids. Intermediacy or mean size can be of two types, *arithmetic* or *geometric*. If the parental size differences are considerable, the difference between the two means is likewise considerable. For instance, if a 6-in. *Cattleya* is crossed to a 1-in. *Sophronitis*, the arithmetic mean or average would be $(1 + 6)/2 = 3.5$ in. On the other hand, the geometric mean would be $\sqrt{1 \times 6} = 2.45$ in. Anyone familiar with *Sophronitis* hybrids knows that such hybrids tend to be smaller than hoped for on the basis of ordinary intermediacy and indeed tend to approach the geometric rather than the arithmetic mean.

This apparent dominance of smallness has received little attention from geneticists and plant breeders, probably because there is little actual difference between the arithmetic and the geometric means unless the parents differ markedly in size. The greater the difference in size of the parental features, the greater the difference in magnitude between the means.

Apparently MacArthur and Butler (1938) were the first to offer definite evidence that in the absence of dominance, the geometric is a better measure of inheritance of size than the arithmetic mean. Their data from intercrossing the very small-fruited currant tomato with those of more normal size clearly indicated that the mean size of the fruit in the hybrids came much closer to the geometric than the arithmetic mean. This was true not only for the first-generation hybrids but for backcrosses to either parent as well.

In 1946 this author pointed out that in *Cymbidium* Eburneo-lowianum, a hybrid between *C. eburneum* which rarely has more than one flower per stem, and *C. lowianum*, with 25-35 flowers per stem, the number of flowers approximated the geometric rather than the arithmetic mean. Since that time the same pattern has been observed in numerous crosses between the small-flowered *C. pumilum* and the large-flowered species and hybrids. In this case it is in the size of flowers where the adherence to the geometric mean is clearly visible. There is, of course, no reason to doubt that the number of flowers of the hybrids follows the geometric mean as well, but this is not readily demonstrated since *C. pumilum* and most of the large-flowered parents involved have so nearly the same number of flowers per stem that the geometric and the arithmetic means would be statistically indistinguishable.

The question still remains why, with the large number of advanced *Sophronitis-Cattleya* crosses that have been made, not a single hybrid has been obtained that combines the large size of a large *Cattleya* with the red color of *Sophronitis grandiflora*. The answer is probably one of different gene alignments coupled with strong genetic linkages. That is, the genes for the red color of *Sophronitis* are located on the same chromosome as the genes for the small size and likewise the genes for the mauve color of *Cattleya* are located on the same chromosome as the genes for large size. At the diploid level the chromosomes from one parent must of necessity pair to some extent with the chromosomes of the other or there would be no seed. During the process of pairing, genetic crossing over between the chromosomes of one parent with those of the other takes place. If, however, the genes are not located in the same order in the chromosomes in the parents, many—perhaps most—of the reconstituted chromosomes would be deficient for a number of genes resulting in nonfunctional gametes or zygotes incapable of normal growth. The result is small numbers of seedlings, which indeed is characteristic of *Sophronitis-Cattleya* crosses in general. The exception are those hybrids which are the result of several backcrosses to *Cattleya*. Some of them are very fertile but they do not segregate for red. Possibly the genes for this color were lost during the backcrossing, or their effect may have been subdued through the accumulation of more genes for nonred. It could also be that the only viable chromosomes are those that possess largely the original gene alignments. If allotetraploid hybrids combining *Sophronitis* and *Cattleya* could be produced that were sufficiently fertile to permit the raising of very large numbers of seedlings, even a small amount of nonhomologous pairing might eventually result in a large, truly red *Sophronitis* hybrid.

One method of achieving this large red hybrid would be to induce tetraploidy in a number of suitable hybrids, that is, *Sophronitis grandiflora* crossed with members of the *Cattleya labiata* complex, and then intercross them in as many different combinations as possible. It might take many crosses and large numbers of seedlings but even one seedling satisfying the objectives could be reproduced in large numbers through mericloning.

Another method would be to make use of some already existing allotetraploid combining these genera and cross this allotetraploid to *Sophronitis grandiflora* or some diploid primary *Sophronitis* hybrid in the hope of obtaining a suitable allotetraploid through nonreduction in the diploid. The best allotetraploid now available is probably *Slc.* Anzac. This hybrid used as pollen parent in crosses to the diploid *Sc.* Doris, *Sl.* Jinn, *Sl.* Psyche, or *Slc.* Marathon might through nonreduction in the diploid parent result in a series of allotetraploids, which in turn might through intercrossing produce the desired results. Although the increased chromosome number would reduce the number of recessive types (red color) in any given population, it would also permit the retention and function of reconstituted chromosomes, which at the diploid level would cause too great an imbalance.

In the genus *Paphiopedilum* fertile tetraploids appeared much later than in the previously mentioned groups and it is not certain at what point they did appear. At any rate more species from different sections of the genus were involved in *Paphiopedilum* breeding than in the other groups (for about 75 years) before cytological information became available. This greater number of different species could be in itself the main reason for the greater sterility encountered here. Not only was seed production low but in addition germination was poor and many seedlings were lost in the early stages of growth. When cytological information became available in the late 1940s and early 1950s it became apparent that not only did many of the species used have different chromosome numbers, from 26 to 38, but also triploids and various aneuploids had been used (Duncan and MacLeod, 1948–1950; Mehlquist, 1947). In fact it is very likely that tetraploidy in the *P. insigne* group came through the use of *P. insigne* 'Harefield Hall,' which has been shown to be a triploid (Duncan and MacLeod, 1948–1950; Mehlquist, 1947). At any rate this form of *P. insigne* occurs in the background of many of the modern tetraploids. At first this collected form was thought to be an autotriploid, but this is by no means certain. It could be an advanced natural hybrid with, perhaps, *P. villosum.* This author has seen other triploids of *P. insigne* which looked and behaved more like pure *P. insigne* than the Harefield Hall form does.

Another nontetraploid polyploid that has been of great importance in the *Paphiopedilum* group is F.C. Puddle, F.C.C. variety. Several attempts at determining the exact chromosome number of this outstanding breeder have so far failed, but it now seems reasonably certain that it is a hypo-tetraploid with 47-50 chromosomes. At any rate this hypotetraploid, what-ever its exact chromosome number, has been the most important pro-ducer of white paphiopedilums, producing for the most part vigorous whites which unfortunately tend to be rather more sterile than F. C. Puddle itself. This is probably due to the fact that to get clear whites it is usually crossed to so-called green or yellow paphiopedilums, most of which apparently are diploids or at most triploids. However, at least one white hybrid, a variety of *P. Susan Tucker* (Shalimar × F. C. Puddle) is known to be tetraploid and undoubtedly other white tetraploids will be detected in this and other crosses as time goes on.

As of now tetraploids are scarce in the yellow-green group but abun-dant in the so-called colored group whether spotted or not. This is re-flected in the greater ease with which good seed pods are now being produced in this group and the relatively good seed germination being obtained. However, difficulties are still being encountered where hybrids between species of different chromosome numbers are used in further breeding.

The first cross (1869) raised in this genus, *P. Harrisianum* (*P. bar-batum* $2n = 38$ × *P. villosum* $2n = 26$), is of interest in this connection, for at least four different chromosome numbers are encountered in the hybrids. The majority of seedlings have been shown to be diploids with 32 chromosomes (Duncan and MacLeod, 1948–1950; Mehlquist, 1947), but there are at least three triploids with 45, 45, and 51 chromosomes (Mehlquist, 1947) and one tetraploid with 64 chromosomes.[*] One of the triploids, *P. Harrisianum* 'Superbum' ($2n = 45$), is almost indistinguish-able from the tetraploid *P. Harrisianum* 'G. S. Ball,' but when grown side by side the latter is larger and more fertile than the former and is prob-ably the one that has been used wherever *P. Harrisianum* figures in the parentage of advanced hybrids.

In the so-called Maudiae group (*P. lawrencianum* $2n = 36$ × *P. cal-losum* $2n = 32$) differences in chromosome numbers in the parent species have also considerably slowed the progress in raising advanced hybrids. It is not until recently that such hybrids have been produced in relatively large numbers, probably due to better methods of seed sowing and growing, resulting in the germination and retention of many seedlings which in earlier days would have been lost. New primary hybrids are now being produced in large numbers between many of the recently in-

---

[*] Unpublished data by Mehlquist. Duncan (in Withner, 1959) found it to have 70 chromosomes.

troduced or reintroduced species without much difficulty. Whether second-generation hybrids will be as readily produced or whether polyploidy will appear to render otherwise sterile hybrids fertile remains to be seen.

One of the main objections to cytological studies being done in conjunction with orchid breeding is the difficulty in getting it done in time, except where the breeding is being done near cytological laboratories. Expense is also cited as a barrier. The first objection is true but the second ordinarily is not, for it costs much less to obtain a cytological determination than to grow hybrid populations to the point that they will indicate by their behavior whether the parents were of suitable chromosome constitution. However, the first objection is still valid, and experienced growers have had to resort to a logical, albeit somewhat wasteful, method for getting around the lack of cytological data.

This method which has been particularly useful in the *Paphiopedilum* group is to intercross a number of selected plants in as many combinations as possible and then plant the seed preferably only from those combinations that produced good seed pods. The next step in the selection process usually comes after germination and sometimes not until after the first replanting or transplanting into community pots. Whether the crosses showing poor germination or poor growth are discarded or not, further breeding is concentrated with those parents and combinations that gave good germination and growth. The third step in the selection process comes when the seedlings reach maturity. Then some crosses will be found which despite their good germination and growth are not good enough for further breeding. It goes without saying that the individual plants used in breeding should be properly labeled so that a good cross can be reproduced with exactly the same parent plants if desired, or the same parents used in different combinations.

It is probably just a matter of time before seedlings of paphiopedilums will be produced in as large quantities as cattleyas and cymbidiums.

As of now the greatest challenge seems to be in those fields where hybrids are produced with difficulty and where sterility hampers further work. Cytological investigations and systematic chromosome-doubling of sterile hybrids, in the hope of producing fertile amphiploids, should pay off in these areas.

As mericloning is extended to increasing numbers of different orchids, the very best seedlings obtained with any method can be effectively spread throughout the orchid world at reasonable prices.

When that time comes we shall look for other worlds to conquer.

Bibliography

ALLARD, R. W. 1960. *Principles of Plant Breeding.* John Wiley & Sons, New York.

BURNHAM, C. R. 1962. *Discussions in Cytogenetics.* Burgess Publishing Company, Minneapolis, Minn.

DE TOMASI, J. A. 1954. An Introductory Note. *Kiesewetter Orchid Gardens, 1954 Catalogue.*

DUNCAN, R. E. 1947. The hybrid lady slipper. *Orchid Digest:* Sept.-Oct. 199–207.

DUNCAN, R. E., and MACLEOD, R. A. 1948–1950. Appendix II, in C. Withner, Ed., *The Orchids.* Ronald Press, New York.

FRANCINI, E. 1934. Ibridazione interspecifica nel genera *Paphiopedilum. Nuovo Giornale Bot. Ital.* 41:189–237.

KAMEMOTO, H. 1950. Polyploidy in cattlyas. *Am. Orchid Soc. Bull.* 19:366–373.

———. 1952. Further studies on polyploid cattleyas. *Bull. Pac. Orchid Soc.* 10:141–149.

MACARTHUR, J. W., and BUTLER, L. 1938. Size inheritance and geometric growth processes in the tomato fruit. *Genetics* 23:253–268.

MEHLQUIST, G. A. L. 1946. Ancestors of our present-day Cymbidium. *Bull. Mo. Bot. Gard.* 34:1–26.

———. 1947. Polyploidy in the genus *Paphiopedilum* Pfitz. (*Cypripedium* Hort.) and its practical implications. *Bull. Mo. Bot. Gard.* 35:211–228.

———. 1949. The importance of chromosome numbers in orchid breeding. *Am. Orchid Soc. Bull.* 18:284–293.

———. 1952. Chromosome numbers in the genus *Cymbidium. Cymbidium Soc. News* 7.

NIIMOTO, D. H., and RANDULPH, L. F. 1058. Chromosome inheritance in *Cattleya. Am. Orchid Soc. Bull.* 27:157–162; 240–247.

STEBBINS, G. L., JR. 1950. *Variation and Evolution in Plants.* Columbia University Press, New York.

WITHNER, C. L., Ed. 1959. *The Orchids.* Ronald Press, New York.

ZUCK, T. T. 1957. Pentaploid *Cattleya* hybrids and their successful breeding. *Am. Orchid Soc. Bull.* 26:477–479; 503–504.

W.G.S.

10

List of Chromosome Numbers in Species of the Orchidaceae

RYUSO TANAKA AND HARUYUKI KAMEMOTO

Several years have passed since the appearance of Duncan's comprehensive tabulation of chromosome numbers of orchids in *The Orchids: A Scientific Survey* (ref. 41). During this period chromosome number determinations have nearly doubled. Numbers for over 990 species in about 170 genera representing nearly all subtribes and tribes and over 1000 hybrids, including over 40 intergeneric combinations, have been recorded. This list is limited to only the species, and they are arranged alphabetically regardless of their taxonomic position within the family.

Chromosome numbers of orchid species range from $2n = 10$ to as high as $2n = \pm 200$, with relatively high frequencies at $2n = 28$, $2n = 38-42$, and $2n = 56$. Polyploidy is common, particularly among groups that have been subjected to intensive natural hybridization. Numerous cultivars exhibiting excellent horticultural qualities have been established as polyploids. Hybrids exhibit a considerable degree of aneuploidy and polyploidy.

Genus or species synonyms are indicated in parentheses. Valid species and varietal names are italicized. Cultivars are designated by quotation marks.

Chromosome numbers are listed in two columns. In the first column the haploid or gametophytic numbers listed were generally obtained from analysis of meiosis in pollen mother cells or from pollen mitosis. In the second column the sporophytic or somatic chromosome number is given. The majority of these counts were derived by various workers from meristematic cells in root tips. In a few instances counts were made from shoot apex, young petal, embryo, or protocorm.

411

The authority and year of publication of chromosome counts are presented in the last column as a number matching the references that appear in the bibliography.

Chromosome List

Plant Name	Gamete Chromosomes	Meristem Chromosomes	Reference
Acampe longifolia		38	102
A. ochracea		38	102
A. papillosa		38	102
		38	226
Acanthophippium pictum		48	220
A. striatum	20		141
A. sylhetense	20		141
Aceras anthropophora	21	42	13
		42	80
		42	106
Acineta superba		42, 40	29
Acroanthes monophyllos			
(= *Malaxis monophylla*)	15–17		202
Aerangis biloba		50	185
		50	92
A. citrata		50	92
A. compta		51	92
A. kotschyana		50, *ca.* 50*	92
A. rhodosticta		42	92
A. ugandensis		*ca.* 50	92
A. species		50, *ca.* 50	92
		200	92
Aerides biswasianum	19		141
A. crassifolium		38	102
A. crispum	19		23
A. falcatum		38	102
	19		23
A. fieldingii		38	102
A. flabellatum		38	102
A. hitchongii		40	23
A. houllettianum		38	102
		38	226
A. japonica		38	147
	19		209
		38	248

* *ca.* = about.

Plant Name	Gamete Chromosomes	Meristem Chromosomes	Reference
		38	98
		38	162
		38	194
		38	220
A. *lawrenceana*		40	49
	19	38	186
A. *longicornu*	19		141
A. *maculosa*		38	92
A. *mitratum*		38	102
		38	226
A. *multiflorum*	19		8
A. *odoratum*		38	184
	19	38	186
		38	102
		38	92
	19		8
A. *odoratum* var. *immaculatum*		76	102
A. *suavissima*		*ca.* 38	92
A. *vandarum*		38	23
A. *williamsii*		38	92
Agrostophyllum brevipes	20		141
A. *callosum*	20		141
A. *khasianum*	20		141
A. *myrianthum*	20		141
Amblostoma tridactylum		40	18
Amitostigma gracile		42	220
A. *keiskei*		42	160
		42	220
A. *lepidum*		44	220
Anacamptis pyramidalis	18		13
	18	36	80
		20, 40	60
		36	106
Ancistrorhynchus recurvus		*ca.* 50	92
Angraecopsis breviloba		50	92
Angraecum snocentrum		38	92
A. *arachnites*		38	92
A. *bilobum*	25	50	23
A. *calceolus*		38	92
A. *chevalieri*		*ca.* 38	92
A. *compressicaule*		42–48	23

413

Plant Name	Gamete Chromosomes	Meristem Chromosomes	Reference
A. eburneum		40	49
		38	184
A. eichlerianum		38	23
		38	92
		38	226
A. erectum		63	92
A. giryamae		38	92
A. guillaumini		50	23
A. infundibulare		38	92
A. leonis		40	49
A. multinominatum		42	92
A. sacciferum		ca. 76	92
A. sanderianum	25		23
A. scottianum		38	92
A. sesquipedale		38	41
		42	23
Anoectochilus sikkimensis	15		141
A. tetsuoi		40	220
Ansellia nilotica		42	217
Anthogonium gracile		42	25
	20		141
Arachnis flos-aeris		38	184
Arundina graminfolia		32	170
	20		141
A. sinensis		40	220
A. species	16	32	178
Ascocentrum ampullaceum		38	41
		38	102
		38	226
A. curvifolium		38	184
		38	102
		38	226
A. micranthum		38	102
		38	226
A. miniatum		38	102
		38	170
		ca. 38	92
		38	226
Ascotainia laxiflora		36	147
Aspasia principissa		58	194

Plant Name	Gamete Chromosomes	Meristem Chromosomes	Reference
A. pusilla		56	194
Bifrenaria harrisoniae		40	82
		38	214
Blephariglottis lacera		42	128
Bletia rodriguesii		40	18
B. verecunda		60	220
Bletilla formosana		36	145
	18		146
		36	220
B. hyacinthina	16		5
B. striata		32	145
	16	32	146
		32	160
		32	214, 220
		32	213
B. striata var. albomarginata		32	145
B. striata var. gebina		32	145
	16	32	146
		32	213
Brassavola cucullata		40	18
B. digbyana		40	23
B. (Laelia) grandiflora	20		43
	20		5
B. nodosa		40	95
		40	18
B. nodosa var. gigas		40	104
B. (Laelia) perrinii	20		5
		40	18
Brassia caudata		60	191
		60	193
B. chloroleuca		60	191
		60	193
B. gireoudiana		60	191
		60	193
B. lawrenceana var. longissima		52–56	23
B. longissima		60	191
		60	193
B. maculata		60	191
		60	193
B. verrucosa		60	191
		60	214
		52–58	23

415

Plant Name	Gamete Chromosomes	Meristem Chromosomes	Reference
		60	193
B. verrucosa var. grandiflora		56	35
Broughtonia sanquinea	20		177
Bulbophyllum adenopetalum		40	170, 171
		38	28
B. aeolium		40	170
B. affinii	20		141
B. alagense		40	170
B. antenniferum		40	171
B. apodum		38	28
B. auratum	20		170, 171
B. barbigerum		38	28
B. braccatum	20		170, 171
B. calamarium	19		28
B. canlanoense		40	170, 171
B. careyanum	19+B		141
B. clarkeyanum	19		141
B. cocoinum		38	28
B. congolanum		38	28
B. cornutum		38	23
B. cumingii		40	170, 171
B. cylindraceum	20		141
B. distans		57	28
B. drymoglossum		40	162
B. elatius		38	28
B. emiliorum		40	170, 171
B. eublepharum	19		141
B. evrardii		38–42	23
B. falcatum		38	28
B. frostii		38	28
B. fuscoides		ca. 38	28
B. gamblei	20		141
B. grandiflorum		38	28
B. imbricatum		38	28
B. inconspicuum		38	220
B. intertextum		38	28
B. japonicum		40	162
		40	220
B. lacerata		80±2	170, 171
B. leopardinum	19		141
B. levanae var. giganteum		60	170, 171
B. lobbii		38–42	23
		39	28

416

Plant Name	Gamete Chromosomes	Meristem Chromosomes	Reference
B. lupulinum		38	28
B. makinoanum		38	220
B. micholitzii		38	28
B. minutipetalum		38	28
B. nutans		38	28
B. odoratissimum		38	23
B. phaeopogon		38	28
B. reflexiflorum		38	23
B. saurocephalum	20		81, 82
B. secundum	20		141
B. sociale		38	28
B. tenuicaule	19	38	23
B. veluntinum	19		28
B. virescens		38	28
Calanthe alpina	20		141
C. aristrulifera (kirishimensis)		40	147
		40	89, 90
		40	160
		40	220
C. biloba	40		141
C. brevicornu	20		141
C. chevalieri	20		23
C. discolor		40	147
		40	155
		40	160
		40	90
		40	220
		40	117
C. discolor var. kanashiroi		40	220
C. fauriei		40	220
C. furcata		40	147
		40	89, 90
		40	160
		40	220
		40	222
C. gracilis	20		141
C. herbacea		40+2B	141
C. japonica		40	89, 90
		40	160
		40	220
C. kirishimensis (aristrulifera)		40	147
C. liukiuensis		40	220
C. longicalcarata		40	220
C. masuca	20		141

417

Plant Name	Gamete Chromosomes	Meristem Chromosomes	Reference
C. nipponica		40	160
		40	90
		38	220
C. plantaginea		40	8
C. reflexa		40	147
		42	89, 90
		42	160
		42	158
		40	220
C. rubens		44	220
C. schlechteri		44	160
		44	90
		44	158
		40	220
C. striata		40	89, 90
		40	160
		40	220
C. striata var. *sieboldii*		40	147
		40	89
		40	160
		40	220
C. sylvataica	20		23
C. torifera (*tricarinata*)		40	147
		40	220
C. tricarinata		40, 60	89
		40, 60	160
		40	90
		40, 60	158
	20		139
	20		140
C. trulliformis	20		141
C. venusta		40	89, 90
		40	160
		40	220
C. veratrifolia		40	170, 171
C. vestita	20		82
Calopogon pulchellus	*ca.* 13	*ca.* 26	168
Calypso bulbosa	14	28	68
		32	84, 85
Camarotis apiculata		38	102
C. manii	19		141
C. obtusa	19		141

Chromosome List—continued

Plant Name	Gamete Chromosomes	Meristem Chromosomes	Reference
Catasetum atratum		56	18
		ca. 108	93
C. callosum		54	93
C. cassidum		54	93
C. cernuum		56	18
		54	93
C. deltoideum		*ca.* 54	93
C. discolor		108	93
C. fimbriatum		108	93
C. fimbriatum var. *inconstans*		108	93
C. fimbriatum var. *morrenianum*		*ca.* 108	93
C. hookeri		56	18
C. integerrimum		54	93
C. luridum		*ca.* 54	93
C. macrocarpum		56	18
		54	93
C. pileatum		*ca.* 108	93
C. planiceps		*ca.* 162	93
C. planiceps (peloric form)		*ca.* 108	93
C. russellianum		54	93
C. thylaciochilum		54	93
C. trulla		54	93
C. viridiflavum		54	93
C. warscewiczii		54	93
Cattleya aurantiaca		40	95
C. bicolor		40	95
		40, 80	18
		40	19
C. bicolor var. *measuresiana*			
(= *C. measuresiana*)		80	19
C. bowringiana	20, 21	41	95
		40	49
		40–42	23
C. bowringiana 'Katayama'		40	222
C. bowringiana 'Labor Day'		40	177
C. bowringiana 'Splendens'		61±1	104
C. citrina		40	95
		40	23
C. dormaniana		40	18
C. dowiana	20	40	95
C. dowiana var. *aurea*		40	175
C. elongata 'No.1'		80	104
C.forbesii		54–60	23
C. gaskelliana	20	40	95

419

Plant Name	Gamete Chromosomes	Meristem Chromosomes	Reference
C. gigas (*warscewiczii*)	20	40	95
		40	164
		40	23
C. guttata		40	95
		40	18
C. harrisoniana	20	40	95
C. intermedia		40	95
C. intermedia var. *alba*		41+1 f.*	175
C. intermedia 'Aquinii'		40	175
C. intermedia 'Graham'		40	95
		40	18
		40	236
C. labiata	20	40, 42	233
		40	95
		40	49
		40	236
C. labiata 'Alba'		Tetraploid	233
C. labiata 'Amesiana'	20, 21	40, 41	95
C. labiata 'Harefield Hall'		Tetraploid	233
		40	177
C. labiata 'Westonbirt'		40	104
C. leopoldii		40	18
C. lueddemanniana	20	40	95
C. measuresiana (= *C. bicolor* var. *measuresiana*)		80	19
C. mossiae	20	40	99
C. mossiae 'Mrs. Butterworth'		ca. 60	95
C. mossiae var. *reineckiana* 'Young's'		41	164
C. mossiae 'Verna'		40	95
C. mossiae var. *wageneri*		41	164
C. percivaliana	20	40	95
		40	99
C. rex		40	95
C. skinneri	20	40	95
		40	99
		40	23
C. trianaei		40	95
		40	49
	20		82
C. trianaei 'Alba'		40	177
C. trianaei 'A. C. Burrage'		ca. 60	96
		60±1	104

* f. = fragment.

Plant Name	Gamete Chromosomes	Meristem Chromosomes	Reference
C. trianaei 'Broomhills'		40	217
C. trianaei 'Grand Monarch'		40	98
C. trianaei 'Joan'		40	95
C. trianaei 'Jungle Queen'		60	175
C. trianaei 'Llewellyn'		83	95
C. trianaei 'Mary Fennell'		60	175
C. trianaei 'Mooreana'		*ca.* 60	98
C. trianaei 'Naranja'		59	175
C. velutina		40	18
C. walkeriana		40	18
C. warneri		40	95
		40	99
C. warscewiczii (*C. gigas*)	20	40	95
		40	164
		40	23
C. warscewiczii 'Firmin Lambeau'		40	96, 104
C. warscewiczii 'Frau Melanie Beyrodt'		Diploid	233
		40	164
Caularthron bicornutum	20	40	1
Cephalanthera damasonium		32	71
		36	13
		36	125
		36	241
C. ensifolia		34	139
		34	140
		32	141
C. erecta		34	144
C. falcata		34	143, 144
		34	161
C. grandiflora	18		13
C. longifolia	16	32	71
	16		5
C. rubra		36	241
C. schizuoi		32	143, 144
Chamaeangis odoratissima		50	92
C. vesicata		95–100	92
Chaemorchis alpina		42	80
Chilochista luniferus		38	102
C. usnoides		38	226
Cirrhopetalum acuminatum		38	28
C. andersonii		38	28

Plant Name	Gamete Chromosomes	Meristem Chromosomes	Reference
C. caespitosum	20		141
C. caudatum	19		141
C. cuminghii		38	23
		38	28
C. gracillimum		*ca.* 38	28
C. lasiochilum		38	28
C. lepidum		38	28
C. longiflorum		38–40	23
C. maculosum		38	25
C. makoyanum		38–40	23
C. mastersianum		38–40	23
		38	28
C. mundulum		38	28
C. mysorense		*ca.* 38	28
C. ornatissimum		38–40	23
		ca. 38	28
C. parvulum	19		141
C. picturatum		48	49
C. pulchrum		38	28
C. robustum		38	28
C. stramineum		38–40	23
C. thouarsii		38	23
C. umbellatum		38	28
C. vaginatum		38	28
C. viridiflorum	19		141
C. wallichii	19		141
Cleistes divaricata		18	10
Cleistostoma brevipes	20		141
C. gemmatum	20		141
C. micranthum	19		141
Coeloglossum viride (= *Platanthera bracteatum*)		40	122, 124
	20	40, 80, 85	5
		40	197, 198
		40	80
	20	40	173
		40	162
		40	125
		40	62
		40	106
		ca. 40–42	196
		40	108
		40	112

Plant Name	Gamete Chromosomes	Meristem Chromosomes	Reference
C. viride ssp. bracteatum		40	128
C. viride ssp. islandicum		40	122, 124
Coelogyne barbata	20		23
	20		141
C. candoonensis		40	170, 171
C. chloroptera		40	170, 171
C. corymbosa	20		141
C. cristata		40	217
	20		141
C. eberhardtii		40–44	23
C. elata	20		141
C. elmeri		40	170, 171
C. fimbriata	20		81, 82
C. flaccida	20		141
C. flavida	20		141
C. flexuosa	20		81, 82
C. fuliginosa	20		81, 82
C. fuscescens	20		141
C. hettneriana	20		23
C. longipes	20		141
C. merrittii		40	170, 171
C. nigrofurfuracea	20		23
C. occultata	20		141
C. ochracea	20		141
C. ovalis	20		141
C. uniflora	20		141
Comparettia falcata		42	191
C. speciosa		42	191
Corallorrhiza innata	21		68
		42	147
C. maculata	42		128
C. maculata ssp. mertensiana	20		228
C. striata		42	128
C. trifida		42	123
		42	68
		42	147
		42	124
	21		94
		ca. 42	200
		42	112
C. trifida var. verna		42	124
Coryanthes maculata		40	29
Corymborchis confusa		40	170, 171

Plant Name	Gamete Chromosomes	Meristem Chromosomes	Reference
Cremastra unguiculata		48	161
		50	220
C. variabilis	24	48	165
		48	160
		46	194
		48	220
C. wallichiana	26		141
Cryptochilus lutea	19		141
C. sanguinea	19		141
Cryptopus elatus		95	92
Cycnoches chlorochilon		68	93
C. egertonianum		ca. 68	93
C. loddigesii		64	93
C. ventricosum		68	93
Cymbidiella rhodochila		54	243
Cymbidium aloifolium		40	138
	16	32	178
	20		141
		40	170, 171
C. bicolor	20		210
C. cochleare		40	243
C. dayanum		40	162
		40	220
C. devonianum		40	243
	20		141
C. eburneum		40	243
		40	138
C. ensifolium		40	243
		40	214, 220
C. erythrostylum		40	243
		40	137
		40	138
C. faberi		40	127
C. finlaysonianum		40	138
		40	170, 171
C. forrestii		40	243
C. giganteum		40	243
		40	138
C. grandiflorum (*hookerianum* var.)		40	243
		40	138
C. grandiflorum 'Westonbirt'		40	138

424

Plant Name	Gamete Chromosomes	Meristem Chromosomes	Reference
C. gyokuchin		40	98
C. hoosai		40	98
C. i'ansoni		40	243
		40	138
C. insigne		40	243
		40	138
C. insigne var. *albens*		40	138
C. insigne 'Album'		40	138
C. insigne f. *album*	20		247
C. insigne var. *atrosanguinea*		40	214, 217
C. insigne 'Rodochilum'		40	138
C. insigne 'Westonbirt'		40	138
C. iridifolium		40	243
C. kanran		40	143
		40	220
C. lancifolium		40	243
	20		23
C. lowianum		40	243
		40	138
		Diploid	137
	20		81, 82
	9–10		206
		40	98
C. lowianum 'Concolor'		40	138
C. lowianum 'Fir Grange'		40	138
C. lowianum 'McBeans'		40	138
C. lowianum 'Pitt's'		40	138
C. lowianum 'St. Denis'		40	138
C. mastersii		40	243
C. nagifolium		40	161
		38	220
C. parishii 'Sanderae'		40	138
C. pendulum	20		141
C. pumilum		40	243
		40	138
		40	98
		40	214, 217
C. pumilum 'Folia Albo marginalis'		40	138
C. pumilum 'Gessho'		40	214, 217
C. pumilum 'Gesshohen'		80	62
C. schroederi		40	243
		40	138
C. sikkhimense	19		23
C. simonsianum		40	243

Plant Name	Gamete Chromosomes	Meristem Chromosomes	Reference
C. sinense		40	243
	20		209
		40	217, 220
C. tracyanum		40	243
		40	138
		Diploid	137
C. virescens		40	160
		40	98
		40	158
		40	220
C. virescens forma		46	162
C. whiteae		40	243
Cynorkis anacamptoides		14	229
Cyperorchis cochleare	20		141
C. elegans	20		141
C. eubernea	20		141
C. grandiflora	20		141
C. longifolia	20		141
C. mastersii	20		141
Cypripedium acaule	10	20	84, 85
	10	20	14, 16
		20	128
C. arietinum		20	128
C. calceolus		20	84, 85
		22	55
		20	120
		20	125
C. candidum	10	20	84, 85
		20	128
C. cordigerum		20	139
		20	140
	10		141
C. debile	10	20	150
		20	160
	10	20	246
C. fasciculatum		20	128
C. hirsutum	10	20	86
C. japonicum		20	160
		20	220
C. jatabeanum		20	196
C. marcanthum var. rebunense		20	160

426

Plant Name	Gamete Chromosomes	Meristem Chromosomes	Reference
C. parviflorum	10	20	22
	11		167
C. passerinum		20	126
		20	128
C. pubescens		20	84, 85
	10		16
	11	22	167
		20	128
C. reginae (spectabile)	10		41
		20	86
		20	125
		20	128
C. spectabile	11		81
		22	82
	11	22	167
	10		41
Cyrtorchis arcuata ssp. variabilis		ca. 150	92
C. chailluana	22–23		23
C. species		46, 50	92
Dactylorchis cordigera (see also Orchis)		80	238
D. cruenta		40	238
		40	122
		40	76
		40	80
D. foliosa		40	237, 238
D. fuchsii		40	119
(= Orchis maculata var. meyeri)		40	77
		20, 40	238
		20, 40	69
	20		67
		40	13
		40	124
		40	125
		40	7
	20	40	106
		40	108
D. fuchsii ssp. hebridensis		40	75
D. fuchsii ssp. okellyi		40	125
D. fuchsii ssp. psychrophila		80	241
D. fuchsii ssp. rheumensis		40	77
D. fuchsii ssp. typica		40	238
D. hybrid	12	36	61

Plant Name	Gamete Chromosomes	Meristem Chromosomes	Reference
D. (*Orchis*) *incarnata*	20	40	238
(= *Orchis latifolia*)		40	75
(= *Orchis elodes*)		79, 80	133
		40	67
		40	80
	10	20	60
		40	197
		40	83
	20		200
	20	40	106
D. (*Orchis*) *maculata*		40	122
	20		13
		40, 42, 80	67, 69
		40, 80	80
	10	20	60
	16		205
		80	238
		80	31, 32
		80	52
		80	77
		80	83
		ca. 80	199
		80	200
		80	108
D. (*Orchis*) *maculata* ssp. *elodes*		80	119
		80	238
		80	77
	40		67, 69
		80	52
		80	124
	40	80, 100, 120	106
D. (*Orchis*) *maculata* ssp. *ericetorum*		80	75, 77
(= *Orchis maculata* var. *meyeri*)		80	238
(= *Orchis maculata* var. *ericetorum*)	20		67
		80 (40)	69
D. (*Orchis*) *maculata* var. *genuina*		40	67
D. *maculata* ssp. *islandica*		80	119
		80	124
D. *maculata* var. *o'kelly*		80	133
D. *maculata* ssp. *typica*		80	238
D. (*Orchis*) *majalis* (*latifolia*)		80	237, 238
		80	67
		80	80
		80	49
		80	125
	40	80	106

Plant Name	Gamete Chromosomes	Meristem Chromosomes	Reference
D. majalis ssp. *occidentalis*		80	237, 238
		80	125
D. (*Orchis*) *munbyana*		80	237, 238
D. (*Orchis*) *praetermissa*		80	237, 238
		80, 82	133
		80	125
	40	80	106
D. praetermissa ssp. *junialis*		80	237
	40	80	106
D. (*Orchis*) *praetermissa* var. pulchella		80, 82	133
D. pseudotraunsteineri	6, 8, 9, 10,	16, 18, 20,	
(= *Orchis traunsteineri*)	11, 12, 15	25, 28	61
		80	238
		80	75
		80	133
		80	133
D. russovii		120, 122	237, 238
D. saccifera		80	238
D. (*Orchis*) *sambucina*		40	238, 239
	21		67
	20	40	80
		42	200
D. sesquipedale		80	237, 238
D. (*Orchis*) *traunsteineri*		40, 80	238
		80	80
	9, 10, 11, 12,	16, 17, 18,	
	15, 16, 22	20, 24	61
		80	83
		80	198
D. (*Orchis*) *traunsteineri* var. *gigas*	20		60, 61
D. traunsteineri ssp. *russowii*		122	237
D. traunsteineri ssp. *typica*		80	239
Dactylorhiza aristata		42	128
Dactylostalix ringens		42	189
Dendrobium acuminatissimum		40	170, 171
D. acuminatum (= *D. lyonii*)		40	110
D. aggregatum	19	38	235
	19	38	110
		32–35	23
		38	100
		38	252
D. aggregatum var. *majus*		38	109
D. albayense		40	170, 171

Plant Name	Gamete Chromosomes	Meristem Chromosomes	Reference
D. alpestre	20		8
D. amoenum		38	91
	20		141
D. ampulum	20		141
D. anceps	19+(0-2B)		141
D. anosmum (= D. superbum)		40	49
		40	88
	19		109
	19		235
	19		110
		38	252
D. aphyllum		38	91
D. aporoides		38±2	170, 171
D. aqueum		38	91
D. arachnites		38	170, 171
D. atroviolaceum		38	252
D. bicallosum		40	170, 171
D. bicameratum		38	91
		40	8
D. bigibbum		38	91
		38	252
D. bigibbum var. compactum		38, ca. 57	91
D. bronckartii		40	88
		40	49
D. brymerianum		40	88
		38	91
D. bullenianum		38	252
D. canaliculatum		2x	91
		38	252
D. candidum		38	91
D. capillipes		38	91
D. capituliflorum		38	91
D. cariniferum		38	110
		38	100
D. cathcartii	19		141
D. chameleon		38	170, 171
D. chrysanthum	19	38	235
		38	110
		38	91
		38	100
D. chrysocrepis		ca. 76	91
D. chrysotoxum	20	40	88
	20 + 5 f.	40	81, 82
		38	110
		40	23

Plant Name	Gamete Chromosomes	Meristem Chromosomes	Reference
		38	91
		38	217
		38	100
		38	252
D. chrysotoxum var. suavissimum		38	91
D. chrysotropis		38	91
D. clavatum	19		141
D. cobbianum		38	170, 171
D. crassinode		38	100
D. crepidatum		38	91
		38	100
D. cretaceum		38 + 1 f.	91
D. cruentum		40	100
D. crumenatum		38 + 1 f.	91
		40	170, 171
		38	100
		38	252
D. crystallinum		38	91
		38	100
D. curranii		38	170, 171
D. d'albertsii	19	38	110
		38	252
D. decuphum		08	91
D. delacourii (ciliatum)		38	100
		38	252
D. delicatum		ca. 57	91
D. densiflorum		40 + 2 f.	109
	20+(1-2B)		141
D. denudans		40	91
D. devonianum	19		23
D. distichum		57	235
		38	170, 171
		38	252
D. dixanthum		41	91
		40	100
		40	252
D. draconis		38	187
		38	100
		38	252
D. falconeri		2x	91
D. farmeri		40	100
D. farmeri var. aureoflava		40	100
D. fimbriatum		38	88
		38	91
		38	100

431

Plant Name	Gamete Chromosomes	Meristem Chromosomes	Reference
D. fimbriatum var. oculatum		38	88
		38	110
D. findlayanum		38	91
		38	100
D. flammula		38	91
D. formosum		38	88
		38	187
D. formosum var. giganteum		38	110
		38	100
		38	252
D. friedericksianum		38	91
		38	23
		38	100
D. fusiforme		38	109
D. gamblei		2x	91
D. gibsonii		38	235
D. gordonii		38	110
D. gouldii		38	109
		38	110
		38	252
D. gracilicaule		38	91
D. gracilicaule var. howeanum		38	91
D. graminifolium		38	170, 171
D. grantii		38	109
		38	110
		38	252
D. gratiosissimum		38	91
D. hendersonii		38	91
D. heterocarpum (aureum)		38	109
		38	110
		38	91
		38	170, 171
		38	100
		38	252
D. hildebrandii		38, 38 + 1 f.	109
		38, 38 + 1 f.	110
		38	91
		38	100
D. hookerianum		40	91
D. infundibulum	20		81
		40	82
		38	217
		38	100
D. infundibulum var. jamesianum		38	91

432

Plant Name	Gamete Chromosomes	Meristem Chromosomes	Reference
D. jamesianum		38	217
J. jenkinsii		38	91
D. johannis	19		110
D. kingianum		76	235
		38, 112–114	91
		76	217
D. kingianum var. album		ca. 57, 76	91
D. kingianum var. silcockii		ca. 76	91
D. kwashotense		38	220
D. leonis		40	252
D. leucorhodum		38	91
D. lindleyi		38	91
D. linguella (hercoglossum)		38	100
		38	252
D. linguiforme		38	91
D. lituiflorum		38	91
		38	100
D. loddigesii		40	88
		38	91
		38	23
D. longicornu		38	91
D. longispicatum		00	170, 171
D. macraei	19		141
D. macrophyllum		38	109
		38	110
D. macrostachyum		38	91
D. microchilum		38	170, 171
D. mirabelianum		38	252
D. monile (moniliforme)		38	147
		38	88
		38	160
		38	110
		38, 38 + 1–3 f.	91
		38	220
		38	252
D. moniliforme 'Ginryu'		38	214
D. moniliforme 'Pink Flower'		48	162
D. moschatum		40	91
		38	23
		38, 39	100
		38	252
D. moschatum var. cupreum		38 + 3 f.	109
		38, 38 + 3 f.	110
D. mutabile		2x	91

Plant Name	Gamete Chromosomes	Meristem Chromosomes	Reference
D. nobile	19	38	88
		40	49
	19	38	147
		ca. 20	81, 82
	19	38	235
	19		23
		38, 57	91
		38	100
D. nobile 'Cooksonianum'	19	38	88
		38	91
D. nobile 'King George'	38	76	88
		76	214
D. nobile 'Nobilius'		*ca.* 57	91
D. nobile 'Sanderianum'		38–40	23
D. nobile 'Virginale'		57	91
D. nobile 'Wallichianum'		38	91
D. ochreatum		2x	91
D. ovatum		40	91
D. palpebrae		40	91
D. paravalum		38	170, 171
D. parcoides	20		23
D. parishii		40	88
		38	91
		38	100
D. pendulum		2x	91
D. phalaenopsis		38	109
	19	38	110
		38	252
D. phalaenopsis 'Extra'		76	104
D. phalaenopsis 'Giganteum'		*ca.* 76	109
		ca. 76	110
		76±1	104
D. phalaenopsis 'Hololeucum'		38	235
D. phalaenopsis 'Lyons Light'	19,38	38	110
		38	104
	Variable	38	37
D. phalaenopsis 'Ruby'		*ca.* 76	109
		ca. 76	110
		76	104
D. phalaenopsis 'Schroederianum'	19	38	235
D. phalaenopsis 'Shibata'		38	104
D. philippinensis		38	110
D. pierardii	19		109
		38	235
	19		110
	19–20		23
		38	100

Plant Name	Gamete Chromosomes	Meristem Chromosomes	Reference
D. pitcherianum		2x	91
D. playcaulon		38	170, 171
D. porphyrophyllum	19		23
D. primulinum		38	88
		38	91
		38	100
	19		141
D. prostratum		2x	91
D. pulchellum (dalhousianum)		38	110
		2x	91
		40	100
D. ramosum		40	91
D. regium		38	91
D. revolutum		40	100
D. rotundatum	20		141
D. sanderae	20	40	187
D. scabrilingue		38	100
D. schuetzei	20	40	187
D. schulleri		38	235
D. secundum		40	91
	20		23
		40	100
D. senile		38	100
		38	252
D. signatum		2x	91
D. smilliae		38	91
D. sophronites		ca. 80	91
D. speciosum		38	18
D. speciosum var. fusiforme		38	110
D. speciosum var. hillii		38	91
D. spectabile		38	109
		38	110
		38	252
D. spurium		40	170, 171
D. stratiotes		38	109
		38	235
		38	110
		38	91
D. stratiotes var. giganteum		38	252
D. strebloceras		38	91
		38	252
D. superbiens	19	38	235
		38	91
D. superbiens 'Daeng Yai'		38	235

Plant Name	Gamete Chromosomes	Meristem Chromosomes	Reference
D. superbum (anosmum)	19		109
		40	88
		40	49
	19		235
	19		110
D. sutepense		2x	91
		38	252
D. taurinum		38	109
		38	110
D. teretifolium		2x	91
D. thyrsiflorum	20		81, 82
		40	235
		40	110
		40	100
D. toftii		38	235
		38	110
D. tokai		38	235
	19		110
		38	91
D. topaziacum		38	170, 171
D. tortile		38	110
		38	91
		38	100
		38	252
D. tosaense		40	161
		38	220
D. transparens		38	91
	20		141
D. trigonopus		38	252
D. undulatum	19	38	110
		38	91
		38	252
D. undulatum var. broomfieldii	19		109
D. ventricosum		38	170, 171
D. veratrifolium		38	109
		38	235
		38	110
		38	91
D. victoriae-reginae		38	91
		38	252
D. wardianum		2x	91
D. wardianum var. album		ca. 57	91
D. wardianum var. gigantum		40	81, 82
Dendrochilum cobbianum (see also Platyclinis)		40	170, 171

Plant Name	Gamete Chromosomes	Meristem Chromosomes	Reference
D. filiforme		40	170, 171
D. glumaceum	20		81, 82
D. pumilum		40	170, 171
D. tenellum		40	170, 171
Dendrophylax funalis		42	185
Diacrium bilamelatum		40	18
Diaphananthe cuneata		*ca.* 50	92
D. densiflora		50	92
D. plehniana		50	92
D. rutila		100	92
D. species		*ca.* 50	92
Dichaea muricata var. *neglecta*		52	41
Dilomilis montana	21		163
Dimerandra stenopetala		40	18
Diploprora championi		38	102
Doritis buyssoniana			
(= *Phalaenopsis buyssoniana*)		76	244
D. pulcherrima			
(= *Phalaenopsis esmeralda*)		38	244
	19	38	177
		38	188
	19		23
		38	102
D. pulcherrima var. *buyssoniana*			
(= *D. buyssoniana*)		76	102
D. wightii	19		141
Eleorchis conformis		40	145
		40	160
E. japonica		40	145
		40	160
		40	220
Encyclia odoratissima		40	18
(see also *Epidendrum*)			
E. odoratissima var. *serroniana*		40	18
Ephippianthus schmidtii		40	162
		42	189
		36	220
Epidendrum atropurpureum	20	40	95
		80–90	23

Plant Name	Gamete Chromosomes	Meristem Chromosomes	Reference
E. brassavolae	20	40	95
E. campylostalix	20	40	95
		40	99
E. ciliare	20	40	95
		40	49
	20	40 (80, 160)	64
		40	18
E. cochleatum	20	40	95
E. conopseum	20	40	95
		40	23
E. denticulatum		40	18
E. difforme		39–40	23
E. diffusum	20	40	95
E. ellipticum		56	18
E. elongatum		56	18
E. floribundum		40	18
E. lindenii		56	18
E. linkianum	20	40	95
	ca. 20		82
E. loefgrenii		40	18
E. longispathum		40	18
E. mariae		40	217
E. mosenii		24	18
E. munroeanum		40	49
E. nocturnum		ca. 80	95
	20		81, 82
		40, 80	18
		74–85	23
E. nocturnum var. guadeloupense		42–48	23
E. ochraceum	20	40	99
	20	40	95
E. patens		40	23
E. prismatocarpum		40	18
E. propinquum		40	95
E. purpureum		56	18
E. radicans		40, 70	95
		48–57	23
	19		141
E. ramiferum	20		81, 82
		40	18
E. rigidum		40	18
E. tampense	20	40	95
E. xanthinum		ca. 80	99
Epipactis atropurpurea	20		71

Plant Name	Gamete Chromosomes	Meristem Chromosomes	Reference
E. atrorubens		40	122
		40	68, 71
		40	125
		40	106
		40	108
E. confusa		40	70, 71
E. falcata		24	208
E. helleborine		40, 20?	242
		40	70
		38, 40	71
		38, 40	125
		38	13
		38	62
	19	38	106
		38	127
	19	38	140
E. latifolia	19, 20		71
	20	40(20)	70
		38	13
		40	195
		36, 44	118
		40	141
E. leptochila		36	71
E. microphylla		40	71
E. palustris	20	40	122
	20	40	68
	12		59
	12		157
		38	71
		40	62
	20	40	106
		40, 44, 46, 48	118
E. papillosa	20		132
		38–40	196
E. persica	20	40	71
E. phyllanthes		36	71
E. sayekiana		40	143, 144
		40	189
E. thunbergii	20	40	144
		40	143
		40	158
		40	161
		40	220
E. veratrifolia		40	8

Plant Name	Gamete Chromosomes	Meristem Chromosomes	Reference
Epipogium aphyllum	34		54
Eria arisanensis		40	220
E. brachystachya		44	170, 171
E. clemensiae		44	170, 171
E. convallarioides	19		23
	20		141
E. corneri		36	220
E. coronaria	18		141
E. cymbiformis		42 ± 2	170, 171
E. excavata	20		141
E. floribunda		44	170, 171
E. giungii	20		23
E. graminifolia	19		141
E. lagunensis		44	170, 171
E. luchuensis		36	220
E. nudicaulis		40	220
E. ovata		44	170, 171
E. paniculata	19		23
E. pannea	18		141
E. philippinensis		44	170, 171
E. reptans		44	160
		38	220
E. ringens		44	170, 171
E. woodiana		44	170, 171
E. yakushimensis		36	220
Eriopsis biloba		40	29
E. rutidobulbon		40	29
Erycina diaphana		52	191
		56	192
Eulophia geniculata	19	38	24
E. gusukumai		56	220
E hormusfii	27		141
E. macrostachya		32	170, 171
E. nuda		54	24
E. squalida		32	170, 171
E. species	16	32	178
E. stricta		32	170, 171
Eurychone rothschildiana		50	92
Galeola falconeri	15		141
Galeorchis (Orchis) spectabilis		42	84, 85
Gastrochilus calceolaris		38	102
		38	92
G. dasypogon		38	226

Plant Name	Gamete Chromosomes	Meristem Chromosomes	Reference
G. japonicus		40	162
		38	220
G. (Saccolabium) matsuran		34	161
Gastrodia elata	8–9	16, 18	111
	18	36	Unpublished
G. galeata	20		82
Geodorum nutans		36	170, 171
Gomesa recurva		56	191
		56	193
Gongora galeata	20		82
G. quinquenervis		38, 40, 42	29
G. truncata		ca. 38	29
Goodyera boninensis		28	220
G. hachijoensis		28	145
		28	220
G. hachijoensis var. leuconeura		28	220
G. hachijoensis var. yakushimensis		38	220
G. macrantha		30	145
		30	160
		30	220
G. matsumarana		28	145
		28	220
G. maximowicziana		42	145
		56	158
		56	161
		28, 56	219
G. oblongifolia	15		228
		22	128
G. ogatai		22	220
G. ophioides		30	122
G. pendula		28	162
		30	220
G. procera	11		5
		42	145
		42	220
G. repens		30	122
		28, 32	141
	15	30	173
		30	161
		30	120
		30	62
	15	30	106
		32	220
	15		140

Plant Name	Gamete Chromosomes	Meristem Chromosomes	Reference
G. repens ssp. ophioides		30	128
G. schlechtendaliana		30	160
		30	189
		30	220
G. velutina		28	145
		28	158
		28	162
		28	220
G. yaeyamae		44	220
Grammangis ellisii		54	23
Gramotophyllum scriptum		40	243
		40	170, 171
G. speciosum		40	243
		40	170, 171
Grobya galeata		56	18
Gymnadenia camtschatica		40	160
		40	158
G. chidori		42	147
G. clavellata		40	128
G. conopsea	18, 19, 20		5
	20		13
	20	40	53
	60	40, 80	197
		40, 80	80
	20	40	173
	8	16	27
	10	20	60
	16		205
		40	31, 32
		40	160
		38	194
		40	125
		40	62
	20	40	106
		80	241
		40	108
		40	249
G. conopsea ssp. conopsea		40	118
G. conopsea ssp. serotina		40	80
G. cuculata		42	220
G. odoratissima	10	20	80
	20	40	60
		40	125

Plant Name	Gamete Chromosomes	Meristem Chromosomes	Reference
Gymnigritella suaveolens	18		232
Gyrostachys (*Spiranthes*) *cernua*	30		169
G. (*Spiranthes*) *gracilis*	15	30	169
Habenaria aitchisonii	21		141
H. (*Leucorchis*) *albida*		42	73
H. andrewsii		42	17
H. arietina	21		139
		21	140
H. aristata	21		141
H. biermanniana	21		141
H. blephariglottis		42	86
		42	17
H. bracteata		42	84, 85
H. chorisiana	21	42	228
H. (*Platanthera*) *ciliaris*	16		21
		16	173
H. clavellata		42	86
		42	17
H. copelandii		42	170, 171
H. densa	23		141
H. (*Platanthera*) *dilatata*		42	86
	21	42	228
		42	17
H. edgeworthii	21		8
	21	42	140
H. elisbethae		42	139
H. ensifolia	21		139
	21		140
H. fallax	21		141
H. fimbriata		42	17
H. flava		42	17
H. galeandra	21		139
	ca. 21		140
H. geniculata		63	145
	42		141
H. goodyeroides	23		141
H. graveolens		42	8
H. hookeri		42	86
		42	17
H. hyperborea		42	86
	42	84	73
H. hyperborea var. *huronensis*		84	17
H. hystrix		42	170, 171

Plant Name	Gamete Chromosomes	Meristem Chromosomes	Reference
H. intermedia		42	8
	21		140
H. lacera		42	17
H. miersiana		62	145
		64	162
H. obtusata		42	86
H. oldhami		28	147
H. orbiculata		42	84, 85
		42	17
H. pectinata	21		139
		42	8
	21		140
H. plantaginea	21		8
	63		141
H. psycodes		42	17
H. (Pecteilis) radiata (suzannae)	32, (64)		145
	32		160
	32		158
	32		194
	32		220
H. rhodocheila	19		23
H. robinsonii		42	170, 171
H. saccata	21	42	228
H. sagittifera		28	145
		28	160
		28	158
H. species	16	32	178
H. (Leucorchis) straminea	21	42	73
H. (Pecteilis) susannae (radiata)	21		139
	21		140
H. tridactylites	21		141
H. unalascensis ssp. *maritima*	21		228
	20		8
H. viridis	21		141
H. viridis var. *bracteata*		42	17
Haemalia discolor var. *dawsoniana*		44	220
Hammarbya paludosa	14		68
	14		106
Hemipilia cordifolia		44	140
Herminium angustifolium	20		139
	19	38	140
	19, 38		141

Plant Name	Gamete Chromosomes	Meristem Chromosomes	Reference
H. elisabethae		42	140
H. gramineum	20		8
H. jaffereyanum	19		140
H. josephii	19		140
H. monorchis	20	40	80
	12–13		12
		40	62
	20	40	106
		40	140
H. quinquilobium	19		140
Hetaeria rubens	21		140
H. yakushimensis		42	220
H. xenantha		20	220
Himantoglossum (Loroglossum)			
hircinum	18	36	80
	12		79
	16		205
		36	30
	18		106
Hormidium calamarium		40	18
H. fragans		40	18
H. glumaceum		40	18
H. variegatum		40	18
Ionopsis paniculata		46	18
I. utricularioides		46	191
Isotria medeoloides		18	10
I. verticillata		18	10
Jumellea filicornoides		38–40	92
Koellensteinia graminea		ca. 48	81
Laelia albida		40, ca. 63	95
(see also Brassavola)			
L. anceps		40	18
L. anceps 'Brilliant'		40, 60	95
L. anceps 'Sanderiana'		40	95
L. anceps 'Stella'		40	95
L. autumnalis		41, 42	95
L. briegeri		80	18
L. caulescens		80	18
L. cinnabarina		40	18
		40	23

445

Plant Name	Gamete Chromosomes	Meristem Chromosomes	Reference
L. crispilabia		80	18
L. esalqueana		40	18
L. flava		40	18
L. gouldiana		40, 60	95
L. harpophylla		40	18
L. longipes		40, 60, 80	18
L. milleri		40	18
L. mixta		40	18
L. ostermayerii		40	18
L. peduncularis		40–44	23
L. pumila		40	18
		40	23
L. purpurata		40	95
		40	18
L. purpurata 'Semi-alba'		40 ± 1	104
L. rubescens	20	40	95
	20	40	99
L. rupestris		80	18
L. tereticaulis		80	18
Lanium avicula		40	18
Leptotes unicolor		40	18
Leucorchis (Habenaria) albida		42	107
		42	122
		42	80
		42	73
		40	198
		40	125
L. (Habenaria) straminea		42	120, 124
	21	42	73
Limnorchis convallariaefolius		80	196
L. dilatata		42	128
L. dilatata ssp. albiflora		42	128
L. hyperborea		84	128
L. saccata		42	128
Limodorum abortivum		64	130
Liparis amesiana		30	170, 171
L. confusa		30	170, 171
L. duthiei	15		8
	15		141

Plant Name	Gamete Chromosomes	Meristem Chromosomes	Reference
L. formosana		42	220
		42	153
L. formosana var. hachijoensis		42	153
L. gamblei	18		141
L. glossula	10		141
L. haponica		30	220
L. krameri		30	145
		30	160
		30	158
		30	220
L. kumokiri		30	145
	13		165
		30	160
		30	220
L. kurameri		30	153
L. loeselii		32	68
		26	62
		26	106
L. longipes		42	153
L. longipes var. spathulata	15		8
L. makinoana		30	145
		30	160
		00	158
		30	220
L. nepalensis	18		141
L. nervosa	21	42	147
		42	160
		42	158
		42	220
		42	153
L. paradoxa	18		141
L. perpusilla	15		141
L. plicata		42	162
		38	220
		38	153
L. pulverulenta	40		23
L. pusila		40	25
L. resupinata	14		141
L. rostrata	15		8
	14	28	139
L. taiwaniana		38	220
Listera borealis		56, 56 + B	190
L. caurina	17	34	228
L. convallarioides		36	190

447

Plant Name	Gamete Chromosomes	Meristem Chromosomes	Reference
L. cordata		36–38	124
		38	197
		42	133
		42	231
		38	195
		38	196
		38	125
		40	62
		40	106
	19		227
		40	108
		38	228
		42	113
L. cordata var. *japonicum*		42	194
L. cordata ssp. *nephrophylla*		36–38	190
L. makinoana		38	220
		42	189
L. nipponica		38	162
L. ovata		34, 36, 38	123
	17, 18	34, 36, (17)	71
	17		13
		34, 34 + f.	129
		42	41
	17, 18	34, 35	172
	17		81, 82
	17		201
	16, 17, 18	32, 34, 36	234
	16–17	32, 34	157
	16		174
	16		65, 66
		34–40	124
		38	195
		ca. 38	199
	19		200
		34	125
	17	34, 35, 36	106
		34, 35, 36, 37, 38	62
		34	108
		34	118
L. pinetorum	20		141
L. sikokiana		38	160
Lockhartia oerstedii		14	63
Loroglossum hircinum (see also *Himantoglossum*)	12		79

Plant Name	Gamete Chromosomes	Meristem Chromosomes	Reference
Luisia boninensis		40	147
L. inconspicua	19		141
L. liukiuensis		38	220
L. teretifolia (= *L. teres*)		40	162
		38	184
		38	170, 171
		ca. 38	92
	19	38	25
	19		141
		38	226
L. trichorhiza		38	92
	20		8
Lycaste aromatica	20		81, 82
Lysiella obtusata		42	128
Macradenia paraensis		52	18
Malaxis monophylla	15	30	68
(= *Acroanthes monophyllos*)	15–17		202
M. monophylla ssp. *brachypoda*		28	127
M. paludosa		28	220
	14		228
Masdevallia coccinea 'Lindenii'		44	217
M. ochthoides		44	217
Maxillaria picta		40	18
M. tenuifolia		40	222
Microstylis biloba	18		141
M. commelinifolia	19–21		23
M. congesta	21		141
M. cylindrostachya	15		139
	15		141
M. iriomotensis		44	220
M. maximowicziana	21		141
M. muscifera	15		141
M. saprophyta	21		141
M. species	ca. 20		82
M. wallichii	18		141
Microtis parviflora		44	220
Miltonia bluntii		60	191
		60	193

Plant Name	Gamete Chromosomes	Meristem Chromosomes	Reference
M. flavescens		56	18
		60	191
		60	193
M. regnellii		60	191
		60	193
M. roezlii var. alba		56	191
		60	193
M. spectabilis		60	191
		60	193
M. spectabilis var. lineata		56	191
		56	193
M. spectabilis var. morelliana subvar. rosea		86	191
		86	193
M. vexillaria		60	191
		60	193
M. warscewiczii		56	191
		56	193
M. warscewiczii var. panamense		56	193
Mormodes buccinator		54	93
M. buccinator var. citrinum		54	93
M. histrio		54	93
Neofinetia falcata		38	160
		38	98
		38	184
		38	220
Neotinea intacta		40	20
Neottia listeroides	20		141
	20	40	139
N. nidus-avis	18		13
		36	49
	18		154
	16		66
		36	125
	18		106
Nervilia discolor		20	170, 171
Neuwiedea singapureana		ca. 144	116
Nigritella apomictica		40	80
		32	32

450

Plant Name	Gamete Chromosomes	Meristem Chromosomes	Reference
N. nigra		64	107
	32		5
	20	40	80
		32	240
	19	38	26
		64	108
N. nigra var. apomictica	32		4
N. rubra (miniata)		80	80
	19		26
Notylia bicolor		42	191
N. panamensis		42	191
Oberonia caulescens	13		141
O. integerrima		30	23
O. japonica		30	220
O. obcordata	15		141
O. prainiana	15		141
Odontoglossum citrosmum		44–48	23
O. cordatum		56	191
		56	193
O. crispum		56	82
O. grande		44	36
		60?	191
		60?	193
O. insleayi		44	36
O. kegeljani		56	36
O. pendulum		44	36
O. schlieperianum		44	36
Oeoniella polystachys		38	222
Oncidium altissimum		56	191
		56	192
O. ampliatum		44	35
		44	177
		44	191
		44	192
O. anciferum		56	23
		56	191
		56	192
O. anthocrene		56	191
		56	192
O. aurosum		54	23

Plant Name	Gamete Chromosomes	Meristem Chromosomes	Reference
O. bauerii		*ca.* 52	35
		56	191
		56	192
O. barbatum		56	18
O. bicallosum		28	35
	14		81, 82
O. brachyandrum		56	191
		56	192
O. brunleesianum		56	250
O. carthaginense		28	35
		30	191
		30	250
		30	192
O. carthaginense var. *roseum*		30	191
O. cavendishianum		28	35
		28	18
O. cebolleta		36	18
		34	191
		34	192
O. cheirophorum		*ca.* 48	35
		56	191
		56	217
		56	192
O. cordatum		56	35
O. crispum		56	35
		56	23
O. cucullatum		54?	250
O. curtum		52	23
O. desertorum (*intermedium*)		40	191
		40	250
		40	192
O. desertorum (*intermedium*) 'Gigas'		40	192
O. ebrachiatum		28	35
O. excavatum		56	49
		56	191
		56	192
O. flexuosum		56	35
		56	81, 82
O. globuliferum		56	191
		56	192
C. glossomystax		14	250
O. guttatum		28	35
O. guttatum var. *olivaceum*		32	191
		32	192

452

Plant Name	Gamete Chromosomes	Meristem Chromosomes	Reference
O. haematochilum		28, ca. 40	191
O. harrisonianum		42	191
		42	192
O. hastatum		56	191
		56	192
O. henekenii		40	191
		40	192
O. hieroglyphicum		56	250
O. hyphaematicum		56	191
		56	192
O. incurvum		56	191
		56	192
O. isthmi		56	191
		56	192
O. jonesianum		30	250
O. kenscoffii		84	191
		84	192
O. kramerianum		38	34, 35
O. lammeligerum		55–57	250
O. lanceanum		28	35
		28	191
		28	192
O. leucochilum		56	35
O. lieboldii		42	36
		42	192
O. lieboldii 'Alba'		40–42	250
O. lieboldii f. album		42	191
O. longifolium		28	35
O. longipes		56	18
		56	250
O. loxense		56?	250
O. lucayanum		40	250
O. luridum		32	191
		32	250
		32	192
O. macranthum		56	35
		50–57	250
O. maculatum		56	191
		56	192
O. marshallianum		58	250
O. "Miami" (variegatum?) (Moir)		84	192
O. microchilum		36	35
		36, 37	191
		36	250
		36, 37	192

Plant Name	Gamete Chromosomes	Meristem Chromosomes	Reference
O. nanum		26	191
		26	192
O. nebulosum		56	250
O. obryzatoides		56	191
		56	192
O. obryzatum		56	191
		56	192
O. oestlundianum		28	35
O. onustum		56	191
		56	192
O. ornithorhynchum		56	35
		56	191
	28		23
		56	192
O. panamense		56	191
		56	192
O. papilio		38	34, 35
		38	192
O. papilio 'Latour'		38	191
		38	192
O. parviflorum		56	35
O. pentadactylon		40–42	250
O. phalaenopsis		56	250
O. phymatochilum		56	191
		56	192
O. polyadenium		56	191
		56	250
		56	192
O. powellii		28–56 f. 5	35
		56	191
		56	192
O. praetextum	28		2
O. pulchellum		42	36
		42	191
		42	250
		42	192
O. pumilum		30	250
O. pusillum		10	33, 35
		10	191
		10, 14	250
		10	192
O. robustissimum		44	33, 35
O. sarcodes		56	35

Plant Name	Gamete Chromosomes	Meristem Chromosomes	Reference
O. sphacelatum		56	191
		56	23
		56	192
O. splendidum		34	35
		36	191
		36	23
		36	250
		36	192
		56	191
O. stenotis		56, 105–111	23
		56	192
O. stipitatum		28	35
O. stramineum		28	191
		30	250
		28	192
O. sylvestre		84	191
		84	192
O. teres		28	25
O. tetrapetalum		42	36
O. tetraskelidon	28		23
O. tigrinum		54	250
O. triquetrum		42	36
		42	191
		42	192
O. Unnamed from Abaco (Moir)		40	191
		40	192
O. Unnamed from Bahamas (Moir)		40	191
		40	192
O. Unnamed from Cuba (Moir)		133	191
		133	192
O. urophyllum		84	191
		84	192
O. varicosum		56	35
	28		81, 82
		112, 168	191
		112, 168	192
O. varicosum var. *rogersii*		56	23
O. variegatum		42	36
		40	191
		40	192
O. volvox		28	23
O. warmingii		140–150	250
O. wentworthianum		56	191
		56	192

Plant Name	Gamete Chromosomes	Meristem Chromosomes	Reference
Ophrys apifera		36	13
	18	36	80
	18	36	106
		36	62
O. aranifera (*sphegodes*)	18	36	80
O. fuciflora	18	36	80
O. insectifera (*muscifera*)	18	36	5
		36	13
		36	80
	18	36, 38	106
O. muscifera (*myodes*)		36	13
		36	80
	11–12		179
O. sphegodes		36	13
		36	182
O. sphegodes ssp. *litigiosa*			
(ssp. *pseudospeculum*)		36	80
		36	182
Orchis aristata (see also *Dactylorchis*)		40	145
		40–42	196
O. aristata var. *immaculata*		38	194
O. caucasica		40	197
O. chidori		42	147
		42	161
O. coriophora	19	38	80
	10	20	60
		38	239
O. (D.) cruenta		40	238
		40	76
O. drudei	7	14	60
O. elodes			
(= *D. incarnata*)		79, 80	133
		21	132
		42	161
O. fauriei		21	132
		42	161
O. foliosa (*maderensis*)		40	238
O. fuchsii (*maculata*)		40	238
O. globosa	21	42	80
(= *Traunsteinera globosa*)		42	31
O. graminifolia		42	220
O. habenariodes	20		141
O. (D.) incarnata	10		60

456

Plant Name	Gamete Chromosomes	Meristem Chromosomes	Reference
O. latifolia		40	49
(*D. incarnata*)		40, 80	133
(*O. strictifolia*)		80	67
		80	80
		80	237
	10	20	60
		40	75
O. latifolia var. *duensis*		40	133
O. laxiflora		42	238, 239
O. (*D.*) *maculata* (*fuchsii*)	16		205
	10		60
		40	122
		40	238
O. (*D.*) *maculata* ssp. *elodes*		80	77
O. (*D.*) *maculata* var. *ericetorum*		80 (40)	69
O. (*D.*) *maculata* var. *genuina*		40	67
O. maculata var. *meyeri* (*ericetorum*)	20		67
O. maculata var. *meyeri* (*D. fuchsii*)		20	69
O. maderensis (*Or. foliosa*)		40	238
O. (*D.*) *majalis*		80	238
O. mascula	21	42	67
	21		80
		42	238, 239
		42	125
		42	62
		42	106
O. militaris		42	238
	21		67
	21	42	80
	10	20	60
		42	125
	21	42	106
		42	62
O. morio		36	238
	18		67
	18	36	80
	10	20	60
		36	31, 32
		36	239
		36	125
		36, 38	106
		36	62

457

Plant Name	Gamete Chromosomes	Meristem Chromosomes	Reference
O. morio ssp. picta		36	125
O. (D.) munbyana		80	238
O. ochroleuca Boll. (incarnata var. straminea Rchb.)		40	80
O. pallens	20	40	80
		40	239
O. palustris		42	80
		42	239
O. palustris ssp. elegans		42	125
O. papilionacea	16	32	80
O. (D.) praetermissa		80, 82	133
O. (D.) praetermissa var. pulchella		80, 82	133
O. provincialis		42	80
O. purpurea		42	238, 239
		40	49
	21	42	67
	21	42	80
	21	42	106
		42	62
O. (D.) purpurella		80	75
O. rotundifolia		42	84, 85
O. (D.) sambucina	21		67
		40	238
O. sesquipedalis		80	238
O. simia	21	42	80
		42	239
		42	106
O. (Galeorchis) spectabilis		42	84, 85
O. sphaerica		42	197
O. strictifolius	20		67
O. traunsteineri (= D. pseudo-traunsteineri)	6–15	16–28	61
		40, 80, 120, 122	238
O. traunsteineri (= D. traunsteineri)	10	20	60
	9–22	16–24	61
		80	198
O. (D.) traunsteineri var. gigas	20		60
O. tridentata		42	80
		42	67
		42	239
O. ustulata		42	238, 239
	21	42	67
	21	42	80
	10	20	61
	10		60
		42	31, 32

Plant Name	Gamete Chromosomes	Meristem Chromosomes	Reference
Oreorchis patens	24	48	166
		48	161
		50	220
Ornithidium densum	24		81, 82
Otochilus alba	20		141
O. fusca	20		141
O. porrecta	20		141
Paphiopedium argus		26	170, 171
P. barbatum	19	38	47
		38	39
		38	135
		38	57
	16		55
	16	32	205
	19		23
P. bellatulum	13	26	47
		26	39
		26	135
		26	101
P. boxallii	13	26	43
		26	39
		26	135
P. callosum	16	32	47
		32	39
		32	135
		32	101
		32	220
P. callosum 'Sanderae'		32	134
		32	39
		32	135
P. chamberlainianum	16	32	46
		32	39
	16	32	81, 82
P. charlesworthii	13	26	44
		26	39
P. charlesworthii 'Bromilowianum'	13	26	44
		26	39
P. concolor	13	26	42
		26	101
P. curtisii	18	36	47
		36	135
P. curtisii 'Exquisitum'		36	39
P. curtisii 'Sanderae'		36	39
		36	135

Plant Name	Gamete Chromosomes	Meristem Chromosomes	Reference
P. dayanum	17	34	47
		34	39
P. delenatii	13	26	42
		26	39
		26	135
P. druryi	13	26	44
		26	39
P. exul	13	26	44
		26	39
		26	101
P. fairieanum	13	26	44
		26	39
		26	135
P. glaucophyllum	18	36	46
		36	39
P. godefroyae		26	101
P. gratrixianum	13	26	43
		26	39
P. haynaldianum	13	26	45
		26	39
		26	170, 171
P. hirsutissimum	13	26	44
		26	39
P. insigne		26	39
(= *C. insigne*)		26	135
		28	49
	16		55
	12		41
	16		81, 82
	12–13		74
	8–9		206
	ca. 12		2
		26	217, 220
P. insigne 'Ernesti'	13	26	43
		26	39
P. insigne 'Harefield Hall'		39	43
		39	135
P. insigne 'Laura Kimball'		26	135
P. insigne 'Royalty'		26	135
P. insigne 'Sanderae'		26	43
		26	39
		26	135
		26	217, 220
P. insigne 'Sylhetense'		26, 28	39
		26	135

Plant Name	Gamete Chromosomes	Meristem Chromosomes	Reference
P. insigne 'Tonbridgense'		26	135
P. javanicum	18	36	47
		36	39
P. lawrenceanum	18	36, 40(I)	47
		36	134
		36(II)	135
		40	47
P. lawrenceanum 'Hyeanum'		36	134
		36	39
P. lowii	13	26	44
		26	39
P. mastersianum	16	32	47
		32	39
P. niveum	13	26	42
		26	39
		26	135
		26	101
P. parishii	13	26	45
		26	39
		26	101
P. philippinense		26	45
		26	170, 171
P. spiceranum	14	28	44
		28	39
P. purpuratum	ca. 24	ca. 48	82
P. rothschildianum		26, 28	45
		26, 28	39
P. speciosum		26	156
		26	57
P. spiceranum	14	28	44
	14	28	39
	15	30	54–56, 58
P. stonei 'Mary Reginae'	13	26	45
		26	39
P. sublaeve	29	58	47
		57, 58	39
P. superbiens	19	38	47
		38	39
P. tonsum	17	34	47
		34	39
P. venustum	21	42	47
		42	39
	18		55
	20 + 1B		141

461

Plant Name	Gamete Chromosomes	Meristem Chromosomes	Reference
P. villosum	13	26	43
		26	39
		26	135
		19–28	58
		26	57
	14		55
		26	101
P. wardii	20	40	47
		40–45	38
Pecteilis radiata (*Habenaria suzannae*)		32, (64)	145
P. radiata variety		48	147
Pelatantheria ctenoglossa		38	102
		38	226
Peristeria elata var. *gattonensis*		40	29
P. guttata		40	29
Peristylis cordatus		36	115
P. goodyeroides	21		8
Phaius elatus		42	41
P. flavus		42	170, 171
P. grandifolius (*tankervilliae*)		50	220
		38	8
P. mindorensis		42	170, 171
P. minor		42	160
		44	220
P. minor f. *punctatus*		44	220
P. wallichii	21		141
Phalaenopsis amabilis (*Elisabethae*) (see also *Doritis*)		38, 114	244
		38, 69 + 3 f.	175
		38	170, 171
P. amabilis 'Grandiflora'		38	104
		38	176
		38	184
		152 ± 2	170, 171
P. ambilis var. *grandiflora*		38	188
P. amboinensis 'P'		38	176
P. aphrodite		38	244
	19	38	9

Plant Name	Gamete Chromosomes	Meristem Chromosomes	Reference
P. boxallii		38	184
P. buyssoniana			
(Doritis buyssoniana)		76	41
P. cornu-cervi		38	244
		38	176
		38	102
P. decumbens		38	102
P. equestris 'New Type'		38	176
P. equestris (rosea)		38	244
		38	176
		38	188
	19	38	9
P. equestris 'Three Lips'		38	104
P. esmeralda		38	244
(Doritis pulcherrima)	19	38	177
P. lindenii		38	176
		38	188
P. lueddemanniana		38	244
		38	188
	19	38	9
P. lueddemanniana 'Boxalli'		38	176
(= ochracea?)			
P. lueddemanniana 'Division 1'		38	176
P. lueddemanniana 'Division Plant A'		38	176
P. lueddemanniana 'Fennell'		38	176
P. lueddemanniana var. heiroglyphica		38	176
P. lueddemanniana 'Jones'		38	176
P. lueddemanniana 'Majus'		38	176
P. lueddemanniana var. ochracea		38	176
		38	188
P. lueddemanniana var. pulchra		38	176
P. lueddemanniana var. pulchra			
'Plant 2 (red)'		38	176
P. mannii		38	244
		00	176
		38	188
	19	38	9
P. mariae		38	176
P. (Doritis) pulcherrima		38	188
		38	102
P. pulcherrima var. buyssoniana		76	102
P. sanderana		38	176
		38	188
	19	38	9

Plant Name	Gamete Chromosomes	Meristem Chromosomes	Reference
P. schillerana		38	244
		40	49
		38, 76	104
		38	176
		38	188
		38	217
P. schillerana 'Malibu'		65	176
P. schillerana 'No. 1280A'		38	104
P. speciosa 'Orchidglade'		38	176
P. stuartiana	19	38	9
P. violacea		38	188
Pholidota articulata	20		8
	20		141
P. chinensis		40	214
P. conchoidea	20		81, 82
P. imbricata		40	170, 171
		40	8
	20		141
P. protracta	20		141
P. recurva	20		141
Phragmopedium blenheimense		24	81
P. caudatum		32	81
P. longifolium var. *hartwegii*		20	41
P. sedenii	12	24	81
Physosiphon bifolia	21		3
		42	173
	21	42	80
P. carinatus	*ca.* 16		81, 82
P. loddigesii	16, 17, 18		81, 82
Placoglottis javanica	19		5
Platanthera bifolia (see also Coeloglossum; Habenaria)	21	42	80
		42	173
	21		3
		42	31, 32
		42	125
		42	106
	21	42	62
		ca. 42	200
P. chlorantha		42	123
	21	42, (21)	71
		42	80

Plant Name	Gamete Chromosomes	Meristem Chromosomes	Reference
	21		3
		42	173
		42	125
		42	62
	21	42	106
P. (Habenaria) cilaris	16		173
	16		21
P. (Habenaria) dilatata		42	86
P. florenti		42	160
P. hologlottis		42	147
		42	220
P. (Habenaria) hyperborea	21	42	173
		84	124
		42	86
	42	84	73
P. mandarinorum var. brachycentron		42	220
P. mandarinorum var. maximowicziana		42	160
P. metabifolia		21	165
		42	220
P. minor		42	160
		42	220
P. obtusata		63	173
		63	3
P. oligantha		126	3
		126	120
		ca. 126	108
P. ophrydioides		20	165
		42	162
		49	194
P. ophrydioides var. takedai		42	160
P. sachalinensis		42	161
P. sussanae		42	8
P. typuloides		42	161
P. typuloides var. nipponica		42	162
		42	220
Platyclinis (Dendrochilum) glumacea	20		81, 82
		38	217
Pleione formosanum		40	147
		40	214, 220
P. humilis	20		141
P. praecox	20		141

Plant Name	Gamete Chromosomes	Meristem Chromosomes	Reference
P. pricei		20 + 1B	114
Pleurothallis vittata	21		23
Plocoglottis javanica	19		5
Podochilus cultratus	19		141
Pogonia japonica		19–23	148
		24	161
		20	214, 220
		19, 20	152
P. minor		18, 20, 21	152
		18	162
		18	214, 220
P. ophioglossoides		18	10
		18	127
Polystachya cultriformis		38–39	23
P. estrellensis		80	18
P. polychaete	ca. 20		82
P. rhodoptera	19		23
Pomatocalpa spicatum		38	102
Ponerorchis rotundifolia		42	128
Pristiglottis tashiroi		26	220
Promenea citrina		46	217
Pseudorchis albida		42	127
Rangaeris brachyceras		50	92
R. musicola		ca. 100	92
Renanthera coccinea	57 ± 1	38, 114, 115	103
		38	102
		ca. 114	92
R. elongata	19	38	103
R. histrionica	19	38	103
R. imschootiana	19		23
R. matutina	19	38	103
R. monachica		38	41
	19	38	103
		38	184
R. storiei	19	38	103
		38	105
		38	226

Plant Name	Gamete Chromosomes	Meristem Chromosomes	Reference
Rhynchostylis coelestis		38	102
		38	226
R. gigantea		38	102
R. gigantea var. *illustre*		38	102
R. retusa		38	102
	19		8
		38	226
Robiquetia paniculata		38	102
		38	226
R. spathulata		38	102
Rodriguezia batemannii		42	191
R. decora		42	191
R. fragrans		42	191
R. secunda		42	191
R. strobelii		42	191
R. teuscherii		28, 29	191
R. venusta		42	191
Saccolabium albo-lineatum (see also *Gastrochilus*)	19–20		23
S. calceolare	19–20		23
	38		141
S. distichum	19		141
S. eberhardtii	19–20		23
S. longifolium	19		141
S. obtussifolium	19		141
S. papillosum	38		141
S. pseudo-distichum	19		141
S. rubescens	19–20		23
S. tenerum		*ca.* 38	92
S. triflorum	19–20		23
Sarcanthus appendiculatus		38	102
		38	25
S. carinatus		38	102
		38	226
S. crinaceous		38	226
S. dealbatus	19		23
S. filiformis	19		141
S. flagelliformis		38	102
		38	226
S. kunstleri		38	102
S. pallidus		38	25
S. rostratus	18		209
		36	82

Plant Name	Gamete Chromosomes	Meristem Chromosomes	Reference
S. scolopendrifolius		38	161
S. strongyloides		38	92
S. subulatus		38	102
		38	92
		38	226
S. termissus		38	102
		38	226
Sarcochilus japonicus		38	161
		36	189
		38	220
S. longicalcarus		38	170, 171
S. luniferus	19		141
S. palawanensis		38	170, 171
Sarcorhynchus bilobatus		50	92
Satyrium nepalense		41	211
Sauroglossum nitidum		44	18
Schomburgkia crispa		40	18
Serapias longipetala	12		11
S. vomeracea		36	80
Sigmatostalix radicans		60	191
		60	193
Sophronitis cernua		40	18
Spathoglottis plicata		40	220
Spiranthes australis (*sinensis*)	12		212
(see also *Gyrostachys*)	12		11
	15, 16		141
S. autumnalis		35	230
S. (*Gyrostachys*) *cernua*	30		169
		30	220
S. (*Gyrostachys*) *gracilis*	15	30	169
S. romanzoffiana		60	125
	15		228
S. sinensis (*australis*)		30	145
	12	24	166
		30	160, 161
		30	158
		26, 30	194
		30	215, 220, 221
	15	30	68
		30	223, 224

Plant Name	Gamete Chromosomes	Meristem Chromosomes	Reference
S. spiralis	15	30	68
Stanhopea bucephalus		40	29
S. candida		40	29
S. costaricensis		40	29
S. devoniensis		40	29
S. ecornuta		*ca.* 40	29
S. gibbosa		40	29
S. grandiflora		40	29
S. grandiflora 'Alba'		42	29
S. graveolens		40	29
S. inodora		40	29
S. inodora 'Amona'		42	29
S. insignis	20		82
S. oculata	20		5
		40	23
		ca. 40, 42	29
S. peruviana		42	29
S. ruckeri		40	29
S. saccata		*ca.* 40	29
S. tigrina (*hernandezii*)	20		81, 82
		42	214
		40	23
		40	29
S. tigrina 'Superba'		40	29
S. wardii		41, 42	29
Stauropsis undulatus	19		141
Staurochilus dawsonianus		38	102
S. fasciatus			
(= *Trichoglottis fasciata*)		38	102
S. lukyuensis		38	220
Stelis atropurpurea (*ciliaris*)	16		81, 82
S. concaviflora		32	51
S. miersii		32	81, 82
S. pygmaea	16		23
Stenoglottis longifolia		36	217, 220
Taeniophyllum aphyllum		38?	160
		38?	158
		24	159
		38	220
T. crepidiforme	19		141
T. elmeri		40	170, 171
T. philippinensis		40	170, 171

Plant Name	Gamete Chromosomes	Meristem Chromosomes	Reference
Tainia laxiflora		36	147
		36+1B, 36+2B,	220
		36+3B, 36+4B,	
		36+O-9B(mode 2B)	225
Thelymitra longifolia		26	72
Thrixpermum acuminatissimum		38	102
		38	226
T. arachnites		38	102
T. centripeda		38	23
Thunia alba		42	217
	20		141
T. marshalliana	20		141
Traunsteinera globosa	21	42	80
(= *Orchis globosa*)			
		42	31, 32
		42	125
Trichocentrum albo-purpureum		28	35
		24	191
T. maculata		24	191
T. panamense		28	36
		28	191
T. tigrinum		24	191
Trichoglottis cirrhifera		38	226
(see also *Staurochilus*)			
T. fasciata		38	105
		38	226
T. philippinensis		38	184
T. rosea		38	226
Trichopilia suavis			
(presumed from hybrid)		56	191
Tridactyle anthomaniaca		*ca.* 100	92
T. tridactylites		*ca.* 100	92
T. species		*ca.* 50	92
Triphora trianthophora		18–44 f. 5	10
Tropidia nipponica		56	251
Uncifera species		38	217
Uraria neglecta	10		8
Vanda alpina (see also *Vandopsis*)		38	97
	19		141

Plant Name	Gamete Chromosomes	Meristem Chromosomes	Reference
V. amesiana		38	203
		38	245
		38	97
V. amoena		38	104
V. coerulea		38	203
		38	245
		38	102, 104
V. coerulescens		38	102
		ca. 38	92
V. concolor		76	203
V. dearei		38	203
		38	104
V. denisoniana (brown)		38, 76	102
V. denisoniana (green to yellow)		38	102
V. hookerana	19	38	203
V. kimballiana		38	203
V. lamellata		38	184
V. lamellata var. boxallii	19	38	203
V. laotica		38	102
V. luzonica	19	38	203
		38	245
		38–42	23
V. parviflora		ca. 40	92
V. roeblingiana		38	203
V. sanderana	19	38	203
		38	104
V. sanderana '0'		38	104
V. sanderana 'No. 1'		38	104
V. sanderana 'No. 2'		38	104
V. sanderana 'No. 3'		38	104
V. spathulata	38	76	203
		114	104
		114, 115	204
V. stangeana	19		141
V. suavis		38–39	23
V. teres	19	38	203
		38	184
		38	102
V. teres 'Alba'		38	104
V. teres 'Aiba-candida'		38	104
V. tricolor	19	38	203
		38	245
		28	49
	ca. 18, 20		82
	ca. 16		81

Plant Name	Gamete Chromosomes	Meristem Chromosomes	Reference
V. tricolor var. suavis	19	38	203
	ca. 18		82
	ca. 16		81
		38	23
V. tricuspidata		76	203
Vandopsis gigantea		38	102
		38	92
V. lissochiloides		38	102
V. (Vanda) parishii		38	203
		38	245
		38	23
		38	102
	19		141
Vanilla aromatica		32	50
V. barbellata		32	131
V. dilloniana		32	131
V. fragans		30–32	23
V. hartii		32	50
V. imperialis	16		50
V. moonii		32	50
V. papeno		32	50
V. phaeantha		32	131
V. planifolia	16	32	50
		28–32	87
		32	81, 82
		32	131
V. pompona	16	32	50
		32	131
V. thaitii		32	50
Vexillabium fissum		40	220
V. nakaianum		26	220
V. yakushimense		26	220
Warscewiczella (see Zygopetalum)			
Zeuxine straumatica		42	220
	10, 11, 20, 50	50	141
Z. sulcata	22	44	180
		44	181
	10		140
Zygopetalum crinitum		96	217
Z. (Warczewiczella) discolor		ca. 48	23
Z. mackayi		48	82
	ca. 24		207
		96	217
Z. maxillare		48	18
Z. odoratissimum		48–50	23

Bibliography

1. ADAIR, V. L., and SAGAWA, Y. 1969. Cytological and morphological studies of *Caularthron bicornutum*. *Caryologia* 22:369–373.

2. AFZELIUS, K. 1916. Zur Embryosackentwicklung der Orchideen. *Svensk. Bot. Tidskr.* 10:183–227.

3. ———. 1922. Embryosackentwicklung und Chromosomenzahl bei einigen *Platanthera*-Arten. *Svensk. Bot. Tidskr.* 16:371–382.

4. ———. 1932. Zur Kenntnis der Fortpflanzungeverhaltnisse und Chromosomenzahlen bei *Nigritella nigra*. *Svensk. Bot. Tidskr.* 26:365–369.

5. ———. 1943. Zytologische Beobacktungen an einigen Orchidaceen. *Svensk. Bot. Tidskr.* 37:266–276.

6. ———. 1954. Embryo-sac development in *Epigogium aphyllum*. *Svensk. Bot. Tidskr.* 48:513–520.

7. ———. 1958. En engendomling form av *Orchis maculata* L. sens. lat. *Svensk. Bot. Tidskr.* 52:18–22.

8. ARCRA, C. M. 1968, In IOPB chromosome number reports, XVI. *Taxon* 17:199–204.

9. ARENDS, J. C. 1970. Cytological observations on genome homology in eight interspecific hybrids of *Phalaenopsis*. *Genetica* 41:88–100.

10. BALDWIN, J. T., and SPEESE, B. M. 1957. Chromosomes of *Pogonia* and of its allies in the range of Gray's manual. *Am. J. Bot.* 44:651–653.

11. BARANOW, P. 1915. Recherches sur le developpement du sac embryonnaire chez les *Spiranthes australis* Lindl. et *Serapias pseudocordigera* Mor. *Bull. Soc. Imp. Nat. Moscow* 29:74–92.

12. ———. 1925. Ueber die Reduktion des weiblichen Gametophyten in der Familie Orchidaceae. *Bull. Univ. Centralasiens* 10:181–195.

13. BARBER, H. N. 1942. The pollen grain division in the Orchidaceae. *J. Genetics* 43:97–103.

14. BELLING, J. 1924. Detachment (elimination) of chromosomes in *Cypripedium acaule*. *Bot. Gaz.* 78:458–460.

15. ———. 1926a. Iron-acetocarmine method of staining chromosomes. *Biol. Bull.* 50:160–162.

16. ———. 1926b. Structure of chromosomes. *Br. J. Exp. Biol.* 3:145–147.

17. BENT, F. C. 1969. Chromosome studies in *Habenaria*. *Rhodora* 71:541–543.

18. BLUMENSCHEIN, A. 1960. Numero de chromosomas de algunas especies de orquideas. *Publ. Cien. Univ. Sao Paulo, Inst. Genet.* 1:45–50.

19. ———. 1961. Una nova especie de genero *Cattleya* Lindl. *Pull. Cien. Univ. Sao Paulo, Inst. Genet.* 2:23–33.

20. BORGEN, L. 1969. Chromosome numbers of vascular plants from the Canary Islands, with special reference to the occurrence of polyploidy. *Nytt Magasin Botanik* 16:18–121.

21. BROWN, W. H. 1909. The embryo sac of *Habenaria*. *Bot. Gaz.* 48:241–258.

22. CARLSON, M. C. 1945. Megasporogenesis and development of the embryo sac of *Cypripedium parviflorum*. *Bot. Gaz.* 107:107–113.

23. CHARDARD, R. 1963. Contribution a l'etude cytotaxinomique des Orchdees. *Rev. Cyt. Biol. Veg.* 26:1–58.

24. CHATTERJI, A. K. 1965. Chromosomes of *Eulophia*. *Chromosome Inf. Serv.* 6:8–9.

25. ———. 1968. Chromosome numbers and karyotypes of some orchids. *Am. Orchid Soc. Bull.* **37**:202–205.

26. CHIARUGI, A. 1929. Diploidismo con anfimissa e tetraploidismo con apomissa in una medesina specie: *Nigritella nigra*, Rehb. *Bol. Soc. Ital. Biol. Sper.* **4**:659–661.

27. CHODAT, R. 1924. La caryocinese et la reduction chromatique observees sur le vivant. *Compt. Rend. Soc. Phys. et Hist. Nat. Geneve.* (*Ser. 2*) **41**:96–99.

28. DAKER, M. G. 1970. The chromosomes of orchids, IV. Bulbophyllinae Schltr. *Kew Bull.* **24**:179–184.

29. DAKER, M. G., and JONES, K. 1970. The chromosomes of orchids, V. Stanhopeinae Benth. (Gongorinae Auct.). *Kew Bull.* **24**:457–459.

30. DELIOT, M. 1955. Etude structurale du chromosome somatique chez le *Loroglossum hircinum* (L.) Richard. *Botaniste* **39**:315–337.

31. DIANNELIDIS, T. 1948. A study of chromosomes of the Orchidaceae. *Proktika Acad. Athenon.* **23**:352–359.

32. ———. 1955. Chromosomenzahlen einiger Orchidaceen. *Ann. Fac. Sci. Univ. Thessaloniki* **7**:99–105.

33. DODSON, C. H. 1957a. Studies in *Oncidium*, I. *Oncidium pusillum* and its allies. *Am. Orchid Soc. Bull.* **26**:170–172.

34. ———. 1957b. *Oncidium papilio* and its allies. *Am. Orchid Soc. Bull* **26**:240–244.

35. ———. 1957c. Chromosome numbers in *Oncidium* and allied genera. *Am. Orchid Soc. Bull.* **26**:323–330.

36. ———. 1958. Cytogenetics in *Oncidium*. *Proc. of the Second World Orchid Congress.* Harvard University Press, Cambridge, Mass.

37. DORN, E. C., and KAMEMOTO, H. 1962. Chromosome transmission of *Dendrobium phalaenopsis* 'Lyons Light No. 1.' *Am. Orchid Soc. Bull.* **31** (12):997–1006.

38. DUNCAN, R. E. 1945. Production of variable aneuploid numbers of chromosomes within the root tips of *Paphiopedilum wardii*. *Am. J. Bot.* **32**:506–509.

39. ———. 1947. The hybrid lady slipper. *Orchid Digest.* Sept.-Oct.:199–207.

40. ———. 1959a. Orchids and cytology, in C. L. Withner, Ed., *The Orchids*. Ronald Press, New York; *Chronica Bot.*, No. **32**:189–260.

41. ———— 1959b. List of chromosome numbers in orchids, in C. L. Withner, Ed., *The Orchids*. Ronald Press, New York; *Chronica Bot.*, No. **32**:529–587.

42. DUNCAN, R. E., and MacLEOD, R. A. 1948a. Chromosomes of the Brachypetalums. *Am. Orchid Soc. Bull.* **17**:170–174.

43. ———. 1948b. Chromosomes of the *insigne* complex of lady-slippers. *Am. Orchid Soc. Bull.* **17**:424–429.

44. ———. 1949a. The chromosomes of the continental species of *Paphiopedilum* with solid green leaves. *Am. Orchid Soc. Bull.* **18**:84–89.

45. ———. 1949b. The chromosomes of some of the Polyantha. *Am. Orchid Soc. Bull* **18**:159–163.

46. ———. 1949c. The chromosomes of the species of Cochlopetalum Hallier. *Am. Orchid Soc. Bull.* **18**:573–576.

47. ———. 1950a. The chromosomes of Eremantha Tesselata. *Am. Orchid Soc. Bull.* **19**:137–142.

48. ———. 1950b. The chromosomes of *Paphiopedilum sublaeve*. *Am. Orchid Soc. Bull.* **19**:489–492.

49. Eftimiu-Heim, P. 1941. Recherches sur les noyaux des orchidees. *Botaniste* 31:65–111.

50. ———. 1950. Le noyau dans les genre *Vanilla*. *Encyclopedie Mycologique*.

51. FAVARGER, C., and HUYNH, K. L. 1965. IOPB chromosome number reports IV (by A. Löve and O. T. Solbrig). *Taxon* 14(3):86–87.

52. FERNANDES, A. 1950. Sobre a cariologia de algunas plantas da Serra do Geres. *Agron. Lusit* 12:551–600.

53. FRAHM-LELIVELD, J. A. 1941. Some remarks on the formation of the pollinia of *Gymnadaenia conopsea* (L.) R. Br. *Natuurkundig Tijdschr.* 101:242–244.

54. FRANCINI, E. 1930. Primi dati di una revisione critica della sviluppo del gametofito gemmineo del genere *Cypripedium*. *Nuovo Giornale Bot. Ital.* 37:277–278.

55. ———. 1931. Ricerche embriologiche e cariologiche sul genere *Cypripedium*. *Nuovo Giornale Bot. Ital.* 38:155–212.

56. ———. 1932. Un reperto cariologico nella F_2 di *Paphiopedilum leeanum* × (*P. Spicerianum* + *P. insigne*). *Nuovo Giornale Bot. Ital.* 39:251–253.

57. ———. 1934. Ibridazione interspecifica nel genera *Paphiopedilum*. *Nuovo Giornale Bot. Ital.* 41:189–237.

58. ———. 1945. Ibridazione interspecifica nel genere *Paphiopedilum*. *Nuovo Giornale Bot. Ital.* 52:21–29.

59. FRIEMAN, W. 1910. Über die Entwicklung der generativen Zelle der Pollenkorn der monokotylen Pflanzen. Diss., Bonn.

60. FUCHS, A., and ZIEGENSPECK, H. 1923. Aus der Monographie des *Orchis Traunsteineri* Saut, IV. Chromosomen eineger Orchideen. *Bot. Arch.* 5:457–471.

61. ———. 1924. *Naturw. Veretin F. Schwaben u. Neuberg*, 43.

62. GADELLA, T. W. J., and KLIPHUIS, K. 1963. Chromosome numbers of flowering plants in the Netherlands. *Acta Bot. Neerl.* 12:106 130.

63. GARAY, L. A. 1963. *Oliveriana* and its position in the Oncidieae. *Am. Orchid Soc. Bull.*, (January): 18–24.

64. GEITLER, L. 1940. Die Polyploidie der Dunergewebe hoherer Pflanzen. *Ber. Deutsch. Bot. Ges.* 58: 131–142.

65. GUIGNARD, L. 1884. Structure et division du noyan cellulaire. *Ann. Sci. Nat. Bot.* (Ser. VI). 17:5–59.

66. ———. 1886. Sur la pollinisation et ses effets chez les orchidees. *Ann. Sci. Nat. Bot.* (Ser. VII). 4:202–240.

67. HAGERUP, O. 1938. Studies on the significance of polyploidy, II. *Orchis*. *Hereditas* 24:258–264.

68. ———. 1944a. Notes on some boreal polyploids. *Hereditas* 30:152–160.

69. ———. 1944b. On fertilisation, polyploidy and haploidy in *Orchis maculatus* L. sens lat. *Dansk. Bot. Ark.* 11:1–26.

70. ———. 1945. Facultative parthenogenesis and haploidy in *Epipactis latifolia*. *K. Dansk. Videnskab. Biol. Meddelel.* 19:1–13.

71. ———. 1947. The spontaneous formation of haploid, polyploid, and aneuploid embryos in some orchids. *K. Dansk. Videnskab. Selskab. Biol. Meddelel.* 20:1–22.

72. HAIR, J. B. 1942. The chromosome complements of some New Zealand plants, I. *Trans. Roy. Soc. N.Z.* 71:271–276.

73. HARMSEN, L. 1943. Studies on the cytology of Arctic plants, II. *Habenaria*. *Meddelelser om Gronland.* 131:3–15.

74. HEITZ, E. 1926. Der Nachweis der Chromosomen vergleichende Studien über ihre Zahl, Grosse und Form im Pflanzenreich. *Z. Bot.* **18**:625–681.

75. HESLOP-HARRISON, J. 1948. Field studies in *Orchis* L., I. The structure of dactylorchid populations on certain islands in the Inner and Outer Hebrides. *Transact. Proc. Bot. Soc. Edinb.* **35**:26–66.

76. ———. 1950. *Orchis cruenta* Mull. in the British Isles. *Watsonia* **1**:366–375.

77. ———. 1951. A comparison of some Swedish and British forms of *Orchis maculata.* *Svensk. Bot. Tidskr.* **45**:608–635.

78. ———. 1953. Microsporogenesis in some triploid dactylorchid hybrids. *Ann. Bot. N.S.* **17**:539–549.

79. HEUSSER, K. 1915. Die Entwicklung der generativen Organe von *Himantoglossum hircinum* Spr. *Beih. Bot. Centralbl.* **32**:218–277.

80. ———. 1938. Chromosomenverhaltnisse bei schweizerischen basitonen Orchideen. *Ber. schweisz. Bot. Gesell.* **48**:562–605.

81. HOFFMANN, K. 1929. Zytologische Studien der Orchideen. Ber. dtsche. *Bot. Gesell.* **47**:321–326.

82. ———. 1930. Beiträge zur Cytologie der Orchidaceen. *Planta* **10**:523–595.

83. HOLMEN, K., and KAAD, P. 1956. Über *Dactylorchis traunsteineri* auf der Insel Laso. *Bot. Tidssk.* **53**:35–48.

84. HUMPHREY, L. M. 1932a. Somatic chromosomes in certain Minnesota orchids. *Am. Nat.* **66**:471–474.

85. ———. 1932b. The somatic chromosomes of eight species of Orchidaceae. *Proc. Iowa Acad. Sci.* **39**:137.

86. ———. 1933–1934. Somatic chromosomes of *Cyp. hirsutum* and six species of genus *Habenaria. Proc. Iowa Acad. Sci.* **40**:75; *Am. Nat.* **68**:184–186.

87. HUREL-PY, G. 1938. Etude des noyaux vegetatifs de *Vanilla planifolia. Rev. Cytol. Cyto-physiol. Veg.* **3**:129–133.

88. ITO, I., and MUTSURRA, O. 1957. Chromosome numbers of *Dendrobium* species and hybrids. *Jap. Orchid Soc. Bull.* **3**:1–4.

89. ———. 1958. Ebine groups and their chromosome numbers. *Jap. Orchid Soc. Bull.* **4**(2):4–6.

90. ———. 1959. Chromosome numbers of "Ebine" and its allies (*Calanthe* species) native to Japan (II). *Jap. Orchid Soc. Bull.* **5**(2):1–2.

91. JONES, K. 1963. The chromosomes of *Dendrobium. Am. Orchid Soc. Bull.* **32**(8):634–640.

92. ———. 1967. The chromosomes of Orchids, II. *Vandeae* Lindl. *Kew Bull.* **21**:151–156.

93. JONES, K., and DAKER, M. G. 1968. The chromosomes of orchids, III. Catasetinae Schltr. *Kew Bull.* **22**:421–427.

94. JORGENSEN, C. A., SORENSEN, Th., and WESTERGAAD, W. 1958. The flowering plants of Greenland. A taxonomical and cytological survey. *Biol. Skr. Dansk.* (*Vidensk. Selsk.* **9**:1–172).

95. KAMEMOTO, H. 1950. Polyploidy in Cattleyas. *Am. Orchid Soc. Bull.* **19**:366–373.

96. ———. 1952. Further studies on polyploid Cattleyas. *Bull. Pac. Orchid Soc.* **10**:141–149.

97. ———. 1959a. The origin and significance of polyploidy in *Vanda. Pacific Orchid Soc. Bull.* **16**(3-4):77–95.

98. ———. 1959b. Studies on chromosome numbers of orchids in Japan. *Jap. Orchid Soc. Bull.* 5(1):1–4.

99. KAMEMOTO, H., and RANDOLPH, L. F. 1949. Chromosomes of the *Cattleya* tribe. *Am. Orchid Soc. Bull.* 18:366–369.

100. KAMEMOTO, H., and SAGARIK, R. 1967. Chromosome numbers of *Dendrobium* species of Thailand. *Am. Orchid Soc. Bull.* 36:889–894.

101. KAMEMOTO, H., SAGARIK, R., and DIEUTRAKUL, S. 1963. Karyotypes of *Paphiopedilum* species of Thailand. *The kasetsart J.* 3(2):69–78.

102. KAMEMOTO, H., SAGARIK, R., DIEUTRAKUL, S., and KASEMSAP, S. 1964. Chromosome numbers of Sarcanthine orchid species of Thailand. *Nat. Hist. Bull. Siam Soc.* 20(4):235–241.

103. KAMEMOTO, H., and SHINDO, K. 1962. Genome relationships in interspecific and intergeneric hybrids of *Renanthera*. *Am. J. Bot.* 49(7):737–748.

104. KAMEMOTO, H., TANAKA, R., and KOSAKI, K. 1961. Chromosome numbers of orchids in Hawaii. *Univ. Hawaii Agr. Exp. Sta. Bull.* 127. 32 pp.

105. KAMEMOTO, H., and TARA, M. 1969. The relationship of *Renanthera storiei* and *Trichoglottis fasciata*. *Brittonia* 21:126–129.

106. KLIPHUIS, E. 1963. Cytological observations in relation to the taxonomy of the orchids of the Netherlands. *Acta Bot. Neerl.* 12:172–194.

107. KNABEN, G. 1950. Chromosome numbers of Scandinavian arctic-alpine plant species, I. *Blyttia* 8:129–155.

108. KNABEN, G., and ENGELSKJON, T. 1967. Chromosome numbers of Scandinavian arctic-alpine plant species, II. *Act. Boreal. A. Sci.* 21:1–57.

109. KOSAKI, K. 1958. Preliminary investigations on the cytogenetics of *Dendrobium*. *Proo. Sooond World Orchid Congress* Harvard University Press, Cambridge, Mass.

110. KOSAKI, K., and KAMEMOTO, H. 1961. Chromosomes of some *Dendrobium* species and hybrids. *Na Pua Okika o Hawaii Nei*, 11:75–86.

111. KUSANO, S. 1915. Experimental studies on the embryonal development in an Angiosperm. *J. Coll. Agr. Imp. Univ. Tokyo.* 6:7–120.

112. LAANE, M. M. 1967. Kromozomundersokelser i Ost-Finnmarks flora, II. (Chromosome numbers in the flora of eastern Finnmark, II.) *Blyttia* 25:45–54.

113. ———. 1969. Further chromosome studies in Norwegian vascular plants. *Blyttia* 27: 5–17.

114. LA COUR, L. F. 1952. *Rep. John Innes Hort. Instn.* 42:47–.

115. LARSEN, K. 1960. Cytological and experimental studies on the flowering plants of the Canary Islands. *Biol. Skrift. K. Dansk. Vidensk. Selsk.* 11(3):1–60.

116. ———. 1968. Brief notes on *Neuwiedia singapureana* in Thailand. *Nat. Hist. Bull. Siam Soc.* 22:330–331.

117. LEE, Y. N. 1967. Chromosome numbers of flowering plants in Korea (1). *J. Korean Cult. Res. Inst.* 11:455–478.

118. LEVEQUE, M., and GORENFLOT, R. 1969. Prospections caryologiques dans la flore littorale du Boulonnais. *Bull. Soc. Bot. Nord France* 22:27–58.

119. LÖVE, A. 1951. Tofragros (*Dactylorchis Fuchsii*) a Islandi. *Natuurufraedingurinn* 21:91–93.

120. ———. 1954. Cytotaxonomical evaluation of corresponding taxa. *Vegetatio Acta Geobotanica* 5(6):212–224.

121. Löve, A., and Löve, D. 1942. Cyto-taxonomic studies on boreal plants. *K. Fysiogr. Sallsk. i Lund Forhandlingar 12.*

122. ———. 1944. Cyto-taxonomical studies on boreal plants, III. Some new chromosome numbers of Scandinavian plants. *Arkiv. Bot.* **31A**:1–22.

123. ———. 1948. Chromosome numbers of Northern plant species. *University Inst. Applied Sci. Dept. Ag. Reports. Series B,* No. 3, Reykjavik, Iceland.

124. ———. 1956. Cytotaxonomical conspectus of Icelandic flora. *Acta Hort. Gotab.* **20:** 65–290.

125. ———. 1961. Chromosome numbers of central and northwest European plant species. *Opera Botanica,* in *Supplm. Ser. Botaniska Notiiser.* 5:1–581.

126. ———. 1965. IOPB chromosome number reports III and IV. (A. Löve and O. T. Solbrig). *Taxon* **14**(2):50–57; *Taxon* **14**(3):86–87.

127. ———. 1969. In IOPB chromosome number reports XXI. *Taxon* **18**:310–315.

128. Löve, A., and Simon, W. 1968. Cytotaxonomical notes on some American orchids. *Southwestern Naturalist* **13**:335–342.

129. MacMahon, B. 1936. Meiosis in the pollen mother-cells of *Listera ovata. Cellule* **45**:209–262.

130. Malvesin-Fabre, G., and Eyme, J. 1949. Le noyan et la mitose chez *Limodorum abortivum. Compt. Rend. Acad. Sci.* **228**:2050–2057.

131. Martin, 1963. Chromosome number and behavior in a *Vanilla* hybrid and several *Vanilla* species. *Bull. Torrey Bot. Club* **90**:416.

132. Matsuura, H., and Sato, T. 1935. Contributions to the idiogram study in phanerogamous plants. I. *Fac. Sci. J. Hokkaido Imp. Univ.* (Series V). **5**:32–75.

133. Maude, R. F. 1939. The Merton catalogue. A list of the chromosome numerals of species of British flowering plants. *New Phytol.* **38**:7–31.

134. McQuade, H. A. 1949. The cytology of *Paphiopedilum* Maudiae. *Ann. Mo. Bot. Gard.* **36**:433–474.

135. Mehlquist, G. A. L. 1947a. Polyploidy in the genus *Paphiopedilum* Pfitz. (*Cypripedium* Hort.) and its implications. *Mo. Bot. Gard. Bull.* **35**:211–228.

136. ———. 1947b. Some smear technics for counting chromosomes in orchids. *Mo. Bot. Gard. Bull.* **35**:229–231.

137. ———. 1949. The importance of chromosome numbers in orchid breeding. *Am. Orchid Soc. Bull.* **18**:284–293.

138. ———. 1952. Chromosome numbers in the genus *Cymbidium. Cymbidium Soc. News* **7**.

139. Mehra, P. N., and Bawa, K. S. 1962. Chromosome studies in Orchidaceae. *Proc. Ind. Sci. Cong.* **1962**:326–327.

140. ———. 1970. Cytological observations on some northwest Himalayan orchids. *Caryologia* **23**:273–282.

141. Mehra, P. N., and Vij, S. P. 1970. In IOPB chromosome number reports, XXV. *Taxon* **19**:102–113.

142. Menninger, E. D. 1963. Diary of a colchicine induced tetraploid *Cymbidium. Am. Orchid Soc. Bull.* **32**(11):885–887.

143. Miduno, T. 1937. Chromosomenstudien von Orchidazeen. *Jap. J. Genetics* **13**:259.

144. ———. 1938. Chromosomenstudien an Orchidazeen, I. Karyotype und Mixoploidie bei Cephalanthera und Epipactis. *Cytologia* **8**:505–514.

145. ———. 1939. Chromosomenstudien an Orchidazeen, II. Somatischen Chromosomenzahlen einiger Orchideen. *Cytologia* 9:447–451.

146. ———. 1940a. Chromosomenstudien an Orchidazeen, III. Über das Vorkommen von haploiden Pflanzen bei *Bletilla striata* Reichb. f. var. *gebina* Reichb. f. *Cytologia* 11:156–177.

147. ———. 1940b. Chromosomenstudien an Orchidazeen, IV. Chromosomenzahlen einiger Arten und Bastarde bei Orchideen. *Cytologia* 11:179–185.

148. ———. 1953. Die Meiose und die erste Teilung in Pollen Korn bei *Pogonia japonica*. *Jap. J. Genet.* 28:175.

149. ———. 1954. Chromosomenstudien an Orchidazeen, V. Über das zytologischen Verhalten des Artbastardes zwischen *Bletilla striata* (n = 16) und *Bl. formosana* (n = 18). *Cytologia* 19(2–3):239–248.

150. ———. 1955. Karyotypanalyse und differentielle farbung der Chromosomen von *Cypripedium debile*. *Jap. J. Genet.* 30:176. (Abstract.)

151. MIDUNO, T., and YAMAZAKI, N. 1952. Über die verschiedenen Karyotypen und Kinetochorpaarung, beobachtet an den Kernteilung der jeden Wurzelspitze der an die Gattung *Pogonia* gehorenden Pflanzen. *Jap. J. Genet.* 27:210.

152. ———. 1963. Chromosome number and meiosis of *Pogonia*. *28th Ann. Meeting Bot. Soc. of Japan.*

153. MITSUKURI, Y., and KOZUKA, T. 1967. Cytological studies on Japanese Orchidaceae, 1. Chromosome numbers and karyotypes of *Liparis*. *La Kromosomo* 68:2221–2228.

154. MODILEWSKI, J. 1918. Cytological and embryological studies on *Neottia nidus-avis* (L.) Rich. *Verh. Kiewer Ges. Naturf.* 26:1–55.

155. MORINAGA, T., and FUKUSHIMA, E. 1931. Chromosome numbers of cultivated plants, III. *Bot. Mag. Tokyo* 45:140–145.

156. MULAY, B. N., and PANIKKAR, T. K. B. 1953. The chromosome number and morphology in *Cypripedium speciosum* L. *Proc. Rajasthan Acad. Sci.* 4:29–31.

157. MÜLLER, H. A. C. 1912. Kernstudien an Pflanzen. *Arch. Zellfor.* 8:1–52.

158. MUTSUURA, O. 1959. On chromosome counts in orchids native to Jap. *Japan Orchid Soc. Bull.* 5(2):5–7.

159. ———. 1963. Chromosome numbers of orchid species in Japan. *Kyoto Shigaku Kenkyu Ronbun Shu.* 1:123–232.

160. MUTSUURA, O., and NAKAHIRA, R. 1958. Chromosome numbers of the family Orchidaceae in Japan (1). *Scientific Reps. Saikyo Univ.* 2(5):25–30.

161. ———. 1959. Chromosome numbers of the family Orchidaceae in Japan (2). *Scientific Reps. Kyoto Prefectural Univ.* 3(1).27 01.

162. ———. 1960. Chromosome numbers of the family Orchidaceae in Japan (3). *Scientific Reps. Kyoto Prefectural Univ.* 3(2):11–16.

163. NEVLING, L. I. 1969. Ecology of an elfin forest in Puerto Rico, 3. Chromosome numbers of some flowering plants. *J. Arnold Arboretum.* 50:99–103.

164. NIIMOTO, D. H., and RANDOLPH, L. F. 1958. Chromosome inheritance in *Cattleya*. *Am. Orchid Soc. Bull.* 27:157–162; 240–247.

165. ONO, R., and HASHIMOTO, A. 1956. Chromosome studies in Orchidaceae, I. Chromosome numbers in four species of Orchidaceae. *Bot. Mag. Tokyo.* 69:286–288.

166. ONO, R., FUJIYA, T., and OKAMOTO, Y. 1957. Chromosome studies in Orchidacea, II.

Chromosome numbers of *Oreorchis patens* Lindl. and *Spiranthes sinensis* Ames. *J. Hokkaido Gakugei Univ.* 8:32–34.

167. PACE, L. 1907. Fertilization in *Cypripedium*. *Bot. Gaz.* 44:353–374.

168. ———. 1909. The gametophytes of *Calopogon*. *Bot. Gaz.* 48:126–139.

169. ———. 1914. Two species of *Gyrostachys*. *Baylor Univ. Bull.* 17:1–16.

170. PANCHO, J. B. 1965a. IOPB Chromosome number reports, III. (A. Löve and O. T. Solbrig). *Taxon* 14(2):50–57.

171. ———. 1965b. IOBP Chromosome number reports, IV. (A. Löve and O. T. Solbrig). *Taxon* 14(3):86–87.

172. RICHARDSON, M. M. 1933. Chromosome variation in *Listera ovata* R. Br. *Univ. Calif. Publ. Bot.* 17:51–60.

173. ———. 1935. The chromosomes of some British orchids. *Durham Univ. Phil. Soc. Proc.* 9:135–140.

174. ROSENBERG, O. 1905. Zur Kenntnis der Reduktionsteilung in Pflanzen. *Bot. Notiser:* 1–25.

175. SAGAWA, Y. 1962a. Chromosome numbers of some species and hybrids of the *Cattleya* group in Florida. *The Florida Orchidist* 5(4):201–203.

176. ———. 1962b. Cytological studies of the genus *Phalaenopsis*. *Am. Orchid Soc. Bull.* 31:459–465.

177. SAGAWA, Y., and NIIMOTO, D. H. 1961. Cytological studies in the Orchidaceae. *Am. Orchid Soc. Bull.* 30(8):628–632.

178. SAMPATHKUMARAN, M., and SESHAGIRIAH, K. N. 1931. Cytology of the embryo sac in orchids. *Proc. Indian Sci. Congress* 18:277–278.

179. SENIANINOVA, M. 1925. Etude embryologique de l'*Ophrys myodes*. *Z. Russ. Bot. Ges.* 9:10–14.

180. SESHAGIRIAH, K. N. 1934. Pollen sterility in *Zeuxine sulcata* Lindley. *Curr. Sci.* 3:205–206.

181. ———. 1941. Morphological studies in Orchidaceae, I. *Zeuxine sulcata* Lindley. *J. Indian Bot. Soc.* 20:357–365.

182. SHIMOYA, C., and FERLAN, L. 1952. Estudos orquideologicos, III. Determinacoes cromosomicas em *Ophyrus*. *Broteria* 21:171–176.

183. SHINDO, K., and KAMEMOTO, H. 1962. Genome relationships of *Neofinetia* Hu and some allied genera of Orchidaceae. *Cytologia* 27:402–409.

184. ———. 1963a. Karyotype analysis of some Sarcanthine orchids. *Am. J. Bot.* 50:73–79.

185. ———. 1963b. Chromosomes of *Dendrophylax funalis* and *Aerangis biloba*. *Am. Orchid Soc. Bull.* 32(10):821–823.

186. ———. 1963c. Chromosome relationships of *Aerides* and allied genera. *Am. Orchid Soc. Bull.* 32(11):922–926.

187. ———. 1963d. Chromosome numbers and genome relationships of some species in the Nigrohirsutae section of *Dendrobium*. *Cytologia* 28:68–75.

188. ———. 1963e. Karyotype analysis of some species of *Phalaenopsis*. *Cytologia* 28:390–398.

189. SHOJI, T. 1963. Cytological studies on Orchidaceae, II. Chromosome numbers and karytoypes of six Japanese species. *La Kromosomo* 55–56:1823–1828.

190. SIMON, W. 1968. Chromosome numbers and B-chromosome in *Listera*. *Caryologia* 21:181–189.

191. SINOTO, Y. 1962. Chromosome numbers in *Oncidium* alliance. *Cytologia* **27**(3):306–313.

192. ———. 1969a. Chromosomes in *Oncidium* and allied genera, I. Genus *Oncidium*. *Kromosom* **76**:2459–2473.

193. ———. 1969b. Chromosomes in *Oncidium* and allied genera, II. Genera *Erycina, Gomesa, Odontoglossum, Miltonia, Brassia* and *Sigmatostalix*. *Kromosom* **77–78**: 2532–2538.

194. SINOTO, Y., and SHOJI, T. 1962. Cytological studies on Orchidaceae, I. Chromosome numbers and karyotypes of some native orchids in Japan. *International Christian Univ. Publ. VA* **1**:20–27.

195. SKALINSKA, M., PIOTROWIZC, M., SOKOLOWSKA-KULCZYCKA, A., et al. 1961. Further additions to chromosome numbers of Polish Angiosperms. *Acta. Polsk. Towarz. Bot.* **30**:463–489.

196. SOKOLOVSKAYA, A. P. 1963. Geographical distribution of polyploidy in plants. *Best. Lenengrad Univ.* 1963, No. 15, *Ser. Biol.*: 38–52.

197. SOKOLOVSKAYA, A. P., and STRELKOVA, O. S. 1940. Karyological investigations of the alpine flora on the main Caucasus range and the problem of geographical distribution of polyploids. *Compt. Rend. Acad. Sci. U.S.S.R.* **29**:415–418.

198. ———. 1960. Geographical distribution of the polyploid species of plants in the Eurasiatic Arctic. *Bot. Zh. U.S.S.R.* **45**:369–381.

199. SORSA, V. 1962. Chromosomenzahlen Finnischer Kormophyten, 1. *Ann. Acad. Sci. Fenn. Ser. A. IV. Biologica* **58**:1–14.

200. ———. 1963.. Chromosomenzahlen Finnischer Kormophyten, II. *Ann. Acad. Sci. Fenn. Ser. A., IV. Biologica* **68**:1–14.

201. STANER, P. 1929. Preieduction ou postreduction dans *Listora ovata* R. Br. *Cellule* **39**:219–235.

202. STENAR, H. 1937. Om *Acroanthos monophyllos* (L.) Greene dessgeogrophiska utbrenning ach embryologi. *Heimbygdas Tidskr.* (1) *Fornvardaren Uppsala* (*Festskrift Till Erik Modin*) **6**:177–231.

203. STOREY, W. B. 1952. Chromosome numbers of some *Vanda* species and hybrids. *Am. Orchid Soc. Bull.* **21**:801–806.

204. STOREY, W. B., KAMEMOTO, H., and SHINDO, K. 1963. Chromosomes of *Vanda spathulata* and its hybrids. *Am. Orchid Soc. Bull.* **32**:703–709.

205. STRASBURGER, E. 1888. *Über Kern-und Zelltheilung im Pflanzenreich, nebst einem Anhand uber Befruchtung. Histol. Beitrage,* Vol. I. Gustav Fischer, Jena.

206. SUESSENGUTH, K. 1921. Beitrage zur Frage des systematischen Anschlusses der Monokotylen. *Bot. Centbl. Beihefte* **38**:1–80.

207. ———. 1923. Über die Pseudogamie bei Zygopetalum Mackayi Hook. *Ber. deut. Bot. Gesell.* **41**:16–23.

208. SUGIURA, T. 1928. Chromosome numbers in some higher plants, I. *Bot. Mag, Tokyo* **42**:504–506.

209. ———. 1939. Studies on the chromosome numbers in higher plants, III. *Cytologia* **10**:205–212.

210. SWAMY, B. G. L. 1941. The development of the male gamete in *Cymbidium bicolor* Lindl. *Proc. Indian Acad. Sci. B.* **14**:454–460.

211. ———. 1944. The embryo sac and embryo of *Satyrium nepalense. Indian Bot. Soc. J.* **23**:66–70.

212. TAKAMINE, K. 1916. Über die ruhenden und die prasynaptischen Phasen der Reduktionsteilung. *Bot. Mag. Tokyo* **30**:293–303.

213. TAN, K. W. 1969. The systematic status of the genus *Bletilla* (Orchidaceae). *Brittonia* **21**:202–214.

214. TANAKA, R. 1962a. Chromosome count of orchids in Japan, I. *Jap. Orchid Soc. Bull.* **8**(1):1–4.

215. ———. 1962b. Cytological studies on the speciation in *Spiranthes sinensis*. *Proc. 27th Ann. Meet. Jap. Bot. Soc.*

216. ———. 1962c. Differentiation in karyotypes between *Pogonia minor* and *P. japonica*. *Jap. J. Genet.* **37**(5):414.

217. ———. 1964a. Chromosome count of orchids in Japan, II. *Jap. Orchid Soc. Bull.* **10**(1):1–5.

218. ———. 1964b. Differentiation and chromosomes in garden orchids. *Heredity* **18**(7): 20–24.

219. ———. 1965a. Intraspecific polyploidy in *Goodyera maximowicziana* Makino. *La Kromosomo* **60**:1945–1950.

220. ———. 1965b. Chromosome numbers of some species of Orchidaceae from Japan and its neighbouring area. *J. Jap. Bot.* **40**(3):65–77.

221. ———. 1965c. H³-thymidine autoradiographic studies on the heteropycnosis, heterochromatin and euchromatin in *Spiranthes sinensis*. *Bot. Mag. Tokyo* **78**(920): 50–62.

222. ———. 1966. Chromosome count of orchids in Japan, III. *Jap. Orchid Soc. Bull.* **12** (2):2–4.

223. ———. 1969a. Speciation and karyotypes in *Spiranthes sinensis*. *J. Sci. Hiroshima Univ. Ser. B. Div. 2.* **12**:165–198.

224. ———. 1969b. Deheterochromatinization of the chromosomes in *Spiranthes sinensis*. *Jap. J. Genet.* **44**:291–296.

225. TANAKA, R., and MATSUDA, T. 1972. A high occurrence of accessory chromosomal types in *Tainia laxiflora*, Orchidaceae. *Bot. Mag. Tokyo* **85**:43–49.

226. TARA, M., and KAMEMOTO, H. 1970. Karyotype relationships in the Sarcanthinae (Orchidaceae). *Am. J. Bot.* **57**:176–182.

227. TAYLOR, R. L. 1967. In IOPB chromosome number reports, XIII. *Taxon* **16**:445–461.

228. TAYLOR, R. L., and MULLIGAN, G. A. 1968. *Flora of the Queen Charlotte Islands, Part 2. Cytological Aspects of the Vascular Plants*. Queen's Printer, Ottawa, p. 148.

229. THULIN, M. 1970. Chromosome numbers of some vascular plants from East Africa. *Bot. Notiser* **123**:488–494.

230. TISCHLER, C. 1934. Die Bedeutung der Polyploidie fur die Verbreitung der Angiospermen. *Bot. Jahrb.* **67**:1–36.

231. ———. 1935–1936. Pflanzliche Chromosomen-Zahlen. *Tabulae Biol. Periodicae* **11-12**: 1–83; 109–226.

232. TITZ, W. 1969. Zur Cytotaxonomie von *Arabis hirsuta* agg. (Cruciferae). IV. Chromosomenzahlen von *A. sagittata* (Bertol.) DC. und *A. hirsuta* (L.) Scop. s. str. aus Europa. *Oesterr. Bot. Zeits.* **117**:197–200.

233. TOMASI, J. A. 1954. An Introductory Note. *Kiesewetter Orchid Gardens 1954 Catalogue.*

234. Tuschnjakova, M. 1929. Embryologische und zytologische Beobachtungen über *Listera ovata*. (Orchidaceae). *Planta* 7:29–44.

235. Vajrabhaya, T., and Randolph, L. F. 1960. Chromosome studies in *Dendrobium*. *Am. Orchid Soc. Bull.* 29:507–517.

236. ———. 1961. Chromosome inheritance in pentaploid and aneuploid cattleyas. *Am. Orchid. Soc. Bull.* 20:209–213.

237. Vermeulen, P. 1938. Chromosomes in *Orchis*. *Chron. Bot.* 4:107–108.

238. ———. 1947. *Studies on Dactylorchids*. F. Schoturius and Jens, Utrecht.

239. ———. 1949. Varieties and forms of Dutch orchids. *Nederlandsche Bot. Vereenegeng, Leyden* 56:204–242.

240. Vis, J. D. 1933. Iets over de Cytologie der Orchideen. *Verh. 24. Neerl. Natuur en Geneesk Congress*, pp. 186–189.

241. Wegener, K. A. 1966. Ein Beltrag zur Zytologie von Orchideen aus dem Gebiet der DDR. *Wiss. Zschr. Ernst-Mortiz-Arndt-Univ. Greifswald, Mat.-nat.* 15:1–7.

242. Weijer, J. 1952. The colour-differences in *Epipactis helleboriae* (Cr. Wats.) Coult. and the selection of the genetical varieties by environment. *Genetica* 26:1–32.

243. Wimber, D. E. 1957a. Cytogenetic studies in the genus *Cymbidium*. Chromosome numbers within the genus and related genera, I. *Am. Orchid Soc. Bull.* 26:636–639.

244. Woodard, J. W. 1951. Some chromosome numbers in *Phalaenopsis*. *Am. Orchid Soc. Bull.* 20:356–358.

245. ———. 1952. Some chromosome numbers in *Vanda*. *Am. Orchid Soc. Bull.* 21:247–249.

246. Yamasaki, N. 1959. Differentielle Farbung der chromosomen der ersten meiotischen Metaphase von *Cypripedium debile*. *Chromosoma* 10:454–460.

247. Yeh, J. C. C. 1962. A cytological study of selected *Cymbidium pumilum* hybrids. *Am. Orchid Soc. Bull.* 31(11):904–915.

248. Yuasa, A. 1936. Nagoran no Sensyokutaisu (Chromosomenahl von *Aerides japonicum*). *Bot. Zool.* 4:953.

249. Zhukova, P. G. 1967. Karyology of some plants, cultivated in the Arctic-Alpine Botanical Garden, in N. A. Avrorin, ed., *Plantarum in Zonam Polarem Transportatio*, Vol. II. Leningrad, 1967, pp. 139–149, (In Russia.).

250. Kugust, K. 1966. Hybridizing with oncidiums. Proceedings of the Fifth World Orchid Conference. Long Beach, Calif.

251. Tanaka, R. 1973. Unpublished.

252. Wilfret, G., and Kamemoto, H. 1971. Genome and karyotype relationships in the genus *Dendrobium* (Orhcidaceae), II. Karyotype relationships. *Cytologia* 36: 604–613.

II

Natural and Artificial Hybrid Generic Names of Orchids

1887-1973

LESLIE A. GARAY AND HERMAN R. SWEET

No other plant family is comparable to the ORCHIDACEAE in the number of known hybrids, both at the natural and artificial level. It is agreed by virtually everybody that, since the time when Dominy flowered his first artificial cross, *Calanthe Dominyi* in 1856, the number of man made hybrids far exceeds that of the described species, which is estimated between 30,000 and 35,000.

There are 577 hybrid generic names recorded in this compendium, most of which were published since 1887, when Rolfe proposed following the precedent established by Masters in *Gardeners' Chronicle* 358 (1872): hybrid generic names are to be compounded from those of the parent genera. Rolfe's proposal was officially adopted in 1910, during the International Horticultural Congress in Brussels. During that Congress it was also resolved that bigeneric and trigeneric hybrid names were to be compounded from the parental generic names, while quadrigeneric names were to be named for some person distinguished either in botany or in horticulture, and that the suffix "-ara" be attached to the name of the person so honored. These general principles are still applicable and strictly adhered to by both the *International Code of Botanical Nomenclature* and the *International Code of Nomenclature of Cultivated Plants*. Both of these Codes derive their authority from the International Union of Biological Sciences.

Because some of the practices that became established during the past hundred years of orchid hybridization are still running contrary to the articles of both Codes, in 1964, during the Tenth International Botanical Congress, the Committee for Hybrids of the Nomenclatorial Section completely redrafted the wordings of Articles 40, H.3 and H.4 of the Botanical Code. These reworded articles embody the following new principles: "that (1) 'generic names' of hybrid genera should be regarded as condensed formulae and should be validly published by an accompanying statement of their parentage, without any Latin diagnosis or other description, and (2) that as a consequence, such 'generic names' should be applicable only to the plants which are accepted taxonomically as derived from the parent genera named."

To clarify the meaning and because of the application of these new principles to hybrid generic names of orchids, we are quoting here the full text of Article 40.

Article 40.   For purposes of valid publication, the name of a hybrid group of generic, subgeneric or sectional rank, which is a condensed formula or equivalent to a condensed formula (see H.3 and H.4), must be published with a statement of the names of the parent genera, subgenera, or sections respectively, but a Latin diagnosis or other description is not necessary . . . .

For purposes of valid publication, names of hybrids of specific or lower rank with Latin epithets are subject to the same rules as are those of non-hybrid taxa of the same rank . . . .

For purposes of priority, names and epithets in Latin form given to hybrids are subject to the same rules as are those of non-hybrid taxa of corresponding rank . . . .

The correctness of the interpretation of these articles is clearly echoed in *An Annotated Glossary of Botanical Nomenclature*, prepared on the authorization of the Nomenclature Section of the Tenth International Botanical Congress in 1964. In this glossary, under the heading Formula we find the following statement: "Condensed formulae formed from parts of generic names and applied to intergeneric hybrids are treated as 'generic names' (Articles 4.3, 4.4) and are subject to the rule of priority and the homonym rule."

In 1969 the International Orchid Commission on Classification, Nomenclature and Registration, a self-elected body, published a *Handbook on Orchid Nomenclature and Registration*. In this handbook many useful principles were borrowed from both the Botanical and Horticultural codes, except the strict adherence to their cornerstone, the *principle of priority*. The priority rule has been only arbitrarily employed and completely subjected to the personal likes and dislikes of the Handbook Committee and its self-selected Advisory Board.

In this compendium, as in the original presentations in 1966 and 1969, there is a faithful adherence to the requirements of both codes, especially to Articles 40, H.3 and H.4 of the International Code of Botanical Nomenclature. The rule of priority is strictly observed, an action most important if stabilization in orchid nomenclature is to be achieved.

For convenience, a list of nomenclatorial changes which are incorporated in this compendium of hybrid generic names is presented here. The column on the left represents those names which are currently appearing or which have been used in horticultural literature; the column on the right gives their correct botanical equivalents.

Aerides Biswasiana	= Papilionanthe Biswasiana
Aerides cylindrica	= Papilionanthe subulata
Aeriles Greenii	= Papilionanthe Greenii
Aerides mitrata	= Seidenfadenia mitrata
Aerides pedunculata	= Papilionanthe pedunculata
Aerides uniflora	= Papilionanthe uniflora
Aerides vandarum	= Papilionanthe vandarum
Aganisia lepida	= Otostylis alba
Angraecum falcatum	= Neofinetia falcata
Angraecum philippinense	= Amesiella philippinensis
Arachnis Cathcartii	= Esmeralda Cathcartii
Arachnis Clarkei	= Esmeralda Clarkei
Arachnis Sulingii	= Armodorum Sulingii
Chondrorhyncha discolor	— Cochleanthes discolor
Cochlioda sanguinea	= Symphyglossum sanguineum
Colax jugosus	= Pabstia jugosa
Colax modestior	= Pabstia modestior
Colax placanthera	= Pabstia placanthera
Cymbidium elegans	= Cyperorchis elegans
Habenaria Susannae	= Pecteilis Susannae
Haemaria discolor	= Ludisia discolor
Kingiella philippinensis	= Kingidium deliciosum
Kingiella taenialis	= Kingidium taeniale
Leucorchis albida	= Pseudorchis albida
Lycaste Skinneri	= Lycaste virginalis
Miltonia Schroederiana	= Odontoglossum confusum
Parasarcochilus spathulatus	= Pteroceras spathulatus
Phaius Humblotii	= Gastrorchis Humblotii
Phaius simulans	= Gastrorchis simulans
Phaius tuberculosus	= Gastrorchis tuberculosa
Phalaenopsis Denevei	= Paraphalaenopsis Denevei
Phalaenopsis Laycockii	= Paraphalaenopsis Laycockii
Phalaenopsis serpentilingua	= Paraphalaenopsis serpentilingua
Renanthera elongata	= Porphyrodesme elongata
Renanthera histrionica	= Renantherella histrionica
Saccolabium giganteum	= Rhynchostylis gigantea
Sophronitis violacea	= Sophronitella violacea

Stauropsis fasciata	= Trichoglottis fasciata
Vanda Hookerana	= Papilionanthe Hookerana
Vanda Sanderana	= Euanthe Sanderana
Vanda teres	= Papilionanthe teres
Vanda tricuspidata	= Papilionanthe tricuspidata
Vandopsis Warocqueana	= Sarcanthopsis Warocqueana
Zygopetalum Jorisianum	= Mendoncella Jorisiana
Zygopetalum rostratum	= Zygosepalum labiosum

Finally, attention is called to the arrangements of present genera in Part II. The generic names involved in the makeup of a given hybrid generic name are arranged in an alphabetical sequence, and accordingly appear only once in the list.

Part I

List of Hybrid Generic Names

Aceraherminium in Camus, Ic. Orch. Eur. 2:366, 1929
Aceras × *Herminium*
1st hybr.: unnamed
Parentage: *Aceras anthropophora* × *Herminium monorchis*
    Syn.: **Aceras-Herminium** in Gremli, Neue Beitr. 3:35, 1883

Aceras-Herminium in Gremli, Neue Beitr. 3:35, 1883
Observation: See **Aceraherminium**

Adaglossum in Orch. Rev. 21:298, 1913
Ada × *Odontoglossum*
1st hybr.: **A.** Juno
Parentage: *Ada aurantiaca* × *Odontoglossum Edwardii*

Adamara in Bull. Roy. Soc. Bot. Belg. 47:402, 1911
Brassavola × *Cattleya* × *Epidendrum* × *Laelia*
1st hybr.: **A.** Fuchsia (as **Yamadara** Fuchsia)
Parentage: **Brasssolaeliocattleya** Eudetta × *Epidendrum Mariae*
    Syn.: **Yamadara** in Orch. Rev. 68:404, 1960

Adioda in Orch. Rev. 19:258, 1911
Ada × *Cochlioda*
1st hybr.: **A.** St. Fuscien
Parentage: *Ada aurantiaca* × *Cochlioda Noezliana*

Aeridachnanathe nom. hybr. gen nov.
Aerides × *Arachnis* × *Papilionanthe*
1st hybr.: **A.** Margaret Ede (as **Burkillara** Margaret Ede)
Parentage: **Aeridachnis** Bogor × *Papilionanthe Cooperi*

Aeridachnis in The Orch. J. 3:165, 1954
Aerides × *Arachnis*
1st hybr.: **A.** Bogor
Parentage: *Arachnis Hookerana* × *Ariedes odorata*

Aeridanthe in Bot. Mus. Leafl. Harv. Univ. 21:148, 1966
Aerides × *Euanthe*
1st hybr.: **A.** Tsuruko Iwasaki (as **Aeridovanda** Tsuruko Iwasaki)
Parentage: *Aerides Lawrenceae* × *Euanthe Sanderana*

Aeriditis in Orch. Rev. 81: March 1973
Aerides × *Doritis*
1st hybr.: **A.** Hermon Slade
Parentage: *Aerides Fieldingii* × *Doritis pulcherrima*

Aeridocentrum in Orch. Rev. 75: February, 1967
Aerides × *Ascocentrum*
1st hybr.: **A.** Luke Nok
Parentage: *Aerides flabellata* × *Ascocentrum curvifolium*

Aeridofinetia in Orch. Rev. 69:267, 1961
Aerides × *Neofinetia*
1st hybr.: **A.** Pink Pearl
Parentage: *Aerides Jarckiana* × *Neofinetia falcata*
    Syn.: **Holcorides** in Bot. Mus. Leafl. Harv. Univ. 23:188, 1972

Aeridoglossum in Orch. Rev. 71: September 1963
Aerides × *Ascoglossum*
1st hybr.: **A.** Peach Blossom
Parentage: *Aerides Lawrenceae* × *Ascoglossum caloplerum*

Aeridolabium in Orch. Rev. 67:329, 1959
Aerides × *Saccolabium*
1st hybr.: not yet reported
Observation: For the hybrid **Aeridolabium** Springtime see **Aeridostylis**

Aeridopsis in Orch. Rev. 46:200, 1938
Aerides × *Phalaenopsis*
1st hybr.: **A.** Shinjiku
Parentage: *Aerides japonica* × *Phalaenopsis* Leda

Aeridopsisanthe, nom. hybr. gen. nov.
Aerides × Euanthe × Vandopsis
1st hybr.: **A.** Hazel (as **Maccoyara** Hazel)
Parentage: **Vandopsides** Apple Blossom × *Euanthe Sanderana*

Aeridostylis in Hawkes, Orchids 242, 1961
Aerides × Rhynchostylis
1st hybr.: **A.** Springtime (as **Aeridolabium** Springtime)
Parentage: *Aerides Lawrenceae × Rhynchostylis gigantea*
  Syn.: **Rhynchorides** in Orch. Rev. 70: October 1962

Aeridovanda in Gard. Chron. ser. 3, 63:93, 1918
Aerides × Vanda
1st hybr.: **A.** Ruth
Parentage: *Aerides crassifolia × Vanda cristata*
Observation: For the hybrid **Aeridovanda** Elizabeth Young see **Eupapilio**
  and for **Aeridovanda** *Mundyi* see **Papilionanthe**
  Syn.: **Aerovanda** in Sander, List Orch. Hybr. Add. 319, 1949–1951
    Aeriovanda in Gartenfl. 86:252, 1937

Aeriovanda in Gartenfl. 86:252, 1937
Observation: see **Aeridovanda**

Aerovanda in Sander, List Orch. Hybr. Add. 319, 1949–1951
Observation: see **Aeridovanda**

Aliceara in Orch. Rev. 72: July 1964
Brassia × Miltonia × Oncidium
1st hybr.: **A.** Pacesetter
Parentage: **Brassidium** Coronet × **Miltonidium** Lustre

Amesangis nom. hybr. gen nov.
Aërangis × Amesiella
1st hybr.: **A.** Snow Nymph (as **Angrangis** Snow Nymph)
Parentage: *Aërangis fastuosa × Amesiella philippinensis*

Amesara in Bot. Mus. Leafl. Harv. Univ. 21:149, 1966
Euanthe × Renanthera × Vanda
1st hybr.: **A.** Donald McIntyre (as **Renantanda** Donald McIntyre)
Parentage: **Vandanthe** Clara Shipman Fisher × *Renanthera Storiei*

Anacamptiplatanthera in Fourn., Brev. Bot. 512, 1927
Anacamptis × Platanthera
1st hybr.: **A.** *Payoti*
Parentage: *Anacamptis pyramidalis × Platanthera bifolia*

Anacamptorchis in J. Bot. Fr. 6:113, 1892
Anacamptis × *Orchis*
1st hybr.: **A.** *Duquesnei* (as *Aceras Duquesnei*, 1851)
Parentage: *Anacamptis pyramidalis* × *Orchis palustris*
 Syn.: **Orchidanacamptis** in Guétrot, Pl. Hybr. Fr. II, 51, 1926

Angrangis in Orch. Rev. 80:161, 1972
Aërangis × *Angraecum*
1st hybr.: not yet reported
Observation: For **Angrangis** Snow Nymph see **Amesangis**

Angulocaste in Rev. Hort. Belg. 32:172, 1906
Anguloa × *Lycaste*
1st hybr.: **A.** *Bièvreana*
Parentage: *Lycaste virginalis* × *Anguloa Rueckeri*

Anoectogoodyera in Gard. Chron. ser. 3, 1:646, 1887
Anoectochilus × *Goodyera*
1st hybr.: not yet reported

Anoectomaria in J. Linn. Soc. Bot. 24:170, 1887
Observation: See **Ludochilus** including the hybrid **Anoectomaria**
 Dominyi. Haemaria is a synonym of *Ludisia*

Ansidium in Orch. Rev. 74: May 1966
Ansellia × *Cymbidium*
1st hybr.: **A.** Bess Waldon
Parentage: *Ansellia africana* × *Cymbidium* Dunster Castle

Anthechostylis nom. hybr. gen. nov.
Euanthe × *Rhynchostylis*
1st hybr.: **A.** Bangkok Sky (as **Vandachostylis** Bangkok Sky)
Parentage: *Rhynchostylis coelestis* × *Euanthe Sanderana*

Antheglottis in Bot. Mus. Leafl. Harv. Univ. 21:150, 1966
Euanthe × *Trichoglottis*
1st hybr.: **A.** Ulaula (as **Trichovanda** Ulaula)
Parentage: *Trichoglottis brachiata* × *Euanthe Sanderana*

Antheranthe in Bot. Mus. Leafl. Harv. Univ. 21:150, 1966
Euanthe × *Renanthera*
1st hybr. **A.** Titan (as **Renantanda** Titan)
Parentage: *Renanthera Imschootiana* × *Euanthe Sanderana*

Arachnadenia nom. hybr. gen. nov.
Arachnis × *Seidenfadenia*
1st hybr.: **A.** May Woo (as **Aeridachnis** May Woo)
Parentage: *Arachnis* Maggie Oei × *Seidenfadenia mitrata*

Arachnoglossum in Orch. Rev. 80:220, 1972
Arachnis × *Ascoglossum*
1st hybr.: **A.** Calobel
Parentage: *Arachnis* Ishbel × *Ascoglossum calopterum*

Arachnoglottis in Orch. Rev. 66:86, 1958
Observation: See **Trichachnis** including the hybrid **Arachnoglottis** Brown
   Bars
Arachnopsis in Anggrek Boelan 1:83, 1939
Arachnis × *Phalaenopsis*
1st hybr.: **A.** *Rosea*
Parentage: *Phalaenopsis Schillerana* × *Arachnis Maingayi*
Observation: For the hybrid **Arachnopsis** Eric Holttum see **Pararachnis**

Arachnostylis in Orch. Rev. 74: April 1966
Arachnis × *Rhynchostylis*
1st hybr.: **A.** Chorchalood
Parentage: *Arachnis Hookerana* × *Rhynchostylis gigantea*

Aranda in Orcrideeën 4:70, 1937
Arachnis × *Vanda*
1st hybr.: **A.** Edwina Tollens
Parentage: *Arachnis Maingayi* × *Vanda tricolor*
Observation: For the hybrid **Aranda** *Jacoba Louisa* see **Papilachnis**
   Syn: **Vandachnanthe** in Anggrek Boelan 1:67, 1939
      Vandarachnis in Orchideeën 6:107, 1939

Arandanthe in Malay. Orch. Rev. 5:13, 1957
Arachnis × *Euanthe* × *Vanda*
1st hybr.: **A.** Wendy Scott
Parentage: *Arachnis Hookerana* × **Vandanthe** *Rothschildiana*

Aranthera in Malay. Orch. Rev. 2:109, 1936
Arachnis × *Renanthera*
1st hybr.: **A**: Mohamed Haniff
Parentage: *Arachnis Hookerana* × *Renanthera coccinea*
Observation: For the hybrid **Aranthera** Star Orange see **Renaradorum**

Arizara in Orch. Rev. 73: October 1965
Cattleya × *Domingoa* × *Epidendrum*
1st hybr. **A.** Luis
Parentage: **Epigoa** Olivine × *Cattleya guttata*

Armodachnis in Na Pua Okika o Hawaii Nei 7:154, 1957
Arachnis × *Armodorum*
1st hybr.: **A.** Catherine (as *Arachnis* Catherine)
Parentage: *Armodorum Sulingi* × *Archnis Hookerana* var. *luteola*

Ascocenda in Orch. Rev. 57:172, 1949
Ascocentrum × *Vanda*
1st hybr.: **A.** Portia Doolittle
Parentage: *Ascocentrum curvifolium* × *Vanda lamellata*
Observation: For the hybrid **Ascocenda** Meda Arnold see **Schlechterara**

Ascodenia nom. hybr. gen. nov.
Ascocentrum × *Seidenfadenia*
1st hybr.: **A.** Cholratana (as **Aeridocentrum** Cholratana)
Parentage: *Ascocentrum curvifolium* × *Seidenfadenia mitrata*

Ascofinetia in Orch. Rev. 69:32, 1961
Ascocentrum × *Neofinetia*
1st hybr.: **A.** Twinkle
Parentage: *Neofinetia falcata* × *Ascocentrum miniatum*
    Syn.: **Holcocentrum** in Bot. Mus. Leafl. Harv. Univ. 23:188, 1972

Asconopsis in Orch. Rev. 76: February 1968
Ascocentrum × *Phalaenopsis*
1st hybr.: **A.** Mini-Coral
Parentage: *Phalaenopsis Schillerana* × *Ascocentrum miniatum*

Ascorachnis in Orch. Rev. 75: August 1967
Arachnis × *Ascocentrum*
1st hybr.: **A.** George Neo
Parentage: *Arachnis* Ishbel × *Ascocentrum miniatum*

Ascorella in Bot. Mus. Leafl. Harv. Univ. 21:151, 1966
Ascocentrum × *Renantherella*
1st hybr.: **A.** Curvionica (as **Renancentrum** Curvionicum)
Parentage: *Renantherella histrionica* × *Ascocentrum curvifolium*

Ascovandoritis in Orch. Rev. 77: September 1969
Ascocentrum × Doritis × Vanda
1st hybr.: **A.** Sonnhild Kitts
Parentage: *Doritis pulcherrima* × **Ascocenda** Red Gem

Aspasium in Orch. Rev. 6:161, 1958
Observation: See **Oncidasia** including the hybrid **Aspasium** Regal

Aspoglossum in Orch. Rev. 70: September 1962
Aspasia × Odontoglossum
1st hybr.: **A.** Nuuanu
Parentage: *Aspasia principissa* × *Odontoglossum cordatum*

Athertonara in Orch. Rev. 56:26, 1948
Observation: See **Renanopsis** including the hybrid **Athertonara** Lena
  Rowold

Bardendrum in Orch. Rev. 70: September 1962
Barkeria × Epidendrum
1st hybr.: **B.** Elvena
Parentage: *Barkeria Lindleyana × Epidendrum Schomburgkii*
  Syn.: **Barkidendrum** in Am. Orch. Soc. Bull. 31:667, 1962

Barkidendrum in Am. Orch. Soc. Bull. 31:667, 1962
Observation: See **Bardendrum**

Barlaceras in Riviera Scientif. 11:62, 1924
Aceras × Barlia
1st hybr.: **B.** *Terraccianoi*
Parentage: *Aceras anthropophora × Barlia longibracteata*
  Syn.: **Barliaceras** in Ciferri and Giacomini, Nomencl. Fl. Ital. pt. 1:167,
  1950

Barliaceras in Ciferri and Giacomini, Nomencl. Fl. Ital. pt. 1:167, 1950
Observation: See **Barlaceras**

Bateostylis in Orch. Rev. 75: November 1967
Batemannia × Otostylis
1st hybr.: **B.** Silver Star
Parentage: *Batemannia Colleyi × Otostylis brachystalix*

Beallara in Orch. Rev. 78: November 1970
Brassia × Cochlioda × Miltonia × Odontoglossum
1st hybr.: **B.** Vashon
Parentage: **Miltassia** Charles M. Fitch × **Odontioda** Carmine

Beardara in Orch. Rev. 78: July 1970
Ascocentrum × *Doritis* × *Phalaenopsis*
1st hybr.: **B.** Charles Beard
Parentage: **Doritaenopsis** Red Coral × *Ascocentrum miniatum*

Beaumontara in Orch. Rev. 69:198, 1961
Observation: See **Recchara** including the hybrid **Beaumontara** Herb

Benthamara in Bot. Mus. Leafl. Harv. Univ. 21:152, 1966
Arachnis × *Euanhte* × *Paraphalaenopsis*
1st hybr.: **B.** Manoa (as **Trevorara** Manoa)
Parentage: **Pararachnis** Eric Holttum × *Euanthe Sanderana*

Bifrillaria in Lager & Hurrell Seedling Listing no. 866-S, p. 1, September
    1966 (as **Bifrinlaria**)
Bifrenaria × *Maxillaria*
1st hybr.: unnamed
Parentage: *Maxillaria Sanderana* × *Bifrenaria Harrisoniae*

Bleteleorchis in Recent Advances in Breedings 12:92, 1971
Bletilla × *Eleorchis*
1st hybr.: unnamed
Parentage: *Blotilla striata* × *Eleorchis japonica*

Bletundina in Recent Advances in Breeding 12:92, 1971
Arundina × *Bletilla*
1st hybr.: unnamed
Parentage: *Arundina sinensis* × *Bletilla formosana*

Bloomara in Orch. Rev. 74: June 1966
Broughtonia × *Laeliopsis* × *Tetramicra*
1st hybr.: **B.** Jim
Parentage: *Tetramicra canaliculata* × **Broughtopsis** Kingston (as
    Lioponia Kingston)

Bolleo-Chondrorhyncha in Gard. Chron. ser. 3, 32:243, 1902
Observation: See **Chondrobollea**

Bradeara in Orch. Rev. 81: July 1973
Comparettia × *Gomesa* × *Rodriguezia*
1st hybr.: **B.** Brasil
Parentage: **Rodrettia** Henry Teuscher × *Gomesa recurva*

Bradriguezia in Bull. Pac. Orch. Soc. Hawaii 14:85, 1957
Brassia × *Rodriguezia*
1st hybr.: **B.** Angellitos (as **Rodrassia** Angellitos)
Parentage: *Brassia Gireoudiana* × *Rodriguezia (venusta) bracteata*
    Syn.: **Rodrassia** in Orch. Rev. 68:404, 1960

Brapasia in Bull. Pac. Orch. Soc. Hawaii 14:85, 1957
Aspasia × *Brassia*
1st hybr.: **B.** Panama
Parentage: *Aspasia principissa* × *Brassia longissima*

Brassada in Orch. Rev. 78: April 1970
Ada × *Brassia*
1st hybr.: **B.** Mem. Bert Field
Parentage: *Ada aurantiaca* × *Brassia verrucosa*

Brassidium in Orch. Rev. 56:186, 1948
Brassia × *Oncidium*
1st hybr.: **B.** Coronet
Parentage: *Oncidium anthocrene* × *Brassia brachiata*

Brassocatlaelia in Gard. Chron. ser. 3, 21:438, 1897
Observation: See **Brassolaeliocattleya**

Brassocattleya in Gard. Chron. ser. 3, 5:438, 1889
Brassavola × *Cattleya*
1st hybr.: **B.** *Lindleyana* (as *Cattleya Lindleyana*, 1857)
Parentage: *Brassavola tuberculata* × *Cattleya intermedia*
    Syn.: **Brassoleya** in Hansen, Orch. Hybr. 81, 1895
        Correvonia in Jardin 240, 1898
        Cattleyovola in Proc. 3rd World Orch. Conf. 323, 1960

Brasso-Cattleya-Laelia in Gard. Chron. ser. 3, 41:259, 1907
Observation: See **Brassolaeliocattleya**

Brassodiacrium in Orchid World 6:62, 1915
Brassavola × *Diacrium*
1st hybr.: **B.** *Colmaniae*
Parentage: *Diacrium bicornutum* × *Brassavola nodosa*

Brassoepidendrum in Gard. Chron. ser. 3, 40:298, 1906
Brassavola × *Epidendrum*
1st hybr.: **B.** *Stamfordiense*

Parentage: *Brassavola glauca* × *Epidendrum Parkinsonianum*
  Syn.: **Epivola** in Orch. Rev. 16:83, 1908
     Epibrassavola in Roanele Manor Coll. Orch. 38, 1926

Brassolaelia in Orch. Rev. 10:85, 1902
Brassavola × *Laelia*
1st hybr.: **B.** *Veitchii* (as *Laelia Digbyano-purpurata*)
Parentage: *Laelia purpurata* × *Brassavola Digbyana*
  Syn.: **Brassavolaelia** in Proc. 3rd World Orch. Conf. 325, 1960
     Laeliavola in Proc. 3rd World Orch. Conf. 325, 1960

Brassolaeliocattleya in Gard. Chron. ser. 3, 40:201, 1906
Brassavola × *Cattleya* × *Laelia*
1st hybr.: **B.** *Lawrencei* (as **Brassocatlaelia** *Lindleyano-elegans*)
Parentage: **Brassocattleya** *Lindleyana* × **Laeliocattleya** *elegans*
  Syn.: **Brassocatlaelia** in Gard. Chron. ser. 3, 21:438, 1897
     Laelio-Brasso-Cattleya in Gard. Chron. ser. 3, 39:254, 1906
     Brasso-Cattleya-Laelia in Gard. Chron. ser. 3, 41:259, 1907

Brassoleya in Hansen, Orch. Hybr. 81, 1895
Observation: See **Brassocattleya**

Brassonotis in Orch. Rev. 70: December 1962
Observation: See **Sophrovola** including the hybrid **Brassonotis** Edna

Brassophronitis in Die Orchidee 5:41, 1954
Observation: See **Sophrovola** including the hybrid **Brassophronitis**
  Waipuna

Brassosophrolaeliocattleya in Bol. Cric. Paul. Orch. 1:191, 1944
Observation: See **Potinara**

Brassotonia in Orch. Rev. 68:223, 1960
Brassavola × *Broughtonia*
1st hybr.: **B.** John H. Miller
Parentage: *Brassavola nodosa* × *Broughtonia sanguinea*

Brassovolaelia in Proc. 3rd World Orch. Conf. 325, 1960
Observation: See **Brassolaelia**

Bratonia in Bull. Pac. Orch. Soc. Hawaii 14:85, 1957
Brassia × *Miltonia*
1st hybr.: **B.** Premier (as **Miltassia** Premier)
Parentage: *Miltonia spectabilis* × *Brassia caudata*
  Syn.: **Miltassia** in Orch. Rev. 66:25, 1958

Broughtopsis in Bull. Pac. Orch. Soc. Hawaii 14:85, 1957
Broughtonia × *Laeliopsis*
1st hybr.: **B.** Kingston (as **Lioponia** Kingston)
Parentage: *Broughtonia sanguinea* × *Laeliopsis domingensis*
   Syn.: **Lioponia** in Orch. Rev. 67:259, 1959

Burkillara in Orch. Rev. 75: December 1967
Aerides × *Arachnis* × *Vanda*
1st hybr.: **B.** August Rose (Orch. Rev. 80:48, 1972)
Parentage: **Aeridachnis** Mandai × *Vanda tricolor*
Observation: For the hybrids **Burkillara** Henry see **Wrefordara,** for **B.**
   Margaret Ede see **Aeridachnanthe,** for **B.** Sanderling see
   Euarachnides

Burrageara in Gard. Chron. ser. 3, 81:309, 1927
Cochlioda × *Miltonia* × *Odontoglosum* × *Oncidium*
1st hybr.: **B.** Windsor
Parentage: **Odontonia** *Firminii* × **Oncidioda** *Cooksoniae*

Calanthidio-preptanthe in Kerchov. Le Liv. Orch. 465, 1894
Observation: See *Calanthe*

Calanthophaius in Plauszew., Orch. Pl. Serr. t. 11, 1899
Observation: See **Phaiocalanthe**

Carrara in Bot. Mus. Leafl. Harv. Univ. 21:154, 1966
Ascocentrum × *Euanthe* × *Rhynchostylis* × *Vanda*
1st hybr.: **C.** Blue Fairy (as **Vascostylis** Blue Fairy)
Parentage: **Schlechterara** Meda Arnold × *Rhynchostylis coelestis*

Carterara in Orch. Rev. 77: January 1969
Aerides × *Renanthera* × *Vandopsis*
1st hybr.: **C.** Evening Glow
Parentage: **Renanopsis** Lena Rowold × *Aerides Lawrenceae*

Catamodes in Orch. Rev. 75:205, 1967
Catasetum × *Mormodes*
1st hybr.: **C.** Brown Derby
Parentage: *Catasetum Warscewiczii* × *Mormodes atropurpurea*

Catanoches in Orch. Rev. 75:205, 1967
Catasetum × *Cycnoches*
1st hybr.: **C.** Green Beret
Parentage: *Catasetum Warscewiczii* × *Cycnoches ventricosum*

Catlaelia in Hansen, Orch. Hybr. 85, 1895
Observation: See **Laeliocattleya**

Catlaenitis in Hansen, Orch. Hybr. 100, 1895
Observation: See **Sophrolaeliocattleya**

Cattleyodendrum in Chron. Orch. 1:115, 1898
Observation: See **Epicattleya**

Cattleyopsisgoa in Orch. Rev. 75: November 1967
Cattleyopsis × *Domingoa*
1st hybr.: **C.** Little Fellow
Parentage: *Domingoa hymenodes* × *Cattleyopsis Ortgiesiana*

Cattleyopsistonia in Orch. Rev. 74: July 1966
Broughtonia × *Cattleyopsis*
1st hybr.: **C.** Leona
Parentage: *Cattleyopsis Ortgiesiana* × *Broughtonia sanguinea*

Cattleyovola in Proc. 3rd World Orch. Conf. 323, 1960
Observation: See **Brassocattleya**

Cattleytonia in Orch. Rev. 67:69, 1959
Broughtonia × *Cattleya*
1st hybr. **C.** Rosy Jewel (as **Cattleytonia** Rosy Gem)
Parentage: *Broughtonia sanguinea* × *Cattleya Bowringiana*

Cephalepipactis in Camus, Monogr. Orch. Europ. 424, 1908
Observation: See **Cephalopactis**

Cephalopactis in Aschers. & Graebn., Syn. 3:883, 1907
Cephalanthera × *Epipactis*
1st hybr.: **C.** *speciosa* (as *Epipactis speciosa,* 1889)
Parentage: *Cephalanthera alba* × *Epipactis rubiginosa*
    Syn.: **Cephalepipactis** in Camus, Monogr. Orch. Europ. 424, 1908

Cephalophrys in Orchid World 2:114, 1912
Cephalanthera × *Ophrys*
1st hybr.: **C.** *integra*
Parentage: *Ophrys apifera* × *Cephalanthera rubra*

Chamodenia in Dhauner Echo 35:23, 1970
Chamorchis × *Gymnadenia*
1st hybr.: **C.** *heteroglossa* (as *Gymnadenia heteroglossa*)
Parentage: *Chamorchis alpina* × *Gymnadenia odoratissima*

Charlesworthara in Orch. Rev. 27:143, 1919 (as **Charlesworthiara**)
Cochlioda × *Miltonia* × *Oncidium*
1st hybr.: **C.** Alpha
Parentage: **Miltonioda** Ajax × **Oncidioda** *Cooksoniae*

Chewara in Orch. Rev. 81: March 1973
Aerides × Renanthera × Rhynchostylis
1st hybr.: **C.** Ruth Wong
Parentage: **Renades** Mahani × *Rhynchostylis retusa*

Chondrobollea in Orch. Rev. 10:347, 1902
Bollea × *Chondrorhyncha*
1st hybr.: **C.** *Froebeliana*
Parentage: *Bollea coelestis* × *Chondrorhyncha Chestertonii*
    Syn.: **Bolleo-Chondrorhyncha** in Gard. Chron. ser. 3, 32:243, 1902

Chondropetalum in Orch. Rev. 16:56, 1908
Observation: See **Zygorhyncha** including the hybrid **Chondropetalum**
    Fletcheri. This name is preempted by *Chondropetalum* Rttb., 1773

Christieara in Orch. Rev. 77: November 1969
Aerides × *Ascocentrum* × *Vanda*
1st hybr.: not yet reported
Observation: For the hybrid **Christieara** Mem. Lillian Arnold see
    Reinikkaara

Cirrhophyllum in Orch. Rev. 73: January 1965
Bulbophyllum × *Cirrhopetalum*
1st hybr.: not yet reported
Observation: For the proposed hybrid **Cirrhophyllum** *Mariae* see *Bulbo-
    phyllum lasioglossum* (Orch. Rev. 77: December 1969, corrigenda)

Cochleatorea in Orch. Rev. 73: May 1965
Observation: See **Pescoranthes** including the hybrid **Cochleatorea** Sunny-
    bank

Cochlenia in Orch. Rev. 75: November 1967
Cochleanthes × *Stenia*
1st hybr.: **C.** Bryn Mawr
Parentage: *Stenia pallida* × *Cochleanthes discolor*

Coeloglossgymnadenia in Rep. Bot. Exch. Cl. Br. Isl. 8:698, 1928
Observation: See **Gymnaglossum**

Coeloglosshabenaria in Rep Bot. Exch. Cl. Br. Isl. 8:698, 1928
Observation: See **Gymnaglossum**

Coeloglossogymnadenia in Camus, Ic. Orch. Eur. 2:377, 1929
Observation: See **Gymnaglossum**

Coeloglossorchis in Guétrot, Pl. Hybr. Fr. II:57, 1926
Observation: See **Orchicoeloglossum**

Coeloplatanthera in Ciferri and Giacomini, Nomencl. Fl. Ital. pt. 1:169, 1950
Coeloglossum × *Platanthera*
1st hybr.: **C.** *Brueggeri*
Parentage: *Coeloglossum viride* × *Platanthera chlorantha*

Cogniauxara in Bot. Mus. Leafl. Harv. Univ. 21:156, 1966
Arachnis × *Euanthe* × *Renanthera* × *Vanda*
1st hybr.: **C.** Bintang Timor (as **Holttumara** Bintang Timor)
Parentage: **Amesara** Palolo × *Arachnis Hookerana*

Colmanara in Gard. Chron. ser. 3, 94:33, 1936
Miltonia × *Odontoglossum* × *Oncidium*
1st hybr.: **C.** Sir Jeremiah (in Orch. Rev. 71: November 1963)
Parentage: *Odontoglossum bictoniense* × **Miltonidium** Lee Hirsch
   Syn.: **Hatcherara** in Gard. Chron. ser. 3, 94:33, 1936

Correvonia in Jardin 240, 1898
Observation: See **Brassocattleya**

Cycnodes in Orch. Rev. 69:402, 1961
Cycnoches × *Mormodes*
1st hybr.: **C.** L. Sherman Adams
Parentage: *Cycnoches chlorochilon* × *Mormodes Wendlandi*

Cymbiphyllum in Orch. Rev. 75: March 1967
Observation: See **Grammatocymbidium**

Cyperocymbidium in Orch. Rev. 72:420, 1964
Cymbidium × *Cyperorchis*
1st hybr.: **C.** *Gammieanum* (as *Cymbidium Gammieanum*)
Parentage: *Cyperorcris elegans* × *Cymbidium longifolium*

Cysepedium in Hansen, Orch. Hybr. 187, 1895
Cypripedium × *Selenipedium*
1st hybr.: not yet reported

Observation: For the hybrid **Cysepedium** *Corndeanei* see **Phragmipaphium**
Syn.: **Selenocypripedium** in J. Hort. Soc. Fr. ser. 4, 13:706, 1912

Dactylanthera in Willis, Dict. Fl. Plants and Ferns, ed. 7; 327, 1966
Observation: See **Rhizanthera**

Dactylella in Ann. Univ. Eötvös, Budapest 8:318, 1966
Observation: See **Dactylitella**

Dactyleucorchis in Ann. Univ. Eötvös, Budapest 8:319, 1966
Observation: See **Pseudorhiza**

Dactylitella in Watsonia 6:132, 1965
Dactylorhiza × *Nigritella*
1st hybr.: **D.** *Tourensis* (as *Nigrorchis Tourensis*)
Parentage: *Nigritella nigra* × *Dactylorhiza maculata*
   Syn.: **Dactylella** in Ann. Univ. Eötvös, Budapest 8:318, 1966

Dactylocamptis in Watsonia 6:132, 1965
Anacamptis × *Dactylorhiza*
1st hybr.: **D.** *Weberi* (as **Anacamptorchis** *Weberi*)
Parentage: *Anacamptis pyramidalis* × *Dactylorhiza maculata*

Dactyloceras in Bot. Mus. Leafl. Harv. Univ. 22:278, 1969
Aceras × *Dactylorhiza*
1st hybr.: **D.** *helvetica*
Parentage: *Aceras anthropophora* × *Dactylorhiza latifolia*

Dactylodenia in Bot. Mus. Leafl. Harv. Univ. 21:157, 1966
Dactylorhiza × *Gymnadenia*
1st hybr.: **D.** *Heinzeliana*
Parentage: *Dactylorhiza maculata* × *Gymnadenia conopea*
   Syn.: **Dactylogymnadenia** in Ann. Univ. Eötvös, Budapest 8:318, 1966

Dactyloglossum in Watsonia 6:132, 1965
Coeloglossum × *Dactylorhiza*
1st hybr.: **D.** *Erdingeri* (as *Platanthera Erdingeri*)
Parentage: *Coeloglossum viride* × *Dactylorhiza sambucina*

Dactylogymandenia in Ann. Univ. Eötvös, Budapest 8:318, 1966
Observation: See **Dactylodenia**

Debruyneara in Orch. Rev. 80:141, 1971
Ascocentrum × Luisia × Vanda
1st hybr.: not yet reported
Observation: For the hybrid **Debruyneara** Victoria de Bruyne see **Neo-debruyneara**

Degarmoara in Orch. Rev. 76: February 1968
Brassia × Miltonia × Odontoglossum
1st hybr.: **D.** Agnes
Parentage: **Odontonia** Debutante × *Brassia antherotes*

Dekensara in Orch. Rev. 63:107, 1955
Brassavola × Cattleya × Schomburgkia
1st hybr.: **D.** Flandria
Parenatge: **Brassocattleya** Helena × *Schomburgkia undulata*

Dendrocattleya in Schultes & Pease, Gen. Names Orch. 329, 1963
Cattleya × Dendrobium
1st hybr.: unnamed
Parentage: *Cattleya Bowringiana × Dendrobium Phalaenopsis*

Devereuxara in Orch. Rev. 78: November 1970
Ascocentrum × Phalaenopsis × Vanda
1st hybr · not yet reported
Observation: For the hybrid **Devereuxara** Ellis see **Neodevereuxara**, and for **D.** Marjorie Wreford see **Eudevereuxara**

DeWolfara in Bot. Mus. Leafl. Harv. Univ. 22:279, 1969
Ascocenrtum × Ascoglosum × Euanthe × Renanthera × Vanda
1st hybr.: **D.** Cassino (as **Shigeuraara** Cassino)
Parentage: **Renanthoglossum** Red Delight × **Schlechterara** Meda Arnold

Diabroughtonia in Orch. Rev. 64:209, 1956
Broughtonia × Diacrium
1st hybr.: **D.** Alice Hart
Parentage: *Broughtonia sanguinea × Diacrium bicornutum*

Diacatlaelia in Orch. Rev. 18:110, 1910
Observation: See **Dialaeliocattleya**

Diacattleya in The Garden 72:95, 1908
Cattleya × Diacrium
1st hybr.: **D.** Colmaniae
Parentage: *Cattleya intermedia × Diacrium bicornutum*
   Syn.: **Diacrocattleya** in Gard. Chron. ser. 3, 43:108, 1908

Diacrocattleya in Gard. Chron. ser. 3, 43:108, 1908
Observation: See **Diacattleya**

Dialaelia in Gard. Chron. ser. 3, 37:174, 1905
Diacrium × *Laelia*
1st hybr.: **D.** *Veitchii*
Parentage: *Diacrium bicornutum* × *Laelia cinnabarina*

Dialaeliocattleya in Orch. World 6:61, 1915
Cattleya × *Diacrium* × *Laelia*
1st hybr.: **D.** Gatton Rose
Parentage: *Diacrium bicornutum* × **Laeliocattleya** *Cappei*
    Syn.: **Diacatlaelia** in Orch. Rev. 18:110, 1910

Dialaeliopsis in Orch. Rev. 74: November 1966
Diacrium × *Laeliopsis*
1st hybr.: **D.** Tobago
Parentage: *Diacrium bicornutum* × *Laeliopsis domingensis*

Diaschomburgkia in Bull. Pac. Orch. Soc. Hawaii 14:84, 1957
Diacrium × *Schomburgkia*
1st hybr.: **D.** Ipo (as **Schombodiacrium** Ipo)
Parentage: *Schomburgkia tibicinis* × *Diacrium bicornutum*
    Syn.: **Schombodiacrium** in Orch. Rev. 66:137, 1958

Dillonara in Orch. Rev. 74: December 1966
Epidendrum × *Laelia* × *Schomburgkia*
1st hybr.: **D.** Bronze Kahili
Parentage: **Schombolaelia** Maunalani × *Epidendrum diurnum*

Domindendrum in Bull. Pac. Orch. Soc. Hawaii 14:85, 1957
Observation: See **Epigoa**

Domindesmia in Orch. Rev. 72: November 1964
Domingoa × *Hexadesmia*
1st hybr.: **D.** Little Gem
Parentage: *Domingoa hymenodes* × *Hexadesmia pulchella*

Domintonia in Bull. Pac. Orch. Soc. Hawaii 14:85, 1957
Broughtonia × *Domingoa*
1st hybr.: not yet reported

Domliopsis in Orch. Rev. 73: August 1965
Domingoa × *Laeliopsis*

1st hybr.: **D.** Lavender Mist
Parentage: *Domingoa hymenodes* × *Laeliopsis domingensis*

Doricentrum in Orch. Rev. 77: October 1969
Ascocentrum × *Doritis*
1st hybr.: **D.** Merrilee Wallbrunn
Parentage: *Doritis pulcherrima* × *Ascocentrum curvifolium*

Doridium nom. hybr. gen. nov.
Doritis × *Kingidium*
1st hybr.: **D.** Tiny (as **Doriella** Tiny)
Parentage: *Doritis pulcherrima* × *Kingidium deliciosum* (*Kingiella philippinensis*)
  Syn.: **Doriella** in Orch. Rev. 74: May 1966

Doriella in Orch. Rev. 74: May 1966
Observation: See **Doridium** including **Doriella** Tiny

Doriellaopsis in Orch. Rev. 76: December 1968
Observation: See **Doriopsisium** including **Doriellaopsis** Burma

Doriopsisium nom. hybr. gen. nov.
Doritis × *Kingidium* × *Phalaenopsis*
1st hybr.: **D.** Burma (as **Doriellaopsis** Burma)
Parentage: **Doritaenopsis** Purple Gem × *Kingidium taenialis* (*Kinglella taenialis*)
  Syn.: **Doriellaopsis** in Orch. Rev. 76: December 1968

Doritaenopsis in Arch. Mus. Nat. Paris ser. 6, 12, pt. 2:613, 1935
Doritis × *Phalaenopsis*
1st hybr.: **D.** Asahi (as *Phalaenopsis* Asahi)
Parentage: *Phalaenopsis Lindeni* × *Doritis pulcherrima*
  Syn.: **Doritopsis** in Rev. Circ. Paul. Orch. 7:218, 1950

Doritopsis in Rev. Circ. Paul. Orch. 7:218, 1950
Observation: See **Doritaenopsis**

Dossinimaria in J. Linn. Soc. Bot. 24:170, 1887
Observation: See **Dossisia** including the hybrid **Dossinimaria** *Dominyi*.
  Haemaria is a synonym of *Ludisia*

Dossisia in Bot. Mus. Leafl. Harv. Univ. 21:159, 1966
Dossinia × *Ludisia*
1st hybr.: **D.** *Dominyi* (as *Anoectochilus Dominyi*, 1861)

Parentage: *Dossinia marmorata* × *Ludisia discolor*
    Syn.: **Dossinimaria** in J. Linn. Soc. Bot. 24:170, 1887

Ellanthera in Orch. Rev. 71:137, 1963
Renanthera × *Renantherella*
1st hybr.: **E.** Histrimona (as *Renanthera* Histrimona)
Parentage: *Renantherella histrionica* × *Renanthera monachica*

Encyclipedium in Am. Orch. Soc. Bull. 38:676, 1969
Cypripedium × *Encyclia*
1st hybr.: not yet reported

Epibrassavola in Roanele Manor Coll. Orch. 38, 1926
Observation: See **Brassoepidendrum**

Epibroughtonia in Bull. Pac. Orch. Soc. Hawaii 14:85, 1957
Broughtonia × *Epidendrum*
1st hybr.: **E.** Lilac (as **Epitonia** Lilac)
Parentage: *Epidendrum cochleatum* × *Broughtonia sanguinea*
    Syn.: **Epitonia** in Orch. Rev. 68:371, 1960

Epicattleya in Gard. Chron. ser. 3, 5:491, 1889
Cattleya × *Epidendrum*
1st hybr.: **E.** *matutina*
Parentage: *Cattleya Bowringiana* × *Epidendrum ibaguense*
    Syn: **Epileya** in Hansen, Orch. Hybr. 203, 1895
        Cattleyodendrum in Chron. Orch. 1:115, 1898

Epidiacrium in Orch. Rev. 16:82, 1908
Diacrium × *Epidendrum*
1st hybr.: **E.** *gattonense* (in Rolfe, Stud-book 268, 1909)
Parentage: *Diacrium bicornutum* × *Epidendrum ibaguense*

Epidrobium in Hansen, Orch. Hybr. 203, 1895
Dendrobium × *Epidendrum*
1st hybr.: not yet reported

Epigoa in Orch. Rev. 65:137, 1957
Domingoa × *Epidendrum*
1st hybr.: **E.** Olivine
Parentage: *Domingoa hymenodes* × *Epidendrum Mariae*
    Syn.: **Domindendrum** in Bull. Pac. Orch. Soc. Hawaii 14:85, 1957

Epilaelia in Gard. Chron. ser. 3, 16:605, 1894
Epidendrum × *Laelia*
1st hybr.: **E.** *Hardyana*
Parentage: *Epidendrum ciliare* × *Laelia anceps*
  Syn.: **Laeliodendrum** in J. Hort. Soc. Fr. ser. 3, 19:602, 1897

Epilaeliocattleya in Orch. Rev. 68:193, 1960
Cattleya × *Epidendrum* × *Laelia*
1st hybr.: **E.** Mint
Parentage: **Laeliocattleya** Kahili Kea × *Epidendrum Mariae*

Epilaeliopsis in Orch. Rev. 67:405, 1959 (as **Epilaelopsis**)
Observation: See **Epilopsis** including the hybrid **Epilaeliopsis** Ariza-Julia

Epileya in Hansen, Orch. Hybr. 203, 1895
Observation: See **Epicattleya**

Epilopsis in Bull. Pac. Orch. Soc. Hawaii 14:85, 1957
Epidendrum × *Laeliopsis*
1st hybr.: **E.** Ariza-Julia (as **Epilaelia** Ariza-Julia)
Parentage: *Laeliopsis domingensis* × *Epidedrum Eggersii*
  Syn.: **Epilaeliopsis** in Orch. Rev. 67:405, 1959 (as **Epilaelopsis**)

Epiphaius in Hansen, Orch. Hybr. Suppl. II, 322, 1897
Epidondrum × *Phaius*
1st hybr.: not yet reported

Epiphronitella in Hawkes, Orchids 244: 1961
Epidendrum × *Sophronitella*
1st hybr.: **E.** *Orpeti* (as **Epiphronitis** *Orpeti*)
Parentage: *Epidendrum O'Brienianum* × *Sophronitella violacea*

Epiphronitis in Gard. Chron. ser. 3, 7:799, 1890
Epidendrum × *Sophronitis*
1st hybr.: **E.** *Veitchii*
Parentage: *Epidendrum ibaguense* × *Sophronitis grandiflora*

Epitonia in Orch. Rev. 68:371, 1960
Observation: See **Epibroughtonia** including the hybrid **Epitonia** Lilac

Epivola in Orch. Rev. 16:83, 1908
Observation: See **Brassoepidendrum**

Ernestara in Orch. Rev. 76: December 1968
Phalaenopsis × *Renanthera* × *Vandopsis*

1st hybr.: **E.** Helga Reuter
Parentage: **Renanopsis** Cape Sable × *Phalaenopsis* Dos Pueblos

Esmeranda in Vacherot, Les Orchidées 150, 1954
Esmeralda × *Vanda*
1st hybr.: **E.** Würzburg
Parentage: *Esmeralda Clarkei* × *Vanda coerulea*

Esmenanthera nom. hybr. gen. nov.
Esmeralda × *Renanthera*
1st hybr.: **E.** Ruben (as **Aranthera** Ruben)
Parentage: *Esmeralda Cathcartii* × *Renanthera Imschootiana*

Euarachnides nom, hybr. gen. nov.
Aerides × *Arachnis* × *Euanthe*
1st hybr.: **E.** Sanderling (as **Burkillara** Sanderling)
Parentage: **Aeridachnis** Bogor × *Euanthe Sanderana*

Eucentrum nom. hybr. gen. nov.
Ascocentrum × *Euanthe*
1st hybr.: **E.** Sagarik (as **Ascocenda** Sagarik)
Parentage: *Ascocentrum miniatum* × *Euanthe Sanderana*

Eudevereuxara nom. hybr. gen. nov.
Ascocentrum × *Euanthe* × *Phalaenopsis* × *Vanda*
1st hybr.: **E.** Marjorie Wreford (as **Devereuxara** Marjorie Wreford)
Parentage: *Phalaenopsis* Zada × **Schlechterara** Meda Arnold

Eupapilanda nom. hybr. gen. nov.
Euanthe × *Papilionanthe* × *Vanda*
1st hybr.: **E.** Betty Wright (as *Vanda* Betty Wright)
Parentage: *Papilionanthe teres* × **Vandanthe** *Burgeffii*

Eupapilio nom. hybr. gen. nov.
Euanthe × *Papilionanthe*
1st hybr.: **E.** Maurice Restrepo (as *Vanda* Maurice Restrepo)
Parentage: *Euanthe Sanderana* × *Papilionanthe teres*

Euporphyranda nom. hybr. gen. nov.
Euanthe × *Porphyrodesme* × *Vanda*
1st hybr.: **E.** Rothschongata (as **Amesara** Rothschongata)
Parentage: **Vandanthe** *Rothschildiana* × *Porphyrodesme elongata*

Eurachnis in Bok Choon, List Malay. Orch. Hybr. III, 1960
Arachnis × *Euanthe*

1st hybr.: **E.** Helen Gagan (as **Aranda** Helen Gagan, 1957)
Parentage: *Arachnis* Maggie Oei × *Euanthe Sanderana*

Forgetara in Orch. Rev. 80:141, 1972
Aspasia × *Brassia* × *Miltonia*
1st hybr.: **F.** Mexico
Parentage: **Brapasia** Serene × *Miltonia* Fortaleza

Fujiwarara in Orch. Rev. 71: April 1963
Brassavola × *Cattleya* × *Laeliopsis*
1st hybr.: **F.** Frolic (as **Tenranara** Frolic)
Parentage: **Brassocattleya** Kinipopo × *Laeliopsis domingensis*
  Syn.: **Tenranara** in Orch. Rev. 70: December 1962

Garayara in Am. Orch. Soc. Bull. 38:676, 1969
Arachnis × *Paraphalaenopsis* × *Vandopsis*
1st hybr.: **G.** Lee Kim Hong (as **Laycockara** Lee Kim Hong)
Parentage: **Pararachnis** Lee Siew Chin × *Vandopsis lissochiloides*

Gastrocalanthe in The Orch. J. 1:245, 1952
Calanthe × *Gastrorchis*
1st hybr.: **G.** Berryana (as **Phaiocalanthe** Berryana)
Parentage: *Calanthe masuca* × *Gastrorchis Humblotii*

Gastrophaius in The Orch. J. 1:245, 1952
Gastrorchis × *Phaius*
1st hybr.: **G.** Cooksoni (as *Phaius Cooksoni*)
Parentage: *Gastrorchis tuberculosa* × *Phaius Wallichii*

Gauntlettara in Orch. Rev. 74: April 1966
Broughtonia × *Cattleyopsis* × *Laeliopsis*
1st hybr.: **G.** Noel
Parentage: *Cattleyopsis Ortgiesiana* × **Broughtopsis** Kingston (as **Lioponia** Kingston)

Giddingsara in Am. Orch. Soc. Bull. 38:676, 1969
Ascocentrum × *Euanthe* × *Renanthera* × *Vanda* × *Vandopsis*
1st hybr.: **G.** Sapphire (as **Onoara** Sapphire)
Parentage: **Schlechterara** Meda Arnold × **Renanopsis** Lena Rowold

Gilmourara in Bot. Mus. Leafl. Harv. Univ. 22:280, 1969
Aerides × *Arachnis* × *Ascocentrum* × *Euanthe* × *Vanda*
1st hybr.: **G.** Gracia (as **Lewisara** Gracia)
Parentage: **Aeridachnis** Bogor × **Schlechterara** Ophelia

Goffara in Orch. Rev. 81: July 1973
Luisia × *Rhynchostylis* × *Vanda*
1st hybr.: not yet reported
Observation: For the hybrid **Goffara** Eva Goff see **Rhynchopapilisia**

Grammatocymbidium in Orch. Rev. 30:38, 1922
Cymbidium × *Grammatophyllum*
1st hybr.: **G.** Emil Anderson (as **Cymbiphyllum** Emil Anderson)
Parentage: *Cymbidium pendulum* × *Grammatophyllum Measuresianum*
 Syn.: **Cymbiphyllum** in Orch. Rev. 75: March 1967

Greatwoodara in Bot. Mus. Leafl. Harv. Univ. 22:28, 1969
Ascocentrum × *Euanthe* × *Renanthera* × *Vanda*
1st hybr.: **G.** William Doi (as **Kagawara** William Doi)
Parentage: *Renanthera* Kilauea × **Schlechterara** Meda Arnold

Gymleucorchis in Lond. Cat. Br. Fl., ed. 11; 43, 1925
Observation: See **Pseudodenia**

Gymnabicchia in Camus, Monogr. Orch. Eur. 315, 1908
Observation: See **Pseudodenia**

Gymnacamptis in Roy. Hort. Soc. Dict. 2:938, 1951
Observation: See **Gymnanacamptis**

Gymnadeniorchis in Hawkes, Encycl. Cult. Orch. 340, 1965
Observation: See **Orchigymnadenia**

Gymnaglossum in Orch. Rev. 27:171, 1919
Coeloglossum × *Gymnadenia*
1st hybr.: **G.** *Jacksonii* (as **Gymnplatanthera** *Jacksonii*, 1911)
Parentage: *Gymnadenia conopea* × *Coeloglossum viride*
 Syn.: **Coeloglossgymnadenia** in Rep. Bot. Exch. Cl. Br. Isl. 8:698, 1928
  Coeloglosshabenaria in Rep. Bot. Exch. Cl. Br. Isl. 8:698, 1928
  Coeloglossogymnadenia in Camus, Ic. Orch. Eur. 2:377, 1929

Gymnanacamptis in Aschers. & Graebn., Syn. 3:854, 1907
Anacamptis × *Gymnadenia*
1st hybr.: **G.** *Anacamptis* (as *Gymnadenia Anacamptis*, 1868)
Parentage: *Anacamptis pyramidalis* × *Gymnadenia conopea*
 Syn.: **Gymnacamptis** in Roy. Hort. Soc. Dict. 2:938, 1951

Gymnaplatanthera in Lambert, Notes Orch. Hybr. 9, 1907
Gymnadenia × *Platanthera*
1st hybr.: **G.** *Chodati* (as *Gymnadenia Chodati*)

Parentage: *Gymnadenia conopea* × *Platanthera bifolia*
  Syn.: **Gymnplatanthera** in Camus, Monogr. Orch. Eur. 337, 1908
  Gymplatanthera in Winchester Coll. N.H. Rep. 33, 1911

Gymnigritella in J. Bot. Fr. 6:484, 1892
Gymnadenia × *Nigritella*
1st hybr.: **G.** *suaveolens* (as *Orchis suaveolens*, 1787)
Parentage: *Nigritella nigra* × *Gymnadenia conopea*

Gymnorchis in Dostál, Fl. Czechosl. (Květena ČSR) ed. 2, 2101, 1950
Observation: See **Pseudodenia**

Gymnotraunsteinera in Ciferri and Giacomini, Nomencl. Fl. Ital. pt. 1:
  171, 1950
Gymnadenia × *Traunsteinera*
1st hybr.: **G.** *Vallesiaca* (as *Orchis Vallesiaca*)
Parentage: *Gymnadenia conopea* × *Traunsteinera globosa*

Gymnplatanthera in Camus, Monogr. Orch. Eur. 337, 1908
Observation: See **Gymnaplatanthera**

Gymplatanthera in Winchester Coll. N.H. Rep. 33, 1911
Observation: See **Gymnaplatanthera**

Habenari-orchis in Ann. Bot. 6:325, 1892
Observation: See **Orchicoeloglossum**

Haemari-anoectochilus in Kerchov, Le Liv. Orch. 468, 1894
Observation: See **Ludochilus**

Haemari-macodes in Kerchov, Le Liv. Orch. 468, 1894
Observation: See **Macodisia**

Hagerara in Orch. Rev. 81: March 1973
Doritis × *Phalaenopsis* × *Vanda*
1st hybr.: **H.** Herb
Parentage: **Doritaenopsis** Mem. Clarence Schubert × **Vandaenopsis** Henrietta Fujiwara

Hartara in Orch. Rev. 73: August 1965
Broughtonia × *Laelia* × *Sophronitis*
1st. hybr.: **H.** George
Parentage: **Sophrolaelia** Valda × *Broughtonia sanguinea*

Hatcherara in Gard. Chron. ser. 3, 94:33, 1936
Observation: See **Colmanara**

Hawaiiara in Orch. Rev. 67:405, 1959
Renanthera × *Vanda* × *Vandopsis*
1st hybr.: **H.** Sunglow
Parentage: **Renanopsis** Lena Rowold × *Vanda spathulata*
Observation: For the hybrid **Hawaiiara** Copper Coin see **Lindleyara**

Hawkesara in Orch. Rev. 76: November 1968
Cattleya × *Cattleyopsis* × *Epidendrum*
1st hybr.: **H.** Alex
Parentage: **Epicattleya** Frances Dyer × *Cattleyopsis Ortgiesiana*

Herbertara in Orch. Rev. 76: December 1968
Cattleya × *Laelia* × *Schomburgkia* × *Sophronitis*
1st hybr.: **H.** Thelma
Parentage: **Schombocattleya** Harry Dunn × **Sophrolaeliocattleya** Radians

Hermibicchia in Camus, Monogr. Orch. Eur. 312, 1908
Observation: See **Pseudinium**

Hermileucorchis in Ciferri and Giacomini, Nomencl. Fl. Ital. pt. 1:169, 1950
Observation: See **Pseudinium**

Herminorchis in Fourn., Quatre Fl. Fr. 201, 1935
Observation: See **Pseudinium**

Holcanthera in Bot. Mus. Leafl. Harv. Univ. 23:188, 1972
Observation: See **Renanetia**

Holcocentrum in Bot. Mus. Leafl. Harv. Univ. 23:188, 1972
Observation: See **Ascofinetia**

Holconopsis in Bot. Mus. Leafl. Harv. Univ. 23:188, 1972
Observation: See **Phalaenetia**

Holcorides in Bot. Mus. Leafl. Harv. Univ. 23:188, 1972
Observation: See **Aeridofinetia**

Holcostylis in Bot. Mus. Leafl. Harv. Univ. 23:188, 1972
Observation: See **Neostylis**

Holttumara in Malay. Orch. Rev. 5:75, 1958
Arachnis × *Renanthera* × *Vanda*
1st hybr.: **H.** Cochineal

Parentage: **Aranda** Hilda Galistan × *Renanthera coccinea*
Observation for the hybrid **Holttumara** Bintang Timor see **Cogniauxara**
  Syn.: **Renanda** in Orch. Rev. 69:63, 1961

Honoluluara nom. hybr. gen. nov.
Papilionanthe × *Rhynchostylis* × *Vanda*
1st hybr.: **H.** Dawn (as **Saccovanda** Dawn)
Parentage: **Papilionanda** Colorful × *Rhynchostylis gigantea*

Hookerara in Orch. Rev. 71: October 1963
Brassavola × *Cattleya* × *Diacrium*
1st hybr.: **H.** Fragrance
Parentage: **Discattleya** Chastity × *Brassavola Digbyana*

Hueylihara in Orch. Rev. 79: October 1971
Neofinetia × *Renanthera* × *Rhynchostylis*
1st hybr.: **H.** Hueylih Jane
Parentage: **Renanetia** Sunrise × *Rhynchostylis gigantea*

Huntara in Bot. Mus. Leafl. Harv. Univ. 22:282, 1969
Arachnis × *Euanthe* × *Renanthera* × *Vanda* × *Vandopsis*
1st hybr.: **H.** Teoh Cheng Swee (as **Teohara** Teoh Cheng Swee)
Parentage: **Arandanthe** Kian Kee × **Renanopsis** Lena Rowold

Huntleanthes in Orch. Rev. 74: January 1900
Cochleanthes × *Huntleya*
1st hybr.: **H.** Narberth
Parentage: *Cochleanthes discolor* × *Huntleya Burtii* (*meleagris*)

Ioncidium in Orch. Rev. 76: December 1968
Observation: See **Ionocidium** including the hybrid, **Ioncidium** Little Bit

Ionettia in Orch. Rev. 76: May 1968
Comparettia × *Ionopsis*
1st hybr.: **I.** Rose Petal
Parentage: *Comparettia falcata* × *Ionopsis paniculata*

Ionocidium in Orch. Rev. 76: May 1968
Ionopsis × *Oncidium*
1st hybr.: **I.** Ressie Toy
Parentage: *Ionopsis paniculata* × *Oncidium pulchellum*
  Syn: **Ioncidium** in Orch. Rev. 76: December 1968

Isanitella in Am. Orch. Soc. Bull. 40:710, 1971
Isabelia × *Sophronitella*

1st hybr.: **I.** *Pabstii*
Parentage: *Isabelia pulchella* × *Sophronitella violacea*

Iwanagara in Orch. Rev. 68:223, 1960
Observation: See **Linneara** including the hybrid **Iwanagara** Frontier

Jacquinparis in Am. Orch. Soc. Bull. 38:676, 1969
Jacquiniella × *Liparis*
1st hybr.: not yet reported

Jimenezara in Orch. Rev. 81: October 1973
Broughtonia × *Laelia* × *Laeliopsis*
1st hybr.: **J.** Jose
Parentage: *Laelia autumnalis* × **Broughtopsis** Kingston (as **Lioponia**
    Kingston)

Joannara in Orch. Rev. 81: October 1973
Renanthera × *Rhynchostylis* × *Vanda*
1st hybr.: not yet reported
Observation: For the hybrid **Joannara** Jetstar see **Neojoannara**

Kagawara in Orch. Rev. 76: June 1968
Ascocentrum × *Renanthera* × *Vanda*
1st hybr.: **K.** Firebird (Orch. Rev. 77: March 1969)
Parentage: *Renanthera Storiei* × **Ascocenda** Red Gem
Observation: For the hybrid **Kagawara** William Doi see **Greatwoodara**

Kamemotoara in Bot. Mus. Leafl. Harv. Univ. 22:283, 1969
Aerides × *Euanthe* × *Rhynchostylis* × *Vanda*
1st hybr.: not yet reported
Observation: For the hybrid **Kamemotoara** Porchina Blue see **Thaiara**

Kawanishiara nom. hybr. gen. nov.
Euanthe × *Papilionanthe* × *Vanda* × *Vandopsis*
1st hybr.: **K.** Fascination (as **Opsisanda** Fascination)
Parentage: *Papilionanthe* Josephine van Brero × **Opsisanthe** May M.
    Kawanishi

Kirchara in Orch. Rev. 67:33, 1959
Cattleya × *Epidendrum* × *Laelia* × *Sophronitis*
1st hybr.: **K.** Topaz
Parentage: **Sophrolaeliocattleya** Firefly × *Epidendrum Mariae*

Kraenzlinara in Bot. Mus. Leafl. Harv. Univ. 21:164, 1966
Euanthe × *Trichoglottis* × *Vanda*
1st hybr.: **K.** Richard Emery (as **Trichovanda** Richard Emery)
Parentage: *Trichoglottis brachiata* × **Vandanthe** *Rothschildiana*

Laelio-Brasso-Cattleya in Gard. Chron. ser. 3, 39:254, 1906
Observation: See **Brassolaeliocattleya**

Laeliocatonia in Orch. Rev. 75: March 1967
Broughtonia × *Cattleya* × *Laelia*
1st hybr.: **L.** Betty Holley
Parentage: **Laeliocattleya** Bright Night × *Broughtonia sanguinea*

Laeliocattkeria in Orch. Rev. 73: October 1965
Barkeria × *Cattleya* × *Laelia*
1st hybr.: **L.** Serendipity
Parentage: **Laeliocattleya** Ibbie × *Barkeria Lindleyana*

Laeliocattleya in J. Linn. Soc. Bot. 24:168, 1887
Cattleya × *Laelia*
1st hybr.: **L.** *elegans* (as *Cattleya elegans*, 1848)
Parentage: *Cattleya Leopoldii* × *Laelia purpurata*
   Syn.: **Catlaelia** in Hansen, Orch. Hybr. 85, 1895

Laeliodendrum in J. Hort. Soc. Fr. ser. 3, 19:602, 1897
Observation: See **Epilaelia**

Laeliokeria in Orch. Rev. 78: August 1970
Barkeria × *Laelia*
1st hybr.: **L.** Elnora
Parentage: *Laelia anceps* × *Barkeria Skinneri*

Laeliopleya in Orch. Rev. 74: March 1966
Cattleya × *Laeliopsis*
1st hybr.: **L.** Orange Glow
Parentage: *Cattleya aurantiaca* × *Laeliopsis domingensis*

Laeliovola in Proc. 3rd World Orch. Conf. 325, 1960
Observation: See **Brassolaelia**

Laelonia in Orch. Rev. 65:231, 1957
Broughtonia × *Laelia*
1st hybr.: **L.** Ruby
Parentage: *Broughtonia sanguinea* × *Laelia autumnalis*
Observation: For the hybrid **Laelonia** Federation see **Laeopsis**

Laeopsis in Bull. Pac. Orch. Soc. Hawaii 14:85, 1957
Laelia × Laeliopsis
1st hybr.: **L.** Federation (as **Laelonia** Federation)
Parentage: *Laelia rubescens × Laeliopsis domingensis*
  Syn.: **Liaopsis** in Orch. Rev. 67:147, 1959
    Opsilaelia in Hawkes, Orchids 244, 1961

Lagerara in Orch. Rev. 80:141, 1972
Aspasia × Cochlioda × Odontoglossum
1st hybr.: **L.** Printaw
Parentage: *Aspasia principissa ×* **Odontioda** Taw

Laycockara in Orch. Rev. 74: August 1966
Arachnis × Phalaenopsis × Vandopsis
1st hybr.: not yet reported
Observation: For the hybrids **Laycockara** Lee Kim Hong and **L.** Ian
  Trevor see **Garayara**

Leeara in Orch. Rev. 80:220, 1972
Arachnis × Vanda × Vandopsis
1st hybr.: **L.** Lissom Lucy
Parentage: **Aranda** Lucy Laycock × *Vandopsis lissochiloides*

Leocidium in Orch. Rev. 80:182, 1972
Leochilus × Oncidium
1st hybr.: **L.** Anne Borden
Parentage: *Oncidium pulchellum × Leochilus labiatus*

Leptolaelia in Gard. Chron. ser. 3, 31:280, 1902
Laelia × Leptotes
1st hybr.: **L.** *Veitchii*
Parentage: *Leptotes bicolor × Laelia cinnabarina*

Leucadenia in Fedde Rep 16:290, 1920
Observation: See **Pseudodenia**

Leuerminium in Gartenfl. 85:253, 1936
Observation: See **Pseudinium**

Leucororchis in Ciferri and Giacomini, Nomencl. Fl. Ital. pt. 1:170, 1950
Observation: See **Pseudorhiza** including the hybrids **Leucororchis** *Bru-niana* and **L.** *albucina*

Leucotella in Fedde Rep. 16:272, 1920
Observation: See **Pseuditella**

Lewisara in Orch. Rev. 76: February 1968
Aerides × *Arachnis* × *Ascocentrum* × *Vanda*
1st hybr.: **L. Max**
Parentage: **Aeridachnis** Bogor × **Ascocenda** Charm
Observation: For the hybrid **Lewisara** Gracia see **Gilmourara**

Liaopsis in Orch. Rev. 67:147, 1959
Observation: See **Laeopsis** including the hybrid **Liaopsis** Federation

Limara in Orch. Rev. 68:403, 1960
Arachnis × *Renanthera* × *Vandopsis*
1st hybr.: **L. Lim Lean Teng**
Parentage: **Renanopsis** Lena Rowold × *Arachnis* Maggie Oei

Limatopreptanthe in Kerchov, Le Liv. Orch. 471, 1894
Observation: See *Calanthe*

Lindleyara in Bot. Mus. Leafl. Harv. Univ. 21:165, 1966
Euanthe × *Renanthera* × *Vanda* × *Vandopsis*
1st hybr.: **L.** Copper Coin (as **Hawaiiara** Copper Coin)
Parentage: **Vandanthe** Ellen Noa × **Renanopsis** Lena Rowold

Linneara in Bull. Soc. Roy. Bot. Belg. 47:402, 1911
Brassavola × *Cattleya* × *Diacrium* × *Laelia*
1st hybr.: **L.** Frontier (as **Iwanagara** Frontier)
Parentage: **Diacattleya** Chastity × **Brassolaeliocattleya** Hodaco
  Syn: **Iwanagara** in Orch. Rev. 68:223, 1960

Lioponia in Orch. Rev. 67:259, 1959
Observation: See **Broughtopsis** including the hybrid **Lioponia** Kingston

Loroglorchis in J. Bot. Fr. 6:110, 1892
Observation: See **Orchimantoglossum**

Lowara in Orch. Rev. 20:360, 1912 (as **Lowiara**)
Brassavola × *Laelia* × *Sophronitis*
1st hybr.: **L.** *insignis*
Parentage: *Sophronitis grandiflora* × **Brassolaelia** Helen

Ludochilus in Bot. Mus. Leafl. Harv. Univ. 21:166, 1966
Anoectochilus × *Ludisia*
1st hybr.: **L.** *Dominyi* (as *Anoectochilus Dominyi*)
Parentage: *Anoectochilus Roxburghii* × *Ludisia discolor*
Observation: *Haemaria* is a synonym of *Ludisia*
  Syn.: **Anoectomaria** in J. Linn. Soc. Bot. 24:170, 1887
        Haemari-anoectochilus in Kerchov, Le Liv. Orch. 468, 1894

Luisaerides in Maekawa, Wild Orch. Jap. Color 480, 1971
Aerides × *Luisia*
1st hybr.: **L.** Furusei Maekawa
Parentage: *Aerides japonica* × *Luisia teres*

Luisanda in Orch. Rev. 60:180, 1952
Luisia × *Vanda*
1st hybr.: **L.** Rippa (see Orch. Rev. 75: November 1967)
Parentage: *Luisia Jonesii* × *Vanda coerulea*
Observation: For the hybrid **Luisanda** Uniwai see **Papilisia**

Lutherara in Orch. Rev. 81: March 1973
Phalaenopsis × *Renanthera* × *Rhynchostylis*
1st hybr.: **L.** Ruth Takemoto
Parentage: **Renanthopsis** Jan Goo × *Rhynchostylis gigantea*

Lycastenaria in Colman, Hybr. Orch. 80, 1933
Bifrenaria × *Lycaste*
1st hybr.: **L.** Darius (as **Lyfrenaria** Darius, 1954)
Parentage: *Bifrenaria Harrisoniae* × *Lycaste virginalis* var. *hellemense*
   Syn.: **Lyfrenaria** in Gard. Chron. ser. 3, 135:175, 1954
      Lycasteria in Orch. Rev. 62:92, 1954

Lycasteria in Orch. Rev. 62:92, 1954
Observation: See **Lycastenaria** including the hybrid **Lycasteria** Darius

Lyfrenaria in Gard. Chron. ser. 3, 135:175, 1954
Observation: See **Lycastenaria** including the hybrid **Lyfrenaria** Darius

Lymanara in Orch. Rev. 75: August 1967
Aerides × *Arachnis* × *Renanthera*
1st hybr.: **L.** Mary Ann
Parentage: **Aeridachnis** Bogor × *Renanthera Storiei*

Lyonara in Orch. Rev. 56:94, 1948
Observation: See **Trichovanda;** for the hybrid **Lyonara** Ulaula see **Antheglottis**

Lyonara in Orch. Rev. 67:405, 1959
Observation: See **Schombolaeliocattleya** including the hybrid **Lyonara** Fiesta

Maccoyara in Orch. Rev. 80:246, 1972
Aerides × *Vanda* × *Vandopsis*

1st hybr.: not yet reported
Observation: For the hybrid **Maccoyara** Hazel see **Aeridopsisanda**

Macodisia in Bot. Mus. Leafl. Harv. Univ. 21:167, 1966
Ludisia × *Macodes*
1st hybr.: **M.** *Veitchii* (as *Goodyera Veitchii*)
Parentage: *Macodes Petola* × *Ludisia discolor*
Observation: *Haemaria* is a synonym of *Ludisia*
   Syn.: **Macomaria** in J. Linn. Soc. Bot. 24:170, 1887
      Haemari-macodes in Kerchov, Le Liv. Orch. 468, 1894

Macomaria in J. Linn. Soc. Bot. 24:170, 1887
Observation: See **Macodisia** including the hybrid **Macomaria** *Veitchii*

Macradesa in Orch. Rev. 76: December 1968
Gomesa × *Macradenia*
1st hybr.: **M.** Brown Baby
Parentage: *Gomesa recurva* × *Macradenia brassavolae*

Macrangraecum in Cost., La Vie des Orch. 180, 1917
Angraecum × *Macroplectrum*
1st hybr.: **M.** *Veitchii* (as *Angraecum Veitchii*)
Parentage: *Macroplectrum sesquipedale* × *Angraecum superbum*
Observation: *Macroplectrum* is a synonym of *Angraecum*. If the two
   genera are kept separate, **Macrangraecum** should be used

Maxillacaste in La Sem. Hortic. 1:350, 1897
Lycaste × *Maxillaria*
1st hybr.: not yet reported

Mayara nom. hybr. gen. nov.
Papilionanthe × *Renanthera* × *Vanda*
1st hybr.: **M.** Gold Nugget (as **Renantanda** Gold Nugget)
Parentage: **Papilionanda** Cobber Kain × *Renanthera Storiei*

Milpasia in Orch. Rev. 67:33, 1959
Observation: See **Miltonpasia** including the hybrid **Milpasia** Candissa

Milpilia in Orch. Rev. 69:33, 1961
Observation: See **Miltonpilia** including the hybrid **Milpilia** Magic

Miltassia in Orch. Rev. 66:255, 1958
Observation: See **Bratonia** including the hybrid **Miltassia** Premier

Miltoglossum in Tribune Hortic. 5:241, 1910
Observation: See **Odontonia**

Miltoncidium in Am. Orch. Soc. Bull. 25:186, 1956
Observation: See **Miltonidium**

Miltonguezia in Bull. Pac. Orch. Soc. Hawaii 14:85, 1957
Miltonia × *Rodriguezia*
1st hybr.: **M.** Freckles (as **Rodritonia** Freckles, 1959)
Parentage: *Miltonia Bluntii* × *Rodriguezia secunda*
   Syn: **Rodritonia** in Orch. Rev. 67:33, 1959

Miltonidium in Gard. Chron. ser. 3, 94:33, 1936
Miltonia × *Oncidium*
1st hybr.: **M.** Lee Hirsch (Orch. Rev. 61:27, 1953)
Parentage: *Oncidium varicosum* × *Miltonia spectabilis*
Observation: For the hybrid **Miltonidium** Aristocrat see **Odontocidium**
   Syn.: **Miltoncidium** in Am. Orch. Soc. Bull. 25:186, 1956

Miltonioda in Orch. Rev. 17:57, 1909
Cochlioda × *Miltonia*
1st hybr.: **M.** *Lindenii*
Parentage: *Cochlioda vulcanica* × *Miltonia Phalaenopsis*

Miltoniopsis in Orchidophile 9:145, 1889
Observation: See *Miltonia*

Miltonpasia in Bull. Pac. Orch. Soc. Hawaii 14:85, 1957
Aspasia × *Miltonia*
1st hybr.: **M.** Candissa (as **Milpasia** Candissa)
Parentage: *Aspasia principissa* × *Miltonia candida*
   Syn.: **Milpasia** in Orch. Rev. 67:33, 1959

Miltonpilia in Bull. Pac. Orch. Soc. Hawaii 14:85, 1957
Miltonia × *Trichopilia*
1st hybr.: **M.** Magic (as **Milpilia** Magic)
Parentage: *Miltonia spectabilis* × *Trichopilia suavis*
   Syn.: **Milpilia** in Orch. Rev. 69:33, 1961

Mizutara in Orch. Rev. 74: August 1966
Cattleya × *Diacrium* × *Schomburgkia*
1st hybr.: **M.** Pink Kahili
Parentage: **Schombocattleya** Diamond Head × *Diacrium bicornutum*

Moirara in Orch. Rev. 71: June 1963
Phalaenopsis × Renanthera × Vanda
1st hybr.: not yet reported
Observation: For the hybrid **Moirara** Sunshine see **Paramayara** and for
    M. Sunbeam see **Neomoirara**

Mokara in Orch. Rev. 77: December 1969
Arachnis × Ascocentrum × Vanda
1st hybr.: **M.** Wai Liang
Parentage: *Arachnis* Ishbel × **Ascocenda** Red Gem
Observation: For the hybrid **Mokara** Ooi Leng Sun see **Neomokara**

Moscosoara in Orch. Rev. 77: September 1969
Broughtonia × Epidendrum × Laeliopsis
1st hybr.: **M.** Santo Domingo
Parentage: **Broughtopsis** Kingston × *Epidendrum olivaceum*

Myrmecocattleya in Orch. Rev. 28:50, 1920
Observation: See **Schombocattleya**

Myrmecolaelia in Orch. Rev. 28:50, 1920
Observation: See **Schombolaelia**

Nakamotoara in Orch. Rev. 72: August 1964
Ascocentrum × Neofinetia × Vanda
1st hybr.: **N.** Blanc (in Orch. Rev. 73: March 1965)
Parentage: **Ascocenda** Charm × *Neofinetia falcata*
Observation: For the hybrid **Nakamotoara** Wendy see **Smithara**

Neochristieara nom. hybr. gen. nov.
Aerides × Ascocentrum × Papilionanthe × Vanda
1st hybr.: **N.** Malibu Gold (as **Christieara** Malibu Gold)
Parentage: **Papilionanda** Josephine van Brero × **Aeridocentrum** Luke
    Nok

Neodebruyneara nom. hybr. gen. nov.
Ascocentrum × Euanthe × Luisia × Vanda
1st hybr.: **N.** Victoria de Bruyne (as **Debruyneara** Victoria de Bruyne)
Parentage: **Schlechterara** Ophelia × *Luisia Jonesii*

Neodevereuxara nom. hybr. gen. nov.
Ascocentrum × Euanthe × Paraphalaenopsis × Vanda
1st hybr.: **N.** Ellis (as **Devereuxara** Ellis)
Parentage: **Schlechterara** Ophelia × **Paranthe** Jawaii

Neojoannara nom. hybr. gen. nov.
Euanthe × *Renanthera* × *Rhynchostylis* × *Vanda*
1st hybr.: **N.** Jetstar (as **Joannara** Jetstar)
Parentage: **Renantanda** Violet × **Rhynchovandanthe** Blue Angel

Neokagawara nom. hybr. gen. nov.
Ascocentrum × *Euanthe* × *Porphyrodesme* × *Vanda*
1st hybr.: **N.** Mok Yeow Seng (as **Kagawara** Mok Yeow Seng)
Parentage: *Porphyrodesme elongata* × **Schlechterara** Ophelia

Neomoirara nom. hybr. gen. nov.
Papilionanthe × *Phalaenopsis* × *Renanthera* × *Vanda*
1st hybr.: **N.** Sunbeam (as **Moirara** Sunbeam)
Parentage: **Mayara** Gold Nugget × *Phalaenopsis* Doris

Neomokara nom. hybr. gen. nov.
Arachnis × *Ascocentrum* × *Euanthe* × *Vanda*
1st hybr.: **N.** Ooi Leng Sun (as **Mokara** Ooi Leng Sun)
Parentage: *Arachnis* Ishbel × **Schlechterara** Meda Arnold

Neorobinara nom. hybr. gen. nov.
Aerides × *Ascocentrum* × *Euanthe* × *Renanthera* × *Vanda*
1st hybr.: **N.** Kosher Red (as **Robinara** Kosher Red)
Parentage: **Renades** Red Jewel × **Schlechterara** Meda Arnold

Neostylis in Orch. Rev. 73: August 1965
Neofinetia × *Rhynchostylis*
1st hybr.: **N.** Dainty
Parentage: *Neofinetia falcata* × *Rhynchostylis retusa*
    Syn.: **Holcostylis** in Bot. Mus. Leafl. Harv. Univ. 23:188, 1972

Neoyusofara nom. hybr. gen. nov.
Arachnis × *Ascocentrum* × *Euanthe* × *Renanthera* × *Vanda*
1st hybr.: **N.** Nong (as **Yusofara** Nong)
Parentage: **Aranthera** Anne Black × **Schlechterara** Meda Arnold

Nigribicchia in Camus, Monogr. Orch. Eur. 360, 1908
Observation: See **Pseuditella**

Nigrorchis in J. Bot. 63:313, 1925
Nigritella × *Orchis*
1st hybr.: not yet reported
Observation: For the hybrid **Nigrorchis** *Tourensis* see **Dactylitella**

Nobleara in Orch. Rev. 77: September 1969
Aerides × *Renanthera* × *Vanda*
1st hybr.: **N.** Royal Monarch
Parentage: **Renantanda** Jukichi Murata × *Aerides odorata*

Odontioda in Gard. Chron. ser. 3, 35:360, 1904
Cochlioda × *Odontoglossum*
1st hybr.: **O.** *Vuylstekeae*
Parentage: *Cochlioda Noezliana* × *Odontoglossum Pescatorei*

Odontiodonia in Orch. World 1:84, 1911
Observation: See **Vuylstekeara**

Odontobrassia in Gartenfl. 84:121, 1935
Brassia × *Odontoglossum*
1st hybr.: **O.** Alice
Parentage: *Brassia brachiata* × *Odontoglossum* Tagus

Odontocidium in Gard. Chron. ser. 3, 50:343, 1911
Odontoglossum × *Oncidium*
1st hybr.: **O.** *Fowlerianum*
Parentage: *Odontoglossum cirrhosum* × *Oncidium Forbesii*

Odontonia in Gard. Chron. ser. 3, 37:398, 1905
Miltonia × *Odontoglossum*
1st hybr.: **O.** *Lairesseae*
Parentage: *Miltonia Warscewiczii* × *Odontoglossum crispum*
  Syn.: **Miltoglossum** in Tribune Hortic. 5:241, 1910

Odopetalum in Hansen, Orch. Hybr. 227, 1895
Odontoglossum × *Zygopetalum*
1st hybr.: **O.** *Heathii* (in Hansen, Orch. Hybr. Suppl. II, 329, 1897)
Parentage: *Zygopetalum Mackayi* × *Odontoglossum* sp.
Observation: This is a very doubtful hybrid

Oncidarettia in Bull. Pac. Orch. Soc. Hawaii 14·85, 1957 (as **Oncidaretia**)
Comparettia × *Oncidium*
1st hybr.: **O.** Valentine (as **Oncidettia** Valentine)
Parentage: *Oncidium Wydleri* (*O. altissimum auct.*) × *Comparettia falcata*
  Syn.: **Oncidettia** in Orch. Rev. 71: June 1963

Oncidasia in Bull. Pac. Orch. Soc. Hawaii 14:85, 1957
Aspasia × *Oncidium*

1st hybr.: **O.** Regal (as **Aspasium** Regal)
Parentage: *Aspasia epidendroides* × *Oncidium Wydleri*
  Syn.: **Aspasium** in Orch. Rev. 66:161, 1958

Oncidenia in Orch. Rev. 74: February 1966
Macradenia × *Oncidium*
1st hybr.: **O.** Helen
Parentage: *Oncidium* Helen Brown × *Macradenia brassavolae*

Oncidesa in Orch. Rev. 72: December 1964
Gomesa × *Oncidium*
1st hybr.: **O.** America
Parentage: *Oncidium triquetrum* × *Gomesa recurva*

Oncidettia in Orch. Rev. 71: June 1963
Observation: See **Oncidarettia** including the hybrid **Oncidettia** Valentine

Oncidguezia in Bull. Pac. Orch. Soc. Hawaii 14:85, 1957
Observation: See **Rodricidium**

Oncidioda in Orch. Rev. 18:266, 1910
Cochlioda × *Oncidium*
1st hybr.: **O.** *Charlesworthii*
Parentage: *Cochlioda Noezliana* × *Oncidium incurvum*

Oncidophora in Orch. Rev. 78: July 1970
Observation: See **Ornithocidium**

Oncidpilia in Orch. Rev. 74: February 1966
Oncidium × *Trichopilia*
1st hybr.: **O.** Don Carlos
Parentage: *Oncidium Papilio* × *Trichopilia coccinea*

Onoara in Orch. Rev. 75: August 1967
Ascocentrum × *Renanthera* × *Vanda* × *Vandopsis*
1st hybr.: not yet reported
Observation: For the hybrid **Onoara** Sapphire see **Giddingsara**

Opsilaelia in Hawkes, Orchids 244, 1961
Observation: See **Laeopsis**

Opsisanda in Orch. Rev. 57:24, 1949
Vanda × *Vandopsis*
1st hybr.: **O.** Colombo
Parentage: *Vanda Dearei* × *Vandopsis lissochiloides*

Observation: For the hybrid **Opsisanda** Helen Miyamoto see **Opsisanthe;**
for the hybrid **Opsisanda** Kimo Cardus see **Reichenbachara**
Syn.: **Tanakara** in Orch. Rev. 55:120, 1947 (not **Tanakara,** 1952)
Vandopsisvanda in Rev. Circ. Paul Orch. 7:219, 1950

Opsisanthe in Schultes & Pease, Gen. Names Orch. 330, 1963
Euanthe × *Vandopsis*
1st hybr.: **O.** Helen Miyamoto (as **Opsisanda** Helen Miyamoto)
Parentage: *Vandopsis lissochiloides* × *Euanthe Sanderana*

Opsistylis in Orch. Rev. 78: October 1970
Rhynchostylis × *Vandopsis*
1st hybr.: **O.** Lanna Thai
Parentage: *Rhynchostylis gigantea* × *Vandopsis Parishii*

Orchiaceras in J. Bot. Fr. 6:107, 1892
Aceras × *Orchis*
1st hybr.: **O.** *Bergoni*
Parentage: *Orchis simia* × *Aceras anthropophora*

Orchicoeloglossum in Aschers. & Graebn., Syn. 3:849, 1907
Coeloglossum × *Orchis*
1st hybr.: not yet reported
Observation: For the hybrid **Orchicoeloglossum** *Erdingeri* see **Dactylo-**
glossum
Syn.: **Habenari-orchis** in Ann. Bot. Fr. 6:325, 1892
Coeloglossorchis in Guétrot, Pl. Hybr. Fr. II, 57, 1926

Orchidactyla in Watsonia 6:133, 1965
Dactylorhiza × *Orchis*
1st hybr.: **O.** *Schulzei* (as *Orchis Schulzei,* 1882)
Parentage: *Orchis coriophora* × *Dactylorhiza latifolia*
Syn.: **Orchidactylorhiza** in Ann. Univ. Eötvös, Budapest 8:315, 1966

Orchidactylorhiza in Ann. Univ. Etövös, Budapest 8:315, 1966
Observation: See **Orchidactyla**

Orchidanacamptis in Guétrot. Pl. Hybr. Fr. II, 51, 1926
Observation: See **Anacamptorchis**

Orchigymnadenia in J. Bot. Fr. 6:477, 1892
Gymnadenia × *Orchis*
1st hybr.: **O.** *Evequei* (as *Orchis Evequei,* 1905)
Parentage: *Gymnadenia odoratissima* × *Orchis laxiflora*
Syn.: **Gymnadeniorchis** in Hawkes, Encycl. Cult. Orch. 340, 1965

Orchimantoglossum in Aschers. & Graebn., Syn. 3:799, 1907
Himantoglossum × *Orchis*
1st hybr.: **O.** *Lacasei*
Parentage: *Orchis simia* × *Himantoglossum hircinum*
  Syn.: **Loroglorchis** in J. Bot. Fr. 6:110, 1892

Orchiplatanthera in J. Bot. Fr. 6:474, 1892
Orchis × *Platanthera*
1st hybr.: not yet reported
Observation: For the hybrid **Orchiplatanthera** *Chevallierana* see **Rhizanthera**

Ochiserapias in J. Bot. Fr. 6:31, 1892
Orchis × *Serapias*
1st hybr.: **O.** *triloba* (as *Serapias triloba*)
Parentage: *Orchis ensifolia* × *Serapias cordigera*

Ornithocidium in Orquidea 29:181, 1969
Oncidium × *Ornithophora*
1st hybr.: **O.** *Roczonii*
Parentage: *Oncidium riograndense* × *Ornithophora radicans*
  Syn.: **Oncidophora** in Orch. Rev. 78: July 1970

Osmentara in Orch. Rev. 74: June 1966
Broughtonia × *Cattleya* × *Laeliopsis*
1st hybr.: **O.** Bill
Parentage: *Cattleya* R. Prowe × **Broughtopsis** Kingston (as **Lioponia** Kingston)

Otocolax in Orch. Rev. 78: September 1970
Observation: See **Otopabstia** including the hybrid **Otocolax** Pat Duruty

Otonisia in Orch. Rev. 77: January 1969
Aganisia × *Otostylis*
1st hybr.: **O.** Broadway
Parentage: *Otostylis brachystalix* × *Aganisia pulchella*

Otopabstia in Bradea 1:308, 1973
Otostylis × *Pabstia*
1st hybr.: **O.** Pat Duruty (as **Otocolax** Pat Duruty)
Parentage: *Pabstia jugosa* × *Otostylis brachystalix*
  Syn.: **Otocolax** in Orch. Rev. 78: September 1970

Otosepalum in Orch. Rev. 78: December 1970
Otostylis × *Zygosepalum*
1st hybr. **O.** Tommy Aitken
Parentage: *Otostylis brachystalix* × *Zygosepalum labiosum*

Palmerara in Orch. Rev. 81: May 1973
Batemannia × *Otostylis* × *Zygosepalum*
1st hybr.: **P.** Raymond Palmer
Parentage: **Bateostylis** Silver Star × *Zygosepalum labiosum*

Papilachnis nom. hybr. gen. nov.
Arachnis × *Papilionanthe*
1st hybr.: **P.** Nagrok (as **Aranda** Nagrok)
Parentage: *Papilionanthe teres* × *Arachnis alba*

Papilandachnis nom. hybr. gen. nov.
Arachnis × *Papilionanthe* × *Vanda*
1st hybr.: **P.** Mars (as **Aranda** Mars)
Parentage: **Papilionanda** Marguerite Maron × *Arachnis alba*

Papilanthera nom. hybr. gen. nov.
Papilionanthe × *Renanthera*
1st hybr.: **P.** Meteore (as **Renantanda** Meteore)
Parentage: *Papilionanthe teres* × *Renanthera Imschootiana*

Papiliodes nom. hybr. gen. nov.
Ascocentrum × *Papilionanthe*
1st hybr.: **P.** Pink Elf (as **Ascocenda** Pink Elf)
Parentage: *Papilionanthe* Miss Joaquim × *Ascocentrum curvifolium*

Papiliodes nom. hybr. gen. nov.
Aerides × *Papilionanthe*
1st hybr.: **P.** Moana (as **Aeridovanda** Moana)
Parentage: *Papilionanthe teres* × *Aerides Lawrenceae*

Papilionanda in Schultes & Pease, Gen. Names Orch. 330, 1963
Papilionanthe × *Vanda*
1st hybr.: **P.** Marguerite Maron (as *Vanda* Marguerite Maron)
Parentage: *Vanda suavis* × *Papilionanthe teres*

Papilionetia nom. hybr. gen. nov.
Neofinetia × *Papilionanthe*
1st hybr.: **P.** Little Blossom (as **Vandofinetia** Little Blossom)
Parentage: *Neofinetia* falcata × *Papilionanthe* Miss Joaquim

Papiliopsis nom. hybr. gen. nov.
Papilionanthe × *Vandopsis*
1st hybr.: **P.** Hilo (as **Opsisanda** Hilo)
Parentage: *Papilionanthe teres* × *Vandopsis lissochiloides*

Papilisia nom. hybr. gen. nov.
Luisia × *Papilionanthe*
1st hybr.: **P.** Uniwai (as **Luisanda** Uniwai)
Parenatge: *Papilionanthe* Miss Joaquim × *Luisia teretifolia*

Parachilus in Orch. Rev. 80:63, 1972
Observation: *Parasarcochilus* is a synonym of *Pteroceras*. For the hybrid
    Parachilus Perky see **Sarcoceras**

Paramayara nom. hybr. gen. nov.
Papilionanthe × *Paraphalaenopsis* × *Renanthera* × *Vanda*
1st hybr.: **P.** Sunshine (as **Moirara** Sunshine)
Parentage: **Mayara** Gold Nugget × *Paraphalaenopsis Denevei*

Parandachnis in Bot. Mus. Leafl. Harv. Univ. 21:171, 1966
Arachnis × *Paraphalaenopsis* × *Vanda*
1st hybr.: **P.** Hong Trevor (as **Trevorara** Hong Trevor)
Parentage: **Pararachnis** Eric Holttum × *Vanda Dearei*

Parandanthe in Bot. Mus. Laefl. Harv. Univ. 21:172, 1966
Euanthe × *Paraphalaenopsis* × *Vanda*
1st hybr.: **P.** Pang Nyuk Yin (as **Vandaenopsis** Pang Nyuk Yin)
Parentage: **Vandanthe** Ellen Noa × *Paraphalaenopsis Denevei*

Paranthe in Bot. Mus. Leafl. Harv. Univ. 21:172, 1966
Euanthe × *Paraphalaenopsis*
1st hybr.: **P.** *Jawaii* (as **Vandaenopsis** *Jawaii*)
Parentage: *Euanthe Sanderana* × *Paraphalaenopsis Denevei*

Paranthera in Bot. Mus. Leafl. Harv. Univ. 21:172, 1966
Arachnis × *Paraphalaenopsis* × *Renanthera*
1st hybr.: **P.** Ahmad Zahab (as **Sappanara** Ahmad Zahab)
Parentage: **Pararachnis** Eric Holttum × *Renanthera Storiei*

Parapapilio nom. hybr. gen. nov.
Papilionanthe × *Paraphalaenopsis*
1st hybr.: **P.** *superba* (as **Vandaenopsis** *superba*)
Parentage: *Papilionanthe teres* × *Paraphalaenopsis Denevei*

Pararachnis in Orquidea 25:215, 1963
Arachnis × *Paraphalaenopsis*
1st hybr.: **P.** Eric Holttum (as **Arachnopsis** Eric Holttum)
Parentage: *Arachnis* Maggie Oei × *Paraphalaenopsis Denevei*

Pararenanthera in Orquidea 25:215, 1963
Paraphalaenopsis × *Renanthera*
1st hybr.: **P.** Firefly (as **Renanthopsis** Firefly)
Parentage: *Paraphalaenopsis Denevei* × *Renanthera Storiei*

Pararides nom. hybr. gen. nov.
Aerides × *Paraphalaenopsis*
1st hybr.: **P.** Trengganu (as **Aeridopsis** Trengganu in Orch. Rev. 78: October 1970)
Parentage: *Aerides odorata* × *Paraphalaenopsis serpentilingua*

Paravanda in Orquidea 25:215, 1963
Paraphalaenopsis × *Vanda*
1st hybr.: **P.** *bogoriana* (as **Vandaenopsis** *bogoriana*, 1939)
Parentage: *Paraphalaenopsis Denevei* × *Vanda coerulea*

Paravandanthera in Bot. Mus. Leafl. Harv. Univ. 21:172, 1966
Paraphalaenopsis × *Renanthera* × *Vanda*
1st hybr.: not yet reported
Observation: For the hybrid **Paravandanthera** Sunshine see **Paramayara**

Pattoniheadia in Am. Orch. Soc. Bull. 38:676, 1969
Bromheadia × *Pattonia*
1st hybr.: not yet reported

Paulsenara in Orch. Rev. 80:247, 1972
Aerides × *Arachnis* × *Trichoglottis*
1st hybr.: **P.** Medellin
Parentage: *Trichoglottis fasciata* × **Aeridachnis** Colombia

Pectabenaria in Hawkes, Orchids 244, 1961
Habenaria × *Pecteilis*
1st hybr.: **P.** Original (as *Habenaria* Original)
Parentage: *Habenaria militaris* × *Pecteilis Susannae*

Perreiraara in Orch. Rev. 77: September 1969
Aerides × *Rhynchostylis* × *Vanda*
1st hybr.: not yet reported
Observation: For the hybrid **Perreiraara** Porchina Blue see **Thaiara**

Pescarhyncha in Orch. Rev. 69:33, 1961
Chondrorhyncha × Pescatoria
1st hybr.: not yet reported
Observation: For the hybrid **Pescarhyncha** Painted Lady see **Pesco-ranthes**

Pescatobollea in Orch. Rev. 10:347, 1902
Bollea × Pescatoria
1st. hybr.: **P.** *bella* (as *Pescatoria bella*)
Parentage: *Pescatoria Klabochorum × Bollea coelestis*

Pescoranthes in Orch. Rev. 69:403, 1961
Cochleanthes × Pescatoria
1st hybr.: **P.** Painted Lady (as **Pescarhyncha** Painted Lady)
Parentage: *Pescatoria cerina × Cochleanthes discolor*
    Syn.: **Cochleatorea** in Orch. Rev. 73: May 1965

Phabletia in Hansen, Orch, Hybr. Suppl. II, 330, 1897
Bletia × Phaius
1st hybr.: not yet reported

Phaiocalanthe in J. Linn. Soc. Bot. 24:168, 1887
Calanthe × Phaius
1st hybr.: **P.** *irrorata* (as *Phaius irroratus*)
Parentage: *Phaius grandifolius × Calanthe vestita*
    Syn.: **Phaiopreptanthe** in Kerchov, Le Liv. Orch. 485, 1894
        Phaiolimatopreptanthe in Kerchov, Le Liv. Orch. 485, 1894
        Phalanthe in Hansen, Orch. Hybr. 233, 1895
        Calanthophaius in Plauszew., Orch. Pl. Serr. t. 11, 1899

Phaiocymbidium in Gard. Chron. ser. 3, 31:219, 1902
Cymbidium × Phaius
1st hybr.: **P.** *Chardwarense*
Parentage: *Cymbidium giganteum × Phaius Wallichii*
Observation: This is a very doubtful hybrid

Phaiolimatopreptanthe in Kerchov, Le Liv. Orch. 485, 1894
Observation: See **Phaiocalanthe**

Phaiopreptanthe in Kerchov, Le Liv. Orch. 485, 1894
Observation: See **Phaiocalanthe**

Phalaenetia in Orch. Rev. 72: July 1964 (as **Phalanetia**)
Neofinetia × Phalaenopsis

1st hybr.: **P.** Pacifica
Parentage: *Neofinetia falcata* × *Phalaenopsis* Chieftain
  Syn.: **Holconopsis** in Bot. Mus. Leafl. Harv. Univ. 23:188, 1972

Phalaenidium nom. hybr. gen. nov.
Kingidium × *Phalaenopsis*
1st hybr.: **P.** Pale Face (as **Phaliella** Pale Face)
Parentage: *Phalaenopsis Lindenii* × *Kingidium deliciosum* (*Kingella phi-lippinensis*)

Phalaenopapilio nom. hybr. gen. nov.
Papilionanthe × *Phalaenopsis*
1st hybr.: **P.** Explorer (as **Vandaenopsis** Explorer)
Parentage: *Papilionanthe* Miss Joaquim × *Phalaenopsis Aphrodite*

Phalaerianda in Orch. Rev. 59:124, 1951
Aerides × *Phalaenopsis* × *Vanda*
1st hybr. **P.** Honolulu
Parentage: **Aeridovanda** Ruth × *Phalaenopsis Schillerana*
  Syn.: **Tanakara** in Orch. Rev. 60:13, 1952 (as **Tanakaria**)

Phalandopsis in Orch. Rev. 68:224, 1960
Phalaenopsis × *Vandopsis*
1st hybr.: **P.** Milena Joot (in Orch. Rev. 77· June 1969)
Parentage: *Phalaenopsis Veitchiana* × *Vandopsis undulata*
Observation: For the hybrid **Phalandopsis** Star of Hawaii see **Sarca-laenopsis**

Phalanthe in Hansen, Orch. Hybr. 233, 1895
Observation: See **Phaiocalanthe**

Phaliella in Sander's List of Orchid Hybrids. Addendum 1961–1970:561, 1972
Observation: See **Phalaenidium** including the hybrid **Phaliella** Pale Face

Phragmipaphiopedilum in Hawkes, Orchids 244, 1961
Observation: See **Phragmipaphium**

Phragmipaphium in Gartenfl. 85:253, 1936
Paphiopedilum × *Phragmipedium*
1st hybr.: **P.** *Corndeani* (as **Cysepedium** *Corndeanii* in Hansen, Orch. Hybr. 189, 1895)
Parentage: *Phragmipedium Sedeni* × *Paphiopedilum gigas*
  Syn.: **Phragmipaphiopedilum** in Hawkes, Orchids 244, 1961

Pomatisia in Orch. Rev. 81: October 1973
Luisia × *Pomatocalpa*
1st hybr., **P.** Rumrill
Parentage: *Pomatocalpa latifolia* × *Luisia teretifolia*

Porphyrachnis nom. hybr. gen. nov.
Arachnis × *Porphyrodesme*
1st hybr.: **P.** Lilleput (as **Aranthera** Lilleput)
Parentage: *Arachnis Hookerana* × *Porphyrodesme elongata*

Porphyranda nom. hybr. gen. nom.
Porphyrodesme × *Vanda*
1st hybr.: **P.** Sawadi George Neo (as **Renantanda** Sawadi George Neo)
Parentage: *Porphyrodesme elongata* × *Vanda limbata*

Porphyrandachnis nom. hybr. gen. nov.
Arachnis × *Porphyrodesme* × *Vanda*
1st hybr.: **P.** Bright Eyes (as **Holttumara** Bright Eyes)
Parentage: **Porphyrachnis** Lilleput × *Vanda tricolor*

Porphyranthera nom. hybr. gen. nov.
Porphyrodesme × *Renanthera*
1st hybr.: **P.** Mok York-Seng (as *Renanthera* Mok York-Seng)
Parentage: *Porphyrodesme elongata* × *Renanthera coccinea*

Potinara in Gard. Chron. ser. 3, 71:98, 1922
Brassavola × *Cattleya* × *Laelia* × *Sophronitis*
1st hybr.: **P.** *Juliettae*
Parentage: **Brassocattleya** Ena × **Sophrolaeliocattleya** Marathon
   Syn.: **Brassosophrolaeliocattleya** in Bol. Circ. Paul. Orch. 1:191, 1944

Pseudinium in Orch. Rev. 79:141, 1971
Herminium × *Pseudorchis*
1st hybr.: **P.** Aschersonianum (as *Gymnadenia Aschersoniana*, 1888)
Parentage: *Pseudorchis albida* × *Herminium monorchis*
   Syn.: **Hermibicchia** in Camus, Monogr. Orch. Eur. 312, 1908
         Herminorchis in Fourn., Quatre Fl. Fr. 201, 1935
         Leucerminium in Gartenfl. 85:253, 1936
         Hermileucorchis in Ciferri and Giacomini, Nomencl. Fl. Ital. pt.
           1:169, 1950

Pseuditella in Orch. Rev. 79:142, 1971
Nigritella × *Pseudorchis*

1st hybr.: **P.** *micrantha* (as *Nigritella micrantha*)
Parentage: *Nigritella nigra* × *Pseudorchis albida*
   Syn.: **Nigribicchia** in Camus, Monogr. Orch. Eur. 360, 1908
      Leucotella in Fedde Rep. 16:272, 1920

Pseudodenia in Orch. Rev. 79:141, 1971
Gymnadenia × *Pseudorchis*
1st hybr.: **P.** *Schweinfurthii* (as *Gymnadenia Schweinfurthii*, 1865)
Parentage: *Gymnadenia conopea* × *Pseudorchis albida*
   Syn.: **Gymnabicchia** in Camus, Monogr. Orch. Eur. 315, 1908
      Leucadenia in Fedde Rep. 16:290, 1920, not Klotzsch, 1864
      Gymleucorchis in Lond. Cat. Br. Fl., ed. 11; 43, 1925
      Gymnorchis in Dostál, Fl. Czechosl. (Květena ČSR) ed. 2, 2101,
        1950

Pseudorhiza in Orch. Rev. 79:142, 1971
Dactylorhiza × *Pseudorchis*
1st hybr.: **P.** *Bruniana* (as *Orchis Bruniana*)
Parentage: *Pseudorchis albida* × *Dactylorhiza maculata*
   Syn.: **Leucororchis** in Ciferri and Giacomini, Nomencl. Fl. Ital. pt.
      1:170, 1950
      Dactyleucorchis in Ann. Univ. Eötvös, Budapest 8:319, 1966

Pterocottia in Am. Orch. Soc. Bull. 38:676, 1969
Prescottia × *Pterostylis*
1st hybr.: not yet reported

Quisumbingara nom. hybr. gen. nov.
Aerides × *Papilionanthe* × *Vanda*
1st hybr.: **Q.** Florence Loving (as **Aeridovanda** Florence Loving)
Parentage: **Papilionanda** Poepoe × *Aerides odorata*

Recchara in Rev. Circ. Paul. Orch. 7:165, 1950 (as **Recchiara**)
Brassavola × *Cattleya* × *Laelia* × *Schomburgkia*
1st hybr.: **R.** Amelia
Parentage: **Brassolaelia** Brasil × **Schombocattleya** *crispo-Loddigesii*
   Syn.: **Beaumontara** in Orch. Rev. 69:198, 1961

Reichenbachara in Bot. Mus. Leafl. Harv. Univ. 21:175, 1966
Euanthe × *Vanda* × *Vandopsis*
1st hybr.: **R.** Kimo Cardus (as **Opsisanda** Kimo Cardus)
Parentage: **Vandanthe** *Burgeffii* × *Vandopsis lissochiloides*

Reinikkaara in Bot. Mus. Leafl. Harv. Univ. 22:285, 1969
Aerides × *Ascocentrum* × *Euanthe* × *Vanda*
1st hybr.: **R.** Mem. Lillian Arnold (as **Christieara** Mem. Lillian Arnold)
Parentage: *Aerides Lawrenceae* × **Schlechterara** Meda Arnold

Renades in Orch. Rev. 63:108, 1955
Aerides × *Renanthera*
1st hybr.: **R.** Kaiulani
Parentage: *Renanthera monachica* × *Aerides Fieldingii*

Renaglottis in Bull. Pac. Orch. Soc. Hawaii 14:85, 1957
Renanthera × *Trichoglottis*
1st hybr.: **R.** Lone Warrior (in Orch. Rev. 74: November 1966)
Parentage: *Renanthera Storiei* × *Trichoglottis fasciata*

Renancentrum in Orch. Rev. 70: September 1962
Ascocentrum × *Renanthera*
1st hybr.: **R.** Yap Sin Yee (in Orch. Rev. 73: May 1965)
Parentage: *Ascocentrum curvifolium* × *Renanthera* Brookie Chandler
Observation: For the hybrid **Renancentrum** Curvionica see **Ascorella**

Renanda in Orch. Rev. 69:63, 1961
Observation: See **Holttumara** including the hybrid **Renanda** Ruby Star

Renanetia in Orch. Rev. 70: September 1962
Neofinetia × *Renanthera*
1st hybr.: **R.** Bali
Parentage: *Renanthera* Brookie Chandler × *Neofinetia falcata*
  Syn.: **Holcanthera** in Bot. Mus. Leafl. Harv. Univ. 23:188, 1972

Renanopsis in Orch. Rev. 57:24, 1949
Renanthera × *Vandopsis*
1st hybr.: **R.** Lena Rowold (as **Athertonara** Lena Rowold)
Parentage: *Renanthera Storiei* × *Vandopsis lissochiloides*
  Syn.: **Athertonara** in Orch. Rev. 56:26, 1948
      Renopsis in Malay. Orch. Rev. 4:36, 1949

Renanstylis in Orch. Rev. 68:224, 1960
Renanthera × *Rhynchostylis*
1st hybr.: **R.** Jo Ann
Parentage: *Renanthera* Brookie Chandler × *Rhynchostylis retusa*
  Syn.: **Rhynchanthera** in Hawkes, Encycl. Cult. Orch. 529, 1965

Randactyle in Harrison, Epiphytic Orchids of Southern Africa 98, 1972
Rangaeris × Tridactyle
1st hybr.: not yet named
Parentage: *Rangaeris muscicola × Tridactyle bicaudata*

Renantanda in Bull. Soc. Hort. Fr. ser. 6, 2:92, 1935
Renanthera × Vanda
1st hybr.: **R.** *Sanderi*
Parentage: *Renanthera Imschootiana × Vanda suavis*
Observation: For the hybrid **Renantanda** Titan see **Antheranthe**
    Syn.: **Renantheranda** in Malay. Orch. Rev. 2:139, 1938
        Vandathera in De Orchidee 8:159, 1939

Renantheranda in Malay. Orch. Rev. 2:139, 1938
Observation: See **Renantanda**

Renanthoceras in Bot. Mus. Leafl. Harv. Univ. 22:286, 1969
Pteroceras × Renanthera
1st hybr.: **R.** Kona (as **Sarcothera** Kona)
Parentage: *Pteroceras pallidus × Renanthera monachica*

Renanthoglossum in Orch. Rev. 71: September 1963
Ascoglossum × Renanthera
1st hybr.: **R.** Red Delight
Parentage: *Renanthera Storiei × Ascoglossum calopterum*

Renanthopsis in Bull. Soc. Hort. Fr. ser. 5, 4:342, 1931
Phalaenopsis × Renanthera
1st hybr.: **R.** Premier
Parentage: *Renanthera Imschootiana × Phalaenopsis Sanderana*
Observation: For the hybrid **Renanthopsis** Firefly see **Pararenanthera**

Renaradorum in Bot. Mus. Leafl. Harv. Univ. 21:176, 1966
Arachnis × Armodorum × Renanthera
1st hybr.: **R** Star Orange (as **Aranthera** Star Orange)
Parentage: **Armodachnis** Catherine × *Renanthera coccinea*

Renopsis in Malay. Orch. Rev. 4:36, 1949
Observation: See **Renanopsis**

Restesia in Am. Orch. Soc. Bull. 38:676, 1969
Orleanesia × Restrepia
1st hybr.: not yet reported

Rhinochilus in Orch. Rev. 80:142, 1972
Rhinerrhiza × *Sarcochilus*
1st hybr.: **R.** Dorothy (as **Sarcorhiza** Dorothy)
Parentage: *Sarcochilus Hartmannii* × *Rhinerrhiza divitiflora*
  Syn.: **Sarcorhiza** in Orch. Rev. 74: January 1966, not *Sarcorrhiza* Bullock, 1962

Rhizanthera in Watsonia 6:133, 1965
Dactylorhiza × *Platanthera*
1st hybr.: **R.** *Chevallieriana* (as *Orchis Chevallieriana*, 1891)
Parentage: *Dactylorhiza maculata* var. *elodes* × *Platanthera bifolia*
  Syn.: **Dactylanthera** in Willis, Dict. Fl. Plants and Ferns, ed. 7; 327, 1966

Rhynchanthera in Hawkes, Encycl. Cult. Orch. 529, 1965
Observation: See **Renanstylis**

Rhynchocentrum in Orch. Rev. 71: April 1963
Ascocentrum × *Rhynchostylis*
1st hybr.: **R.** Sagarik
Parentage: *Rhynchostylis coelestis* × *Ascocentrum curvifolium*

Rhynchonopsis in Orch. Rev. 73: November 1965
Phalaenopsis × *Rhynchostylis*
1st hybr.: **R.** Winona Jordan
Parentage: *Rhynchostylis gigantea* × *Phalaenopsis* Doris

Rhynchopapilisia nom. hybr. gen. nov.
Luisia × *Papilionanthe* × *Rhynchostylis*
1st hybr.: **R.** Eva Goff (as **Goffara** Eva Goff)
Parentage: **Papilisia** Uniwai × *Rhynchostylis coelestis*

Rhynchorides in Orch. Rev. 70: October 1962 (as **Rhynchorades**)
Observation: See **Aeridostylis** including the hybrid **Rhynchorides** Springtime

Rhynchovanda in Orch. Rev. 66:231, 1958
Observation: See **Vandachostylis** including the hybrid **Rhynchovanda** Fantasy For the hybrid **Rhynchovanda** Blue Angel see **Rhynchovandanthe**

Rhynchovandanthe in Bot. Mus. Leafl. Harv. Univ. 21:177, 1966
Euanthe × *Rhynchostylis* × *Vanda*
1st hybr.: **R.** Blue Angel (as **Rhynchovanda** Blue Angle)
Parentage: **Vandanthe** *Rothschildiana* × *Rhynchostylis coelestis*

Rhynchovola in Proc. 3rd World Orch. Conf. 326, 1960
Brassavola × *Rhyncholaelia*
1st hybr.: **R.** David Sander (as *Brassavola* David Sander)
Parentage: *Brassavola cucullata* × *Rhyncholaelia Digbyana*
Observation: The genera *Rhyncholaelia* and *Brassavola* are not kept
   separate in horticulture

Rhyndoropsis in Orch. Rev. 74: October 1966
Doritis × *Phalaenopsis* × *Rhynchostylis*
1st hybr.: **R.** Florida
Parentage: **Doritaenopsis** Dorette × *Rhynchostylis retusa*

Ridleyara in Malay. Orch. Rev. 5:2, 1957
Arachnis × *Trichoglottis* × *Vanda*
1st hybr.: **R.** Fascad
Parentage: **Aranda** Eileen Addison × *Trichoglottis fasciata*

Robinara in Orch. Rev. 80:247, 1972
Aerides × *Ascocentrum* × *Renanthera* × *Vanda*
1st hybr.: not yet reported
Observation: For the hybrid **Robinara** Kosher Red see **Neorobinara**

Rodrassia in Orch. Rev. 68:404, 1960
Observation: See **Bradriguezia** including the hybrid **Rodrassia** Angollitos

Rodrenia in Am. Orch. Soc. Bull. 31:357, 1962
Observation: See **Rodridenia**

Rodrettia in Orch. Rev. 66:231, 1958
Comparettia × *Rodriguezia*
1st hybr.: **R.** Hawaii
Parentage: *Comparettia falcata* × *Rodriguezia secunda*

Rodricidium in Orch. Rev. 65:89, 1957
Oncidium × *Rodriguezia*
1st hybr.: **R.** Twyla
Parentage: *Oncidium tetrapetalum* × *Rodriguezia secunda*
   Syn.: **Oncidguezia** in Bull. Pac. Orch. Soc. Hawaii 14:85, 1957

Rodridenia in Orch. Rev. 70: April 1962
Macradenia × *Rodriguezia*
1st hybr.: **R.** Red Gem
Parentage: *Rodriguezia secunda* × *Macradenia brassavolae*
   Syn.: **Rodrenia** in Am. Orch. Soc. Bull. 31:357, 1962

Rodriglossum in Orch. Rev. 81: April 1973
Odontoglossum × Rodriguezia
1st hybr.: **R.** Dolly
Parentage: *Odontogolssum bictoniense × Rodriguezia Strobelii*

Rodriopsis in Orch. Rev. 77: September 1969
Ionopsis × Rodriguezia
1st hybr.: **R.** Edwardine Klemm
Parentage: *Rodriguezia secunda × Ionopsis paniculata*

Rodritonia in Orch. Rev. 67:33, 1959
Observation: See **Miltonguezia** including the hybrid **Rodritonia** Freckles

Rolfeara in Orch. Rev. 27:3, 1919
Brassavola × Cattleya × Sophronitis
1st hybr.: **R.** *rubescens*
Parentage: **Sophrocattleya** Blackii × **Brassocattleya** Mrs. J. Leemann

Rosakirschara in Orch. Rev. 80:162, 1972
Ascocentrum × Neofinetia × Renanthera
1st hybr.: **R.** Liliput
Parentage: *Renanthera monachica ×* **Ascofinetia** Peaches

Rothara in Orch. Rev. 78: December 1970
Brassavola × Cattleya × Epidendrum × Laelia × Sophronitis
1st hybr.: **R.** Richard Roth
Parentage: **Epicattleya** Rosita × **Potinara** Estelle Smith

Rumrillara in Orch. Rev. 77: September 1969
Ascocentrum × Neofinetia × Rhynchostylis
1st hybr.: **R.** Rosyleen
Parentage: **Rhynchocentrum** Lilac Blossom × **Ascofinetia** Peaches

Saccanthera in Orch. Rev. 69:269, 1961
Renanthera × Saccolabium
1st hybr.: not yet reported
Observation: For the hybrid **Saccanthera** Queen Emma see **Renanstylis**

Saccovanda in Orch. Rev. 67:330, 1959
Observation: See **Sanda.** For the hybrid **Saccovanda** Dawn see **Vanda-chostylis**

Sanda in Sander, List. Orch. Hybr. Add. III, x, 1955
Saccolabium × Vanda
1st hybr.: not yet reported
   Syn.: **Saccovanda** in Orch. Rev. 67:330, 1959

Sanderara in Orch. Rev. 45:257, 1937
Brassia × *Cochlioda* × *Odontoglossum*
1st hybr.: **S.** Alpha
Parentage: *Brassia Lawrenceana* × **Odontioda** Grenadier

Sappanara in Orch. Rev. 73: June 1965
Arachnis × *Phalaenopsis* × *Renanthera*
1st hybr.: not yet reported
Observation: For the hybrid **Sappanara** Ahmad Zahab see **Paranthera**

Sarcalaenopsis nom. hybr. gen. nov.
Phalaenopsis × *Sarcanthopsis*
1st hybr.: **S.** Star of Hawaii as **Phalandopsis** Star of Hawaii)
Parentage: *Sarcanthopsis Warocqueana* × *Phalaenopsis* Grace Palm

Sarcocentrum in Orch. Rev. 79: March 1971
Ascocentrum × *Sarcochilus*
1st hybr.: **S.** Little Sue
Parentage: *Sarcochilus Hartmannii* × *Ascocentrum curvifolium*

Sarcoceras nom. hybr. gen. nov.
Pteroceras × *Sarcochilus*
1st hybr.: **S.** Perky (as **Parachilus** Perky)
Parentage: *Pteroceras spathulatus* × *Sarcochilus Hartmannii*
    Syn.: **Parachilus** in Orch. Rev. 80:63, 1972

Sarconopsis in Orch. Rev. 79: March 1971
Phalaenopsis × *Sarcochilus*
1st hybr.: **S.** Jean Cannons
Parentage: *Phalaenopsis Schillerana* × *Sarcochilus Hartmannii*

Sarcopapilionanda nom. hybr. gen. nov.
Papilionanthe × *Sarcochilus* × *Vanda*
1st hybr.: **S.** Suzanne (as **Sarcovanda** Suzanne)
Parentage: *Sarcochilus Hartmannii* × **Papilionanda** Poepoe

Sarcorhiza in Orch. Rev. 74: January 1966, not *Sarcorrhiza* Bullock, 1962
Observation: See **Rhinochilus** including the hybrid **Sarcorhiza** Dorothy

Sarcothera in Orch. Rev. 62:92, 1954
Renanthera × *Sarcochilus*
1st hybr.: not yet reported
Observation: For the hybrid **Sarcothera** Kona see **Renanthoceras**

Sarcovanda in Orch. Rev. 79: March 1971
Sarcochilus × *Vanda*
1st hybr.: not yet reported
Observation: For the hybrid **Sarcovanda** Suzanne see **Sarcopapilionanda**

Sartylis in Orch. Rev. 81: October 1973
Rhynchostylis × *Sarcochilus*
1st hybr.: **S.** Blue Knob
Parentage: *Sarcochilus Hartmannii* × *Rhynchostylis retusa*

Schlechterara in Bot. Mus. Leafl. Harv. Univ. 21:179, 1966
Ascocentrum × *Euanthe* × *Vanda*
1st hybr.: **S.** Meda Arnold (as **Ascocenda** Meda Arnold)
Parentage: *Ascocentrum curvifolium* × **Vandanthe** *Rothschildiana*

Schombavola in Orch. Rev. 72: January 1964
Observation: See **Schombobrassavola** including the hybrid **Schombavola**
    Purple Star

Schombletia in Hansen, Orch. Hybr. 234, 1895
Bletia × *Schomburgkia*
1st hybr.: not yet reported

Schombobrassavola in Bull. Pac. Orch. Soc. Hawaii 14:85, 1957
Brassavola × *Schomburgkia*
1st hybr.: **S.** Purple Star (as **Schombavola** Purple Star)
Parentage: *Schomburgkia* Kalihi × *Brassavola glauca*
    Syn.: **Schombavola** in Orch. Rev. 72: January 1964

Schombocattleya in Orch. Rev. 13:245, 1905
Cattleya × *Schomburgkia*
1st hybr.: **S.** *spiralis*
Parentage: *Cattleya Mossiae* × *Schomburgkia tibicinis*
    Syn.: **Schomburgkio-Cattleya** in J. Hort. Soc. Fr. ser. 4, 4:534, 1903
        Schomcattleya in Gard. Chron. ser. 3, 38:53, 1905
        Schomocattleya in Orch. Rev. 28:50, 1920
        Myrmecocattleya in Orch. Rev. 28:50, 1920

Schombodiacrium in Orch. Rev. 66:137, 1958
Observation: See **Diaschomburgkia** including the hybrid **Schombodia-crium** Ipo

Schomboepidendrum in Orch. Rev. 65:90, 1957
Epidendrum × *Schomburgkia*

1st hybr.: **S.** Crispa-Glow
Parentage: *Schomburgkia crispa* × *Epidedrum* Orange Glow

Schombolaelia in Orch. Rev. 21:254, 1913
Laelia × *Schomburgkia*
1st hybr.: **S.** tibibrosa
Parentage: *Laelia tenebrosa* × *Schomburgkia tibicinis*
  Syn.: **Myrmecolaelia** in Orch. Rev. 28:50, 1920

Schombolaeliocattleya in Roy. Hort. Soc. Dict. Gard. 4:1905, 1951
Cattleya × *Laelia* × *Schomburgkia*
1st hybr.: **S.** Fiesta (ás **Lyonara** Fiesta)
Parentage: **Laeliocattleya** Issy × *Schomburgkia Thomsoniana*
  Syn.: **Lyonara** in Orch. Rev. 67:405, 1959

Schombonia in Orch. Rev. 70: January 1962
Observation: See **Schombotonia** including the hybrid **Schombonia** Firefly

Schombonitis in Orch. Rev. 28:50, 1920
Schomburgkia × *Sophronitis*
1st hybr.: **S.** Stella
Parentage: *Schomburgkia superbiens* × *Sophronitis grandiflora*

Schombotonia in Bull. Pac. Orch. Soc. Hawaii 14:85, 1957
Broughtonia × *Schomburgkia*
1st hybr.: **S.** Firefly (as **Schombonia** Firefly)
Parentage: *Schomburgkia Thomsoniana* × *Broughtonia sanguinea*
  Syn.: **Schombonia** in Orch. Rev. 70: January 1962

Schomburgkio-Cattleya in J. Hort. Soc. Fr. ser. 4,4:534, 1903
Observation: See **Schombocattleya**

Schomcattleya in Gard. Chron. ser. 3, 38:53, 1905
Observation: See **Schombocattleya**

Schomocattleya in Orch. Rev. 28:50, 1920
Observation: See **Schombocattleya**

Scullyara in Jones & Scully Orchidglade Catalog, p. 120, October 1973
Cattleya × *Epidendrum* × *Schomburgkia*
1st hybr.: not yet named
Parentage: **Schombocattleya** Snow White × *Epidendrum diurnum*

Sealara nom. hybr. gen. nov.
Papilionanthe × *Paraphalaenopsis* × *Vanda*

1st hybr.: **S.** Khoo Kay Ann (as **Vandaenopsis** Khoo Kay Ann)
Parentage: **Papilionanda** Prolific × *Paraphalaenopsis Denevei*

Selenipanthes in Am. Orch. Soc. Bull. 38:676, 1969
Lepanthes × Selenipedium
1st hybr.: not yet reported

Selenocypripedium in J. Hort. Soc. Fr. ser. 4, 13:706, 1912
Observation: See **Cysepedium.** For the hybrids **Selenocypripedium** *Mal-houitri* and **Selenocypripedium** Confusion see **Phragmipaphium**

Serapicamptis in J. Bot. 59:57, 1921
Anacamptis × Serapias
1st hybr.: **S.** *Forbesii*
Parentage: *Serapias lingua × Anacamptis pyramidalis*

Serapirhiza in Jahresb. Naturwiss. Ver. Wuppertal, Heft 21 22:103, 1968
Dactylorhiza × Serapias
1st hybr.: **S.** *Sambucino-lingua*
Parentage: *Serapias Lingua × Dactylorhiza sambucina*

Shigeuraara in Orch. Rev. 77: November 1969
Ascocentrum × Ascoglossum × Renanthera × Vanda
1st hybr.: not yet reported
Observation: For the hybrid **Shigeuraara** Cassino see **DeWolfara**

Shipmanara in Orch. Rev. 71: April 1963
Broughtonia × Diacrium × Schomburgkia
1st hybr.: **S.** Pink Angel
Parentage: **Diaschomburgkia** Ipo × **Diabroughtonia** Alice Hart

Singaporeara nom. hybr. gen. nov.
Ascocentrum × Papilionanthe × Vanda
1st hybr.: **S.** Singapore Beauty (as **Ascocenda** Singapore Beauty)
Parentage: **Papilionanda** Josephine van Brero × **Ascocenda** Charm

Smithara in Bot. Mus. Leafl. Harv. Univ. 21:180, 1966
Ascocentrum × Euanthe × Neofinetia × Vanda
1st hybr.: **S.** Wendy (as **Nakamotoara** Wendy)
Parenatge: *Neofinetia falcata ×* **Schlechterara** Meda Arnold

Sobraleya in Hansen, Orch. Hybr. 242, 1895
Cattleya × Sobralia
1st hybr.: not yet reported

Sophrobroughtonia in Bull. Pac. Orch. Soc. Hawaii 14:85, 1957
Broughtonia × *Sophronitis*
1st hybr.: not yet reported

Sophrocatlaelia in Orch. Rev. 8:354, 1900
Observation: See **Sophrolaeliocattleya**

Sophrocattleya in J. Linn. Soc. Bot. 24:169, 1887
Cattleya × *Sophronitis*
1st hybr.: **S.** *Batemaniana* (as *Laelia Batemaniana*)
Parentage: *Sophronitis grandiflora* × *Cattleya intermedia*
   Syn.: **Sophroleya** in Hansen, Orch. Hybr. 242, 1895

Sophrolaelia in Orch. Rev. 2:333, 1894
Laelia × *Sophronitis*
1st hybr.: **S.** *laeta*
Parentage: *Laelia pumila* var. *Dayana* × *Sophronitis grandiflora*

Sophrolaeliocattleya in J. Roy. Hort. Soc. 21:468, 1897
Cattleya × *Laelia* × *Sophronitis*
1st hybr.: **S.** *Veitchii*
Parentage: *Sophronitis grandiflora* × **Laeliocattleya** *elegans* (*Schillerana*)
   Syn.: **Catlaenltis** in Hansen, Orch. Hybr. 100, 1895
         Sophrocatlaelia in Orch. Rev. 8:354, 1900

Sophroleya in Hansen, Orch. Hybr. 242, 1895
Observation: See **Sophrocattleya**

Sophrovola in Hansen, Orch. Hybr. 243, 1895
Brassavola × *Sophronitis*
1st hybr.: **S.** Edna (as **Brassonitis** Edna, 1962)
Parentage: *Sophronitis coccinea* × *Brassavola nodosa*
   Syn.: **Brassophronitis** in Die Orchidee 5:41, 1954
         Brassonotis in Orch. Rev. 70: December 1962

Spathophaius in Orchid Weekly 1:268, 1959
Phaius × *Spathoglottis*
1st hybr. not yet reported

Stacyara in Orch. Rev. 81: April 1973
Cattleya × *Epidendrum* × *Sophronitis*
1st hybr.: **S.** Sam
Parentage: *Epidendrum gracile* × **Sophrocattleya** Cleopatra

Stanfieldara in Orch. Rev. 77: November 1969
Epidendrum × Laelia × Sophronitis
1st hybr.: **S.** Will Bates
Parentage: **Sophrolaelia** Psyche × *Epidendrum vitellinum*

Staurachnis in Orch. Rev. 58:65, 1950
Observation: See **Trichachnis** including the hybrid **Staurachnis N.** Sorapure

Stauranda in The Orchid J. 1:300, 1952
Observation: See **Trichovanda.** For the hybrid **Stauranda** Ulaula see
Antheglottis

Stylisanthe nom. hybr. gen. nov.
Papilionanthe × Rhynchostylis
1st hybr.: **S.** *Bernardii* (as **Vandachostylis** *Bernardii*)
Parentage: *Papilionanthe teres × Rhynchostylis retusa*

Sweetara in Am. Orch. Soc. Bull. 38:676, 1969
Paraphalaenopsis × Rhynchostylis × Vanda
1st hybr.: **S.** Oi Yee (as **Yapara** Oi Yee)
Parentage: **Paravanda** Suavei × **Vandachostylis** Tan Great Leng (as
Rhynchovanda Tan Geat Leng)

Symphodontioda in Orchid Weekly 4:121, 1963
Cochlioda × Odontoglossum × Symphyglossum
1st hybr.: **S.** Hermione (as **Odontioda** Hermione, 1910)
Parentage: **Symphodontoglossum** *heatonensis × Cochlioda vulcanica*

Symphodontoglossum in Orchid Weekly 4:121, 1963
Odontoglossum × Symphyglossum
1st hybr.: **S.** *heatonensis* (as **Odontioda** *heatonensis,* 1906)
Parentage: *Symphyglossum sanguineum × Odontoglossum cirrhosum*

Symphodontonia in Bot. Mus. Leafl. Harv. Univ. 21:182, 1966
Miltonia × Odontoglossum × Symphyglossum
1st hybr.: **S.** Felicia (as **Vuylstekeara** Felicia, 1921)
Parentage: *Miltonia Warscewiczii ×* **Symphodontoglossum** (**Odontioda**)
Felicia

Symphyglossonia in Orchid Weekly 4:121, 1963
Miltonia × Symphyglossum
1st hybr. **S.** Pink Pearl (as **Miltonioda** Pink Pearl)
Parentage: *Miltonia* St. Andre × *Symphyglossum sanguineum*

Tanakara in Orch. Rev. 55:120, 1947
Observation: See **Opsisanda** including the hybrid **Tanakara** Colombo

Tanakara in Orch. Rev. 60:13, 1952 (as **Tanakaria**)
Observation: See **Phalaerianda** including the hybrid **Tanakara** Honolulu

Tenranara in Orch. Rev. 70: December 1962
Observation: See **Fujiwarara** including the hybrid **Tenranara** Frolic

Teohara in Orch. Rev. 76: February 1968
Arachnis × *Renanthera* × *Vanda* × *Vandopsis*
1st hybr.: not yet reported
Observation: For the hybrid **Teohara** Teo Cheng Swee see **Huntara**

Tetralaelia in Gard. Chron. ser. 3, 31:280, 1902 (as Tetralaenia)
Laelia × *Tetramicra*
1st hybr.: not yet reported

Tetraliopsis in Orch. Rev. 73: July 1965
Laeliopsis × *Tetramicra*
1st hybr.: **T.** Candystripe
Parentage: *Tetramicra canaliculata* × *Laeliopsis domingensis*

Tetratonia in Orch. Rev. 73: August 1965
Broughtonia × *Tetramicra*
1st hybr. **T.** Dark Prince
Parentage: *Tetramicra canaliculata* × *Broughtonia sanguinea*

Thaiara nom. hybr. gen. nov.
Euanthe × *Rhynchostylis* × *Seidenfadenia* × *Vanda*
1st hybr.: **T.** Porchina Blue (as **Perreiraara** Porchina Blue)
Parentage: **Rhynchovandanthe** Blue Angel × *Seidenfadenia mitrata*

Thesaëra in Orch. Rev. 78: September 1970
Aërangis × *Aëranthes*
1st hybr. **T.** Rex van Delden
Parentage: *Aërangis Kotschyana* × *Aëranthes ramosus*

Thorntonara nom. hybr. gen. nov.
Ascocentrum × *Doritis* × *Euanthe* × *Vanda*
1st hybr.: **T.** John Miller (as **Ascovandoritis** John Miller)
Parentage: *Doritis pulcherrima* × **Schlechterara** Ophelia

Trevorara in Orch. Rev. 71: March 1963
Arachnis × *Phalaenopsis* × *Vanda*

1st hybr.: **T.** Inggraini (in Orch. Rev. 72: June 1964)
Parentage: *Vanda tricolor* × **Arachnopsis** Rosea
Observation: For the hybrid **Trevorara** Hong Trevor see **Parandachnis.**
   For the hybrid **Trevorara** Manoa see **Benthamara**

Trichachnis in Na Pua Okika o Hawaii Nei 7:154, 1957
Arachnis × *Trichoglottis*
1st hybr.: **T. N.** Sorapure (as **Staurachnis** N. Sorapure)
Parentage: *Arachnis flos-aeris* × *Trichoglottis fasciata*
   Syn.: **Staurachnis** in Orch. Rev. 58:65, 1950
          Arachnoglottis in Orch. Rev. 66:86, 1958

Trichocidium in Orch. Rev. 63:155, 1955
Oncidium × *Trichocentrum*
1st hybr.: **T.** Elvena
Parentage: *Oncidium Lanceanum* × *Trichocentrum albopurpureum*

Trichopasia in Bull. Pac. Orch. Soc. Hawaii 14:85, 1957
Aspasia × *Trichopilia*
1st hybr.: not yet reported

Trichopsis in Orch. Rev. 78: September 1970
Trichoglottis × *Vandopsis*
1st hybr.: **T.** Clare Booth Luce
Parentage: *Trichoglottis Guibertii* × *Vandopsis lissochiloides*

Trichovanda in Orch. Rev. 57:24, 1949
Trichoglottis × *Vanda*
1st hybr. **T.** Bonfire
Parentage: *Vanda* Herziana × *Trichoglottis brachiata*
Observation: For the hybrid **Trichovanda** Ulaula see **Antheglottis.** For
   the hybrid **Trichovanda** Richard Emery see **Kraenzlinara**
   Syn.: **Lyonara** in Orch. Rev. 56:94, 1948
          Stauranda in The Orchid J. 1:300, 1952

Vancampe in Bull. Pac. Orch. Soc. Hawaii 14:85, 1957
Acampe × *Vanda*
1st hybr.: **V.** Beans
Parentage: *Vanda* Frank Scudder × *Acampe longifolia*

Vandachnanthe in Anggrek Boelan 1:67, 1939
Observation: See **Aranda**

Vandachnis in Orch. Rev. 57:66, 1949
Arachnis × Vandopsis
1st hybr.: **V.** Premier
Parentage: *Arachnis flos-aeris × Vandopsis lissochiloides*

Vandachostylis in Arch. Mus. Hist. Nat. Paris ser. 6, 12 pt. 2:608, 1935 (as **Vandacostylis**)
Rhynchostylis × Vanda
1st hybr.: **V.** Tan Geat Leng (as **Rhynchovanda** Tan Geat Leng)
Parentage: *Rhynchostylis coelestis × Vanda coerulescens*
Observation: For the hybrid **Vandachostylis** *Bernardii* see *Stylisanthe*
Syn.: **Rhynchovanda** in Orch. Rev. 66:231, 1958

Vandaecum in Orch. Rev. 68:224, 1960
Angraecum × Vanda
1st hybr.: not yet reported
Observation: For the hybrid **Vandaecum** Premier see **Vandofinetia**

Vandaenopsis in Arch. Mus. Nat. Paris ser. 6, 12 pt. 2:607, 1935
Phalaenopsis × Vanda
1st hybr.: **V.** *ferrierensis* (as × **Vandopsis** *ferrierensis*)
Parentage: *Vanda suavis × Phalaenopsis amabilis (Rimestadiana)*
Observation. For the hybrid **Vandaenopsis** Jawaii see **Paranthe**; for the hybrid **V.** Pang Nyuk Yin see **Parandanthe**; for the hybrid **V.** Boguriana see **Paravanda**
Syn.: **Vandaeopsis** in Gartenfl. 86:252, 1937
Vandanopsis in De Orchidee 8:186, 1939

Vandaeopsis in Gartenfl. 86:252, 1937
Observation: See **Vandaenopsis**

Vandanopsis in De Orchidee 8:186, 1939
Observation: See **Vandaenopsis**

Vandanthe in Orchis 13:52, 1919
Euanthe × Vanda
1st hybr.: **V.** *Tatzeri*
Parentage: *Vanda tricolor × Euanthe Sanderana*

Vandathera in De Orchidee 8:159, 1939
Observation: See **Renantanda**

Vandantherella nom. hybr. gen. nov.
Renantherella × Vanda

1st hybr.: **V.** Chang Min-Tat (as **Renantanda** Chang Min-Tat)
Parentage: *Renantherella histrionica* × *Vanda lamellata*

Vandantherides in Bot. Mus. Leafl. Harv. Univ. 21:184, 1966
Aerides × *Euanthe* × *Vanda*
1st hybr.: **V.** Grace Dunn (as **Aeridovanda** Grace Dunn)
Parentage: *Aerides crassifolia* × **Vandanthe** *Tatzeri*
Observation: For the hybrid **Vandantherides** Elizabeth Young see **Eupa-pilio**

Vandarachnis in Orchideeen 6:107, 1939
Observation: See **Aranda**

Vandofinetia in Orch. Rev. 68:404, 1960
Neofinetia × *Vanda*
1st hybr.: **V.** Premier (as **Vandaecum** Premier)
Parentage: *Neofinetia falcata* × *Vanda lamellata*
   Syn.: **Vandoglossum** in Bot. Mus. Leafl. Harv. Univ. 23:188, 1972

Vandoglossum in Bot. Mus. Leafl. Harv. Univ. 23:188, 1972
Observation: See **Vandofinetia**

Vandopsides in Orch. Rev. 66:231, 1958
Aerides × *Vandopsis*
1st hybr.: **V.** Apple Blossom
Parentage: *Aerides Lawrenceae* × *Vandopsis lissochiloides*

Vandopsisvanda in Rev. Circ. Paul. Orch. 7:219, 1950
Observation: See **Opsisanda**

Vandoritis in Orch. Rev. 73: August 1965
Doritis × *Vanda*
1st hybr.: **V.** Malaysia
Parentage: *Vanda* Lanikea × *Doritis pulcherrima*

Vanglossum in Orch. Rev. 77: March 1969
Ascoglossum × *Vanda*
1st hybr.: **V.** Oriental Jewel
Parentage: *Vanda Merrillii* × *Ascoglossum calopterum*

Vascostylis in Orch. Rev. 72: January 1964
Ascocentrum × *Rhynchostylis* × *Vanda*
1st hybr.: not yet reported
Observation: For the hybrid **Vascostylis** Blue Fairy see **Carrara**

Vaughnara in Orch. Rev. 73: March 1965
Brassavola × *Cattleya* × *Epidendrum*
1st hybr.: **V.** Sparklet
Parentage: **Brassocattleya** *Cliftonii* × *Epidendrum vitellinum*

Vuylstekeara in Orch. Rev. 19:60, 1911
Cochlioda × *Miltonia* × *Odontoglossum*
1st hybr.: **V.** *insignis*
Parentage: *Miltonia vexillaria* × **Odontioda** *Vuylstekeae*
  Syn.: **Odontiodonia** in Orchid World 1:84, 1911

Warneara in Orch. Rev. 72: July 1964
Comparettia × *Oncidium* × *Rodriguezia*
1st hybr.: **W.** Robert
Parentage: *Oncidium* Agnes Ann × **Rodrettia** Hawaii

Wilkinsara in Orch. Rev. 81: April 1973
Ascocentrum × *Vanda* × *Vandopsis*
1st hybr.: **W.** Gemini
Parentage: *Vandopsis Parishii* × **Ascocenda** Red Gem

Wilsonara in Gard. Chron. ser. 3, 59:315, 1916
Cochlioda × *Odontoglossum* × *Oncidium*
1st hybr.: **W.** *insignis*
Parentage: *Odontoglossum illustrissimum* × **Oncidioda** *Charlesworthii*

Withnerara in Orch. Rev. 74: November 1966
Aspasia × *Miltonia* × *Odontoglossum* × *Oncidium*
1st hybr.: **W.** Moon Glow
Parentage: **Oncidasia** Starlight × **Odontonia** Wonder

Wrefordara in Am. Orch. Soc. Bull. 38:676, 1969
Aerides × *Arachnis* × *Euanthe* × *Vanda*
1st hybr.: **W.** Henry (as **Burkillara** Henry)
Parentage: **Aeridachnis** Bogor × **Vandanthe** Ellen Noa

Yamadara in Orch. Rev. 68:404, 1960
Observation: See **Adamara** including the hybrid **Yamadara** Fuchsia

Yapara in Orch. Rev. 74: November 1966
Phalaenopsis × *Rhynchostylis* × *Vanda*
1st hybr., not yet reported
Observation: For the hybrid **Yapara** Oi Yee see **Sweetara**

Yoneoara in Orch. Rev. 80:162, 1972
Renanthera × Rhynchostylis × Vandopsis
1st hybr.: **Y.** Hadrian
Parentage: **Renanopsis** Lena Rowold × *Rhynchostylis gigantea*

Yusofara in Orch. Rev. 80:247, 1972
Arachnis × Ascocentrum × Renanthera × Vanda
1st hybr.: not yet reported
Observation: For the hybrid **Yusofara** Nong see **Neoyusofara**

Zygobatemannia in Semaine Hortic. 3:76, 1899
Batemanniana × Zygopetalum
1st hybr.: **Z.** *Mastersii*
Parentage: *Zygopetalum crinitum × Batemannia Colleyi*

Zygocaste in Orch. Rev. 54:41, 1946
Lycaste × Zygopetalum
1st hybr.: **Z.** Van Belle
Parentage: *Lycaste virginalis × Zygopetalum Mackayi*

Zygocella in Bot. Mus. Leafl. Harv. Univ. 21:186, 1966
Mendoncella × Zygopetalum
1st hybr.: **Z.** *Max-Jorisii* (as *Zygopetalum Max-Jorisii*)
Parentage: *Zygopetalum maxillare × Mendoncella Jorisiana*

Zygocidium in Hansen, Orch. Hybr. 244, 1895
Oncidium × Zygopetalum
1st hybr.: not yet reported

Zygocolax in Gard. Chron. ser. 3, 1:756, 1887
Observation: See **Zygopabstia** including the hybrid **Zygocolax** *leopardinus*

Zygodendrum in Hansen, Orch. Hybr. 244, 1895
Epidendrum × Zygopetalum
1st hybr.: not yet reported

Zygolax in Hansen, Orch. Hybr. 244, 1895
Observation: See **Zygopabstia**

Zygolum nom. hybr. gen. nov.
Zygopetalum × Zygosepalum
1st hybr.: **Z.** *Roeblingianum* (as *Zygopetalum Roeblingianum*)
Parentage: *Zygopetalum maxillare × Zygosepalum labiosum*
    Syn.: **Zygomena** in Die Natuerl. Pflanzenfam. Erg.-heft II, 92, 1908

Zygomena in Die Natuerl. Pflanzenfam. Erg.-heft II, 92, 1908
Observation: For the hybrid Zygomena *Roeblingiana* see Zygolum

Zygonisia in Gard. Chron. ser. 3, 31:443, 1902
Aganisia × *Zygopetalum*
1st hybr.; not yet reported
Observation: For the hybrids Zygonisia *Rolfeana* and Z. *Sanderi* see
  Zygostylis

Zygopabstia in Bradea 1:308, 1973
Pabstia × *Zygopetalum*
1st hybr.: Z. *leopardina* (as *Zygopetalum leopardinum*)
Parentage: *Pabstia jugosa* × *Zygopetalum maxillare*
  Syn.: Zygocolax in Gard. Chron. ser. 3, 1:756, 1887
      Zygolax in Hansen, Orch. Hybr. 244, 1895

Zygorhyncha in Gartenfl. 85:254, 1936
Chondrorhyncha × *Zygopetalum*
1st hybr. Z. *Fletcheri* (as Chondropetalum *Fletcheri*)
Parentage: *Zygopetalum Mackayi* × *Chondrorhyncha Chestertonii*
Observaiton: This is a very doubtful hybrid
  Syn.: Chondropetalum in Orch. Rev. 16:56, 1908, not Rttb., 1773

Zygostylis in Handbook Orch. Nomoncl. and Reg., 47, 1965
Otostylis × *Zygopetalum*
1st hybr.: Z. *Rolfeana* (as *Zygonisia Rolfeana*)
Parentage: *Zygopetalum maxillare* × *Otostylis alba*
Observation: See Orch. Rev. 74: April 1966

Part II

Parent Genera	Hybrid Genus
Acampe × Vanda	= Vancampe
Aceras × Barlia	= Barlaceras
Aceras × Dactylorhiza	= Dactyloceras
Aceras × Herminium	= Aceraherminium
Aceras × Orchis	= Orchiaceras
Ada × Brassia	= Brassada
Ada × Cochlioda	= Adioda
Ada × Odontoglossum	= Adaglossum
Aërangis × Aëranthes	= Thesaëra
Aërangis × Amesiella	= Amesangis
Aërangis × Angraecum	= Angrangis

Parent Genera	Hybrid Genus
Aerides × Arachnis	= Aeridachnis
Aerides × Arachnis × Ascocentrum × Euanthe × Vanda	= Gilmourara
Aerides × Arachnis × Ascocentrum × Vanda	= Lewisara
Aerides × Arachnis × Euanthe	= Euarachnides
Aerides × Arachnis × Euanthe × Vanda	= Wrefordara
Aerides × Arachnis × Papilionanthe	= Aeridachnanthe
Aerides × Arachnis × Renanthera	= Lymanara
Aerides × Arachnis × Trichoglottis	= Paulsenara
Aerides × Arachnis × Vanda	= Burkillara
Aerides × Ascocentrum	= Aeridocentrum
Aerides × Ascocentrum × Euanthe × Renanthera × Vanda	= Neorobinara
Aerides × Ascocentrum × Euanthe × Vanda	= Reinikkaara
Aerides × Ascocentrum × Papilionanthe × Vanda	= Neochristieara
Aerides × Ascocentrum × Renanthera × Vanda	= Robinara
Aerides × Ascocentrum × Vanda	= Christieara
Aerides × Ascoglossum	= Aeridoglossum
Aerides × Doritis	= Aeriditis
Aerides × Euanthe	= Aeridanthe
Aerides × Euanthe × Rhynchostylis × Vanda	= Kamemotoara
Aerides × Euanthe × Vanda	= Vandantherides
Aerides × Euanthe × Vandopsis	= Aeridopsisanthe
Aerides × Luisia	= Luisaerides
Aerides × Neofinetia	= Aeridofinetia
Aerides × Papilionanthe	= Papiliodes
Aerides × Papilionanthe × Vanda	= Quisumbingara
Aerides × Paraphalaenopsis	= Pararides
Aerides × Phalaenopsis	= Aeridopsis
Aerides × Phalaenopsis × Vanda	= Phalaerianda
Aerides × Renanthera	= Renades
Aerides × Renanthera × Rhynchostylis	= Chewara
Aerides × Renanthera × Vanda	= Nobleara
Aerides × Renanthera × Vandopsis	= Carterara
Aerides × Rhynchostylis	= Aeridostylis
Aerides × Rhynchostylis × Vanda	= Perreiraara
Aerides × Saccolabium	= Aeridolabium
Aerides × Vanda	= Aeridovanda
Aerides × Vanda × Vandopsis	= Maccoyara
Aerides × Vandopsis	= Vandopsides
Aganisia × Otostylis	= Otonisia
Aganisia × Zygopetalum	= Zygonisia
Anacamptis × Dactylorhiza	= Dactylocamptis
Anacamptis × Gymnadenia	= Gymnanacamptis
Anacamptis × Orchis	= Anacamptorchis
Anacamptis × Platanthera	= Anacamptiplatanthera
Anacamptis × Serapias	= Serapicamptis
Angraecum × Macroplectrum	= Macrangraecum
Angraecum × Vanda	= Vandaecum

Parent Genera	Hybrid Genus
Anguloa × Lycaste	= Angulocaste
Anoectochilus × Goodyera	= Anoectogoodyera
Anoectochilus × Ludisia	= Ludochilus
Ansellia × Cymbidium	= Ansidium
Arachnis × Armodorum	= Armodachnis
Arachnis × Armodorum × Renanthera	= Renaradorum
Arachnis × Ascocentrum	= Ascorachnis
Arachnis × Ascocentrum × Euanthe × Renanthera × Vanda	= Neoyusofara
Arachnis × Ascocentrum × Euanthe × Vanda	= Neomokara
Arachnis × Ascocentrum × Renanthera × Vanda	= Yusofara
Arachnis × Ascocentrum × Vanda	= Mokara
Arachnis × Ascoglossum	= Arachnoglossum
Arachnis × Euanthe	= Eurachnis
Arachnis × Euanthe × Paraphalaenopsis	= Benthamara
Arachnis × Euanthe × Renanthera × Vanda	= Cogniauxara
Arachnis × Euanthe × Renanthera × Vanda × Vandopsis	= Huntara
Arachnis × Euanthe × Vanda	= Arandanthe
Arachnis × Papilionanthe	= Papilachnis
Arachnis × Papilionanthe × Vanda	= Papilandachnis
Arachnis × Paraphalaenopsis	= Pararachnis
Arachnis × Paraphalaenopsis × Renanthera	= Paranthera
Arachnis × Paraphalaenopsis × Vanda	= Parandachnis
Arachnis × Paraphalaenopsis × Vandopsis	= Garayara
Arachnis × Phalaenopsis	= Arachnopsis
Arachnis × Phalaenopsis × Renanthera	= Sappanara
Arachnis × Phalaenopsis × Vanda	= Trevorara
Arachnis × Phalaenopsis × Vandopsis	= Laycockara
Arachnis × Porphyrodesme	= Porphyrachnis
Arachnis × Porphyrodesme × Vanda	= Porphyrandachnis
Arachnis × Renanthera	= Aranthera
Arachnis × Renanthera × Vanda	= Holttumara
Arachnis × Renanthera × Vanda × Vandopsis	= Teohara
Arachnis × Renanthera × Vandopsis	= Limara
Arachnis × Rhynchostylis	= Arachnostylis
Arachnis × Seidenfadenia	= Arachnadenia
Arachnis × Trichoglottis	= Trichachnis
Arachnis × Trichoglottis × Vanda	= Ridleyara
Arachnis × Vanda	= Aranda
Arachnis × Vanda × Vandopsis	= Leeara
Arachnis × Vandopsis	= Vandachnis
Arundina × Bletilla	= Bletundina
Ascocentrum × Ascoglossum × Euanthe × Renanthera × Vanda	= DeWolfara
Ascocentrum × Ascoglossum × Renanthera × Vanda	= Shigueraara
Ascocentrum × Doritis	= Doricentrum
Ascocentrum × Doritis × Euanthe × Vanda	= Thorntonara

Parent Genera	Hybrid Genus
Ascocentrum × Doritis × Phalaenopsis	= Beardara
Ascocentrum × Doritis × Vanda	= Ascovandoritis
Ascocentrum × Euanthe	= Eucentrum
Ascocentrum × Euanthe × Luisia × Vanda	= Neodebruyneara
Ascocentrum × Euanthe × Neofinetia × Vanda	= Smithara
Ascocentrum × Euanthe × Paraphalaenopsis × Vanda	= Neodevereuxara
Ascocentrum × Euanthe × Phalaenopsis × Vanda	= Eudevereuxara
Ascocentrum × Euanthe × Porphyrodesme × Vanda	= Neokagawara
Ascocentrum × Euanthe × Renanthera × Vanda	= Greatwoodara
Ascocentrum × Euanthe × Renanthera × Vanda × Vandopsis	= Giddingsara
Ascocentrum × Euanthe × Rhynchostylis × Vanda	= Carrara
Ascocentrum × Euanthe × Vanda	= Schlechterara
Ascocentrum × Luisia × Vanda	= Debruyneara
Ascocentrum × Neofinetia	= Ascofinetia
Ascocentrum × Neofinetia × Renanthera	= Rosakirschara
Ascocentrum × Neofinetia × Rhynchostylis	= Rumrillara
Ascocentrum × Neofinetia × Vanda	= Nakamotoara
Ascocentrum × Papilionanthe	= Papiliocentrum
Ascocentrum × Papilionanthe × Vanda	= Singaporeara
Ascocentrum × Phalaenopsis	= Asconopsis
Ascocentrum × Phalaenopsis × Vanda	= Devereuxara
Ascocentrum × Renanthera	= Renancentrum
Ascocentrum × Renanthera × Vanda	= Kagawara
Ascocentrum × Renanthera × Vanda × Vandopsis	= Onoara
Ascocentrum × Renantherella	= Ascorella
Ascocentrum × Rhynchostylis	= Rhynchocentrum
Ascocentrum × Rhynchostylis × Vanda	= Vascostylis
Ascocentrum × Sarcochilus	= Sarcocentrum
Ascocentrum × Seidenfadenia	= Ascodenia
Ascocentrum × Vanda	= Ascocenda
Ascocentrum × Vanda × Vandopsis	= Wilkinsara
Ascoglossum × Renanthera	= Renanthoglossum
Ascoglossum × Vanda	= Vanglossum
Aspasia × Brassia	= Brapasia
Aspasia × Brassia × Miltonia	= Forgetara
Aspasia × Cochlioda × Odontoglossum	= Lagerara
Aspasia × Miltonia	= Miltonpasia
Aspasia × Miltonia × Odontoglossum × Oncidium	= Withnerara
Aspasia × Odontoglossum	= Aspoglossum
Aspasia × Oncidium	= Oncidasia
Aspasia × Trichopilia	= Trichopasia
Barkeria × Cattleya × Laelia	= Laeliocattkeria
Barkeria × Epidendrum	= Bardendrum
Barkeria × Laelia	= Laeliokeria
Batemannia × Otostylis	= Bateostylis
Batemannia × Otostylis × Zygosepalum	= Palmerara

Parent Genera	Hybrid Genus
Batemannia × Zygopetalum	= Zygobatemannia
Bifrenaria × Lycaste	= Lycastenaria
Bifrenaria × Maxillaria	= Bifrillaria
Bletia × Phaius	= Phabletia
Bletia × Schomburgkia	= Schombletia
Bletilla × Eleorchis	= Bleteleorchis
Bollea × Chondrorhyncha	= Chondrobollea
Bollea × Pescatoria	= Pescatobollea
Brassavola × Broughtonia	= Brassotonia
Brassavola × Cattleya	= Brassocattleya
Brassavola × Cattleya × Diacrium	= Hookerara
Brassavola × Cattleya × Diacrium × Laelia	= Linneara
Brassavola × Cattleya × Epidendrum	= Vaughnara
Brassavola × Cattleya × Epidendrum × Laelia	= Adamara
Brassavola × Cattleya × Epidendrum × Laelia × Sophronitis	= Rothara
Brassavola × Cattleya × Laelia	= Brassolaeliocattleya
Brassavola × Cattleya × Laelia × Schomburgkia	= Recchara
Brassavola × Cattleya × Laelia × Sophronitis	= Potinara
Brassavola × Cattleya × Laeliopsis	= Fujiwarara
Brassavola × Cattleya × Schomburgkia	= Dekensara
Brassavola × Cattleya × Sophronitis	= Rolfeara
Brassavola × Diacrium	= Brassodiacrium
Brassavola × Epidendrum	= Brassoepidendrum
Brassavola × Laelia	= Brassolaelia
Brassavola × Laelia × Sophronitis	= Lowara
Brassavola × Rhyncholaelia	= Rhynchovola
Brassavola × Schomburgkia	= Schombobrassavola
Brassavola × Sophronitis	= Sophrovola
Brassia × Cochlioda × Miltonia × Odontoglossum	= Beallara
Brassia × Cochioda × Odontoglossum	= Sanderara
Brassia × Miltonia	= Bratonia
Brassia × Miltonia × Odontoglossum	= Degarmoara
Brassia × Miltonia × Oncidium	= Aliceara
Brassia × Odontoglossum	= Odontobrassia
Brassia × Oncidium	= Brassidium
Brassia × Rodriguezia	− Bradriguezia
Bromheadia × Pattonia	= Pattoniheadia
Broughtonia × Cattleya	= Cattleytonia
Broughtonia × Cattleya × Laelia	= Laeliocatonia
Broughtonia × Cattleya × Laeliopsis	= Osmentara
Broughtonia × Cattleyopsis	= Cattleyopsistonia
Broughtonia × Cattleyopsis × Laeliopsis	= Gauntlettara
Broughtonia × Diacrium	= Diabroughtonia
Broughtonia × Diacrium × Schomburgkia	= Shipmanara
Broughtonia × Domingoa	= Domintonia
Broughtonia × Epidendrum	= Epibroughtonia

Parent Genera	Hybrid Genus
Broughtonia × Epidendrum × Laeliopsis	= Moscosoara
Broughtonia × Laelia	= Laelonia
Broughtonia × Laelia × Laeliopsis	= Jimenezara
Broughtonia × Laelia × Sophronitis	= Hartara
Broughtonia × Laeliopsis	= Broughtopsis
Broughtonia × Laeliopsis × Tetramicra	= Bloomara
Broughtonia × Schomburgkia	= Schombotonia
Broughtonia × Sophronitis	= Sophrobroughtonia
Broughtonia × Tetramicra	= Tetratonia
Bulbophyllum × Cirrhopetalum	= Cirrophyllum
Calanthe × Gastrorchis	= Gastrocalanthe
Calanthe × Limatodes	= Calanthe
Calanthe × Limatodes × Phaius	= Phaiocalanthe
Calanthe × Phaius	= Phaiocalanthe
Calanthidium × Preptanthe	= Calanthe
Catasetum × Cycnoches	= Catanoches
Catasetum × Mormodes	= Catamodes
Cattelya × Cattleyopsis × Epidendrum	= Hawkesara
Cattleya × Dendrobium	= Dendrocattleya
Cattleya × Diacrium	= Diacattleya
Cattleya × Diacrium × Laelia	= Dialaeliocattleya
Cattleya × Diacrium × Schomburgkia	= Mizutara
Cattleya × Domingoa × Epidendrum	= Arizara
Cattleya × Epidendrum	= Epicattleya
Cattleya × Epidendrum × Laelia	= Epilaeliocattleya
Cattleya × Epidendrum × Laelia × Sophronitis	= Kirchara
Cattleya × Epidendrum × Schomburgkia	= Scullyara
Cattleya × Epidendrum × Sophronitis	= Stacyara
Cattleya × Laelia	= Laeliocattleya
Cattleya × Laelia × Schomburgkia	= Schombolaeliocattleya
Cattleya × Laelia × Schomburgkia × Sophronitis	= Herbertara
Cattleya × Laelia × Sophronitis	= Sophrolaeliocattleya
Cattleya × Laeliopsis	= Laeliopleya
Cattleya × Schomburgkia	= Schombocattleya
Cattleya × Sobralia	= Sobraleya
Cattleya × Sophronitis	= Sophrocattlyea
Cattleyopsis × Domingoa	= Cattleyopsisgoa
Cephalanthera × Epipactis	= Cephalopactis
Cephalanthera × Ophrys	= Cephalophrys
Chamorchis × Gymnadenia	= Chamodenia
Chrondrorhyncha × Pescatoria	= Pescarhyncha
Chondrorhyncha × Zygopetalum	= Zygorhyncha
Cochleanthes × Huntleya	= Huntleanthes
Cochleanthes × Pescatoria	= Pescoranthes
Cochleanthes × Stenia	= Cochlenia
Cochlioda × Miltonia	= Miltonioda
Cochlioda × Miltonia × Odontoglossum	= Vuylstekeara

Parent Genera	Hybrid Genus
Cochlioda × Miltonia × Odontoglossum × Oncidium	= Burrageara
Cochlioda × Miltonia × Oncidium	= Charlesworthara
Cochlioda × Odontoglossum	= Odontioda
Cochlioda × Odontoglossum × Oncidium	= Wilsonara
Cochlioda × Odontoglossum × Symphyglossum	= Symphodontioda
Cochlioda × Oncidium	= Oncidioda
Coeloglossum × Dactylorhiza	= Dactyloglossum
Coeloglossum × Gymnadenia	= Gymnaglossum
Coeloglossum × Orchis	= Orchicoeloglossum
Coeloglossum × Platanthera	= Coeloplatanthera
Comparettia × Gomesa × Rodriguezia	= Bradeara
Comparettia × Ionopsis	= Ionettia
Comparettia × Oncidium	= Oncidarettia
Comparettia × Oncidium × Rodriguezia	= Warneara
Comparettia × Rodriguezia	= Rodrettia
Cycnoches × Mormodes	= Cycnodes
Cymbidium × Cyperorchis	= Cyperocymbidium
Cymbidium × Grammatophyllum	= Grammatocymbidium
Cymbidium × Phaius	= Phaiocymbidium
Cypripedium × Encyclia	= Encyclipedium
Cypripedium × Selenipedium	= Cysepedium
Dactylorhiza × Gymnadenia	= Dactylodenia
Dactylorhiza × Nigritella	= Dactylitella
Dactylorhiza × Orchis	— Orchidactyla
Dactylorhiza × Platanthera	= Rhizanthera
Dactylorhiza × Pseudorchis	= Pseudorhiza
Dactylorhiza × Serapias	= Serapirhiza
Dendrobium × Epidendrum	= Epidrobium
Diacrium × Epidendrum	= Epidiacrium
Diacrium × Laelia	= Dialaelia
Diacrium × Laeliopsis	= Dialaeliopsis
Diacrium × Schomburgkia	= Diaschomburgkia
Domingoa × Epidendrum	= Epigoa
Domingoa × Hexadesmia	= Domindesmia
Domingoa × Laeliopsis	= Domliopsis
Doritis × Kingidium	= Doridium
Doritis × Kingidium × Phalaenopsis	= Doriopsisium
Doritis × Phalaenopsis	= Doritaenopsis
Doritis × Phalaenopsis × Rhynchostylis	= Rhyndoropsis
Doritis × Phalaenopsis × Vanda	= Hagerara
Doritis × Vanda	= Vandoritis
Dossinia × Ludisia	= Dossisia
Epidendrum × Laelia	= Epilaelia
Epidendrum × Laelia × Schomburgkia	= Dillonara
Epidendrum × Laelia × Sophronitis	= Stanfieldara
Epidendrum × Laeliopsis	= Epilopsis

Parent Genera	Hybrid Genus
Epidendrum × Phaius	= Epiphaius
Epidendrum × Scohmburgkia	= Schomboepidendrum
Epidendrum × Sophronitella	= Epiphronitella
Epidendrum × Sophronitis	= Epiphronitis
Epidendrum × Zygopetalum	= Zygodendrum
Esmeralda × Renanthera	= Esmeranthera
Esmeralda × Vanda	= Esmeranda
Euanthe × Papilionanthe	= Eupapilio
Euanthe × Papilionanthe × Vanda	= Eupapilanda
Euanthe × Papilionanthe × Vanda × Vandopsis	= Kawanishiara
Euanthe × Paraphalaenopsis	= Paranthe
Euanthe × Paraphalaenopsis × Vanda	= Parandanthe
Euanthe × Porphyrodesme × Vanda	= Euporphyranda
Euanthe × Renanthera	= Antheranthe
Euanthe × Renanthera × Rhynchostylis × Vanda	= Neojoannara
Euanthe × Renanthera × Vanda	= Amesara
Euanthe × Renanthera × Vanda × Vandopsis	= Lindleyara
Euanthe × Rhynchostylis	= Anthechostylis
Euanthe × Rhynchostylis × Seidenfadenia × Vanda	= Thaiara
Euanthe × Rhynchostylis × Vanda	= Rhynchovandanthe
Euanthe × Trichoglottis	= Antheglottis
Euanthe × Trichoglottis × Vanda	= Kraenzlinara
Euanthe × Vanda	= Vandanthe
Euanthe × Vanda × Vandopsis	= Reichenbachara
Euanthe × Vandopsis	= Opsisanthe
Gastrorchis × Phaius	= Gastrophaius
Gomesa × Macradenia	= Macradesa
Gomesa × Oncidium	= Oncidesa
Gymnadenia × Nigritella	= Gymnigritella
Gymnadenia × Orchis	= Orchigymnadenia
Gymnadenia × Platanthera	= Gymnaplatanthera
Gymnadenia × Pseudorchis	= Pseudodenia
Gymnadenia × Traunsteinera	= Gymnotraunsteinera
Habenaria × Pecteilis	= Pectabenaria
Herminium × Pseudorchis	= Pseudinium
Himantoglossum × Orchis	= Orchimantoglossum
Ionopsis × Oncidium	= Ionocidium
Ionopsis × Rodriguezia	= Rodriopsis
Isabelia × Sophronitella	= Isanitella
Jacquiniella × Liparis	= Jacquinparis
Kingidium × Phalaenopsis	= Phalaenidium
Laelia × Laeliposis	= Laeopsis
Laelia × Leptotes	= Leptolaelia
Laelia × Schomburgkia	= Schombolaelia

Parent Genera	Hybrid Genus
Laelia × Sophronitis	= Sophrolaelia
Laelia × Tetramicra	= Tetralaelia
Laeliopsis × Tetramicra	= Tetraliopsis
Leochilus × Oncidium	= Leocidium
Lepanthes × Selenipedium	= Selenipanthes
Limatodes × Preptanthe	= Calanthe
Ludisia × Macodes	= Macodisia
Luisia × Papilionanthe	= Papilisa
Luisia × Paplionanthe × Rhynchostylis	= Rhynchopapilisia
Luisia × Pomatocalpa	= Pomatisia
Luisia × Rhynchostylis × Vanda	= Goffara
Luisia × Vanda	= Luisanda
Lycaste × Maxillaria	= Maxillacaste
Lycaste × Zygopetalum	= Zygocaste
Macradenia × Oncidium	= Oncidenia
Macradenia × Rodriguezia	= Rodridenia
Mendoncella × Zygopetalum	= Zygocella
Miltonia × Odontoglossum	= Odontonia
Miltonia × Odontoglossum × Oncidium	= Colmanara
Miltonia × Odontoglossum × Symphyglossum	= Symphodontonia
Miltonia × Oncidium	= Miltonidium
Miltonia × Rodriguezia	= Miltonguezia
Miltonia × Symphyglossum	= Symphyglossonia
Miltonia × Trichopilia	= Miltonpilia
Neofinetia × Papilionanthe	= Papilionetia
Neofinetia × Phalaenopsis	= Phalaenetia
Neofinetia × Renanthera	= Renanetia
Neofinetia × Renanthera × Rhynchostylis	= Hueylihara
Neofinetia × Rhynchostylis	= Neostylis
Neofinetia × Vanda	= Vandofinetia
Nigritella × Orchis	= Nigrorchis
Nigritella × Pseudorchis	= Pseuditella
Odontoglossum × Oncidium	= Odontocidium
Odontoglossum × Rodriguezia	= Rodriglossum
Odontoglosum × Symphyglossum	= Symphodontoglossum
Odontoglossum × Zygopetalum	= Odopetalum
Oncidium × Ornithophora	= Ornithocidium
Oncidium × Rodriguezia	= Rodricidium
Oncidium × Trichocentrum	= Trichocidium
Oncidium × Trichopilia	= Oncidpilia
Oncidium × Zygopetalum	= Zygocidium
Orchis × Platanthera	= Orchiplatanthera
Orchis × Serapias	= Orchiserapias
Orleansia × Restrepia	= Restesia
Otostylis × Pabstia	= Otopabstia

Parent Genera	Hybrid Genus
Otostylis × Zygopetalum	= Zygostylis
Otostylis × Zygosepalum	= Otosepalum
Pabstia × Zygopetalum	= Zygopabstia
Paphiopedilum × Phragmipedium	= Phragmipaphium
Papilionanthe × Paraphalaenopsis	= Parapapilio
Papilionanthe × Paraphalaenopsis × Renanthera × Vanda	= Paramayara
Papilionanthe × Paraphalaenopsis × Vanda	= Sealara
Papilionanthe × Phalaenopsis	= Phalaenopapilio
Papilionanthe × Phalaenopsis × Renanthera × Vanda	= Neomoirara
Papilionanthe × Renanthera	= Papilanthera
Papilionanthe × Renanthera × Vanda	= Mayara
Papilionanthe × Rhynchostylis	= Stylisanthe
Papilionanthe × Rhynchostylis × Vanda	= Honoluluara
Papilionanthe × Sarcochilus × Vanda	= Sarcopapilionanda
Papilionanthe × Vanda	= Papilionanda
Papilionanthe × Vandopsis	= Papiliopsis
Paraphalaenopsis × Renanthera	= Pararenanthera
Paraphalaenopsis × Renanthera × Vanda	= Paravandanthera
Paraphalaenopsis × Rhynchostylis × Vanda	= Sweetara
Paraphalaenopsis × Vanda	= Paravanda
Phaius × Spathoglottis	= Spathophaius
Phalaenopsis × Renanthera	= Renanthopsis
Phalaenopsis × Renanthera × Rhynchostylis	= Lutherara
Phalaenopsis × Renanthera × Vanda	= Moirara
Phalaenopsis × Renanthera × Vandopsis	= Ernestara
Phalaenopsis × Rhynchostylis	= Rhynchonopsis
Phalaenopsis × Rhynchostylis × Vanda	= Yapara
Phalaenopsis × Sarcanthopsis	= Sarcalaenopsis
Phalaenopsis × Sarcochilus	= Sarconopsis
Phalaenopsis × Vanda	= Vandaenopsis
Phalaenopsis × Vandopsis	= Phalandopsis
Porphyrodesme × Renanthera	= Porphyranthera
Porphyrodesme × Vanda	= Porphyranda
Prescottia × Pterostylis	= Pterocottia
Pteroceras × Renanthera	= Renanthoceras
Pteroceras × Sarcochilus	= Sarcoceras
Rangaeris × Tridactyle	= Randactyle
Renanthera × Renantherella	= Ellanthera
Renanthera × Rhynchostylis	= Renanstylis
Renanthera × Rhynchostylis × Vanda	= Joannara
Renanthera × Rhynchostylis × Vandopsis	= Yoneoara
Renanthera × Saccolabium	= Saccanthera
Renanthera × Sarcochilus	= Sarcothera
Renanthera × Trichoglottis	= Renaglottis
Renanthera × Vanda	= Renantanda

Parent Genera	Hybrid Genus
Renanthera × Vanda × Vandopsis	= Hawaiiara
Renanthera × Vandopsis	= Renanopsis
Renantherella × Vanda	= Vandantherella
Rhinerrhiza × Sarcochilus	= Rhinochilus
Rhynchostylis × Sarcochilus	= Sartylis
Rhynchostylis × Vanda	= Vandachostylis
Rhynchostylis × Vandopsis	= Opsistylis
Saccolabium × Vanda	= Sanda
Sarcochilus × Vanda	= Sarcovanda
Schomburgkia × Sophronitis	= Schombonitis
Trichoglottis × Vanda	= Trichovanda
Trichoglottis × Vandopsis	= Trichopsis
Vanda × Vandopsis	= Opsisanda
Zygopetalum × Zygosepalum	= Zygolum

Index of Persons

Index of Plant Names

For list of alkaloid-containing species, see pages 353-359.
For list of hybrid generic names, see pages 488-561.
For list of species with chromosome numbers reported, see pages 412-472.
For list of horticultural nomenclatorial changes of species used in hybrid combinations, see pages 487 and 488.

Index of Subjects

sectile, 293
separation of, 294
Pollution, air, 72
effects, 158
Polyacetate chains, 350
Polyembryony, 244, 245, 248, 249
Polyphenoloxidase, 146
Polyploid, 23, 384-386
ancient, 385, 389
Cattleya, 402-405
cells, 342
Cymbidium, 394, 395, 401, 402
definition of, 394
through hybridization, 395
influence of, 393
occurrence, 390
Paphiopedilum, 406-408
Sophrocattleya, 404-406
Polystele structure, 341
Population, breeding, 292
and pollinia size, 298
Postpollination changes, anthocyanin
formation, 145
ethylene, 145
flower, 144
naphthalene acetic acid, 145
stigma, 145
wilting of sepals and petals, 145
Potassium, deficiency, 154
traces in rainwater, 63
Potomageton natans, 257
Potting media, bark mixes, 155
deficiencies, 156
fertilization, 155
fungal content, 155
mineral content, 154
nutrient levels, 154
over fertilization, 156
Primary roots, 121
Primordia, 256
Production ecology, 80, 81
Proembryo, 224, 258
stages of, 238
Propagation, clonal, 169
meristem culture, 169
nodal, 159
structures in, 123
Propagules, 4, 9, 10, 11, 15. *See also*
Dispersal; Gemmae; and Seeds
Proteinase, 146

Protocorm, 131, 133, 249
budding of, 255
definition, 180
development, 29, 180-182
effect of, amino acids, 189
light, 189
vitamins, 189
differentiation, 188, 256
DNA content, 187
hairs, mycorrhizal infection, 108
in vitro cultivation, 185, 186, 211-
215
Cattleya, 212, 214, 215
Cymbidium, 186, 211, 212
Lycaste, 215
Miltonia, 215
Odontoglossum, 215
Phaius, 215
mode of infection, epidermal hairs,
139
number formed from meristem, 181
Cymbidium, 186
photosynthetic and non-photosynthetic, 110
stage, 131, 138
subterranean development, 110
survival, parasitism, 142
Pseudobulb, 305, 331, 334
ant-inhabited, 335
epidermis of, 334
heteroblastic, 326, 331, 334
homoblastic, 326, 334
Pseudocopulation, 287
Pseudomonas, 158
Pterocarpus, 36
Pyridoxine, 130, 143
Pyrrolizidine alkaloids, 368, 378
Pythium ultimum, 158
Python, 72

Quercus, castanea, 135
magnoliaefolia, 135
peduncularis, 135
petraea, 63
scytophylla, 135
vicentensis, 135
Rabbits, 72
Radiation, 158
Radicle, 256
Raffinose, 135, 147